THE SHAPE OF LIFE

THE SHAPE OF LIFE

Genes, Development, and
the Evolution of Animal Form

Rudolf A. Raff

THE UNIVERSITY OF CHICAGO PRESS

Chicago and London

Rudolf A. Raff is professor of biology at Indiana University and director of the Institute for Molecular and Cellular Biology.

The University of Chicago Press, Chicago 60637
The University of Chicago Press, Ltd., London
© 1996 by The University of Chicago
All rights reserved. Published 1996
Printed in the United States of America
05 04 03 02 01 00 99 98 97   2 3 4 5

ISBN 0–226–70265–0 (cloth)
      0–226–70266–9 (paper)

Library of Congress Cataloging-in-Publication Data

Raff, Rudolf A.
    The shape of life: genes, development, and the evolution of animal
form / Rudolf A. Raff.
       p.   cm.
    Includes bibliographical references and index.
    1. Evolutionary genetics.   I. Title.
QH390.R325   1996
575.1—dc20   95-49224
                                                      CIP

⊗ The paper used in this publication meets the minimum requirements of the American National Standard for Information Sciences—Permanence of Paper for Printed Library Materials, ANSI Z39.48–1984.

In memory of
Rudolf August Viktor Raff
1908–1981
my first scientific mentor

Contents

Preface   xiii
Acknowledgments   xxi

1   Over the Ice for Ontogeny and Phylogeny   1
The worst journey in the world
Old roots and tangled branches
Recapitulation, transformation, and von Baer's laws
Haeckel: Metaphor as mechanism
The embryological strand
The conflict between heredity and development
Rate genes: An attempted synthesis
Ontogeny meets the operon
An almost meeting of the minds
A deep intellectual divide
Issues for an evolutionary developmental biology

2   Metazoan Phyla and Body Plans   30
Questions of macroevolution and body plans
Homology and body plan
The living phyla
The need for phylogeny
Phylogenetic tools
Outgroups and primitive characters
Cladograms and phylogenetic trees
Diversity and disparity

3   Deep Time and Metazoan Origins   63
Deep time
The first animals
Precambrian life and environments
Are the Ediacaran fossils animals at all?

If not animals, what?
Bodies and behaviors on the Cambrian boundary
Small shelly fossils
Unearthing the unimaginable
More body plans, faster evolution?
Evolution and progress
A summary: The importance of the fossils

4   **Molecular Phylogeny: Dissecting the Metazoan**   **103**
           Radiation
Inferring molecular phylogenies
Outlining a molecular phylogeny of the phyla
The base of the metazoan radiation
The base of the radiation of the Bilateria
Strange territory: The pseudocoelomates
Origin of the coelom
Coelomate protostomes
Lophophorates
Arthropod monophyly versus polyphyly
Arthropod molecular phylogenies
The disparity of Cambrian arthropods
Deuterostomes
Molecular biology and the metazoan radiation

5   **Recovering Data from the Past**   **142**
The Crystal Palace dinosaurs
Loss and recovery of data from the past
Interpreting lost body plans
The Neanderthal's missing voice and DNA's forgotten
      bases
Can our methods recover phylogenies from genes?
Gene trees versus species trees
Setting the molecular "clock"
Extinct lineages affect molecular phylogenies
Fossil genes
Summing up: Phylogeny and the evolution of
      development

6   **The Developmental Basis of Body Plans**   **173**
Body plans and developmental biology
Why have no new phyla appeared since the Cambrian?
Mass extinctions and big opportunities

Invasion of the land
Hypotheses on the stability of body plans
Patterns of development in the metazoan radiation
Hox genes and body plans
Evolutionary stability of early development
Radical evolutionary changes in early development
Body plans and how they develop
Fiddling with the rules
Stability of phylotypic stages
Evolution after the phylotypic stage
The developmental hourglass

7   Building Similar Animals in Different Ways   211
The evolutionary significance of early development
Dichotomies and model systems
The limits of model systems
Rules for evolutionary developmental biology
Life history and developmental strategies
Conservation and change in early development
Developmental modes in sea urchins
Radically reorganizing development
Changes in morphogenesis
Similar genes, different embryos
Parallel and divergent mechanisms
Experimental approaches to the evolution of early
      development
   Nematodes
   A molluscan diversion
   Evolution of development in primitive chordates:
      Ascidians
   Vertebrate life histories and embryos: Frogs and
      salamanders
How general are evolutionary changes in early
      development?
Experimental evolution: Manipulating egg size
Regulation: The underlying flexibility of development

8   It's Not All Heterochrony   255
What price immortality?
Defining heterochronies
Local versus global heterochronies

Heterochronies need not be limited to late
    development
Development of the dead
The meaning of paedomorphosis
The meaning of peramorphosis
Pre-displacement and post-displacement
Genetics of heterochrony
Time and growth control
Heterochrony as a result, not a process
Human evolution
To recapitulate

9   **Developmental Constraints**   **292**
The divine watchmaker
Constraints
Why are there no centaurs or six-legged greyhounds?
Is selection adequate?
Whence constraints?
On the other hand
Genomic constraints
Constraint by number of cell types
Frozen controls?
Constraints imposed by the limits of structure
Constraint by complexity
Allometry: Done in by big antlers?

10   **Modularity, Dissociation, and Co-option**   **321**
Mosaic bodies
Evolutionary mechanisms
Principles of evolvability
Modularity
    The genetic organization of modules
    Standard parts
    Physical location within the developing system
    Connectivity to other modules
    Temporal transformations
Dissociation
Duplication and divergence
Co-option
A co-option event at morphological and gene levels
Duplication, co-option, and redundancy
Limb buds: Evolution of a module

Identities of molecules signaling between limb bud
    modules
Genes and the execution of pattern
Commonalities in limb development across phyla
Limbs as serial homologues
Why not eight toes?
Modules set up by gene switches
Summing up

11   **Opportunistic Genomes**   **362**
A termite history of the world
Sex and "polymorphic technology"
Fluidity of sex determination mechanisms
Evolution of eyes
Co-option of eye structures and genes
Dollo's law
Turning back the clock: Evolutionary reversals
Consequences of fluid genomes

12   **Evolving New Body Plans**   **397**
Patagonian thinkers
Whales and the return to the sea
Wings and the insect body plan
Genes for legs and wings
Differentiating the flies from the butterflies
The echinoderm radial body plan
Rates and fossils
Exploring body plan evolution
Genes, homology, and evolution

References   435
Index   493

Preface

Some of the most commonplace items in our world are also paradoxically the most improbable and wonderful: the physical properties of water, the existence of sex, the fact that the millions of animal species have only thirty-five basic body plans, the fact of evolutionary transformation through time. The latter two items, the objects of this book, are related in a seemingly antithetical way. The basic animal body plans are half a billion years old. Arthropods, chordates, mollusks, echinoderms, and others can all be recognized from the first appearance of the animal fossil record. Even so, the striking diversity of living animals shows that dramatic evolutionary transformations have taken place within these ancient patterns. Body form evolves through geologic time, but it also arises in each generation through development. The time frames and apparent processes couldn't be more different, but explanations of the evolution of form have to consider how form is generated, and they must account for both underlying stability and immense change in design.

My awakening to the scope and significance of the relationship between development and evolution came during my first year as a graduate student at Duke University, when I read de Beer's book, *Embryos and Ancestors*. I no longer remember how I found out about the book, but I got hold of a copy and set out to study it. This was not as simple a task as vacation novel reading. I was engaged in my first year of heavy coursework in biochemistry and was expected to read the *Journal of Biological Chemistry* in my spare time. Consequently, on Saturday mornings I sat with de Beer in the laundromat across the street from an old Durham eatery, the Little Pigs Barbecue, while I added coins to the dryers and waited for the endless drying cycles to finish. At that time I knew nothing about embryology, and so the terminology was simply mind-boggling; germ layers, neural plates, and archenterons swam before my mystified eyes until I got it all straight. But it was a stimulating experience. Until that time, my views of evolutionary change were largely of transformations between features of adult body plans. I had never considered the embryological roots of those transformations. *Embryos and Ances-*

tors provided the connection between the construction of an individual animal and the course of its evolutionary history. De Beer also presented a mechanistic explanation by which changes in the timing of developmental events in the descendant as contrasted with the ancestor could yield evolutionary novelties by changing the relations of developmental features. This phenomenon, heterochrony, has motivated most thought on the evolution of development. De Beer offered fascinating scenarios to demonstrate how animal groups, including the vertebrates, could potentially have originated by heterochronic changes. What de Beer didn't do was consider the roles of genes in development, nor did he discuss any experimental approaches. He provided a paradigm for the evolution of morphology, but his approach lacked any means by which one might investigate the hypotheses presented. On reflection, it's hard to blame him. It certainly took me a long time before I was able to design for myself an experimental approach to some problems in this still largely theoretical and anecdotal discipline.

I thought a good deal about the evolution of development during my own first years as a developmental biologist, and finally decided in the early 1970s to offer a graduate course on the subject. That experience led to my writing a book with my colleague Thomas Kaufman. When we wrote *Embryos, Genes, and Evolution,* we sought the basis for a synthesis between the disciplines of evolutionary and developmental biology. It was already clear that the course of development, or ontogeny, of an organism is the product both of selection acting upon mechanisms of development and of the evolutionary history of its particular lineage. However, the great progress then being made in developmental genetics indicated that the evolution of development was far richer in potential mechanisms than heterochrony alone.

Higher taxonomic groups, most notably phyla, possess suites of anatomical features that distinguish them from other groups. Such an underlying anatomical arrangement is called a body plan. Vertebrates, arthropods, and echinoderms have immediately obvious and distinct body plans. There is an extraordinary paradox about body plans. The major animal body plans first appear in the fossil record in early Cambrian rocks, deposited just over half a billion years ago. Body plans arose rapidly during the radiation of the first animals, but have been conserved since their debuts. Despite the enormous amount of developmental and morphological innovation that has occurred since then within body plans, no new phyla appear to have originated since the Cambrian. Nonetheless, enormous evolutionary changes have gone on within the animal kingdom within these long-conserved underlying patterns.

Evolution of the developmental processes that shape the individual in each generation underlies the long-term changes in form observed in the record of life. This book is about the role of developmental processes in the origins of animal body plans and about the mechanistic connections between development and evolution. The evolution of animals has been a process of expanding adaptations to the variety of niches available on Earth. This evolution has generated novel anatomical features that have opened new opportunities to their bearers, and has resulted in the great diversity of living and extinct animals.

Of necessity, an understanding of the relationship between development and evolution requires synthesizing two great and traditionally separate disciplines in biology, developmental biology and evolutionary biology. The relationship between the two disciplines has always been an enterprise in creative inference making. The recognition that an appreciation of embryonic development would be important in understanding evolution appears in *The Origin of Species*. Darwin suggested that we could use embryos to trace evolutionary relationships because "the embryo is the animal in its less modified state; and in so far it reveals the structure of its progenitor." Darwin certainly had a concrete example before him, having earlier carried out a long study of barnacles. Barnacles were long thought to be mollusks, but once discovered, their larvae quickly gave them away as related to shrimp and other crustaceans.

From the 1860s to the end of the nineteenth century, the German evolutionary biologist Ernst Haeckel built a hypothesis as to how ontogeny and phylogeny should interact, and founded a research program that featured the large-scale application of data from embryology to working out phylogenies. Early in the twentieth century, embryologists became interested in how embryos worked rather than how they evolved, abandoned Haeckel's program, and turned to experimental studies of development itself. The current study of the relationship between evolution and development has grown from a rebirth of interest in the subject as well as from the invention of new tools in genetics and molecular biology.

Dramatic changes have occurred within the past decade. When *Embryos, Genes, and Evolution* was published in 1983, there was no organized experimental discipline for uniting the three topics of the book's title. There were very few experimental studies explicitly designed to test ideas generated at the boundary between the disciplines. The few that did exist were given disproportionate weight and were seldom critically repeated. An experimental discipline had not yet coalesced. The synthesis between the two fields is not as obvious as one might like, and it continues to be difficult to unite two very disparate disciplines that have followed divergent paths for a century. Issues crucial to one discipline can be completely misconstrued or even invisible to practitioners of the other. One of the aims of this book will be to point out these disciplinary differences and indicate directions of synthesis.

Developmental biologists must come to appreciate diversity, the comparative method, and the importance of phylogeny. Evolutionary biologists need to understand and incorporate the underlying rules of developing systems into a more comprehensive theory of evolution.

Comparisons between animals often reveal differences in developmental patterns, and traditionally, scenarios to account for the differences and their evolutionary significance are then inferred. A classic and famous example discussed by de Beer is the tadpole larva of ascidians. This larva has the notochord, dorsal nerve cord, and somites characteristic of chordates, which are then lost when it metamorphoses into the filter-feeding sessile adult. Garstang suggested that the ascidian larva, by acquiring the ability to reproduce as a nonmetamorphosed larva, became the ancestor of the chordates. It's a nice easy wrap-up of the whole problem of vertebrate origin. However, the converse is just as likely. The point is that without an unequivocal phylogeny or evolutionary history, we cannot determine the direction of evolutionary change in development. For most situations in which evolutionary conclusions are drawn from comparative data, the direction of change must be inferred. In many of the most interesting cases, such as consideration of the evolution of developmental processes among animal phyla, the problem is compounded by the fact that the phylogenies rest heavily on comparative embryology. Circularity fails as a method.

We are now in a far better position to evaluate polarities of evolutionary change. The growth of cladistics, a robust methodology for doing systematics, allows us to map developmental features on more objectively inferred phylogenies and to distinguish homologies from independently acquired convergences. The growing availability of gene sequences provides a new set of data, making it possible to infer molecular phylogenies independently of the morphological and developmental features that we want to study.

Another element unavailable ten years ago is an understanding of the molecular machinery underlying development. During the past decade, developmental genetics has become one of the liveliest disciplines of biology. Its enormous success has come from the realization on the part of developmental biologists that genes underlie developmental processes and even serve as switches that determine the course of differentiation of cells within the embryo. During the same period, powerful tools became available to dissect developmental processes at the molecular level. The most striking part of that success has come from the joint application of genetics and molecular biology to identifying and isolating the genes that control development. The genetic analysis of development was under way a decade ago, and we could point to the roles of homeotic genes in the evolution of segmental features in insects. But there were major limitations. At that time, none of the genes involved had been cloned, and the functions of their products were not known. Without gene sequence information, gene homology could not be established. Once that information became available, it became possible to trace patterns of gene utilization in development across phylogenetic lines.

The discovery during the past decade that insects and vertebrates use homologous homeobox-containing genes in their development has revealed a deep genetic relationship underlying the development of the two phyla. More than any other single discovery, this finding of a deep commonality of the genetic controls for establishing components of axial polarity in animal development has helped to fix the focus of developmental biologists on how evolutionary data can contribute to our understanding of developmental mechanisms.

I've organized this book around the two related themes of the origins of the major animal body plans in the great Cambrian radiation and their subsequent modification through the evolution of developmental processes during the ensuing 530 million years. The first chapter reviews the historical origins of the relationship between the disciplines of evolutionary biology and developmental biology. Our attitudes toward the great problems of biology are conditioned by historical contingencies. These contingencies have had a particularly powerful impact on the outlooks of the two disciplines that I'm attempting to synthesize here. When history dictates the form experiments take, it's important to take note.

The second chapter introduces animal body plans and the use of phylogenetic methods. I do that because it is impossible to consider the evolution of developmental features and processes without placing them in a phylogenetic context. The use of phylogenetic reasoning has not been a part of developmental biology, and many readers may not have given it much thought, nor realized the pervasiveness of phylogeny in evolutionary studies, nor understood the need to know the direction of evolutionary changes being studied. The following three chapters consider the origins of the animal phyla in the Cambrian radiation and the use of gene sequence data in resolving phylogenetic problems, especially the origins of major groups, that are not amenable to other methods.

Chapters 6 and 7 deal with the developmental problems posed by body plans and their modifications. These chapters present evidence on the evolvability of different stages of development and consider the significance of the freedom of early development to evolve rapidly and radically. Evolutionary observations provide quite a different perspective on the structure of development than do observations from developmental genetics based on single species used as model systems. Chapters 8 through 10 consider the mechanistic relationship between development and evolution. In chapter 8, I propose that heterochrony has been vastly oversold as a universal evolutionary mechanism. The consequence has been the substitution of labels for explanations in all too many studies of evolutionary lineages. In chapters 9 and 10, I present the mechanistic issues posed by the hypothesis that the internal architecture of the genome and of developmental processes and their controls constrains the course of evolution. This issue is a central one in the study of development and evolution. If externally applied natural selection is the only force required to produce evolutionary change, then developmental processes don't matter except as features upon which selection can act. If internal organization and processes govern modes of change, then development must be incorporated into any complete theory of evolution. I propose that internal organization and a set of distinct evolutionary processes acting to sort internal variation produce nonrandom morphological variation in evolution, and I identify and characterize these processes.

The eleventh chapter examines special topics that reveal the rich and unexpected fluidity of genomes and developmental processes in evolution. The examples discussed reveal that genes can be readily co-opted for new functions and that regulatory circuitry can be radically changed while continuing to perform a conserved function. It is important to know as much about function as possible. Apparently identical developmental phenomena can prove to be quite distinct when their mechanistic bases are known. Knowledge of function and of gene sequence homology are both required. Different genetic systems might converge functionally, as they do, for example, in sex determination in flies and nematodes. Conversely, homologous genetic systems can be "wired in" differently and play quite different roles. This chapter explores the extent to which the fluidity of genomes allows evolutionary reversals in developmental processes.

In the final chapter I discuss some examples of the evolution of novelties. The underlying genetic mechanisms are relatively well known for some of these, but not for others. The point to be made is that scenarios about the origins of novel features in evolution can now be cast in terms that will allow their investigation with the increasingly powerful tools provided by molecular biology and developmental genetics. The commonality of genetic controls in animal development has provided an important new impetus to the study of "nonstandard" animals drawn from nature in addition to model systems.

Studying the evolution of development requires more than a focus on development in single species; it also requires using evolutionary approaches to understand the processes of development per se and the roles of the genetic and developmental machinery of organisms in channeling or constraining the course of evolution. Within the context of the time, Kaufman and I were successful in pointing to some of the elements such a synthesis should include. Still, the project shared a bit with Tom Sawyer's trip down the Mississippi. There were a few things dimly lurking in the mists and crouching below the eddies, and it was a big river.

Over a decade has passed since *Embryos, Genes, and Evolution,* and an experimental discipline that integrates developmental and evolutionary biology has begun to coalesce. Most important, the whole emphasis of work on development and evolution has shifted to new ground due to the transformation of our understanding of the genes that regulate development. Our discipline deserves a new look at how the theoretical ideas that have molded it have been utilized and how an experimental approach has shaped our understanding. This book is dedicated to helping that enterprise off to a good start.

Acknowledgments

A large number of colleagues, students, and friends have selflessly helped me by discussing ideas with me, listening to my lectures, sharing data and publications, providing photographs and illustrations, reading drafts of my manuscript, and enthusiastically criticizing any loose thinking and dubious writing they could find. Unfortunately, any errors that remain after all that help are clearly my own.

My first debt is to my wife and colleague Elizabeth C. Raff for her love, encouragement, and help over the course of this project. She has given a great deal to the development of the ideas presented in the book through spirited discussions that helped me discard ill-formed enthusiasms and come to solid (or at least defensible) conclusions, and she has collaborated with me on projects ranging from the molecular phylogeny of the animal kingdom to the reversibility of evolution. The book owes much in a practical way as well to her generous employment of the red pencil in the early versions of the manuscript and to her time, patience, and artistic abilities in doing several of the drawings. This book would not have appeared without her support and inspiration in dark moments and light. I also owe special debts to my long-term friends and co-workers in the research enterprises that contributed to this book. Gregory Wray and Jonathan Henry were my partners in Bloomington and Sydney for a number of years in creating an experimental study of the evolution of early development. Their contributions (some made in hot debates over cold beer) have been immeasurable. Jessica Bolker has shared her intellectual style and thoughts, and has stimulated enjoyable discussions of some major issues. Ellen Popodi has played a vital role in the study of the molecular biology of the transformation of the echinoderm body plan, and has done more than her share to make our Australian expeditions feasible. Charles Marshall brought a cheerful, stimulating, and sophisticated phylogenetic outlook to my laboratory, and also succeeded in proving to me the toughness of Australians by wearing shorts during an Indiana winter. Katharine Field played an instrumental role in our study of the phylogeny of the animal kingdom, and has always lent a sympathetic ear. J. M. Turbeville was my long-term collaborator on projects in molecular phylogeny, and shared his rigorous thinking and unparalleled knowledge of invertebrates. Steven Palumbi helped build an enjoyable and stimulating collaboration, made me appreciate the power of population genetics, and took me to visit Hawaiian sea urchins in their home waters. Scott Gilbert, a card-carrying evolutionary developmental biologist and a historian of this forming discipline, has shared ideas and encouraged my work over the years. He has also generously read parts of the book. I also acknowledge two old friends in this enterprise, Stephen Jay Gould and Per Alberch, whose work has stimulated my thinking.

I am grateful to Jarmila Kukalová-Peck for her sense of humor, for her thoughtful discussions on Paleozoic insects, and for providing some of her original drawings. I also thank her and Stuart Peck for their detailed critique of my discussion of the evolution of insect appendages. I thank Bryan Rogers, Rudolf Turner, and Thomas Kaufman for the use of their remarkable scanning electron micrographs of juvenile insect heads, and James Gehling for his excellent photographs of Ediacaran fossils and his insights on the earliest metazoans.

Several readers generously donated their time to critique the book in progress. Jessica Bolker, Eric Haag, and Jessica Kissinger gave much help in this way, as did William Wimsatt, Michael LaBarbera, George Lauder, Günter Wagner, Nicholas Holland, and James Hanken. Their suggestions have greatly improved the manuscript.

Much of my own research discussed in the book was done on visits to the University of Sydney. I am grateful to Donald T. Anderson for enabling me to start work on the evolution of early development in Australian sea urchins, and for making me feel at home in Sydney. Many other Australian colleagues have welcomed me and my co-workers into their laboratories and homes and have generously provided help, discussions, facilities, and lasting friendship. They include George Barrett, Maria Byrne, Ron Hill, Ove Hoegh-Guldberg, Ian Hume, Guillermo Moreno, Valerie Morris, David Patterson, Angela Low, Frank Rowe, Heather and Craig Sowden, and Noel Tait. Heather and Craig have given freely of their time and expertise as marine biologists to provide us invaluable help in collecting and maintaining sea urchins, and they have shown us some fascinating parts of Oz. I also thank the University of Sydney for hosting me and my students as visiting scholars. Without the assistance of the University and the Sydney Aquarium, my antipodean research on evolution and development would have been far less pleasant and productive.

Many colleagues and friends unstintingly answered my many questions and provided copies of figures, preprints, and papers, and even copies of their books. These include André Adoutte, Michael Akam, Shonan Amemiya, Wallace Arthur, Eckart Bartnik, Stefan Bengtson, Annalisa Berta,

Hans Bode, David Bottjer, Marianne Bronner-Fraser, Ann Burke, Robert Burke, Maria Byrne, Andrew Cameron, Sean Carroll, Peter Cherbas, Michael Coates, Simon Conway Morris, Della Cook, David Crews, Eric Davidson, Eugenia del Pino, Robin Denell, David Dilcher, Richard Elinson, Richard Emlet, Scott Emmons, David Fitch, James Gehling, Scott Gilbert, James Hanken, John Hayes, David Jablonski, William Jeffery, Stuart Kauffman, Thomas Kaufman, Ellen Ketterson, James Lake, George Lauder, Curt Lively, Jane Maienschein, Margaret McFall-Ngai, George Miklos, Jay Mittenthal, Norman Pace, Joram Piatigorsky, Lars Ramsköld, Matthew Ravosa, Gregory Retallack, Lynn Riddiford, Rupert Riedl, Virginia Roth, Bruce Runnegar, Michael Savarese, Philipp Schatt, Einhard Schierenberg, William Schopf, Brian Shea, Rick Shine, Barry Sinervo, Michael Smith, Mitchell Sogin, James Sprinkle, Billie Swalla, Cliff Tabin, Anthony Underwood, Elisabeth Vrba, David Wake, Dieter Walossek, Cathy Wedeen, David Weisblat, and Graeme Wistow.

I thank all my students, past and present, for their friendship and for the intellectual stimulation they have given me, Rhea Freeman for keeping my administrative life organized during the pleasurable ordeal of writing, Connie Taylor for exhaustive and skillful secretarial help, and Christie Henry, biological science editor at the University of Chicago Press, for her cheerful guidance through the labyrinth of publication.

The writing of this book was begun under the auspices of a Guggenheim sabbatical fellowship.

1
Over the Ice for Ontogeny and Phylogeny

In the strait of Magellan, looking due southward from Port Famine, the distant channels between the mountains appeared from their gloominess to lead beyond the confines of the world.

Charles Darwin, *Voyage of the Beagle*

THE WORST JOURNEY IN THE WORLD

As a young man, Apsley Cherry-Garrard spent two and one-half years as an unpaid assistant zoologist on Robert Falcon Scott's fatal 1911 expedition to Antarctica and the South Pole, then returned to England in time to see the start of World War I. He survived military service in Flanders, and found the time to publish in 1922 one of the most extraordinary accounts of polar exploration ever written, *The Worst Journey in the World*. There he vividly and movingly tells of the tragic fates of Scott and his polar party, but within that grander tragedy he records the staggering difficulties of one of the strangest trips ever undertaken, his own appalling winter journey with ornithologist Edward A. Wilson and Royal Navy lieutenant H. R. Bowers to Cape Crozier to collect the eggs of the emperor penguin. The same zeal for scientific discovery that lured Darwin and so many other great scientific voyagers drew Cherry-Garrard and the other members of the party on a journey of scientific exploration that they barely survived. Cherry-Garrard's idealism comes through in the words he wrote many years later: "We traveled for Science. These three small embryos from Cape Crozier, that weight of fossils from Buckley Island, and that mass of material, less spectacular, but gathered just as carefully hour by hour in wind and drift, darkness, and cold, were striven for in order that the world may have a little more knowledge, that it may build on what it knows instead of on what it thinks." Bowers and Wilson were later to die with Scott on their return from the South Pole. Their passion for science may have contributed to their deaths. They had collected several kilograms of fossil plants from some of the few exposed rocks of the Antarctic continent at Beardmore Glacier in the Transantarctic Mountains. Despite their exhaustion, Scott's South Pole party attempted to return with the specimens the three hundred-odd miles from Beardmore Glacier to their base camp on

1

Ross Island. The fossils were found on their sled at their final camp where they died in a blizzard. The fossils were recovered by the rescue party, and much later in the century were to lead to the discovery of fossils of Triassic mammal-like reptiles in Antarctica. The remains of these animals, so akin to the contemporary mammal-like reptiles of Africa, established that about 250 million years ago Antarctica, along with Africa and Australia, made up the southern part of the supercontinent of Pangaea.

The main goal of Scott's expedition was to be the first to reach the South Pole. Nevertheless, in Scott's mind, his expedition was primarily a scientific journey of exploration, not merely a race with Roald Amundsen's Norwegian polar party. Wilson's satellite expedition to Cape Crozier was driven by the desire to test an evolutionary hypothesis about embryonic development and its relationship to evolutionary history. The reasons underlying the six-week-long sled journey to a penguin rookery in the continuous dark of the Antarctic winter seem extraordinary, but they are not so different from the foundations of many other expeditions dispatched across the planet and into space on the basis of what later were shown to be very shaky theoretical concepts.

Darwin's most forceful adherent was the German zoologist Ernst Haeckel, who became the most prominent late-nineteenth-century evolutionary theorist. Haeckel's achievements included the coining of numerous scientific terms (including such enduring favorites as ecology, phylogeny, and heterochrony), the prediction of the discovery of a fossil link between apes and humans (he called it *Pithecanthropus*), and the inauguration of a passionate quest for phylogenetic trees. In 1866 he propounded the famous and overwhelmingly influential biogenetic law, which states that ontogeny (the development of the individual) results from phylogeny (the evolutionary history of the lineage). Haeckel's mechanism of evolution required that new forms appear as a result of the addition of new terminal stages to the ancestral ontogeny. Thus, all animals should recapitulate their phylogenies in an abbreviated form during development, and developmental stages should reveal those histories. In 1911 Haeckel's ideas still held powerful sway in the minds of zoologists, a vision potent enough to send men on a desperate journey through the Antarctic night. It was Wilson's idea that the emperor penguin is the most primitive living bird, an evolutionary relict, a creature pushed to the very farthest reaches of the south polar regions by competition with more recently evolved and more advanced groups of birds spilling out of Eurasia. In the first decades of the twentieth century, Eurasia was thought by zoologists to be the center of many evolutionary radiations, a cauldron of Darwinian competition. It was thought that humanity originated there as well. Less fit species either became extinct or sought refuge in peripheral parts of the world not reached by their superior competitors.

The winter trip to the penguin rookery was aimed at preserving eggs in fixative and bringing them back to England for study. Wilson thought that if the trip took place early enough in the breeding season, various early developmental stages would be represented in the sampled eggs. Because in Wilson's scenario the emperor penguin is a primitive relict, its embryos were expected to recapitulate the reptile-to-bird transition. As Cherry-Garrard put it, "it is because the Emperor is probably the most primitive bird in existence that the working out of his embryology is so important. The embryo shows remains of the development of an animal in former ages and former states; it recapitulates its former lives. The embryo of an Emperor may prove the missing link between birds and the reptiles from which birds have sprung." This is a clear expression of Haeckel's dogma. Modern bird embryos do not reveal much about their reptilian ancestry. But, in Haeckelian recapitulation theory, it would be expected that with addition of more modified late stages in advanced birds, some of the early evolutionary stages would become so compressed as to effectively vanish. Thus, Wilson wanted to look at what he thought was the most primitive bird, because it would have undergone less modification of its development, and still might recapitulate some reptilian stages.

The trip was executed with awesome persistence in the face of the unending darkness, the killing cold, and the broken and barely visible terrain. Each man pulled a sled loaded at the start of the trip with 250 pounds of food and supplies. Condensation from their breath froze in clothing and sleeping bags, converting these items into a reluctant frozen armor that had to be forced on with help. The extreme cold prevented the snow from melting beneath the sled runners and forming a lubricating film of water. The result was a pull over snow that had the friction of sand. Cherry-Garrard's account of the daily routine is painful: "That day we made $3\frac{1}{4}$ miles, and traveled 10 miles to do it. The temperature was $-66°$ when we camped, and we were already pretty badly iced up. . . . For me it was a very bad night: a succession of shivering fits which I was quite unable to stop." The physical toll mounted. Cherry-Garrard records, "I don't know why our tongues never got frozen, but all my teeth, the nerves of which had been killed, split to pieces."

"And then we heard the Emperors calling."

After nineteen days, they reached the rookery at Cape Crozier. Penguins stood with eggs tucked up onto their feet and under their breast feathers. It required a difficult climb down over the 200-foot ice cliffs to get to the rookery, and only five eggs could be collected from the frightened birds. A blizzard lasting several days prevented any further attempts. In the exertions of returning to camp, two of the eggs were broken; the three survivors were preserved and ultimately returned to Britain.

A thorough analysis was finally published by the ornithologist Cosser Ewart in 1921. Ironically, when the eggs were finally studied, the embryos so painfully recovered were found to be fairly late in development. Wilson's hope of examining early stages was not to be fulfilled. Thus, the test made of recapitulation was limited to seeing whether scales arise before feather primordia. That would be expected in a strict recapitulation in the development of a primitive bird because scales are the homologues of feathers and are their evolutionary precursors. The homology of scales and feathers is supported by the existence of mutations in chickens in which the scales that normally cover the legs are converted to feathers. However, Ewart found that, as in other birds, the emperor penguin's feather primordia form prior to the scale primordia. It has since been shown that penguins have a good fossil record dating back over 40 million years and evolved deep in the Southern Hemisphere from a flying seabird ancestor. Wilson's driving idea became an irrelevance, based both on a misreading of the relationship of biogeography to evolution and on a misplaced faith in an oversimplistic theory of the evolutionary meaning of development.

OLD ROOTS AND TANGLED BRANCHES

Many of the ideas we cherish as novel have origins that predate us by embarrassingly long times. Current studies of the relationship between development and evolution also have a long and complex evolutionary history. To place a new discipline of evolutionary developmental biology in its proper context, we must trace that history. What I attempt to do in this chapter is to show that this discipline has deep roots in biology. These roots have nourished a complex tangle of ideas. At times the fusion of developmental biology and evolutionary biology has been central to a generation's understanding of evolution. At other times only a few biologists have had any interest in the subject, and the two disciplines have each advanced along their own internally driven paths with scarcely a polite glance at each other. The story is not a neat one. There has been a tangle of ideas that have derived from quite disparate disciplines.

History is not merely a record of the dead past. The questions and problems that we will deal with in this book in many cases arose long ago, but remain unresolved. They merely appear in new guises and demand new approaches. Furthermore, the paths traveled during the past century by the disciplines that we are trying to fuse have created domains of thought that are in substantial ways different and not easily melded. The invention of new tools and the discovery by developmental geneticists of deep commonalities in the genetic regulatory mechanisms underlying animal development have inevitably led to the experimental study of the evolution of development. However, to be fully meaningful, such a new discipline must truly fuse older disciplines, not merely attempt to impose a dataset derived from developmental or molecular biology upon a superficial idea of evolution.

RECAPITULATION, TRANSFORMATION, AND VON BAER'S LAWS

The idea of developmental recapitulation as a guide to evolutionary history that sent Cherry-Garrard's party over the ice was an outgrowth of one of the grandest synthetic concepts of biology. This idea was at first pervasive in driving research and later soundly derided. For all of the problems in its application, however, recapitulation in part reflects very real phenomena in the evolution of developmental patterns. Pre-Darwinian comparative anatomists in the early nineteenth century noted a very striking thing: in their development, the embryos of higher organisms appeared to pass through, or recapitulate, the features seen in the adults of related lower organisms. Many of the adherents of a literal recapitulation saw the animal kingdom as a linear progression of various invertebrate groups culminating in the vertebrates and humans. As elaborated by Louis Agassiz in 1849, this comparative anatomical discovery could be given a time dimension as well by examination of the fossil record. Agassiz suggested that a threefold parallelism is evident as the embryo of a higher animal passes through its developmental stages. In many cases a series of developmental stages resembles both the adults in a series of more primitive related forms and the progression of fossil representatives of the group through geologic time. For example, the juveniles of modern bony fishes, Agassiz's favorite animals, initially develop tails similar to those characteristic of adults of primitive fossil species, and only later develop modern-style tails. Some biologists saw in this phenomenon support for "transformation," the pre-Darwinian term for evolution. Not Agassiz. To him, the threefold parallelism did not reflect any transformation between forms, but an underlying "plan designed by an intelligent creator." Although he had revolutionized geology through his discovery of the Ice Age and later founded the Museum of Comparative Zoology at Harvard, he stubbornly, and to modern eyes perversely, opposed evolution throughout his entire life. In an 1867 letter to a colleague (cited by Winsor), he wrote of Darwinism, "I trust to outlive this mania." He didn't.

The embryological supports for parallelisms between developmental stages and levels of animal complexity were often speculations based on poorly interpreted data. There are some notorious examples. The famous French anatomist Geoffroy Saint-Hilaire homologized the internal skeleton of vertebrates with the exoskeleton of arthropods (see Appel). As he summed it up, "every animal lives within or without its vertebral column." His contempo-

rary Etienne Serres considered human teratological monsters to be arrested
at the stages of adults of lower forms. Thus a headless human fetus reflected
the clam stage, since these mollusks seemed to be the highest headless animals. Although evolutionary ideas colored the ideas of the biologists of the
transcendental school, the concept of linear recapitulation harked back to the
much more ancient concept of the Great Chain of Being propounded by
Aristotle and later sanctified by the Church because it illustrated the plenitude
of God's creation. In the great chain or ladder, all possible living creatures
had been created by God in order, from the most simple, such as plants,
through various invertebrates, lower vertebrates and higher, man, and finally
angels and other supernatural beings. Superficially, the idea looks evolutionary, but it was not. The scheme was static, not transformational. No one
climbed this ladder except embryos.

In 1828 the great embryologist Karl Ernst von Baer carried out the investigations of vertebrate development that led to his famous empirical "laws,"
which were intended to deal a stunning blow to loose theorizing on the
nature of development. Von Baer's observations showed that there is no strict
recapitulation of the development of a more primitive form by the embryo
of a more advanced one. Instead, organisms diverge from one another in
development. The embryos of higher forms do not duplicate the adults of
more primitive relatives, but in their stages of development they do resemble
the stages of development of the embryos of related lower forms. Von Baer
intended his "laws" to make the theory of recapitulation untenable. His
results showed that fancied resemblances across phyla, such as those proposed by Serres, were unfounded. The body plans revealed by development
were clearly specific to each of the great "embranchements" defined in 1812
by Georges Cuvier, the father of comparative anatomy. These great branches
of the animal kingdom, radiates (coelenterates and echinoderms), mollusks,
articulates (insects, crustaceans, and annelids), and vertebrates, were each
characterized by unique and distinct body plans. There was not one great
chain, but four branches. Cuvier's definitions of these embranchements were
based on characteristic modes of organization of the nervous system, and
they have a rough correspondence to the phyla we currently recognize. Cuvier
considered the nervous system to be the primary functional system, and thus
its organization the critical feature of each body plan. Like the fundamentally
different developmental pathways defined by von Baer, the embranchements
of Cuvier argued against evolution. For Cuvier it was not just that the embranchements were different in morphology; integrated function also was
crucial. Cuvier believed that the radically different organizations of the four
embranchements could not change from one to another without a fatal loss
of integrated function during the transition. This argument is still a forceful
one.

I do not want to leave behind a caricature of the events I have outlined above. Words like "notorious" and "loose theorizing" are highly colored and imply that the activities of the transcendentalist biologists were somehow silly or irrelevant because they did not correspond to our view of things. Nothing could be further from the truth. The first quarter of the nineteenth century was a time of roiling intellectual ferment and methodological innovation in biology. We look at formal portraits of remote scientists like Cuvier with his high-necked brocade jacket and medals, and too easily forget that all the petty jealousies, academic politics, and power struggles, as well as the grander passions of scientists for their viewpoints, existed then as now.

The differences between Cuvier and the transcendentalists came to a head in 1830 in a prolonged dispute, the Cuvier-Geoffroy debates so well captured and dissected by Toby Appel. For Cuvier, the embranchements were absolutely distinct from one another because of integrated function requirements. To him, function and the integration of all features of an animal's anatomy were primary. A carnivore would have sharp teeth because of the need to kill prey and slice it up. The presence of such teeth and associated features such as claws indicated only that this was the appropriate anatomical design for the functions of a carnivore. Geoffroy, on the other hand, had a more evolutionary outlook, and pointed to homologies between parts of animals as the crucial way to understand anatomy. If the role that an animal played determined its anatomical organization for Cuvier, its body plan determined its function for Geoffroy. Homologous structures were not identical, but were essentially the same structure in modified forms. Geoffroy pioneered the use of connections, the relationships between elements, to identify homologues. Thus, the bones of the skull can be identified by their relationship and placement relative to other bones. As long as Geoffroy confined himself to identifying homologues within an embranchement, no problems arose between him and Cuvier. The break came with attempts by Geoffroy's followers to establish homologies between vertebrates and cephalopod mollusks. Geoffroy's attempt to establish homologies between the vertebrate skeleton and the arthropod exoskeleton was likewise offensive to Cuvier. Their debates centered on the existence and nature of homologies. Cuvier was easily able to demolish claims of homology between vertebrate and mollusk, but was on thin ice in holding that complex structures such as the hyoid bones of various vertebrates were called forth independently to serve specific and distinct purposes. An evolutionist would see these hyoid bones as homologous in having a common evolutionary ancestry but having undergone considerable modifications to serve distinct functions. That was an outlook that Cuvier could never accept.

Neither Cuvier nor Geoffroy had sufficient knowledge of embryology to resolve their arguments. Yet contemporary German embryology was already providing the necessary insights. Von Baer provided no support for homologies between body plans because to him, development represented a common theme within each embranchement. Homology within a body plan was powerfully supported by discoveries such as Karl Reichert's demonstration in 1837 that the bones of the mammalian middle ear are developmentally the same bones that are involved in the articulation of the jaw in more primitive vertebrates. At the time of the Geoffroy-Cuvier debates, Darwinian evolution had not yet provided the most basic unification of homology and adaptation through selection and phylogenetic history. Homologies record the inherited contingencies of evolutionary history, and are now understood to be features that share a common inheritance in evolution. Homologies can exist at any level of organization. There are homologous genes, cell types, tissues, organs, and organ systems. Selection molds homologous features to meet the demands of function, but not necessarily the same functions. The bones of the reptilian jaw and those of the mammalian middle ear show the play of adaptation on ancient homologies.

HAECKEL: METAPHOR AS MECHANISM

Von Baer never accepted evolution, even after the publication of *The Origin of Species*. Ironically, because von Baer's concept of development was progressive, his laws proved to be not at all incompatible with a modified form of evolutionary recapitulation. Embryos of animals within any phylum are transformed from the generalized and simple to the specific and complex. Resemblances between embryos of higher forms and adults of more primitive relatives do exist, and are indicators of shared ancestry and descent.

Darwin himself was fascinated by the evolutionary transformation of the threefold parallelism. To him, the reason embryos of advanced species resemble adults of their more primitive living and fossil relatives was that they recapitulate some of these more primitive features during development. He saw this recapitulation as the manifestation of a genetic continuity between ancestor and descendant. As Robert Richards has pointed out in *The Meaning of Evolution*, embryological development served Darwin as a model for the transformation of form in evolution. Analogously, recapitulation within embryological development served as a reliable guide for inferring the shape of the ancestors of living animals. In *The Descent of Man,* Darwin even applied recapitulationist ideas directly to human evolution. Thus, he stated, "the human embryo likewise resembles certain low forms when adult in various points of structure." In the same book Darwin was to further extend this concept to make an evolutionary connection between invertebrates and vertebrates, reaching for an explanation of transitions between body plans that are still an enigma today. The ascidians are marine invertebrate animals that as adults are saclike filter feeders. They were classified early in the nineteenth

century as mollusks, but they were later discovered by Kovalevsky to have
larvae unlike those of any mollusk. Their larvae resemble tiny tadpoles,
complete with a dorsal nerve cord and a notochord. Although much more
simple in structure than vertebrates, these larvae nevertheless are built along
the same overall body plan. Darwin suggested that an ancestor similar to the
ascidian larva gave rise to both ascidians and vertebrates through choice
of quite different pathways of development by the two diverging lines of
descent.

It was with Ernst Haeckel in the 1860s that recapitulation became the
driving force of research on phylogeny. Haeckel codified the first notion of
a mechanistic link between evolution and development. For him, development was an infallible guide to phylogenetic history and the basis for his
dramatic phylogenetic trees. Because heredity was not understood, Haeckel
was forced to provide a metaphor as a mechanism: "Phylogenesis is the
mechanical cause of Ontogenesis." But there was a bit more to it than that.
Haeckel had seized upon a pair of ideas put forward by Fritz Müller in his
1864 book *Für Darwin*. On the basis of his studies of crustacean development, Müller suggested that descendants pass through the stages of development of their ancestors and then add a new stage at the end. His second
contribution was to realize that new stages could not be just added ad infini-
tum; thus former stages were condensed or lost. Haeckel was a Lamarckian;
to him, terminal addition of new stages not only made sense, but provided
the only sure mechanism by which recapitulation could occur: larval stages
were fleeting, and only adult stages persisted sufficiently long for adaptations
to become fixed through the action of life habits. To Haeckel, evolutionary
changes that produced adaptations to the conditions of larval life and had
occurred in nonterminal stages of development were merely inconveniences
that obscured the straightforward reading of phylogeny from development.
Of course, Haeckel was perfectly aware that adaptations to the conditions of
development had also to be acknowledged to avoid being forced into accepting such absurdities as ancestral mammals being connected throughout
life to their mothers by a placenta, as would be required by an inflexible
recapitulatory reading of mammal development. Haeckel also understood that
a condensation of all the recapitulated stages was necessary to prevent ontogeny from becoming impossibly long. Condensation also explained why phylogenies read from ontogeny were imperfect. Condensation inevitably obscured the recapitulatory record by abbreviating or even deleting features and
stages.

The whole edifice of the Haeckelian program became irrelevant when developmental biologists shifted their efforts to understanding mechanisms of
embryonic development. It became explicitly incorrect with the demise of
Lamarckian heredity in the face of Mendelian genetics in the early twentieth

century. To fully savor the power, influence, and demise of recapitulation, one should read Gould's *Ontogeny and Phylogeny*.

The Haeckelian tradition played a powerful role in the training of many of the generation of scientists who founded developmental biology, yet when they shifted paradigms, they did so without a backward glance at recapitulation theory and evolution, as Garland Allen's biography of Thomas Hunt Morgan shows very clearly. Morgan exemplifies the transition. He was arguably the most important of his generation of biologists in America. At progressive stages in his career, he participated in Haeckelian embryological phylogeny, then experimental embryology, and later still pioneered experimental genetics. Morgan did his graduate work at Johns Hopkins University under William Keith Brooks. Brooks was thoroughly committed to the study of marine invertebrate embryology in the Haeckelian mode of seeking phylogenetic histories from early developmental stages. He was convinced that such studies would illuminate evolutionary histories because he felt that marine forms retain more of the primitive features of a group than terrestrial or freshwater forms. Morgan's dissertation followed Brooks's conviction. To this end, he approached the phylogeny of pycnogonids, or sea spiders, an obscure group of arthropods thought to be related to the chelicerates (the group containing *Limulus,* spiders, scorpions, and other less appetizing entities such as ticks and mites). From a detailed examination of their early development, he concluded that pycnogonids are indeed related to spiders. Like Brooks's other students, however, Morgan became disillusioned with the Haeckelian program and became an experimental biologist.

THE EMBRYOLOGICAL STRAND

The parting of the ways between evolution and development began in the last quarter of the nineteenth century with the establishment of marine stations at such attractive seaside sites as Naples in the Mediterranean, Woods Hole in the Atlantic, Misaki on the coast of Japan, and elsewhere, which made marine embryos readily available. The biologists of the 1890s who flocked to those marine stations gave rapt attention to defining the fates and lineages of individual cells within a variety of embryos. Coincidentally, in 1894, Wilhelm Roux published his influential call to developmental biologists to study the mechanisms of development in living embryos by experimental intervention. The attractions of phylogeny evaporated. Morgan made the longest intellectual journey of any of his contemporaries, from his early studies steeped in Haeckel's phylogenetic embryology with Brooks, to his experimental embryology using marine invertebrates at Woods Hole, and finally to his powerful pioneering genetic studies of *Drosophila* and his founding of the American school of genetics. He might have been expected to take the first steps in synthesizing these fields. He did make a weak attempt to bring together genetics and development in his 1934 book *Embryology and Genetics*. However, the book might as well have been called *Oil and Water*, as it turned out to be a curious and unintegrated effort, with some chapters on genetics and others on embryology.

Embryologists rapidly and enthusiastically embraced Roux's approach and transformed embryology into a powerful experimental science. They attempted to reach into the underlying mechanisms of development by making direct biochemical attempts to understand the important embryological phenomena that they had uncovered. The most exciting work of early-twentieth-century embryology was that of Hans Spemann, who first demonstrated the organizer in amphibian embryos and the role of induction in the cascade of events that culminates in a fully differentiated embryo. The basic experiment was conducted in Spemann's laboratory by Hilde Mangold in 1924. She found that a critical portion of the frog embryo, the dorsal lip of the blastopore of the gastrula, when grafted to a host embryo caused the host to form a second axis, a Siamese twin. The transplanted dorsal lip exerted its action by inducing neighboring tissue in the host to carry out a complex set of developmental events. In some cases, the induced twins were complete with eyes, brain, and spinal cord. The action of the transplanted dorsal lip revealed its normal mode of action in situ and showed that induction had a powerful role in organizing the subsequent differentiation of regions of the embryo. According to Viktor Hamburger, who has reviewed this heroic period of embryology, Spemann understood that activation of the genetic material resulted from the epigenetic interactions among inducing and induced tissues. However, Spemann apparently never attempted to follow the problem in genetic terms. Spemann's data were all comprehended in operational terms at the level of tissues and their interactions.

Approaches to a deeper mechanistic analysis of induction were initiated in the 1930s with biochemical attempts to identify the chemical nature of inducers. An impressively productive and innovative researcher of this period was Johannes Holtfreter, who was a pioneer of in vitro culture methods for embryonic explants and of transplantation methods. Holtfreter (as described in a paper by Bautzmann et al., published in 1932) first showed that killed tissues could still cause neural induction. Presumptive inducing tissues were treated with alcohol to kill the cells but leave potential chemical inducers intact. These tissues were then embedded in ectoderm sandwiches or inserted into embryos, and induction of neural structures was scored. This experiment provided the vital clue that induction resulted from the action of a chemical and did not require live cells. However, Holtfreter and others later found that extracts of a number of tissues not normally involved in neural induction (including adult vertebrate liver, kidney, and bone marrow) were active in neural induction. Because alcohol-killed tissues gave the same results as live tissues, a sort of "chicken soup" chemical induction enterprise resulted in which all sorts of "inducers" were tested. In the best zoological tradition, some really outré sources such as viper liver, worms, mollusks, and insects were tried and found to have inducing activity. However, the interpretation of such experiments and the relationship of their results to in vivo events were dubious. Victor Twitty, in his memoir of those times, *Of Scientists and Salamanders,* records that Holtfreter eventually remarked that such analyses were "bringing chaos out of order." It is interesting that other investigators, including Joseph Needham and his collaborators, working in 1934, put a different spin on the widespread distribution of neural inducer. As they suggested, "We may, in fact, conclude that it is a definite chemical substance universally present throughout the animal kingdom." Needham and his co-workers cited the broad distribution of insulin-like substances in animals. The widespread distribution of these powerful regulatory substances acting between cells suggested that the same could be true of inducer substances. It is worth noting that this interpretation is consistent with more recent discoveries. In many instances, a deep-rooted commonality of molecules and processes in development has emerged. Signaling molecules and their receptors are widely conserved in evolution, although their specific developmental functions are frequently modified.

In the first of two studies by Waddington and his collaborators that continued the work of Needham and his co-workers, an attempt was made to test the role of metabolic differences between the inducing and noninducing cells of the embryo. No differences were found, even though a sensitive micro-respirometer was used. As a part of this study, cells were treated with methylene blue to stimulate respiration, and it was found unexpectedly that methylene blue acted as an inducer. Waddington and his co-workers suggested two hypotheses for the methylene blue action. They suspected that it either acted by the same mechanism as the natural inducer, or acted by triggering an inducing substance already present in the neuroectoderm. This observation was to prove very troubling to the field. Insights into biochemical mechanisms were still limited by the prevailing focus of biochemists on metabolism during that period. What is most significant here, however, is that no genetic mechanisms were even alluded to. But, of course, the connection between genes and the molecules and actions involved in developmental processes such as these had not yet been envisaged. The leading developmental biologists of the time were instead limited to vague interpretations of inductive mechanisms. Incidentally, in his 1938 book *Physiological Genetics,* Richard Goldschmidt contemplated the functions of interactions between genes and the cytoplasm. He specifically referred to the induction process as representing timed nuclear gene action that sets up the required cytoplasmic actions, such as "the pro-

duction and flow of the organizer substance in amphibian development.''
The lack of knowledge about the nature of genes (they were still commonly
thought to be analogous to enzymes) and the lack of any clear model of gene
activation prevented this discussion from providing a fruitful basis for research at that time.

Unfortunately, biochemistry was not sufficiently advanced technically in
the 1930s to isolate and unequivocally identify any inducing substance, and
attempts to do so by Needham and Waddington proved frustrating. Some of
their results were consistent with lipoidal compounds, and the newly discovered role of steroids as hormones suggested a steroid as the inducer. Although
these ideas could not be fruitfully developed in the 1930s, it is interesting
that retinoic acid, which is an isoprenoid biosynthetically related to steroids,
was demonstrated by Tickle and her co-workers and by Thaller and Eichele
in the late 1980s to be a morphogen in differentiation of the limb bud.
Studies on retinoic acid have progressed rapidly because there is now a
well-developed understanding of steroid receptors and of their role as transcription factors in directly regulating gene expression.

The direct biochemical approach was also doomed because identification
of a chemical capable of inducing a neural structure would by itself reveal
little about the cellular machinery involved in receiving and transducing such
a signal. The early experiments on the organizer were bound to fail because
too little was understood about the induction process. The final events of
neural induction may require more than a single signal from other cells. The
nonspecific inductions achieved with dead tissues and a variety of chemical
agents might really be the result of the triggering of a second message system
already present in the presumptive neural tissue. Despite its being the most
famous and most intensively studied induction system, neural induction is
still far from completely understood. Substantial progress in understanding
induction phenomena has come only recently with the application of developmental genetic tools that exploit a mutational analysis of inductive events,
and with the growing understanding of signal reception in general. Neither
genetic nor evolutionary approaches were a significant part of mainstream
embryology during its golden age.

THE CONFLICT BETWEEN HEREDITY AND DEVELOPMENT

Because there was such an evident link between the laws that governed
heredity and the presumed actions of natural selection, Mendelian genetics
was avidly incorporated into the thinking of evolutionary biologists early in
the twentieth century. The attention of geneticists was primarily focused on
transmission genetics. The laws of heredity that had so long baffled biologists
could finally be clearly elucidated. With the birth of population genetics,

evolutionary biology was provided with a powerful quantitative tool for predicting the behavior of genes under selection. No knowledge of the nature of genes or their expression was required.

The situation was not so simple for the embryologist. Transmission genetics provided no theoretical foundation for developmental biology. The expression of genes within cells or embryos was little understood, and indeed could not be even experimentally approached before the 1960s. At the same time, the phenomena of primary interest to early-twentieth-century embryologists seemed to be largely cytoplasmic and nongenetic in operation. For example, in 1905 Edwin Conklin showed that the developmental fates of the cells of ascidian embryos could be mapped from the movements and segregation of colored cytoplasmic components to various regions of the embryo shortly after fertilization. Cells containing particular cytoplasms ultimately gave rise to certain predictable cell types. Thus, the yellow crescent region produced the larval tail muscle cells. The determination of muscle cells appeared to be based upon the cytoplasm, not the nucleus. In the 1890s, Hans Driesch had performed an experiment that altered the distribution of nuclei in the sea urchin embryo. He was able to demonstrate no effect of this drastic treatment on development. All nuclei were equivalent.

The brilliant early studies of cytoplasmic determinants and cell lineage were paralleled by studies that directly indicated a genetic role in the regulation of development. Foremost among these studies was the work of Theodor Boveri, published in 1907. Fritz Baltzer, in his biography of Boveri, nicely summarizes his results and conclusions. Boveri's study was simple and elegant. He noted that sea urchins fertilized with two sperm instead of one divide into four cells, not two, in their first division. Boveri realized that in this situation chromosome distribution must be abnormal, because instead of being attached to two mitotic spindle poles, the chromosomes were pulled by four poles. Cytological staining showed that the chromosomes were not equally partitioned to the daughter cells. Boveri quickly showed that the resulting abnormal development was not merely the result of the size of the cells produced in the first division or their cytoplasmic composition. The four cells of a disaggregated normal four-cell embryo all proceed to form complete, small pluteus larvae. Boveri also exploited the statistics of chromosome segregation in dispermic eggs shaken in such a way as to initially divide into three cells. The chance of a correct distribution of chromosomes into each cell of these embryos was significantly higher than in the four-cell cases, and Boveri found the expected increase in the number of embryos showing normal development. The conclusion was clear: Normal development cannot proceed unless a balanced complement of chromosomes is present. The idea of the chromosome as carrier of the genes was to be brilliantly made into the foundation of genetics in the first decades of the twentieth century by Morgan and his school utilizing *Drosophila.* The focus of Morgan's work, however, was transmission genetics, and the developmental role for genes suggested by Boveri's experiments had as little practical impact upon genetics as it did upon embryology.

Morgan, in his important book *Experimental Embryology,* published in 1927, concluded that no decision was possible on the role of genes in development. As he put it, "One of the most important questions for embryology relating to the activity of the genes cannot be answered at present. Whether all the genes are active all the time, or whether some of them are more active at certain stages of development than are others, are questions of profound interest." He doubted that the genes were themselves enzymes, but so little was known of the molecules of what is now the "central dogma" that he was left unable to go beyond. Morgan put the question in modern enough terms of gene activity, but had no tools to approach the concept further.

Many developmental biologists, such as Frank Lillie, concluded that genes could not direct the differentiation of the very different parts of an embryo. To Lillie, "those who desire to make genetics the basis of physiology of development will have to explain how an unchanging complex can direct the course of an ordered developmental stream." That this view could be seriously maintained in 1927 was due to the disparate outlooks of geneticists and developmental biologists in the first half of the century. Those outlooks must have been deeply ingrained, because evidence that genes are involved in development was already available. It would be a distorted and one-dimensional tale, however, to suggest that embryologists of this era failed to recognize the existence of genes. The feeling that genetics was somehow crucial to developmental biology remained as a persistent, if uncomfortable and undeveloped, theme throughout the first half of the twentieth century.

RATE GENES: AN ATTEMPTED SYNTHESIS

Despite the determined separation of embryology from genetics, there were a few out-of-step characters, notably Richard Goldschmidt, who in the 1930s attempted to combine genetics, evolution, and development. Studies by Goldschmidt as well as by Ford and Huxley showed that genes could influence the rates of developmental processes. In the gypsy moth, *Lymantria dispar,* studied by Goldschmidt, the caterpillars of different races differed in the darkness of their markings. In some races, light markings persisted throughout development. In others, a dark pigment was subsequently deposited. The rate of pigmentation differed with race and was intermediate in heterozygotes between races. The other example of rate genes in action was provided by the course of pigment accumulation in the developing eye of *Gammarus chevreuxi,* a small crustacean studied by Ford and Huxley. They

found that eye color in this species existed in red and black Mendelian alternatives. Black eyes were red at first and then became black as melanin was
deposited. In the red-eyed mutation, melanin deposition was observed to
start later and to proceed at a slower rate. Colors, brown and chocolate,
between red and black could also be produced. These studies had something
akin to the modern experimental idea of looking at the phenotypes produced
by mutations of genes affecting development, but their application was
limited.

The concept of a "rate gene" had an enormous appeal to a number of
evolutionary biologists in the 1930s. Haldane explicitly discussed genes
known to act at various stages of development, and proposed that evolutionary changes in timing of developmental processes could be linked to changes
in time of action of rate genes. Huxley at that time was concerned with the
phenomenon of allometry, that is, the control of relative size of body parts
as a function of body size. He suggested that rate genes could be mutated to
yield changes in plus or minus directions, resulting in acceleration or retardation of growth of features relative to the ancestor.

ONTOGENY MEETS THE OPERON

Ultimately, the application of genetics to dissecting development arose from
the study of mutations that produced effects on development. This tradition
had its roots in the 1930s in the discipline then called "physiological genetics." An explicit role for genes was sought in cellular and developmental
processes. Mutations were sought that affected specific developmental events
and produced specific defect phenotypes. Ernst Hadorn, C. H. Waddington,
and Curt Stern carried this approach forward in the 1940s and 1950s. The
results began to win notice by the early 1960s with the work of E. B. Lewis
on the developmental genetics of the fruit fly *Drosophila melanogaster*. Insects are built of specialized segments, each with appendages related to those
on other segments, but specific to the segment that bears them. In his remarkable book *Materials for the Study of Variation,* published in 1894, William
Bateson presented unusual cases in which structures belonging to one body
segment were transformed in identity to those belonging to a different segment. He called this phenomenon "homoeosis." Goldschmidt and other geneticists demonstrated that mutations of certain genes produced homeotic
transformations during development in *Drosophila*. There are two principal
clusters of these genes in *Drosophila*. Lewis performed a genetic and developmental analysis of the Bithorax complex of homeotic genes, and demonstrated that they govern the development of segmental identities in a highly
coordinated manner. Most important, he showed that the action of these
genes resulted in alternate choices being made in development: genes were

acting as switches. Since the 1970s genetic analysis, combined with the direct study of molecular mechanisms of gene action, has become the dominant program for studying development.

It is only within the past few years that any significant interest has arisen in reuniting the old partners development and evolution. A number of events have promoted their rejoining. After almost a century of mutual avoidance, the application of genetics to understanding development finally became possible when the emphasis of genetics switched from its old focus on transmission of traits between generations to a concern for how genes function within a living cell. That shift stems from the remarkable work of Jacob and Monod in the 1950s, which demonstrated how the short-term differentiation of bacterial cells was governed by controls on gene expression. Those controls suggested for the first time how gene expression could be regulated in animal development. The messenger RNA hypothesis gave a powerful impetus for considering how genetic information could operate in eggs and embryos in the face of the constancy of nuclear DNA.

The study of development has been transformed in large part into a study of how gene action underlies and governs developmental processes. A synthesis of genetics with developmental biology required a theory of gene action. The theoretical basis for understanding how genes function to direct development became available from the operon theory of Jacob and Monod. Their hypothesis was constructed to account for control of expression of bacterial genes, but it was to have a powerful effect on the integration of developmental biology with molecular genetics in the 1960s and 1970s.

The effective conjoining of genetics with developmental biology required two fundamental revolutions as well as an acceptance by developmental biologists that gene action was fundamental to the execution of development. The operon theory of Jacob and Monod provided a model for the role of gene expression in development. The revelation of the double helical structure of DNA by Watson and Crick dictated a focus on the role of DNA structure in genetic processes. Molecular biology originated in the middle decades of the twentieth century, and its roots lie in the fusion of two themes represented by these landmarks. The first theme was the working out of the mechanisms of gene expression. Only then could a molecular-based genetics be profitably applied to working out the genetic basis of function in development. The second theme was the understanding of the structure of macromolecules that came from the work of the X-ray crystallographers on proteins and later on DNA. Genes were no longer ineffable entities detectable only by genetic crosses. Now they acquired a chemical and physical reality. The ability to clone genes and determine their base sequences allowed developmentally important genes to be identified and to be considered from an evolutionary perspective. Sequences of different genes can be compared for common motifs or for degree of similarity. Homologous genes can be identified and sought in other developmental contexts. Comparison is fundamental to evolutionary biology, and even in molecular guise, is in principle the same as the comparative anatomy so powerfully employed by Darwin and his contemporaries.

Genetics has completely transformed the questions we ask about the influence of developmental mechanisms on evolution, as well as the kinds of answers we seek. Yet, despite our rejection of the Haeckelian program and its Lamarckian basis, Haeckel remains relevant to us. We need to know the phylogeny of any set of organisms or genes whose evolution we wish to study. Without phylogeny we cannot establish the polarity, the direction of evolutionary change, of the differences that we observe. History is vital in reconstructing evolutionary processes. Considerations of evolutionary history have not been a part of the growing understanding of developmental processes. Evolutionary biology and developmental biology have not only sought different objectives, but have been so different as disciplines that their reintegration is more difficult than most practitioners of one or the other discipline imagine.

AN ALMOST MEETING OF THE MINDS

In 1981, John Tyler Bonner organized a major cross-disciplinary conference to explore effective ways of creating a study of evolution and development. I was at that stimulating meeting in the Berlin suburb of Dahlem. It provided an interesting education in several ways. To start with, Berlin was a schizophrenic city, with its glitzy western half and, behind a line of dead, abandoned buildings along the boundary, its brooding gray eastern half. The Berlin Wall still stood. A trip organized for us took us through the wall at Checkpoint Charlie into a very different world, where spontaneity was replaced by rigid control. The bus was searched for contraband newspapers on the way in, which seemed a little crazy considering that East Germans could watch West German television. But there were other peculiarities. Western attendees of the conference were welcome in East Berlin, but our Polish colleagues were not, because Poland was already developing unhealthy democratic attitudes. The contingent of evolutionary biologists and paleontologists on the bus requested a change in the scheduled tour to include a visit to the Museum für Naturkunde, which has in its possession the famous Berlin *Archaeopteryx* as well as the mounted skeleton of *Brachiosaurus,* then the largest known dinosaur. That proposal caused some turmoil among the tour authorities, but despite the fact that it was unheard of to change a tour schedule, our guide was good-natured. A few telephone calls while we waited in the parking lot of the Soviet War Memorial, and the deed was done. We spent a happy hour among the skeletons. On the way back to West Berlin we had a less reassuring wait in the parking lot while mirrors on wheels were rolled under the bus to look for refugees clinging to the undercarriage. It would have hardly occurred to us then that the Berlin Wall and the whole system that built it would collapse in only a few years.

It developed that the meeting itself had a touch of schizophrenia too. The organizers and participants had not realized just how difficult it really is to combine two fields that have diverged for nearly a century. Group discussions foundered on issues of vocabulary and on very profound differences as to what each discipline thought evolutionary biology was. After all, evolutionary biology has made the understanding of the diversity of life its major goal, whereas developmental biologists seek unifying mechanisms for developmental processes. There was another problem whose resolution has proved to be the watershed in synthesizing the fields: in 1981, there was no experimental discipline in evolutionary developmental biology. The feeling that a synthesis is necessary has persisted and has motivated subsequent meetings, which have revealed that the conceptual rapprochement is slowly coming along. What is crucially different now from a decade ago is that some experimenters have taken steps toward a practicable fusion and have framed problems that can be addressed experimentally with appropriate organisms.

As in the case of developmental biology and genetics, the separation of developmental biology from evolutionary biology occurred at about the turn of the century. There was one important difference. Developmental biology had no connection with genetics until later in the life of the two disciplines. However, it had been happily married to evolutionary biology since within a few years of *The Origin of Species*. The divorce, when it came in 1894, was straightforward. The experimental program proposed by Wilhelm Roux was so powerful that embryologists rapidly shed their interests in evolutionary problems. The quest for the mechanistic controls of development became far more interesting than the quest for phylogenetic histories traced through embryonic resemblances. The shift was already under way among the younger embryologists gathered at the Marine Biological Laboratory at Woods Hole. Their growing ability to trace cell lineages in the embryos of a sea of marine invertebrates let them follow the histories of each of the individual cells of the early embryo. The initial cell lineage studies were descriptive, but they offered the first real insights into the processes of development and led naturally and directly to experiments in which the roles of particular cell lineages could be tested by deleting individual cells. Embryology became an experimental discipline with a very different paradigm from that constructed by evolutionary biology during the same period.

Perhaps surprisingly, the embryologists who wandered away from the Haeckelian program did not do so because they rejected its validity. Haeckel's recapitulation theory remained an acceptable part of zoology. It had merely lost its motivating force for embryologists. They abandoned it simply because it had become irrelevant to the new concerns of their discipline. The inadequacies of Haeckel's ideas came to the fore with the triumph of Mendelian genetics and the rejection of Lamarckian mechanisms by biologists. Even that occurred relatively late. The last gasps of Lamarckian inheritance came in the 1920s with the work of Paul Kammerer, who claimed to have demonstrated transmission of acquired traits in his experiments with amphibians. Kammerer's extraordinary studies are detailed by him, with illustrations, in *The Inheritance of Acquired Characteristics.* When the results of his experiments on the midwife toad were explosively shown to have been falsified, the whole notion was discredited among biologists. Nevertheless, the sympathetic treatment of Kammerer by Arthur Koestler in his 1971 book, *The Case of the Midwife Toad,* shows that among nonbiologists there is still sympathy for the notion of organisms improving themselves. Koestler made it clear in *The Ghost in the Machine* that he was repelled by the random and mechanistic worldview of Darwinian selection. The issue here, of course, is the difficulty in accepting the Darwinian mechanism of selection acting on undirected variation, which Koestler referred to as one of the "four pillars of unwisdom." It seemed too random and heartless. Ironically, booksellers often didn't seem to have a clue as to what *The Case of the Midwife Toad* is all about and often shelved it with murder mysteries.

A DEEP INTELLECTUAL DIVIDE

There are some key distinctions that show that developmental biology and evolutionary biology are now played as rather different games (table 1.1). Causality has two different faces for the two disciplines. Evolutionary biologists are wont to cite selection as the "cause" of the presence of some structure, whereas developmental biologists will cite genetic and developmental mechanisms as the "cause." Hypothesis making in evolutionary biology is quite different from the hypothesizing of developmental biologists. Hypotheses in developmental biology focus on mechanistic details, such as the presumptive functions of particular domains in genes, mechanisms for generating patterns in cell sheets, or cell behaviors. Hypotheses in evolutionary biology often use historical scenarios. They can include hypotheses on the effects of long-term selection, on the effects of environmental change, or on the causes and effects of extinction. Some historical scenarios are clearly "just-so stories." However, historical scenarios often can be tested by tracing their predicted effects in the fossil record. This is clearly a rather different mode of hypothesis testing from that done on hypotheses about the function of living organisms.

TABLE 1.1. Differences between Evolutionary Biologists and Developmental Biologists in Their Views of Major Biological Qualities

Quality	Evolutionary biologists	Developmental biologists
Causality	Selection	Proximate mechanisms
Genes	Source of variation	Directors of function
Variation	Central role of diversity and change	Importance of universality and constancy
History	Phylogeny	Cell lineage
Time scale	10^1–10^9 years	10^{-1}–10^{-7} years

The things that the disciplines see as significant problems also differ. For evolutionary biologists, the original issue was the explanation of the diversity of life. This was as much Darwin's focus in *The Origin of Species* as explaining adaptation. It is still clearly a driving issue in evolutionary biology, but is all but invisible to developmental biologists.

The time frames of the two disciplines are on the whole very different. Although microevolution can be studied in time intervals of a few years, and thus comes near to the second- to month-long spans of developmental studies, macroevolution—evolution above the species level—usually deals with time frames of tens of thousands to tens of millions of years. The questions about rates and modes of evolution that motivated the efforts of Simpson, and the debate on punctuated equilibrium generated by the proposals of Eldredge and Gould and of Stanley, revitalized paleontology. These ideas have also become central concerns to evolutionary biologists who want to understand macroevolution. The role of catastrophes in causing mass extinctions has become a major question for evolutionary biologists, as has the role of climate change and environment in driving evolution. None of these issues has any place in mainstream developmental biology.

Another substantive way in which the disciplines differ can be seen in how genes are regarded. For evolutionary biologists, genes provide the raw material for the generation of diversity, and thus population genetics has been developed as the major tool for relating the behavior of genes in populations to evolutionary events. This link between population genetics and evolution became the hallmark of the "modern synthesis" that has dominated evolutionary biology for the past half century. Explanations for adaptive success have been sought in terms of displacement of alleles by others that yield a higher fitness. This is hardly an outlook that has any place in current developmental genetics, for which genes are the executors of developmental processes.

A final historical focus of evolutionary biology has been on the definition of phylogenetic relationships, which has traditionally used comparative morphological, embryological, and fossil data. Phylogenetic studies have waned and waxed in importance, and are at a peak again now because of the invention of cladistic methods, gene sequencing, and computer tools for constructing phylogenetic trees. The importance of phylogenetic studies goes beyond inferring phylogenies. They also provide a tool for organizing the most basic historical information in terms of origins, rates of change, extinctions, diversity patterns, and evolution of adaptations. The closest approximation to phylogeny in the methodology of developmental biology is the tracing of cell lineages within individual embryos.

A very different set of priorities has occupied developmental biologists. Questions of phylogenetic relationship long ago ceased to be part of mainstream developmental biology. Instead, a small set of species has been utilized, partly for their convenience as experimental organisms in the study of various developmental processes, and partly because it was thought that universal processes could be explicated from suitable model organisms. Organismal diversity has been valued only inasmuch as some embryos are more suitable for the study of particular processes than others. On the whole, diversity is to be avoided. Standard culture conditions, and for some species genetic stocks, provide the consistent developmental stages necessary for reproducible experiments. In a more profound sense, diversity is not merely an inconvenience. It also has been regarded as a sort of epiphenomenon, the frills and digressions surrounding and obscuring the real elements of development. To developmental biologists, there is a mechanistic universality in developmental processes despite any diversity of ultimate outcome. Clearly eggs of different species yield different animals, but all arise from a small set of basic processes deployed in specific ways. This outlook arises from biochemistry and genetics, which have demonstrated that there is an underlying commonality in basic life processes, and that basic processes can be discovered through the use of a reductionist approach. Development can be considered as composed of separable processes such as fertilization, cell cleavage, pattern formation, gastrulation, induction, morphogenetic movements, and differentiation, which have common mechanisms extending across wide phylogenetic distances. Genes have assumed a primacy in developmental biology in the past decade. Genes are not regarded as important in development because they are sources of variation; after all, variation is to be avoided. Genes are important because they are the controllers and executors of developmental processes. Mutations are utilized to dissect these processes by studying their effects upon particular developmental events.

The consequence of a century of the practitioners of developmental and evolutionary biology politely ignoring each other is that when they are brought together, there is sometimes a mutual flummoxing by jargon and, more seriously, an incomprehension of each other's research traditions and paradigms. My favorite episodes are those in which a speaker on the intimate molecular workings of some embryo will pause two minutes before the end of his talk and say, "Now for evolution." With this incantation, he puts up a slide of his favorite gene sequence aligned and compared with similar DNA sequences extracted from a computer database. Although often informative in placing the gene in a functional context by its similarity to other genes of known function in other organisms, this approach does miss some of the subtleties of evolutionary biology. Yet a boundary discipline exists, and its investigations can yield important complementary insights not possible in either discipline alone.

ISSUES FOR AN EVOLUTIONARY DEVELOPMENTAL BIOLOGY

The central problem for evolutionary biologists interested in development has been how morphology is transformed in evolution. In 1922 Walter Garstang made the very basic observation that because the morphology of animals arises anew in each generation, evolution of new animal forms has to be viewed as a problem in the evolution of development. Although this is a simple concept, there was considerable difficulty in actually executing studies based on the idea. Development and evolution certainly offer a facile sort of analogy to each other: both are processes of change. Although this analogy was compelling during the nineteenth century, it is ultimately sterile. Development is a programmed and reproducible process. If we accept Darwinian selection, evolution can be neither. The apparent inevitability of development was daunting. To connect it effectively with evolution, two major ideas had to be accepted. The first, pointed out by Garstang, is that larval stages also face the rigors of life. Mendelian genetics allows new traits to appear at any developmental stage, and natural selection potentially operates upon them as it does upon traits expressed in adults. The second major point is that although any ontogeny looks inevitable and inextricably orchestrated in its flow, it is not a single process. There are a large number of processes at work, some more or less coupled to others. It was Joseph Needham who, in 1933, using an engineering metaphor of shafts, gears, and wheels, suggested the idea of dissociability of elements of the developmental machinery. He pointed out that it is possible experimentally to separate differentiation from growth or cell division, biochemical differentiation from morphogenesis, and some aspects of morphogenesis from one another. The implication of this idea is enormous: developmental processes could be dissociated in evolution to produce novel ontogenies out of existing processes, as long as an integrated developmental program and organismal function could be maintained.

In his book *Embryos and Ancestors,* the British zoologist Gavin de Beer laid out a simple solution to how dissociation might work. De Beer put his consideration of the interface between ontogeny and phylogeny in terms of a unifying developmental mechanism that would explain the relatively easy transformation of form suggested by evolutionary histories. That universal mechanism was heterochrony, the concept that events in development can be shifted in timing relative to one another to produce new ontogenies. Heterochrony is simply an evolutionary dissociation in timing. So many examples of heterochronies have been documented or suggested that the prevalent view is that heterochrony is the most common mechanism for evolutionary changes in animal form. One example is the suggestion that humans resemble young apes more closely than adult ones, and that human evolution might have involved developmental changes resulting in a more juvenilized morphology in sexually mature adults. Unfortunately, the evidence for such a simple and elegant relationship is weak. However, some timing changes appear potentially important, notably the prolongation of the rapid prenatal growth rate of the brain in human infants.

In the absence of other readily applied concepts, heterochrony came to provide the dominant explanation at the developmental level for evolutionary changes in morphology in fossil as well as living organisms. Heterochronies can readily be comprehended in terms of various dissociations between rates of somatic versus gonadal maturation. As shown by de Beer and later by Gould, strikingly different results obtain from accelerating or retarding one or the other. The persistence of the idea that heterochrony is the major mechanism for evolutionary changes is probably due to its appealing simplicity and to the ease with which examples can be fit to the various predicted categories of heterochrony. Although it is applied widely as an explanation, it is not so certain that heterochrony is a universal mechanism. Heterochronic results are inevitable. We measure all developmental events along a time axis. Therefore, anything that happens to alter the course of development has temporal consequences. Nevertheless, heterochrony has been a very important concept because it has provided a simple unifying mechanism around which data can be ordered. It was vital to the creation of evolutionary developmental biology as a discipline because it tied observations on many evolving lineages into a coherent system of explanation operating at the level of ontogeny. We will return to heterochrony in chapter 8.

The evolution of developmental processes should be of compelling interest, but it touches upon only a part of what has been a long-standing debate over a basic issue in evolutionary theory. The argument has raged pretty much from the time of publication of *The Origin of Species* as to whether natural selection can elicit responses in any direction from the apparently random mutations that appear in organisms, or whether evolution is actually driven by the operation of internal factors. Extreme antiselectionist views were held by prominent paleontologists around the turn of the century because they read the fossil record as revealing strong directional trends, which they interpreted as being maintained in the face of selection, even to driving lineages to extinction. Orthogenesis, the directed linear mode of evolution favored by the paleontologists, replaced random variation with internally directed change. With the triumph of Mendelian genetics, such views became more difficult to defend. The reading of the record was also incorrect. The "pull of the present" has exerted a strong influence. In his *Principles of Geology,* Charles Lyell, Darwin's mentor in geology, used the percentage of similarity among fossil mollusk shells to calibrate the Cenozoic time scale. The later evolutionary implication of this work would be that animals evolve in a linear manner, from fossil ancestor through intermediate forms to living descendant. Since linear patterns of evolution were expected, that is what was seen. Thus, there was for a long time in the American Museum of Natural History in New York an exhibit of the linear evolution of the horse, from four-toed *Eohippus* through three-toed intermediates to the single-hoofed modern race-horse. The real record of horse evolution can be portrayed as a bush that at different times sprouted widely varying lineages of browsers and grazers. Most of those lineages became extinct and are not ancestral to the one surviving genus, *Equus.* The pull of the present as well as expectations of a linear pattern of evolution distorted the picture to make it seem that there had been a majestic 50-million-year evolutionary procession to the modern horse.

Although the idea of internally directed evolution is not tenable, the potential role of internal factors or constraints remains a viable one even in a selectionist context. Existing genetic and developmental systems are not neutral features, nor are they necessarily readily dissociable. Existing developmental systems must produce constraints on the degree of freedom with which selection can operate. Thus, a major principle of evolution exists beyond an all-powerful selection working on randomly generated variation, and the inner workings of genetic regulatory systems and developmental processes as well as their histories must be considered as key elements of evolution. Various discussions by Alberch, Jacob, Müller, and Thomson in the past decade have translated the internalist perspective into modern and researchable terms.

Finally, progress does not come only from good ideas generated within a discipline. Just as often, new techniques create previously unimagined possibilities. The tools of molecular biology have opened the diversity of species in nature to our study. It is literally now possible to turn over a stone in the woods and extract and clone the DNA of any obscure creature—pill bug, vinegar worm, or slime mold—that one happens to find there. That fact opens astonishing possibilities for the sampling of living diversity. It is possible to explore any genome by isolation and sequencing of genes. In the most obvious sense, gene sequence data contain phylogenetic information. That is of immense interest in itself, but there is much more. The genomic information that can be extracted from any organism provides a connection to the genes and developmental controls of better-studied model organisms. The most intensely studied animals in developmental genetics are the fruit fly *Drosophila melanogaster* and the nematode *Caenorhabditis elegans*. The genes that control developmental processes in these organisms, their sequences, and their functions have been detailed in thousands of research papers. It would be simply impossible to devote even a fraction of these efforts to representatives of other interesting evolutionary lineages, and it is not necessary. When a homologous gene is cloned from a more obscure species, its presence suggests the potential existence of a related developmental role. Its sequence can reveal structural domains that indicate the kind of function the encoded protein might perform: such potent entities as DNA-binding domains, various extracellular domains involved in ligand binding, transmembrane domains, and potential enzyme active sites can all be recognized. If embryos can be obtained, other techniques can be applied. In situ hybridization, for instance, allows the visualization of specific gene transcripts within the particular cells of the embryo in which they are expressed. That, combined with knowledge of the site of action of the gene and its function in the model, allows very powerful inferences about its role in the unknown species. The roles might or might not be the same; insights about the evolution of developmental controls have been obtained either way.

Because we now have the ability to clone regulatory genes, purely developmental questions have quickly become issues in evolutionary biology as well. The most prominent of the gene domains that has been studied as I outlined above is the homeobox. This is a sequence that encodes a 60-amino-acid-long protein domain rich in basic amino acids that has been shown by Qian and co-workers and Kissinger and co-workers to fold into a helix-turn-helix structure that binds to DNA in a highly specific manner. The proteins that contain homeodomains are transcription factors that act to turn on genes that possess a homeodomain binding site in their control region. The homeobox was discovered by Scott and Weiner and by McGinnis and co-workers in homeotic genes of *Drosophila*. Homeoboxes are highly conserved in evolution and so were very quickly found by McGinnis and his co-workers to be present in other phyla as well. The surprise was that although homeoboxes are associated with genes that control external body segmentation in *Drosophila,* they are also present in nonsegmented phyla such as echinoderms and nematodes, as well as in the vertebrates, which exhibit some internal segmental features. In all organisms in which they have been studied, homeobox-containing genes homologous to those of the homeotic genes of *Drosophila* function in some aspect of developmental specification of aspects of the body axis.

The finding that homeobox-containing genes are major regulators of axial specification in annelid worms, insects, and vertebrates shows that such ancient regulatory molecules are shared by widely divergent animal groups.

These genes have been detected even in the most primitive metazoans, the cnidarians (such as hydras, jellyfishes, and anemones), as well as in many other phyla (by Holland, Murtha and co-workers, and Schierwater and co-workers). The phylogenetic distribution of homeobox-containing genes indicates a pattern of gene duplication and divergence as well as profound conservation. It is also clear that the roles homeobox-containing genes play in development have undergone significant modification. Homeobox-containing genes in vertebrates establish axial polarity in the central nervous system, but do not set up epidermal segmental patterns as in insects. Additional roles can be shown for them in aspects of vertebrate development and co-option for new functions, such as in neural crest cell patterning as defined by Hunt and his co-workers. Other regulatory gene families, such as the steroid receptor family discussed by Evans and by Amero and colleagues, show analogous patterns of evolutionary expansion and co-option to provide genetic raw material for regulatory innovations in the evolution of development. It has happened again and again in evolution within numerous families of regulators.

The revelation of wide-ranging conservation and co-option of regulatory genes is a new and crucial one for two reasons. First, it presents us with the linkage between an evolutionary continuity of gene structure and the changing functions of those genes in the evolution of development. Second, the underlying genomic similarity in the development of highly disparate animals reveals one of the things that makes the evolution of complex forms possible at all. If each new species required the reinvention of control elements, there would not be time enough for much evolution at all, let alone the spectacularly rapid evolution of novel features observed in the phylogenetic record. There is a kind of tinkering at work, in which the same regulatory elements are recombined into new developmental machines. Evolution requires the dissociability of developmental processes. Dissociability of processes requires the dissociability of molecular components and their reassembly. It is our task to unravel the operation of these mechanisms from the concrete evolutionary comparisons that we have available to us.

A comparative approach is inevitable and powerful in understanding anything that has a complex history. One would not bother to ask about the history of a hydrogen atom. Whether it has been hanging around in a water molecule in the deep sea or drifting endlessly in interstellar space gas, its historical past has no influence upon its properties. Nor does history affect the properties of black holes, which are ultimately simple and self-contained. However, as Carl Sagan has pointed out, for the planets, physical objects with complex histories, "comparative planetology" has been brilliantly informative. Completely unexpected phenomena have become the expected as planetary probes launched into the solar system touch planets and moons that have evolutionary histories completely different from that of our own world.

Developmental genetics and evolutionary developmental biology both explicitly compare the properties of a modified descendant organism with the features of its ancestor. In developmental genetics, the ancestors are wild-type individuals of the species, and the modified descendants are organisms that bear mutations that affect some developmental process. The difference between the two disciplines lies in their objectives. Developmental genetics asks its questions of situations in which mutations have been used to knock out the function of a gene. An examination of the resulting phenotype is used to reveal the functional role of that gene in development. Evolutionary developmental biology seeks to explain the evolutionary transition between two different but workable developmental alternatives. The results of the two disciplines frequently converge: key regulatory genes and pathways turn up again in other organisms. Functional analyses of genes illuminate evolution, and evolution displays the range of possible functions for homologous genes. Nevertheless, the perspectives on the overall structure of development that emerge from the two approaches differ in important ways that will be described in other parts of this book.

The study of the role of development in evolution must shift from a focus on theoretical considerations to the framing of experimental questions that can reveal the mechanisms by which developmental processes influence evolutionary change. Subsidiary questions about the evolution of development have arisen from other themes in evolutionary biology. A major goal of evolutionary biology is to understand the origins of novel morphological or life history features that have led to the appearance and radiation of new groups of animals. The demonstration and explication of developmental innovations that underlie the origins of key features must become a major part of the link between the disciplines. Comparative studies have provided a way to trace the directions of evolutionary changes in development, the number of times they might have occurred independently, and even their potential reversibility. Evolutionary data also reveal important perspectives on the large-scale logic of ontogeny itself. All of these issues form a crucial part of the central theme of this book, the origin of body plans and their subsequent modifications. Evolutionary developmental biology offers an experimental approach that can carry us beyond speculation based on untested hypotheses.

Darwin ended *The Origin of Species* with the following statement:

It is interesting to contemplate a tangled bank, clothed with many plants of many kinds, with birds singing on the bushes, with various insects flitting about, and with worms crawling through the damp earth, and to reflect that these elaborately constructed forms, so different from each other, and dependent upon each other in so complex a manner, have all been produced by laws acting around us. These laws, taken in the largest sense, being Growth with Reproduction; Inheritance which is almost implied by reproduction;

Variability from the indirect and direct action of the conditions of life, and from use and disuse: a Ratio of Increase so high as to lead to a struggle for Life, and as a consequence to Natural Selection, entailing Divergence of Character and the Extinction of less-Improved forms. Thus, from the war of nature, from famine and death, the most exalted object which we are capable of conceiving, namely, the production of the higher animals, directly follows. There is a grandeur in this view of life . . .

That grandeur is still there. We are in a position to add to Darwin's synthesis by being able to probe more deeply into what were for him impenetrable laws of growth, reproduction, and inheritance.

2

Metazoan Phyla and Body Plans

In those days we weren't expecting any more surprises,—*Qfwfq nar-
rated,*—by then it was clear how things were going to proceed. Those
who existed, existed; we had to work things out for ourselves: some
would go farther, some would remain where they were, and some
wouldn't manage to survive. The choice had to be made from a limited
number of possibilities.

Italo Calvino, *t Zero*

QUESTIONS OF MACROEVOLUTION AND BODY PLANS

The major concern of this book is to establish the connection between the
developmental processes that produce body structure in each generation and
the evolutionary processes that produce new anatomical features and novel
animal body shapes. Much of evolutionary biology is concerned with microevolution. This level of study deals with small evolutionary changes such as
those observable in laboratory populations and sometimes in the wild, including mutation, the nitty-gritty of gene flow between populations, the close
work of selection, and speciation itself. These aspects of evolution are unde-
niably important. Studies of living populations show that rapid alterations in
morphology can occur within species under selection. Although selection
acting at this level is relevant, these microevolutionary phenomena are not
at issue here. This book will discuss evolution above the species level, macroevolution.

In macroevolution we come face to face with what goes on over the long
time spans that are the theater for the evolution of novel anatomies. Macroevolutionary events lie beyond the short-duration processes of development
of an individual or microevolution in a small population. The difficult question posed here is how the two very different phenomena of the genetically
programmed change of development at an individual level and the unpro-
grammed long-term evolutionary change in a lineage of organisms are interrelated.

I have placed the discussion of the interaction of development and evolution within the context of the idea of body plans. A body plan (sometimes
referred to by writers in its German form, *Bauplan*) is a basic pattern of

anatomical organization shared by a group of animals—at the highest level of distinction, by a phylum. All chordates share a basic body organization distinct from that of any other phylum. Classes of the vertebrates, a subphylum of the chordates, share a basic set of anatomical features characteristic of the phylum Chordata despite the substantial distinctness of the anatomical features of the classes from each other. The concept of a shared body plan recognizes the hierarchical structure of animal form that has been long perceived by systematists engaged in classification, one of nested sets of anatomical organization. Within each class, orders share a set of characteristics that defines their membership in that class, and so on to the shared features of species in a genus. Systematists think of the features that unite clades at any level of the hierarchy as defining membership in those clades. We must ask whether there is any mechanistic meaning to such a coalition of features. It could be that the sum of shared features is merely that, an arbitrary collection of colored balls in a teacup. If there is no functional reason for the red and blue balls to occur together, mere historical accident could account for their association. We surmise that the commonality of anatomy among groups sharing a body plan is not merely an incidental sharing of descriptive features, but rather, that those features are the component parts of a deeply integrated shared pattern of development. The depth of integration will depend on how basic the complex of features is to the group under consideration. Those features that characterize the phylum-level body plan may be the most deeply integrated of all. The sharing of a dorsal nerve cord and notochord indicates a more profound integration than the presence of feathers or hair. Novel features arise in animal evolution as a result of modification of developmental pattern. A new structure, such as a wing or a flipper, requires a change in the rules governing the patterning of the limb. The evolution of new body plans must have required novel ways of integrating very basic developmental processes.

The body plans that I'll be primarily concerned with are those that distinguish the animal phyla, although body plans at other hierarchical levels exist and will be discussed as well. The origin of phylum-level body plans most clearly demonstrates the concept and some of the major problems of macroevolutionary change.

This book is concerned with the ontogenetic significance of three great macroevolutionary patterns. The first is the Cambrian radiation, in which the diverse body plans that we call phyla first make their appearance in the fossil record. This sudden appearance of the phyla seems to correspond to a rapid evolutionary radiation of the animals, not merely to the vagaries of the geologic record. The Cambrian event poses an enormous and dramatic problem because so many phyla arose so rapidly. Because the animals have a single ancestry, the diverse anatomies of the various phyla are related by descent and divergence. The Cambrian radiation poses the question of how developmental mechanisms could have been plastic enough to generate the thirty-five distinct body plans of our living phyla (as well as a few phyla that seem not to have survived beyond the Cambrian).

The pattern of evolutionary history revealed by the fossil record and by the application of molecular biology to phylogenetic questions also raises a second major problem: Why has there been such an extraordinary stasis in phyla since the Cambrian radiation? Despite the spectacular innovations in animal form that have evolved since the Cambrian, no new phyla have appeared. Stasis in body plans may inform us about organizational properties and processes in development that constrain change. This possibility requires us to decide whether stasis is a matter of ecological circumstances or whether there really are intrinsic constraints on developmental mechanisms. Demonstration of the existence of such constraints would allow us to define a set of internally operating processes connecting development with selection.

Finally, evolution of novel forms within the body plans that define the phyla has produced an enormous range of innovations. These are the events of evolution during Phanerozoic time, the origins of the classes of vertebrate land animals, spiders, insects, wings, hooves, trunks, hands, and brains. Many of these changes have been dramatic in morphological result and rapid in evolutionary time. Development has not acted as a brake on these macroevolutionary events. In fact, radical evolutionary alterations in development may underlie macroevolutionary change. Once again, we are faced with deciding whether and how the nature of developmental systems affects evolution, and what internal evolutionary processes connect development with external selective pressures.

This book explores these questions in several ways. To start, I will introduce body plans and the Cambrian radiation. I will also introduce phylogenetic concepts. The importance of phylogeny is not merely one of tracing evolutionary relationships, as interesting as these may be in their own right. Phylogenetic information is crucial for any study of evolutionary changes in development. I stress this point here because phylogenetic considerations have played little part in the experimental study of development that has dominated the twentieth century. Much of the initial part of this book will emphasize the fossil record because it is a fundamental source of information, and yet it is poorly understood by most molecular and developmental biologists. Lest anyone doubt that statement, I can only note that in a 1994 review, a prominent developmental geneticist considering the evolutionary role of homeobox-containing genes lamented that it is unfortunate that invertebrates make poor fossils. In reality, all the phyla but one are invertebrates, and the vast bulk of fossils are of invertebrates.

The second half of the book deals with evolutionary aspects of developmental biology. I consider three overall goals to be important here. The first is to present the issues (and they sometimes appear paradoxical) that tie developmental biology to the evolutionary concept of body plan and to the evolutionary history of body plans. Second, I discuss the mechanisms that connect development to evolution. This is a crucial topic because we are not merely interested in describing how developmental processes and patterns evolve. There is a much more profound issue in that existing developmental patterns and mechanisms influence or constrain what natural selection can elicit in the course of evolution. There are thus aspects of evolution that are controlled by the internal order of organisms, and potentially by evolutionary processes that operate internally on developmental features. These developmental constraints provide an important challenge. The three central chapters of the book revolve around them and the mechanisms that may drive the internal part of evolution. Finally, I will consider the phenomenology that must be explained and the examples by means of which problems are being studied in the newly thriving experimental discipline of evolutionary developmental biology.

I intend this book to provide an approach to the synthesis of the disciplines of developmental and evolutionary biology. This nascent study is only beginning to find its own paradigms and means of investigation. Therefore, I don't confine myself to concepts and results, but also lay out my own methodological prejudices and suggestions of how the enterprise should be conducted.

HOMOLOGY AND BODY PLAN

The concept of body plan is an old one, deriving from two distinct intellectual roots. The first is the notion put forward by Richard Owen in the 1840s that each group of organisms shares an "archetype," an essential design that underlies the diversity of anatomical detail among the species in the group. The second owes to Von Baer's definition of the stages that characterize the development of the members of each animal phylum. Von Baer's developmental studies showed very powerfully that the body plan is deeply immanent in the process of ontogeny for each species. He found that the body plan characteristic of its phylum is blocked out relatively early in the development of any embryo, long before the features of the species appear. That point in development is called the phylotypic stage.

Owen is an interesting case. He was close to offering evolutionary (in his time called "transformationalist") explanations for the existence of creatures in the fossil record that are intermediate in structure between now distinct groups, for example, transitional forms between fishes and amphibians, but he never accepted Darwinian evolution. Owen seems to have been poised on the intellectual edge between the old transcendentalist tradition, in which

Russell places him in his masterful *Form and Function,* and the new evolutionary biology.

Owen proposed that in a given group of animals—in his example, vertebrates—there was a shared basic design. The elements of that design could be extracted to yield an archetype containing the general features that unite all members of the group, primitive or specialized. For vertebrates, the archetype that Owen constructed consisted of a serially repeated set of vertebral structures. The outcome looked a lot like a fish skeleton, except that Owen (incorrectly) hypothesized that the bones of the vertebrate head also were derived from vertebrae. What connected Owen's archetype to real vertebrates, many of which are quite different from fishes, was the concept of homology propounded by both Owen and Geoffroy Saint-Hilaire. Owen defined homologues as "the same organ in different animals under every variety of form and function." To us, Owen's word "same" means "derived by evolution from a shared ancestral structure."

Owen's concept of the archetype was not an explicitly evolutionary one, but it lends itself easily to an evolutionary explanation. In a dualistic way, the archetype came to be visualized as the ancestor, but it is just as readily interpreted as the body plan. The difference between an archetype and a body plan is that the archetype was considered by Owen to reflect the divine concept for the anatomical makeup of a phylum of animals. The evolutionary concept of a body plan is that it presents the evolutionarily shared and modified anatomical features of a group of animals. Those shared anatomical features are recognized as homologues among the members of the group.

Owen's concept of homology was idealistic. Darwin transformed it by providing a new theoretical basis and gave it an explicitly historical definition. Modern investigators have seen the need to add an informational basis for the historical continuity of homologous structures. As Van Valen has put it, the essential property of homology is the evolutionary "continuity of information" that it reveals. The most obvious evolutionary continuity of information is evident in gene sequence homologies. However, morphological structures are hierarchical, and the development of such complex homologues as limbs poses a difficult problem for defining informational homologues. As recognized by de Beer, Shubin, Shubin and Alberch, Hall, Roth, and Wagner, the formation of structures recognized as historical homologues may not follow the same developmental path. There have thus been attempts to define informational homologues based on the expression of genetic information through development (as, for example, by Gilbert and co-authors). As both gene expression and epigenetic events determine outcome, this is not a straightforward operation. No simple answer to the conundrum yet exists. For the moment we must be satisfied that similar patterns may arise from different developmental pathways. Constraints may operate to maintain pattern while allowing process to vary. I'll consider this issue more fully in chapter 6.

The classic concept of homology as founded solely on structure has been questioned in recent discussions. Lauder, for example, has suggested an explicitly phylogenetic basis for homologies, which would be operationally recognized as features shared by a set of evolutionary lineages that share a common ancestor in a phylogenetic tree. Wagner has suggested that there are three concepts of homology. These are the "historical homology concept," which regards phylogenetic continuity as the primary defining feature; the "morphological homology concept," which regards structural identity as primary; and finally, the "biological homology concept," which regards shared developmental constraints as most important. Although I am seeking developmental explanations for continuity and change in morphological features, including homologous ones, I will rely primarily on the traditional historical and morphological concepts of homology: that is, I will define homologous features as those that share a common evolutionary origin and an underlying commonality of morphology, resulting from a continuity of information. Homology can be operationally recognized on the basis of structural similarity criteria. Phylogenetic information can also be used, as suggested by Lauder, to help us recognize homology, but should not be part of the definition. The definition I've used above carries less risk of circularity than, for example, a developmental homology concept would. Regardless of the definition used, the same methods are used, and the same problems confront anyone attempting to identify homologues in practice.

Owen used homology, as evolutionary biologists have done throughout the past century, to establish the correspondence of structures among animals. This approach has been tremendously fruitful in, for example, tracing the evolution of the bones that make up human and other vertebrate skulls. Homology is also crucial for tracing evolutionary transformations of structures, and even for using DNA sequences to trace phylogenetic relationships. Definitions of homology are one thing, but the recognition of homologies in the real world is another. Owen also suggested criteria by which homologies could be established: similarity, position, and connections. I here follow the criteria for the establishment of homology summarized by Rupert Riedl in his exceptional book *Order in Living Organisms*. The first is the "positional criterion," in which homology is inferred from an element's position in a comparable structural system. This criterion works for morphological structures, such as bones in the skull, and just as well for molecular features, such as the position of bases in a gene sequence. The second is the "structural criterion," by which structures can be presumed homologous if they share numerous features in common. This criterion works best if the positional criterion is also met, but does not depend on it. To use Riedl's example, the testes of male vertebrates share numerous common features, but are located in different places in the body, or even outside of it in the mammalian scrotum. Again, this criterion also works for molecular characters, as in DNA sequences. In this case, if the bases at the corresponding positions in the sequences being compared are identical, then the structural criterion is satisfied. The third is the "transitional criterion," by which homology can be established if a set of commonly shared historical transitions can be demonstrated for the origin of a given feature. Homology is clearest when a direct transition between ancestor and descendant states can be observed in the fossil record. In some cases that has been possible. For instance, the evolutionary history of the vertebrate skeleton can be inferred from the thousands of vertebrate fossils that allow us to trace the history of each bone or tooth. A molecular parallel is provided by the tracing of gene sequences derived from lineages produced experimentally in the laboratory. In these cases, the mutational steps can be tracked.

Historical transitions cannot always be directly observed, but there is another potentially important source of transitional information: ontogeny. If we can demonstrate a commonality of development for suspected homologues, it greatly strengthens the inference of homology. Thus, Gaupp's classic description of the development of the bones of the middle ear in mammals has confirmed the paleontological evidence described by Crompton and Jenkins for their derivation from elements of the reptilian jaw. The bones that became the mammalian middle ear were a part of the jaw articulation system of reptiles. As that articulation system was superseded in the evolution of primitive mammals, its component bones became available for co-option to another function. We need to be somewhat cautious, however, in using developmental criteria for homology because, as recognized by de Beer and many other biologists, structures that are clearly homologous do not always arise in the same way in development. There is not an exact one-to-one mapping of genes, or of developmental pathways guided by genes, to morphological homologies.

The recognition of one final and very puzzling type of homology also owes to Owen. Serial homology is the similarity of structures repeated along the body axis. In arthropods, that would include serially reiterated structures such as the body segments and limbs. In vertebrates, the most prominent serially repeated structures are the vertebrae, somites, and the forelimbs and hindlimbs. A fascinating problem is immediately evident. Unless we choose to believe that the first terrestrial vertebrates leapt up on the land like fishy kangaroos, we do not suppose that the forelimbs are homologous to the hindlimbs because they evolved from them. What made Owen see homology between vertebrate forelimbs and hindlimbs is their shared bone and muscle structure. The basis for that commonality is still unresolved. The most satisfactory explanation for serially homologous morphological structures would be an underlying true homology in genetic instructions and developmental programs. Molecular genetic tools now make it possible to obtain the information necessary to test for such underlying relatedness in constructional rules. In the case of forelimbs and hindlimbs, serial homology implies difficult and interesting alternatives. For example, the structural resemblances may not have a shared genetic basis, with correspondence of form reflecting constraints on function. Alternatively, forelimbs and hindlimbs may have evolved independently, but using the same genetic machinery, from the continuous fin fold that ran along each side of the body in primitive vertebrates. Or possibly, once the genetic machinery to specify a limb pair had evolved, it may have been transposed in site of expression to produce duplicate limbs in other positions. In chapter 10 we will examine a suggested resolution of this question based on knowledge of the developmental genetics of the limb. However, few cases of serial homology have really yet been solved. There are many examples of different embryological precursors for well-established homologues, and there are many instances of apparent homologues that arise by different developmental processes. The principle of co-option of gene programs for patterning of repeated structures may apply widely. Each case needs to be considered in light of knowledge of its genetic regulatory rules, and if available, knowledge of transitional criteria obtained from the fossil record.

Homology is the basis for all evolutionary comparisons. Phylogeny, molecular evolution, the evolution of developmental features, all can be understood only through comparisons in which homologies are correctly assessed. The recognition of homology requires application of the positional, structural, and transitional criteria listed above to real cases. In general, similar features are considered good candidates for homologues. However, because similar functional demands require similar solutions, organisms have often evolved similar features independently. This nonhomologous similarity, first recognized by Owen, is called analogy, and is manifested as body parts that have similar functions and convergent structures, but that do not share a common evolutionary origin: legs of arthropods versus legs of vertebrates or legs of pianos. Analogies can be operationally identified by means of incongruities in underlying features, or by means of incongruities in the appearance of seemingly homologous features in phylogenetic trees. Some systematists (pattern cladists) have claimed that homologies should be defined as shared derived features in cladograms. This idea confounds the definition of homology with the criteria for its identification and is a nonstarter. Analogies are important both because they can be confounded with homologies when one is trying to infer phylogenetic relationships and because, when correctly identified, they reveal that important adaptations in evolution have been acquired independently.

THE LIVING PHYLA

The deepest manifestation of body plan among animals is at the taxonomic level of the phyla. The idea of body plans can also be applied at lower taxonomic levels, as, for example, in insects (a class) or in whales (an order), wherever a distinct anatomical organization underlying an animal group can be distinguished.

There are about thirty-five living animal phyla. Some representatives of major phyla are presented in figure 2.1. The uncertainty expressed by "about" comes from a doubtful phylum erected for one group of small, unusual animals seen only once and long ago, as well as disagreement about the distinctness of one or two other groups that belong to either related phyla or the same phylum. The basis for much of the currently accepted metazoan phylogeny comes from the work of Libbie Hyman in the 1940s and 1950s. In her great compendium *The Invertebrates,* she evaluated both anatomical and embryological traits of the phyla in terms of the construction of a coherent phylogenetic tree.

The most primitive metazoans (the mesozoan phyla, placozoans, mono-blastozoans, rhombozoans, and orthonectids, which are all very small animals with few cell types, and the sponges) are multicellular and have differentiated cell types, but their cells are not organized into discrete tissues. The first branch in the metazoan tree is thus between these primitively organized multi-cellular animals and those with tissues and organs. More complex metazoans come in radial versus bilateral symmetries, with the majority bilateral. Radiates (cnidarians and ctenophores) possess only a tissue level of construction, whereas the bilaterian animals have organs. Thus, a second split distinguishes between a more primitive set of radially symmetrical animals and a larger, more advanced group, the Bilateria. There is a second important distinction between these groups. Diploblastic animals such as the cnidarians (i.e., hydras, jellyfishes, anemones, and corals), and ctenophores (comb jellies) have two body layers, an outer ectoderm (skin, and the diffuse nerve net and muscle fibers) and an inner endoderm (gut). Lying between the ectoderm and endoderm is a hydrated supportive connective tissue layer called the meso-glea. The mesoglea is poorly cellularized. All of the bilaterians have a more elaborate triploblastic structure. Like diploblastic animals, they possess an outer ectoderm and an inner endoderm. However, in the evolution of the triploblastic animals, a novel intermediate layer, a cellular mesoderm, appeared between the inner and outer cellular layers. This mesodermal layer has made possible the evolution of more complex tissues, larger organs, and a powerful body musculature.

It is with the most primitive of the bilaterians, the platyhelminth flatworms, that complex nervous systems appear, with a brain and major longitudinal nerve trunks. In cnidarians there is only a nerve net. The flatworms are the most primitive of the bilaterians in lacking two anatomical features seen in other bilaterians. They have no body cavities, and their digestive systems have only a pharynx that serves for both food intake and waste excretion. The coelom and a pass-through gut with separate mouth and anus are features of more advanced bilaterians, and mark the next branching among living metazoans (fig. 2.2).

Hyman discerned three kinds of coeloms, or body cavities in which organs are suspended. In the first and most primitive case, the cavity is absent. Thus flatworms are acoelomate. In the second case, a mesoderm-lined body cavity, or eucoelom, is present. The major bilaterian animal phyla are eucoelomates. They include the phyla with the best fossil records and contain some of the most familiar living animals, including chordates, arthropods, mollusks, annelids, echinoderms, brachiopods, and bryozoans. Finally, a number of generally small and obscure phyla (i.e., the rotifers, gastrotrichs, nematodes, nematomorphs, acanthocephalans, gnathostomulids, kinorhynchs, priapulans, entoprocts, and loriciferans) possess a "pseudocoelom." The pseudocoelom is a kind of coelomic cavity, but it lacks the mesodermal cell lining characteristic of the eucoelom, and the pseudocoelom supposedly arises differently in development as a persistent blastocoel. The pseudocoelomates have been classed as a superphylum group by Hyman and everyone since, and are considered more primitive than the eucoelomates. However, most pseudocoelomates are very small animals. This observation creates a certain suspicion that many of these phyla independently acquired the pseudocoelom from an ancestral eucoelom as a consequence of severe size reduction and consequent simplification of structure during their origins. The mere fact of their simplicity and the absence of a eucoelom are not good uniting features. Pseudocoelomates may be diverse in origins, and thus polyphyletic. Because these animals share even fewer features with other phyla than usual, and in addition (except for priapulids) lack a fossil record, molecular approaches may be the only possible way to trace their phylogenetic relationships.

The major eucoelomate phyla have rich fossil records that extend to the Cambrian. However, because these phyla appear in the fossil record as distinct as they are at present, interconnections between them have been difficult to establish. There is also a suite of less well known eucoelomate phyla, including the urochordates, chaetognaths, sipunculans, echiurians, pogonophorans, nemerteans, tardigrades, and phoronids, that lack significant hard parts and much in the way of fossil records. The sorting of the eucoelomates

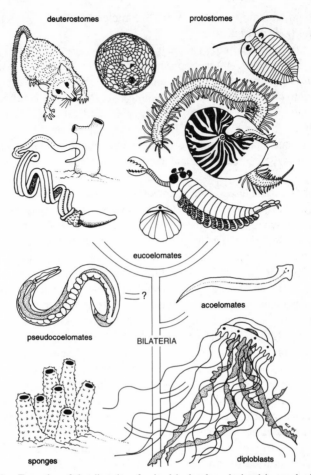

Figure 2.1. Examples of the diversity of animal body plans depicted in rough phylogenetic order. Sponges show the most primitive grade of metazoan body organization. The medusa, a cnidarian, represents the diploblastic grade of organization. The most primitive animals with bilateral symmetry and a triploblastic organization are the platyhelminth flatworms. The phylogenetic position of pseudocoelomates such as the nematode worm is still unknown. The major coelomate phyla are divided into two major superphyla, deuterostomes and protostomes. The deuterostome phyla are represented by an acorn worm (hemichordate), an early Paleozoic edrioasteroid (extinct echinoderm), a saclike ascidian (urochordate), and a mammal (chordate). Protostomes are represented by the cephalopod mollusk *Nautilus,* a polychaete annelid worm, and a trilobite (extinct arthropod). The Cambrian animal *Opabina,* with its peculiar head appendage, has not been assigned to a living phylum, but its anatomy suggests that it is a protostome. A brachiopod is also shown. Molecular sequence data place brachiopods within the protostomes, but on embryonic grounds they have been placed by various investigators in the deuterostomes, the protostomes, or distinct from both, illustrating the uncertainties of phylogenetic relationships among animal phyla. Only about a third of phylum-level body plans are shown. The protostomes and pseudocoelomates include the majority of phyla. (Drawing by E. C. Raff.)

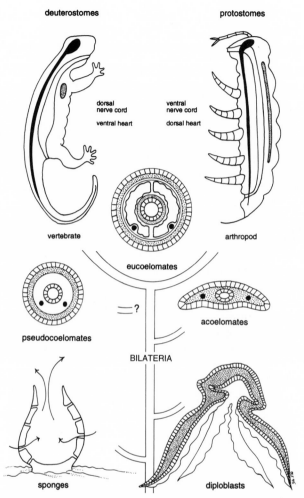

Figure 2.2.   Principal elements of body plan organization. Shared constructional features have been used to infer phylogenetic relationships among metazoan phyla. Sponges possess multiple cell types, but they are not organized into discrete tissues. Diploblastic animals have tissues and are organized into two cellular layers, ectoderm and endoderm, separated by an acellular meso-glea. Most animal phyla exhibit bilateral symmetry combined with a triploblastic arrangement of ectodermal, endodermal, and mesodermal cell layers. The flatworms have no coelomic cavity; the space between body wall and gut is occupied by mesodermal cells. The majority of other phyla have an open body cavity. Eucoelomates have a coelom lined by a mesodermal sheet, the mesentery, which supports the internal organs. Pseudocoelomates, ironically, are defined by a coelomic cavity that lacks a mesentery; this condition may represent a primitively retained blastocoel or a secondary simplification from a eucoelomic ancestry. Two eucoelomate body plans are contrasted to illustrate the profound topological differences that often separate phyla: vertebrates have a dorsal nerve cord and ventral heart, whereas arthropods feature ventral nerve cords and dorsal nervous systems. Ventral nerve cords are the rule among protostome phyla.

into superphylum assemblages has posed a difficult problem. Some shared features of adult anatomy have been taken as offering strong links—for instance, segmentation. In segmented animals the body is divided into repetitive elements that each contain a portion of the coelomic cavity as well as serially repeated organs, musculature, and appendages. Segmentation has been the basis for uniting two very prominent phyla, the annelid worms and the arthropods, as the Articulata. One minor phylum, the pogonophorans, also exhibits segmentation of the rear part of the body. The members of another group of phyla (the phoronids, the ectoproct bryozoans, and the brachiopods) possess a complex feeding structure, the lophophore, that is composed of a more or less elaborate ring of hollow tentacles that surrounds the mouth. This assemblage of phyla is known as the lophophorates. However, adult anatomical features have not proved sufficient to build a coherent phylogeny that maps out the relationships among these supposed superphylum groups. Some phyla, such as the echinoderms, are so distinct that they share only the barest minimum of features with other phyla. Thus, on the sole basis of adult features, about all we could say is that echinoderms are eucoelomate animals. The key to organizing a coherent framework of phylum-level relationships on morphological grounds has proved to lie, at least in part as envisaged by Haeckel and his followers, in the features of early development. To return to the example of echinoderms, despite the uniqueness of their adult anatomies, their embryonic development readily ties them to vertebrates and other phyla that exhibit the "deuterostome" mode of development.

The use of mode of early development in this manner has a powerful influence on our ideas of relationships among phyla. For Haeckel, there was a definite theoretical underpinning for the recording of ancestral relationships in the patterns of early development because he believed that early development recapitulated the adult stages of an animal's ancestors. There is no longer any such theoretical grounding for the use of early developmental stages in phylogeny. However, similar larval forms belonging to otherwise distinct phyla show that early development is generally long conserved in evolution, and that similar larval forms provide homologies that can be used for phylogenetic purposes. In the case of Darwin's favorite animals, barnacles, early development was crucially important in deciding to what phylum they belonged. Adult barnacles were thought to be mollusks until the mid-nineteenth century, when their larvae were observed to be naupli and very similar to the naupli of shrimp and other crustaceans. As development proceeds, morphogenesis in barnacles and shrimp takes very different courses, but the presence of a nauplius in both was a dead giveaway as to their relationship. Barnacles are arthropods, and what's more, crustaceans. The nauplius did not, however, provide a good Haeckelian clue as to the structure of adult primitive arthropods. The nauplius has no segments and only three

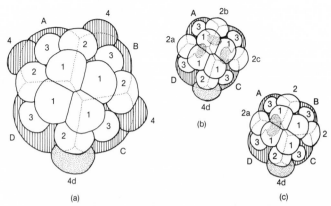

Figure 2.3. Schematic views of spiral cleavage patterns and fate maps shared among several protostome phyla. *(A)* A simplified spiralian embryo showing the spiral arrangement of cells (the smaller cells are micromeres, the large ones macromeres). This diagram simplifies the actual form of the embryo in showing only single cells for first-, second-, and third-tier micromeres in each of the A, B, C, and D quartets. In real embryos these would have continued to divide. The first three tiers of micromeres (unshaded) give rise to the embryonic ectoderm. Some of the second tier of cells also produce ectomesoderm. The mesentoblast cell (4d; dotted) gives rise to the mesoderm, and the macromeres (hatched) give rise to the endoderm. Schematic views of the embryos of two mollusks, *Crepidula (B)* and *Unio (C)*, are also shown to illustrate that despite a great overall similarity, there are differences in cell fate. Embryonic ectomesoderm arises from three second-quartet cells in *(B)*, but from one precursor in *(C)*. (From Raff and Kaufman 1983; drawn by E. C. Raff. Reprinted with permission of Indiana University Press.)

pairs of body appendages; segments and appendages are added during crustacean growth. The most primitive known living and fossil arthropods have long bodies with numerous body segments, each bearing a pair of appendages. Yet direct fossil evidence shows that the nauplius has persisted as the crustacean larval form since at least the late Cambrian, a half billion years ago. The mechanistic grounds for this, as well as for other cases of long-term conservation of larval form in evolution that I discuss in chapter 6, are not well understood. Empirical observations of conservation of early development must serve as the justification for its use in phylogeny.

Early development has offered phylogeneticists a number of apparently solid uniting features. One of these is mode of cleavage. A whole group of phyla (platyhelminths, mollusks, annelids, sipunculans, echiurids, and pogonophorans) exhibits spiral cleavage, in which successive cleavage spindles are oriented at an angle to the preceding cleavage spindle such that each tier of cells lies at an angle to the tier below. As shown in figure 2.3, this characteristic offset of cell tiers gives a spiral twist to the embryo. Spiralian embryos assign cell fates in a very conserved and stereotypic way. The first three tiers of cells give rise to embryonic ectoderm. Some of the second tier of cells also produce ectomesoderm in the larva. The mesentoblast cell produces the definitive mesoderm of the adult animal. The macromeres give rise to the endoderm, the presumptive gut. Both this pattern of cleavage and the assignment of cell fates are remarkably constant across spiralian phyla. These phyla share three other important developmental features. They all produce their coelomic mesoderm by a process called schizocoely, in which two blocks of presumptive mesoderm split to form the cells that line the coelom. Spiralians all have a similar larval form, called the trochophore, and they build a larval mouth in the same manner. The larval mouth in these phyla arises near the blastopore, the site of cell ingression during gastrulation. Thus they have been called protostomes, "first mouth." The protostomes were originally recognized by Grobben at the turn of the century as one great superphylum assemblage, along with the deuterostomes, discussed below. Hyman incorporated Grobben's deuterostome and protostome division into her phylogeny, and these superphylum groups are currently widely accepted.

The protostomes are generally held to include not only the phyla that exhibit the protostome/spiralian mode of development, but the arthropods as well, even though the arthropods have a completely divergent mode of early development. The reason for that has been the similar features shared by arthropod and annelid adults. These include body segmentation, a dorsal heart and circulatory trunk, a ventral nervous system, and a chitin cuticle. Most arthropods do not have a cellular mode of early cleavage. Instead, they undergo multiple nuclear cleavages in a common cytoplasm before nuclei move to the embryo surface and undergo cellularization. This is a derived state, and a few arthropods, notably barnacles, do have a cellular cleavage. Anderson has attempted to homologize that pattern of cleavage with spiral cleavage, but the attempt is forced, and as he has shown, the fate maps of annelids and arthropods are not the same. The closeness of annelids and arthropods is now being questioned. On the basis of a cladistic analysis of morphological and embryological features, Eernisse and co-workers have concluded that these two phyla do not share a close relationship as a monophyletic articulate clade.

The second great superphylum branch, the deuterostomes, includes the chordates, urochordates, hemichordates, echinoderms, and (traditionally, but on molecular grounds evidently erroneously) chaetognaths. At first sight this looks like an impossible grouping of wildly diverse animal body plans. The uniting of these phyla has been entirely based on their embryology and illustrates the powerful influence of comparative embryology in phylogenetic thought. Embryonic cell cleavage in these phyla is primarily radial, in contrast to the cleavage of spiralians. The cells that will form the linings of the coeloms arise from pouches on either side of the gut (a process called enterocoely), and the coeloms that form have a tripartite organization. Finally, the mouth arises at a site distant from the site of gastrulation; thus this assemblage is called the deuterostomes, or "second mouth." Because they also have a tripartite coelom, the lophophorates also have been suggested to be deuterostomes or closely related to them.

Because phyla have been erected on the basis of the uniqueness of their features, a difficult problem emerges in determining their evolutionary relationships to one another. Two classic sources of information, adult morphology and embryology, have been available for a long time, and these are being reevaluated using new phylogenetic tools. The primary method of reconstructing evolutionary patterns is called cladistics. I'll take up its basics later in this chapter. Recent cladistic analyses of the radiation of the animal phyla using morphological features have been carried out by Ax, Brusca and Brusca, Eernisse and co-workers, and Schram. In their agreements and disagreements, these cladograms (diagrams of the pattern of branching of evolutionary lineages) provide a useful exhibit of the state of the art and demonstrate the need for data in addition to the existing morphological features.

Hyman's efforts produced a coherent way to order evolutionary relationships among phyla, but no means of testing the phylogenetic hypotheses she made. There has been no shortage of alternative hypotheses (fig. 2.4). The lack of any means of really testing the correctness of phylogenies is what has generated so much heat in this industry. I'll return to this problem in chapter 5. Eernisse and co-workers recently collected the major phylogenies produced since the 1950s and redrafted them in cladistic format. Including their own, there are thirteen very different trees. Two things strike me as significant and interesting about these conflicting hypotheses. First, they often do not accept at all the reading of the features used by Hyman, and gloriously divergent scenarios have been derived. This dispute over patterns of relationship affects our ideas of how the metazoan radiation proceeded. For instance, as we shall see in chapter 3, the hypothesis that cnidarians are primitive has greatly influenced interpretations of the fossils thought to represent the earliest animals, the Ediacaran fauna. If we were to accept a hypothesis put forward in the 1950s and 1960s by Hadzi that acoelomate flatworms are the most primitive metazoans, followed in origin by pseudocoelomates and then by secondarily radialized and simplified cnidarians and ctenophores, we would have a very different expectation from our current view that cnidarians should be the earliest large metazoans to appear in the fossil record. Hadzi's view of body plan evolution is built on an entirely different hypothesis of diversification than most other views, and he rejects the deuterostome/protostome divergence altogether. Hadzi conceived of metazoans progressing from an undivided body form (the Ameria) to a segmented body form (the Polymeria) and finally to an oligomeric body form (the Oligomeria). The Ameria include

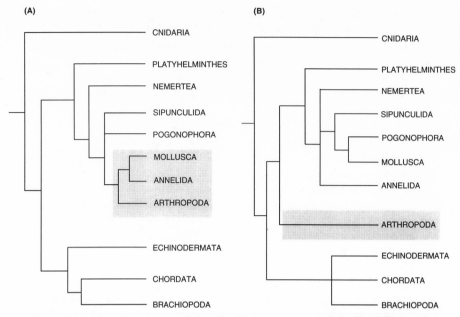

Figure 2.4.   Alternative metazoan phylogenies. These two trees were derived by cladistic analysis of multiple morphological features of animal phyla. The trees have been somewhat simplified for the sake of comparability. In these trees, the platyhelminths are placed in the protostomes. In some other trees (see chapter 4), they are the sister group of all the eucoelomate phyla. Note that in *(A)*, arthropods are the sister group of annelids and mollusks. In *(B)*, the arthropods are no more closely related to annelids than to other protostomes. Both trees place brachiopods in the deuterostomes, a placement inconsistent with results from current gene sequence data. *(A)* is based on the analyses of Brusca and Brusca, *(B)* on those of Eernisse and co-workers.

flatworms, pseudocoelomates, and mollusks. The Polymeria include annelids and arthropods. In the top of this tree, Hadzi placed the oligomeric brachiopods and echinoderms, and finally the chordates.

Hadzi's views have received less than enthusiastic support from zoologists. Most other phylogenies recognize more or less traditional deuterostome/protostome groupings, but disagree substantially about relationships between individual phyla. For example, the mollusks have posed a continuing problem for phylogeneticists as to how they should weigh features in building a phylogeny. Mollusks share many embryological features with annelids, but they lack a distinct coelom and segmentation. Mollusks have a pericardial cavity that has been proposed as homologous to a reduced eucoelom, but this suggestion is rejected by other investigators. The primitive monoplacophoran mollusks, only recently discovered as living fossils in present deep-sea environments, have repeated muscle elements. These have been taken as representing a trace of primitive segmentation, but again, this interpretation is not convincing to all. Clark proposed that the foot of mollusks and their lack of a coelom and segmentation provide a strong tie to flatworms. According to this hypothesis, mollusks do not belong to a coelomate protostome clade, but branch off much earlier from flatworms. This view is controversial, although it is still favored in Willmer's recent book *Invertebrate Relationships*.

The features that Clark used to unite mollusks specifically with flatworms can be evaluated quite differently. Spiral cleavage and the origin of mesoderm from the 4d mesentoblast could be primitive features that unite all spiralians, including flatworms. Shared primitive features do not help us in a cladistic analysis, which must deal with shared derived features. The lack of a coelom in mollusks could be scored as a loss of a feature their ancestors once possessed. The lack of segmentation is moot. Mollusks might have split from the annelid and arthropod lineages before segmentation arose, or they might have had some degree of segmentation early on and lost it.

The second effect of different models of metazoan phylogeny is that they not only lead to strikingly different phylogenetic patterns, but also require very different courses of change in developmental features and underlying genetic evolution. This makes the phylogenetic enterprise a vital one to evolutionary developmental biology. We must know the patterns and directions of evolutionary changes in developmental processes. That knowledge can be derived only from mapping those changes on a phylogenetic tree. However, not only does such a tree need to be reliable, but it must also be built from data other than developmental characters to avoid circularity. Developmental biologists have begun to awaken to the need for phylogenetic information following the discovery that genes that play central roles in development are widespread among the phyla. Shared developmental regulatory genes imply a common origin, and thus homology, for quite different developmental pathways that characterize development among distinct phyla, such as arthropods and vertebrates. If phylogenetic relationships can be inferred reliably, these ideas can be tested. Conversely, shared complex genetic control systems in themselves may prove to be powerful phylogenetic tools. The application of gene sequence data, as discussed in chapter 4, currently offers the most promising approach to producing reliable phylogenies for major animal groups such as arthropods and deuterostomes, and to mapping the course of evolution of developmental features on a reliable phylogenetic foundation.

THE NEED FOR PHYLOGENY

Reliable phylogenies are crucial in three ways for understanding the evolution of body plans and their evolutionary modifications. The first is that we should be able to infer the structure of the Cambrian radiation itself. Without establishing the phylogenetic history of the products of that radiation, we cannot distinguish among the hypotheses that have been proposed for the derivation of body plans, which range from no connectivity among phyla to high connectivity in which many or all phyla are linked in a linear progression. Phylogenies with high linkage derive living phyla from one another, as, for example, in this hypothetical chain: platyhelminth flatworms as the ancestors of annelid segmented worms as the ancestors of arthropods. An extreme low-linkage model was recently proposed by Willmer; in this scenario, most phyla do not share common ancestors after the original metazoan ancestor, and the radiation looks like a porcupine. Most phylogenetic hypotheses lie between these extremes and postulate branching patterns of descent. The trick is in determining which branching pattern is historically correct. That poses a considerable problem. If we have four organisms, and we want to infer a tree rooted at a common ancestor, there are 15 different possible trees. The number grows with distressing rapidity as the number of organisms to be included in a tree increases. There are 105 possible rooted phylogenetic trees for five organisms, 10,000 for seven, 34 million for ten. For the thirty-five living phyla there is an astronomically large number of possible phylogenetic trees. We are saved from having to consider all the possibilities by having some features of organization that eliminate large blocks of possibilities from consideration.

The second need for reliable phylogenies comes from the importance of distinguishing convergent acquisitions that have appeared in separate lineages from homologies. Thus, if we assume a high-linkage phylogeny for the metazoans, the basic structures of animals must be homologous. On the other hand, if the phyla arose independently from protistan ancestors, complex structures characteristic of the various phyla must have originated independently. That would mean that numerous key features (for example, bilateral symmetry, a coelom, segmentation, the central nervous system) that appear on good grounds to be homologous are not, but instead have independent origins in various phyla. We would have very different expectations about shared genetic controls for developmental processes depending on phylogenetic history.

Finally, it is only with a phylogeny in hand that we can decide on the direction of evolution of developmental features. Knowing the polarity of change is crucial for the determination of mechanisms of change. The mapping of derived states from primitive ones must be done on a phylogeny derived independently of the developmental features being studied. The mapping of directional change in developmental features goes beyond simply identifying shared features. Identifying direction provides a powerful tool in the understanding of the evolution of development that allows us to build evolutionary scenarios. These scenarios are important because any living ontogeny has arisen from a combination of selection operating on developmental mechanisms and historical accidents. That is, the ancestral ontogeny is the starting point for modification. Both it and its descendant ontogenies represent local optima in a field of possible ontogenies. As an example, one proposal for the origin of forelimb and hindlimb development builds on a hypothesis that forelimbs and hindlimbs shared a common ancestry in fin folds along the body axis of the primitive vertebrate. This proposal may or may not be correct, but it offers a rational evolutionary hypothesis for studies of mechanism.

Because our central focus is on the evolution of development, there is a problem in accepting the existing approaches to metazoan phylogeny. Because modes of early development are generally conserved in evolution and often shared among phyla, there has been a historical dependency on comparative embryological data for phylum-level metazoan phylogenies. That approach has been fruitful, yet it poses a considerable methodological problem if we are to depend upon phylogeny to trace directions of change in ontogeny. We cannot construct phylogenies based upon the very features we wish to order without introducing more than a little circularity into the enterprise. The need to bring new data into our analyses has thus become painfully evident.

We want to build a phylogenetic tree that presents the actual evolutionary history of the organisms in it. For phylogeny to be correctly inferred from the wealth of data available, it is necessary to identify informative homologous features that allow us to unite organisms in the tree. For example, we unite reptiles, birds, and mammals as one lineage, the amniotes, because they share a complex developmental feature, the amniotic egg. All available evidence supports the amniotic egg as a shared homology; that is, it originated once and is shared by members of a monophyletic lineage. We would not combine whales and fishes into one lineage on the basis of their possessing fins and streamlined body shape. The fossil record and known evolutionary histories show that the similar features of whale and fish shape are convergent. Whales evolved from terrestrial mammals and acquired a fishlike form independently of fishes. When their anatomies are examined in detail, it is apparent that the swimming machinery of fishes and whales is not alike. Fishes utilize their body wall musculature to undulate from side to side. The swimming muscles of whales are located dorsally, and the whale swims by an undulation in the vertical plane. There is a fundamental principle at play here. Whales are built to the mammalian body plan in their deeper features of anatomy and physiology, and adaptations required for an aquatic life are overlaid upon these. This principle of underlying heredity forms the basis for sorting true relationships from convergences.

Convergent similarity is a false homology, termed "homoplasy." The need to distinguish homoplasy from homology applies to molecular evolution as well as to morphological features. There are only four bases in DNA. Although there are biases in substitution frequencies (transitions, which are purine to purine or pyrimidine to pyrimidine substitutions, are generally more frequent than transversions, substitutions between purine and pyrimidine bases), it is not hard to see that convergent substitutions will occur frequently in genes, such that the same base may appear at the same position independently in genes of two distantly related organisms. Sorting out homology from homoplasy is one of the chief pastimes of phylogeneticists. If we get it right, we not only arrive at more valid phylogenies, but also reveal very interesting convergences in evolution.

PHYLOGENETIC TOOLS

A phylogenetic tree, which represents an evolutionary history, can be inferred in a variety of ways from the totality of information available. Until recently, phylogenetic relationships were largely determined on the basis of overall similarity. This phenetic approach seems like a commonsense one because similar organisms should be more closely related than dissimilar ones. However, it does not discriminate between homologous and nonhomologous similarity. Charig has very revealingly contrasted the schools of systematics that do attempt to make that distinction. One school, evolutionary systematics, seeks to use all shared features to define relationships. Homologous features are sought for this purpose, and modifications of shared features are used to group taxa. Because of the intuitive nature of this approach, characters are less rigorously dealt with than by the cladistic methodology described below. Relationships based on characters may be revised by recourse to the fossil record or biogeographic data if these suggest a better reading of phylogeny. This approach has no defined procedure for recognizing parallelisms or convergences other than application of extensive knowledge of the group by the investigator. Although evolutionary systematics has worked remarkably well, a more precise and disciplined approach to systematics has been devised. That method is cladistics. It is now the dominant school of systematics, and I'll present its principles and use here.

    Two aspects of evolution appear in phylogenetic trees: the separation of species or higher taxa from each other, represented by a branching pattern, and the amount of evolutionary change that accumulates with time along any branch. The result is a bush, in which a single ancestral form gives rise to a set of branching and diversifying descendants. Within the bush, we seek monophyletic lineages. A monophyletic lineage, or clade, is one that arose from a single ancestor and that includes all descendants of that ancestor. Similarities within a clade thus indicate true descent with modification. Finding monophyly is one of the most critical aspects of properly understanding phylogenetic relationships. If similarities are incorrectly considered to constitute homologies, a group of animals considered a true clade may actually represent a polyphyletic assemblage. For example, suppose that on the basis of their shared possession of wings evolved from forearms, warm-bloodedness, insulating body coverings, and parental care of young, we decided that bats and birds constituted a hypothetical group of warm-blooded winged animals, the "Thermoptera." That group would not please many zoologists. Based on what we actually know about their evolutionary histories and underlying disparate anatomies, birds and bats derive from separate ancestries. The thermopterans are really a polyphyletic assemblage whose members independently acquired the features we used to unite them. Birds are most similar to archosaurian reptiles, especially dinosaurs, and bats are bona fide mammals. Their distinct wing architectures, brain designs, hearts, body coverings, underlying skull structures, and other anatomical features overwhelm the similar features that support the thermopteran group. Incidentally, some vertebrate phylogeneticists, notably Gardner, have considered warm-bloodedness and a number of characters associated with it to be a major uniting feature, and have seriously proposed that birds and mammals are sister groups. The thorough analysis of relationships among amniotes carried out by Gauthier and his co-workers shows that mammals are far distant from birds in their origins, and that warm-bloodedness is thus convergent. Polyphyletic groups cannot be acceptable if true evolutionary histories are to be sought. Nevertheless, in less obvious cases, false homologies can fool us into inferring untrue relationships.

Phylogenies based on overall similarity utilize all known homologous features of each group. However, there are several different kinds of homologous similarities. The first is primitive features shared by all members of a clade. The second is derived features shared by some lineages within the clade but not by others. The third is features confined to a single lineage within the clade. These distinctions have been recognized as crucial to a more defined cladistic approach to organizing phylogenetic data. The cladistic approach was devised by the German entomologist Willi Hennig. His followers over the past twenty years have converted most phylogeneticists to its methodology (collections of papers illustrating the method and its uses are found in several sources, including those edited by Duncan and Stuessy and by Fernholm, Bremer, and Jörnvall). However, the methodology and terminology of cladistics still rings strange in the ears of developmental and molecular biologists. Basically, monophyletic groups are sought, and the definition of branching patterns between clades is the crucial means of establishing patterns of descent.

Not all shared similarities are used in a cladistic analysis. Primitive traits, those shared by all members of a clade, have long been used by noncladistic systematists to define groups, but are not used in the same manner by cladists. An excellent example is provided by the reptiles. Based on overall similarity, we recognize a group of animals sharing a reptilian "grade" of organization. Yet, within the fossil history of that group, we recognize a number of evolutionary lineages with quite divergent histories. Two reptilian lineages lead to quite nonreptilian descendants, the birds and the mammals. Should we accept the grouping of "reptiles" based on overall similarity, or define distinct lineages based on branching patterns? Most of the features that we score as reptilian are primitive—that is, shared by all of the various lineages within the reptiles. Primitive features shared by all members of the clade are called "symplesiomorphies" by cladists. Although they represent true homologies, they do not inform us about relationships among members within the clade. Therefore, cladists do not accept the concept of "reptile" as a formal taxonomic category. In cladistic parlance, the reptiles are a "paraphyletic" group. Although they have a single ancestor, distinct mammalian and avian lineages are derived from "reptiles" and constitute their own classes. The reptiles as generally defined do not include all daughter clades, and thus are not monophyletic. That is, the reptiles are a grade recognized largely by the primitive features that they share and not by evolutionary branching patterns. The concept of "reptiles" is adequate for everyday use, but is an outdated term if phylogenetic history is to be understood. Instead of reptiles, cladists, such as Gauthier and his co-workers, recognize an inclusive group, the amniotes, on the basis of the amniotic egg shared by all tetrapod vertebrates above amphibians. Within that clade there are distinct lineages such as archosaurs (crocodiles, dinosaurs, and birds), synapsids (mammal-like reptiles and mammals), and others. The point is to find the true historical pattern of descent and to base a classification scheme upon it.

How should lineages be defined? Obviously there are advanced features that occur only in a single lineage within a clade. Thus, birds are very different from other archosaurs such as crocodiles and dinosaurs. Birds have features unique to them, such as feathers, that set them apart from all other groups. Such features are called "autopomorphies." In traditional phylogeny-making, autopomorphies were given important status in setting apart groups. However, because they are unique to a single group, they do not help us in recognizing relationships or the pattern of branching among sister groups. To do that, cladists depend upon the third kind of homologous similarity, shared derived features, called "synapomorphies." These features are shared by sister groups and represent states derived from the primitive features shared by all members of the clade. For example, among the amniotes, crocodiles, dinosaurs, and birds share a suite of anatomical features in the skull. Within that clade, dinosaurs and birds share skull features and locomotor systems that set them apart as a sister group of the crocodilians within the archosaurs. The result is a nested set of lineages: the amniotes contain, among others, the archosaurs, which contain further nested branchings, such as the crocodile branch and its sister, the bird plus dinosaur branch.

To return to the hypothetical "thermopterans" (the bird plus bat clade), we had some features that supported this group, even though they were overwhelmed by a majority of features that disputed thermopteran monophyly. This phenomenon turns out to be the common fate of cladistic efforts: that is, not all traits conform to the favored tree. Convergent evolution confounds phylogenetic analysis. Some features that look like perfectly good homologies, and perfectly good synapomorphies, are seen to be homoplasies once all available features are mapped. The answer to the question of what constitutes the best tree has to be supplied by an outside criterion based on how we think evolution most likely functions. The concept most often applied is parsimony. The concept of parsimony owes much to the medieval English Scholastic philosopher William of Occam. The measure of Occam's abilities is obvious in his rationality in a prescientific era and in his eventual troubles with the Papacy, which charged him with heresy. His principle, known as "Occam's razor," states that in seeking an explanation for some phenomenon, the hypothesis requiring the fewest assumptions is preferred over those requiring additional assumptions.

The concept of parsimony can be applied in two quite distinct ways. The first is the methodological one, and is a good approach to organizing human reasoning in devising hypotheses about natural phenomena. We seek parsimonious hypotheses in our work. The second aspect of parsimony is ontological, and assumes that natural processes—for our purposes, evolution—operate parsimoniously. That is, we assume that the true course of evolutionary history follows the line of fewest changes. Whether nature consistently operates this way is debatable. A law of parsimony cannot be similar to a law of Newtonian physics. Parsimony in nature must be probabilistic rather than deterministic. Many of the reasons cladists give for assuming parsimony are methodological, but their assumption of parsimony is also ontological in requiring parsimony in evolution itself. The validity of that assumption can only be determined empirically.

Sometimes assuming the absurd extreme is helpful. If there were no parsimony in evolutionary processes, the consequences would be very curious. As Farris has suggested, in a nonparsimonious world, each species of insect might be regarded as having evolved segmentation independently, and its offspring might not be expected to closely resemble their parents. Fortunately, this is observably not the case, either for morphological features or for genetic systems. Evolution is descent with modification. More similar forms are likely to be more closely related than more distinct ones, and the

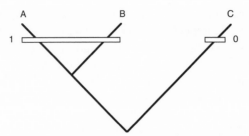

Figure 2.5.   A simple cladogram. State 0 is inferred to be primitive. State 1 is derived from state 0, and is shared by taxa A and B.

hierarchies of living organisms first constructed by Linnaeus should reflect patterns of evolutionary relationship. The problem is, of course, that despite the evidence for an overall parsimony in evolution, nonparsimonious things do happen in the real world. Variable amounts of homoplasy exist. For example, Kettlewell, in his study of industrial melanism in British moths, has recorded that among 259 species, there are 449 known melanic forms, of which 175 evolved independently under selection stemming from industrial causes. That represents a lot of parallel evolution and a lot of homoplasy. However, to make cladistic methods work, one has to be willing to make the parsimony assumption and discard the minority of features that are not congruent with the majority. In some cases the majority tree might be wrong, but in more cases we expect it to be the best choice available, and possibly even true.

The object of cladistics is to produce a diagram of the branching order of clades, that is, a cladogram. The operation is simple in principle, and if not too many lineages are involved, in practice as well. Synapomorphies must be scored to yield nested sets of branches. In order to do that, two preliminary criteria must be satisfied. The first is that there has to be a way of scoring synapomorphies quantitatively, particularly if computer methods of inference are used. And computers must be used if more than a very few lineages are involved, because the number of possible trees rises exponentially with the number of species involved. Generally, the primitive state is assigned a 0 and the derived state a 1. If there is a second derived state, it can be assigned a 2. If organisms being compared share trait 1, it is a synapomorphy for them. Others may share synapomorphy 2. Suppose that there are three organisms to be ordered, A, B, and C. A and B share state 1, and C has state 0. We would thus construct a cladogram, as in figure 2.5, that ties A and B together as sister groups with an A, B branch, with C branching off earlier. If several shared characters are present, all synapomorphies can be mapped and the cladogram requiring the fewest steps selected as the optimal one. That is, we would not choose a cladogram that requires more than the minimum gains or losses of synapomorphies. Second, to order the features, some criterion of polarity is required. This polarity is usually provided by use of an outgroup, a related lineage that can be reasonably inferred to lie outside of the lineages being ordered in the cladogram. Its features are considered to represent the primitive states of the shared derived features inside the cladogram, that is, state 0 for each one.

OUTGROUPS AND PRIMITIVE CHARACTERS

The operation that I have just described for building a cladogram requires that the primitive states of characters be identified. The primitive state of any character can be assigned only if there is a theoretical basis for assigning polarity or if reference can be made to taxa outside the set being considered. If the primitive state is to be assigned based on an outside taxon, that referent is known as an outgroup. The outgroup is chosen from a clade that is outside the set of taxa being treed, but inferred to be related to them. The most useful outgroup would be a closely related taxon that has characters homologous to those being used in inferring the phylogeny, but exhibits them in relatively unmodified (primitive) states. An example of an outgroup for polarizing characters in a phylogeny of placental mammal orders might be the opossum. The choice of the opossum would be justified because it is a mammal, but clearly not a placental mammal. Furthermore, the fossil record suggests that the opossum is a fairly generalized or primitive mammal. Its character states might thus provide a reasonable polarization. That is not to say that the opossum will not have specialized character states of its own, and so might not provide a correct polarization for all characters. Use of a second outgroup might rectify this kind of problem.

The outgroup needs to be selected with some care, and the use of outgroups does not guarantee that characters will be correctly polarized. Furthermore, the choice of outgroup can influence the results of phylogenetic inferences within the taxa being treed. There are no certain criteria for selecting an outgroup as the appropriate closest relative of the group being examined. If an outgroup is too distant, it may contain so many specialized states of the characters it shares with the taxa being treed that it gives a false notion of trait polarity. Another problem is that outgroups might have acquired similar characters independently. If we were to use birds as an outgroup for a phylogenetic study of mammals, we might indeed decide that warm-bloodedness is primitive since the outgroup has it too.

In spite of such problems, outgroups are generally used in cladistic studies. However, not everyone has been satisfied with this method because there are awkward assumptions involved in using outgroups. If an internal criterion for polarization could be found, outgroups would be unnecessary. Ontogeny has been proposed by Nelson as providing just such an internal criterion. The notion supported by Nelson as well as by some other cladists (see Patterson, and Rosen) is that hypotheses about evolution should be avoided in inferring a cladogram. Instead, these researchers cite von Baer's law to justify the use of a so-called ontogenetic criterion, by which the more general features that appear in development are assumed to be more primitive. This proposal has been justifiably criticized (notably by Kluge and Strauss) because von Baer's law is hardly that, and because the ontogenetic criterion is basically a statement that evolution is recapitulatory. As we will see in chapter 8, that is hardly the case. For our purposes in investigating the evolution of development, use of phylogenies inferred by use of the ontogenetic criterion introduces a fatal circularity. A strong empirical test of the ontogenetic criterion has been made by Mabee, and she showed that it did not perform well. It is crucial that we discover the connection between ontogeny and phylogeny, but the ontogenetic criterion doesn't provide it.

CLADOGRAMS AND PHYLOGENETIC TREES

Cladograms and phylogenetic trees are not the same thing, although the data presented in one can be converted to the form of the other. Four major characteristics of cladograms should be noted. First, cladograms seek the branching pattern that connects the organisms included in the analysis. Second, no ancestor-descendant relationship is inferred among the organisms used to construct the cladogram. What is inferred is that the character states split in the order shown. Cladograms can be derived from character states drawn entirely from a set of living organisms, and doing so requires no knowledge or assumptions about ancestral organisms per se. The nodes represent the divergence points and show the inferred character state at each divergence point. Although concrete ancestors are not built into a cladogram, its nodes can be used to infer the features possessed by ancestors of the organisms above any node. Once a large number of features have been incorporated into a cladogram, one can make a rather detailed set of predictions about the characteristics of hypothetical ancestors from the features inferred for each node. Third, fossil organisms can be incorporated into cladograms with living ones. Again, the fossil species are entered as terminal branches just like the living species, and not entered as nodes. In many cases—for example, in human phylogeny, in which only one lineage survives—it is necessary to incorporate fossil species if one is to make any phylogenetic inferences at all. Last, because cladograms deal only in character states and splitting patterns, they make no statements about time beyond the obvious point that the branches lower in the cladogram had to split earlier in time than the higher ones.

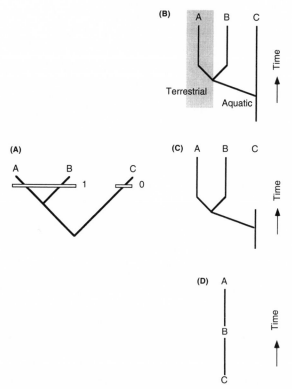

Figure 2.6.   A cladogram *(A)* and three phylogenetic trees consistent with it *(B–D)*. Species C exhibits primitive state 0 of the character used to infer this cladogram. Species A and B share derived state 1. Each phylogenetic tree utilizes the relationships inferred by the cladogram, but adds nonphylogenetic information and hypotheses about evolutionary patterns from other sources. In *(B)* all three taxa are still extant. The tree infers that the ancestor of taxa A and B arose from taxon C, and it presents the time of the transition inferred from the fossil record. It also shows nonphylogenetic information about habitat change in the evolution of taxon A. In *(C)*, taxon C, which is ancestral to taxa A and B, is extinct. The tree gives an estimate of the time of extinction of taxon C. In *(D)*, taxon C evolves into taxon B, which in turn evolves into taxon A. No branching is involved.

Because cladograms look superficially like phylogenetic trees, the two are often confounded. However, they are different and have differing aims. Phylogenetic trees are richer in information than cladograms and generally include inferences not drawn from the data used to construct a cladogram. In phylogenetic trees, inferences are often drawn about ancestors, and because information on the timing of splitting events is often available from the fossil record, a time scale can be presented as well. Figure 2.6 shows a simple cladogram containing three species, A, B, and C. The cladogram is consistent with several possible different phylogenetic trees. Three are shown that include additional information on habitat change, on times of origins and extinctions, and on inferred patterns of evolution. Each species or lineage in a phylogenetic tree is represented as a vertical line. The length of the line indicates the known (but not necessarily the actual) duration of the lineage. Because transitions are uncertain in connection and timing, it is common practice to draw them as dotted lines. It is important to note that because a cladogram deals in characters, more than one hypothesis about ancestors is possible, and several different phylogenetic trees can be consistent with a particular cladogram. Because phylogenetic trees can present data in addition to the character distributions shown in cladograms, phylogenetic trees are more informative than cladograms, but may include more uncertain inferences.

In figure 2.6, the first of the phylogenetic trees shows species C as a still-living ancestor of species A and B. A shift in habitat is inferred to have occurred upon the divergence of species A and B. In the second tree, species C is extinct. In the third tree, species C gives rise to species B, which gives rise to species A, all sequentially and without branching. In that case, species C and B would have become extinct by evolving into the next member of the series, a form of demise called "pseudoextinction." This pattern of straight-line, nonbranching evolution is called "anagenesis." Although the inferred patterns of organismal evolution may differ, the character states can still be readily plotted in the cladogram. The cladogram makes no assumptions about models of evolution, only about the distribution of characters.

Phylogenetic trees are powerful symbolic representations of evolutionary hypotheses. We are so used to this kind of presentation that its underlying assumptions and limitations are usually invisible. The basic aim of phylogenetic trees is to provide a history of a group of organisms. The main elements of such trees are ancestors and descendants and the relationships between them. Trees often also contain information about time: age of ancestor, durations of lineages, times of splits, and times of extinctions of lineages. Occasionally inferences on adaptive zones and geography are presented. As shown in figure 2.6, inferences about evolutionary rates may also be indicated, generally in the slope of the lateral lines linking a descendant to an ancestor. Hypotheses can also be conveyed about patterns of evolution. There is currently a good deal of controversy about whether descendants might arise by splitting of a previous lineage—that is, by cladogenesis—or through progressive evolution within a lineage that did not split—by anagenesis. Both of these hypotheses are expressed in current phylogenetic trees. For example, the mid-Cenozoic horses had a striking flowering of species. A recent analysis of horse phylogeny by Hulbert concludes that some lineages experienced both numerous cladistic splits and direct linear ancestor-descendant evolution.

His phylogenetic tree thus includes large doses of both anagenesis and cladogenesis. In hominid phylogeny, the evolution of the *Homo* lineage from the *Australopithecus* lineage is usually presented as a cladogenic event, whereas the evolution of *Homo erectus* to *Homo sapiens* is usually suggested as anagenetic.

Phylogenetic trees are capacious vehicles for presenting information. Most readers are not sufficiently suspicious and tend to accept their statements as facts, when much of what they present is really complex inferences about evolution. The inferential nature of phylogenetic trees can arise from several sources. Estimates of absolute time come from the fossil record, and their reliability depends on the adequacy of that record. Thus, the known duration of a species will to a greater or lesser extent be an underestimate. Statistical means have been developed to give confidence intervals to time ranges, but uncertainties must remain. Statements about rates of evolutionary processes are similarly uncertain. Ancestor-descendant relationships also represent hypotheses. An organism with the expected evolutionary features and present in the expected time range is generally nominated for the role of ancestor to subsequent species, but genealogical connections are hard to prove among fossils. Nor is it possible to guarantee that a better prospective ancestor will not turn up on the next paleontological expedition to some exotic place. Patterns of evolution presented in phylogenetic trees are also inferential. The diagramming of a direct ancestor-to-descendant line versus a branching pattern of evolution must reflect theoretical bias about evolutionary mechanisms because often the data can be equally well interpreted either way. Choices also will be influenced by the uncertainties of the data themselves. Fossils preserve only part of the anatomical and even less of the genetic structure of extinct species; we cannot be sure of distinguishing similar fossil species even if they have speciated extensively.

Darwin, in *The Origin of Species,* was the first to construct a phylogenetic tree diagram to illustrate a hypothetical evolutionary history. His diagram and pretty much all those following, including Haeckel's hefty Germanic oak-style phylogenetic trees, have illustrated two essential properties of evolution. The first is that lineages split, resulting in speciation, the production of populations with discrete gene pools. The second characteristic of all phylogenetic trees is that lineages become different from each other; they diverge genetically, morphologically, and behaviorally. Splitting of lineages is crucial to generating new lineages, but without significant morphological divergence of lineages, there would be no diversity.

These two aspects of phylogeny are not necessarily correlated. There are species-rich clades in which diversity is high and the accumulation of differences in lineages is obvious. The radiation of antelopes, with their elaborate and highly specific horns and distinct sizes and body forms, is one good example. In other cases, splitting has been extensive, but morphological divergence very low. The peripatids offer a good example of this phenomenon. Peripatids belong to the phylum Onychophora, whose members somewhat resemble velvety caterpillars and are traditionally supposed to represent a link between annelids and arthropods. They are common in eastern Australia, and I've collected them by prying apart rotten gum logs in the sun-dappled eucalyptus forests of the Blue Mountains of New South Wales. Peripatids share their logs with a rich fauna of cryptic invertebrates, including a yellow-and-green-striped land planarian, a handsome pink and blue spider, and a big neotenic roach. They have some less appealing companions in the venomous funnel web spiders with thumb-sized bodies and fangs nearly a centimeter long, aggressive bull ants that sting like wasps, and very hungry land leeches. On wet days the leeches are unendurable as they converge from all directions to search out your armpits and creep unfelt into your socks. The two Australian biologists, Noel Tait and David Briscoe, who have shared these joys of their fieldwork with me have, through detailed study of morphological features and isozyme patterns, found that these Australian peripatids, which were previously thought to be a single species, really represent a complex of several genera and about fifty species.

DIVERSITY AND DISPARITY

As human population growth and the consequent environmental degradation propel ever-increasing numbers of our fellow creatures into the long night of extinction, ecologists and systematists are trying to estimate just how many species we actually have on Earth. Some ecosystems, such as tropical rain forests and coral reefs, are incredibly rich in species. There are some famous quantitative examples. E. O. Wilson has recounted that from a single leguminous tree in Peru, he collected 43 species of ants, belonging to 26 genera. The diversity of the ant fauna of that single tree was about the same as that of the total ant fauna of the British Isles. Surprisingly, despite this late date in the history of biology, the Earth's most species-rich faunas have not been fully sampled. Estimates of undiscovered insect species alone run into the millions, and the total number of species on Earth is estimated to lie between 3 and 30 million. Wilson has tabulated the number of known species for major groups of organisms: animals make up about a million of the currently described 1.4 million living species. This diversity is far from evenly spread among animal groups. There are over three-quarters of a million known insect species, but only 4,000 mammalian species.

The notion of diversity carries with it two attributes. The first is simply numbers of different species. Note that this definition of diversity says nothing about how different the species are from one another. In fact, if we define species as genetic units, that is, as populations of interbreeding individuals (and yes, that is too narrow a definition, because there are organisms that do not reproduce sexually), two species need not differ appreciably from each other in morphology. Most of the diversity discussed by biologists reflects variation upon coherent themes. There are several hundred similar species of flies in the genus *Drosophila* alone, and a single family of beetles, the weevils, contains over 50,000 species.

There is another measure of diversity, however, that is reflected in the existence of higher categories in the hierarchy of Linnaean classification. Jaanusson, as well as Runnegar, has suggested that a separate term, "disparity," be used to distinguish these differences between themes. Disparity is not related to numbers of species, but to some measure of the large-scale morphological evolution that separates them. Disparity is a measure of how fundamentally different organisms are. And indeed, it is the origin of disparity among animals, or macroevolution, with which we are mainly concerned in this book. The degree of disparity is very roughly indicated by the higher Linnaean taxonomic levels, but to be discussed fruitfully, the elements that generate disparity need to be considered aside from ideas of taxonomic level. Although it is not an easy matter to quantify disparity in a systematic way, levels of disparity are easy to grasp by example.

The most disparate divisions of organisms are the three primary kingdoms of deepest evolutionary divergence defined by Karl Woese and his colleagues: archaebacteria (prokaryotes that specialize in hostile environments such as hydrothermal vents and hot springs, and enliven the biosphere with such unusual metabolic pathways as methanogenesis); eubacteria (prokaryotes that include the eubacteria and the photosynthetic cyanobacteria); and eukaryotes (organisms with complex nucleated cells). Based on depth of genetic separation, distinctness of cellular organization, and the Precambrian fossil record, these primary divisions of life must have diverged well over 2 billion years ago.

Within the eukaryotes, there is a vast disparity among the unicellular organisms, the protists, which include such groups as euglenas, slime molds, dinoflagellates, ciliates, and others. Just how distinct these groups are has been shown by the gene sequence phylogenetic studies of Sogin and his colleagues. Protist groups are as morphologically disparate from one another as tulips and truck drivers, and the molecular data show that they are genetically distinct as branches of life. Their separations are evidently deep in time. Animals, plants, and fungi represent the major classic divisions of large, multicellular eukaryotes. Their distinctness of organization is obvious to us because they are organisms on the same size scale as ourselves. The thirty-five phyla supply the basic morphological disparity of the animal kingdom.

Given that we probably know only between one-thirtieth and one-fifth of living species, how complete an accounting of disparity can we have? New discoveries of truly distinct creatures continue to occur. The archaebacteria were first recognized in the early 1970s; a new animal phylum of minute torpedo-shaped creatures with whiskers, the loriciferans, was discovered in 1983. Fortunately, the nature of the widely cast sampling of organisms that we have suggests that most living disparity has been found. A simple "paper" sampling experiment devised by Simpson in 1959 makes it clear why we should expect most disparity to be known. The mammalian fauna of New Guinea includes 352 species belonging to 20 families and 8 orders. Simpson took a random sample of 35 species. In his sample, he recovered only 10% of species diversity, but over half of the families and nearly two-thirds of the orders. Thus, a partial sampling of diversity can yield most of disparity. Of course, our sampling of the Earth's fauna has not been random. Size plays a major role. We are probably close to knowing all large mammal species, but not most insect species. The recent discovery in Vietnam of a new goatlike bovid by Dung and co-workers shows that the occasional discovery of new large mammals is still possible, but unusual enough to be newsworthy. The animal, so far known only from hunter's trophies, is distinct enough to be placed in its own new genus. The wagers seem to be on whether a live one will be seen by zoologists before it becomes extinct.

Some environments are obscure and contain small, nonobvious inhabitants. Some group of organisms, sometimes animals, has taken hold in essentially all environments that have been sampled, including such outré ones as the tops of rain forest trees, in forest floor litter, between grains of sea sand, in reducing mud, in hot springs, in animal digestive tracts, on the flanks of deep-sea hydrothermal vents, beneath Antarctic sea ice, in the interiors of weathering rocks, and even in salt crystals. Newly discovered environments such as deep-sea vents occasionally provide unexplored possibilities, and new animals have been discovered associated with these vents. However, no new animal phyla have been found since the tiny loriciferans were discovered by Kristensen in 1983, unobtrusively living between sand grains. New classes turn up now and then in strange places. One notable example is the recent discovery by Baker and Rowe in 1986 of a new, sixth, class of echinoderms, the concentricycloids, living on sunken wood rotting on the deep-sea floor. The discovery of the concentricycloids adds significantly to both the number of living echinoderm classes and the disparity among living echinoderms. However, the phylum-rich marine environments have been well sampled, even though our knowledge of species diversity for reefs is certainly not complete, nor for the deep sea. The living animal phyla are probably completely tallied up, and their body plans described.

3
Deep Time and Metazoan Origins

It may be objected, that, to assume the world to have been created with fossil skeletons in its crust—skeletons of animals that never really existed—is to charge the Creator with forming objects whose sole purpose was to deceive us.

Philip Henry Gosse, *Omphalos:*
An Attempt to Untie the Geological Knot

DEEP TIME

To understand the processes that affect the evolution of animal body plans, it is necessary to trace their origins back into the deep past, into the rocks that record the first appearance of the animal phyla. The timing and pattern of the radiation of animals with different body plans sets the stage for tracing the kinds of modifications that have occurred in animal form over a long period of evolutionary time. Perhaps paradoxically, observing the attributes of body plans that have not changed over these vast expanses of time will be equally important to the enterprise. Our only direct window into the past history of animal form is the data from paleontology, and these data are thus of central importance to comprehending how development and evolution have interacted to produce the patterns of animals currently found on the Earth.

During the nineteenth century, representations began to be made of extinct beings and past environments. These have grown in sophistication and popularity, and have led both to the scientific reconstruction of past environments and to the thriving dinosaur industry. Representations of events of deep time derive from an earlier artistic conception that created scenes of the remote past based on the biblical record. Dramatic scenes of the destruction of the sinners and of Noah's ark riding on the waters of the Great Flood were a part of this tradition. Few images from Western religious tradition are more dramatic or evocative than the Flood and Noah's ark. The Creation of Genesis was followed by decay (sin), and a wrathful God imposed a shocking purification, a great flood that scoured the Earth of all life, human and animal. But instead of doing what might seem reasonable in cleaning up a bad job, God did not simply wipe the slate clean and start over. Life was not annihilated and recreated. The Earth was repopulated from a small sample, carried in the ark, of each species from the original creation. The physical world of the Creation had been perfect. The present, imperfect world, bearing the manifest scars of the Flood, began with the dispersal of people and animals from the ark as it settled gently onto the sodden top of Mt. Ararat. The Flood was a juncture between the distant mythic world of Genesis and the more immediate and comprehensible world of the historical record. The imagery of the universal flood, the framing of the existing world, and the linear flow of history that it represented exerted a powerful influence on the growth of Western science as well.

As Martin Rudwick has pointed out, the Flood ultimately connected deep time to the human scale of time. The Flood itself became part of early scientific explanations of the form of the planet, very dramatically and visually so in Thomas Burnet's late-seventeenth-century *Sacred Theory of the Earth*. Burnet detailed the sacred cycle of the Earth's history—from its newly created perfection, through the Flood, to its present imperfect state, to its ultimate destruction by fire—as a scientifically valid use of Scripture to understand the long-term history of the planet. The Flood and the ark separated the perfect world from the present and explained the shattered and disrupted nature of the present world's surface. By the eighteenth century, illustrations of the ark drifting in the subsiding flood waters showed stranded shells on the shore, the future fossils of the piece. Later, as the geologic record began to unfold, the connection of time before the Flood with that following was extended to link prehuman time to the historical record of humanity. When the discovery of stone tools in association with fossils of animals such as mammoths revealed the possibility of humans coexisting with extinct prehistoric mammals, the makers of those tools were thought of as "antediluvian man." Eventually, the depth of time was extended even further to include a long period of life existing before the creation of humanity.

The early geologists confronted two very difficult problems with respect to creation and extinction. At first, all fossils could be ascribed to Noah's Flood. This was the approach taken in the 1820s by William Buckland in his book *Reliquiae Diluvianae*, whose full title goes on to read, *Observations on the Organic Remains Contained in the Caves, Fissures, and Diluvial Gravel, and on Other Geological Phenomena, Attesting the Action of an Universal Deluge*. That title pretty much says it all. Buckland conducted excavations in the Kirkdale Caverns in Yorkshire. The sediments in the cave were overwhelmingly rich in the bones of extinct mammals dragged into the cave by its resident hyenas. The hyenas very conveniently left their own remains behind as well. Buckland interpreted the whole ensemble as having been buried by mud brought into the cave by the Great Flood. He dedicated his book to the Bishop of Durham. Science and Scripture were in accord, and the scriptural record could be investigated and confirmed by geologic techniques. However, the agreement was not completely literal, because Buckland, like Cuvier, regarded the Great Flood as only "the last great convulsion" to have affected the planet's surface. Buckland's interpretation was that of a catastrophist, one of the competing schools of geologic thought at that time.

As geologists began to document the geologic record of life, the suspicion grew that just as there was not merely a single Flood, there was not just one past. The fossil record began to show that there could not have been a single Creation followed by a single catastrophic extinction. There were instead a number of past worlds buried one on top of the other. This stupendous revelation had very mundane origins. During the late eighteenth and early nineteenth centuries, canal building was in full swing. Canals provided the transportation web for a growing industrial society before the age of railroads. An engineer involved in canal building, William Smith, made one of the seminal observations in the history of paleontology: he observed that different layers of rock contained different fossils. Smith used specific fossils to trace rock layers over the countryside, and eventually produced the first geologic map of England. The principle he used was simple: If fossil A is found below fossil B at one site, it will be below B at other localities as well. Strata are deposited flat, with the older beneath the younger, as long as the rocks have not suffered some extraordinary folding or overturning in the course of mountain building. The use of such index fossils provided the key to producing a relative dating system for the fossil-bearing rocks of the Earth. Fossils did not recur. They appeared at some level and disappeared forever at a higher one. Therefore there had to have been a succession of species in time. Cuvier observed very much the same thing in his studies of the fossil mammalian faunas of the Paris Basin. A fauna made up of a certain set of species would be present in one stratum, but absent in the stratum above, where a different fauna was present. Cuvier hypothesized that faunas became extinct as the result of catastrophes. The succeeding fauna would migrate in from elsewhere when the dust settled or the water drained away. Thus, according to Cuvier, there had to have been several catastrophes, not just the single Great Flood of the Bible. The abyss of deep time had become apparent. But the concept of evolution had not, and the creation of new forms continued to pose an ever-growing scientific problem.

The realization of the immensity of time and the slowness of geologic processes stemmed from the observations of James Hutton in the late eighteenth century. Hutton saw deep time in the repeated cycles of erosion, deposition, and uplift of rocks evident in the geologic record. The concept that processes currently observable in action had occurred in the past as well, and that these slow processes had sculpted the Earth, became the guiding principle of Charles Lyell, whose *Principles of Geology* shaped nineteenth-century British geology. His was the viewpoint that eventually came to dominate geologic thought: that catastrophes were unnecessary, that slow processes operating at the rates and intensities that we observe today were sufficient to account for the geologic record. No Great Flood was needed if rain alone could wear down a mountain, only time. Geology found no Flood, only a complex and ancient planetary history. Relative dating based on the order of fossils in the geologic record allowed deep time to be ordered into a set of eras, subdivided into periods, each characterized by a unique set of biological events. By the early part of the twentieth century, the basis for absolute dating of rocks using radioactive decay clocks had been established. This method has since been widely applied, and has given us an ever more precise dating of the geologic periods and their events.

The Great Flood continues to hold pride of place with modern creationists, who still, despite its physical impossibility and all of the contrary evidence of geology, biogeography, and archaeology, want to explain the entire geologic record as having been produced by a single year-long flood. According to the classic work of this genre, Whitcomb and Morris's *The Genesis Flood,* the sequence of fossils represents not a record of the order of existence of the fossil beings, but the hydrological sorting of the drowned corpses suspended in the flood waters. Why delicate, lacy trilobites should have settled faster than the heavy shells of oysters or the massive bones of decaying dinosaurs poses one of the wonderful absurdities of this scenario. Expeditions are regularly mounted by true believers, including, incredibly enough, a former astronaut, to seek the remains of the ark on Mount Ararat. Each year success is reported to be almost within reach. Just as the expedition's time is running out, an expedition member sees a boat-shaped shadow looming deep in the ice of a glacier or finds bits of old-looking wood lying high above the tree line on the mountain slope. All that is needed is more financial donations to fund next year's trip, and in no time we'll be stepping aboard the long-lost ship and admiring Noah's dinnerware and reading his log.

Of course, another great scientific mystery, the extinction of the dinosaurs, might be solved by a look at the ark's animal pens. At least one creationist book, Segraves's *The Great Dinosaur Mistake,* proves that there were dinosaurs aboard Noah's ark. They were killed later—by Nimrod the mighty hunter, as I recall. Perhaps there was something to the behemoth of the Book of Job. The behemoth is generally identified by commentators as the hippopotamus, but maybe we should think again.

Noah and his ark have not escaped literary revision. In Stephen Minot's *Surviving the Flood,* Noah's unfavored and persecuted son Ham, later in his long life, tells the unofficial version of what actually transpired, of the mysterious and officially unacknowledged presence of Methuselah, the racy life on the ark, the petty jealousies, Noah's monstrous cruelty, and the ultimate fraud of the whole event. Ham puzzles over why the official version is so untrue, but recalls old Methuselah's words to him so long before: "Men struggle no harder to control the future than they do to control the past."

In Julian Barnes's *A History of the World in 10½ Chapters,* we hear another unofficial history of the ark, from a very undesirable stowaway on a long cruise aboard a wooden ship: a wood-boring insect, an invisible observer lurking in one of the ark's beams. As he tells it, from the animals' point of view, the enterprise was exceedingly unpleasant and incomprehensible. They were not to blame for God's wrath, yet every species of animal stood to be wiped out save for one breeding pair. To start with, not all were given sufficient time to make it to the ark. The giant sloths were just too slow to arrive before sailing day, and missed the boat. "What do you call that— natural selection? I'd call it professional incompetence." Some species, such as our observer's, were not wanted on the voyage. Others were not recognized as separate species by Noah, and so were not allowed aboard. A lot were lost once the voyage began. As the observer tells it, there were several boats in a small convoy, not just a single ark. One went down in a storm, with a fifth of the species aboard. Some species were eaten by Noah's meat-hungry family. All of this left otherwise inexplicable holes in the fauna that got off the ark at the end of the voyage. "I mean, if you look around the animal kingdom nowadays, you don't think this is all there was, do you? A lot of beasts looking more or less the same, and then a gap and another lot of beasts looking more or less the same? I know you've got some theory to make sense of it all—something about relationships to the environment and inherited skills or whatever—but there's a much simpler explanation for the puzzling leaps in the spectrum of creation." This version reveals what we never imagined: It wasn't that Noah loved animals or wanted to preserve the diversity of the Creation; he wanted tasty animals around to eat once life returned to normal after the Flood.

Somehow, this tale serves as a good parable for the fates of the animal body plans in the world following the Cambrian radiation of multicellular animal life, the metazoan explosion. The ark, our planet, not only nurtured its passengers but indifferently killed them as well. Not only were some judged and found wanting—Darwin's survival of the fittest—but more were unlucky in the large and small catastrophes that have befallen the biosphere and in their consequent no-fault extinctions. We are left with holes, missing beings that, if still present, would link now distinct body plans. As Stephen Jay Gould has argued, there is a great contingency evident in the history of animal life. Success was not guaranteed for any lineage, not even the verte-brates. Although vertebrates are diverse and dominate modern ecosystems, they were neither diverse nor dominant in the Cambrian. We were not inevitable. The largest extinctions often removed lineages that seem to have been as successful in normal times as the survivors. It can be argued that if the whole show were run over again, other lineages might have survived, and the pattern of life's development might have been quite different. But the effects of extinctions were not entirely negative, even those that struck on a massive scale. Large-scale extinctions reset the evolutionary stage for the survivors. Some spectacular radiations resulted, such as that of the mammals after the extinction of the dinosaurs.

THE FIRST ANIMALS

The Earth is very ancient indeed, and most of the history of life on it occurred during the first 4 billion years, before the first record of animals. The Cambrian fossil record marks the unequivocal appearance of the animal phyla that inhabit the world today. Its date is arguably the second most important in geology (the age of the Earth is the obvious first). The base of the Cambrian is defined by the stratum in the geologic record where the first fossils of complex metazoan burrows occur. This assemblage of burrows and feeding traces is called the *Phycodes pedum* zone. The earliest metazoan skeletal fossil faunas follow in slightly younger rocks (see Landing for a discussion of the establishment of this geologic boundary). Until recently, there has been considerable uncertainty as to the absolute date of the Precambrian-Cambrian boundary, with variable results of radiometric dating of igneous rocks from that time placing it variously between 540 and 600 million years ago. That has led to waffling and the use of a sort of average, what Compston and co-workers refer to as a "damage control" age of 570 million years. This date is often cited in discussions of the Cambrian radiation but, as the average of two uncertainties, cannot be taken very seriously. A consensus between the radiometric date found by Odin and co-workers and more recent studies by Compston and co-workers suggests 540–530 million years ago as a reasonable estimate for the beginning of the Cambrian. The most recent work, that of Bowring and co-workers, places the boundary at 544 million years ago, and better defines the dates of the stages within the early Cambrian. Since the Cambrian radiation of animals with preservable hard body parts was under way by 530 million years ago, the age of the boundary sets important constraints on the rates of evolution being considered. Within a few million years, the Cambrian radiation had produced a diverse fauna of animals, many of which had acquired the innovation of readily fossilized skeletons. These innovations occurred during a relatively modest interval of time in the Cambrian, which was overall about 40 million years in length. By the end of the Cambrian nearly all recognizable modern phyla had come into being.

There was an earlier radiation as well, of what has been called the Ediacaran fauna. The remains of that fauna consist of various medium-sized to large soft-bodied creatures preserved as flattened casts in fine sandstone (fig. 3.1). Some of these fossils resemble simple medusoid jellyfishes, sea pen fronds, polychaete worms, and simple arthropods, and some are more enigmatic sculptured discoid forms. They were first discovered at Ediacara in the Flinders Range of South Australia, but occur at many sites worldwide in rocks of the uppermost Precambrian. Sites as far apart as Australia, England, and Russia have very similar faunas, and some species are found in all localities, indicating that similar environments prevailed and that species were widespread rather than provincial in distribution. The world had not yet been finely divided into ecological or geographic life zones. One somewhat different fauna has been described from the Avalon Peninsula of Newfoundland by Anderson and Conway Morris. It contains some large (up to a meter long) and truly difficult to interpret frondlike organisms. According to Jenkins, this fauna was deposited in much deeper water than other Ediacaran assemblages and is different in composition. Runnegar, as of 1992, records about 50 well-documented species in the Ediacaran fauna, although its real diversity may have been higher. This terminal Precambrian age is referred to as the Vendian. The absolute age of these fossils often has been estimated as about 600 million years, but Benus, as well as Harris and Glover, has recently presented radiometric data supporting an even younger age, about 565 million years old. Given the present best estimate for the start of the Cambrian, perhaps as little as 20 million years elapsed between the Ediacaran radiation and the Cambrian radiation.

The Ediacaran fauna was discovered in the 1940s, but its significance was widely recognized only in the 1960s with the efforts of Martin Glaessner, who first presented a reconstruction of the Ediacaran animals and their environment. Glaessner's work was tremendously significant because it began to fill the terrible void between the rich metazoan fossil record found in Cambrian and younger rocks and the lack of any traces of ancestral Precambrian animals. The apparently sudden appearance of animals had presented Darwin with a problem of origins. He suggested in *The Origin of Species* that the Precambrian must have been of long duration and "swarmed with living creatures." As to why there was no trace of this swarm, he could only allude to the imperfection of the fossil record and suggest that either Precambrian animal evolution proceeded in regions of open seas that left no depositional record on the present continents, or the likely metamorphosis of such old rocks destroyed any fossils they contained. The problem was not lessened by subsequent paleontological work. Charles Walcott was the pioneer student of the metazoan radiation and the discoverer of the mid-Cambrian Burgess Shale fauna that has played such a significant role in the past decade in our understanding of animal origins. By 1910, Walcott was pressed to suggest

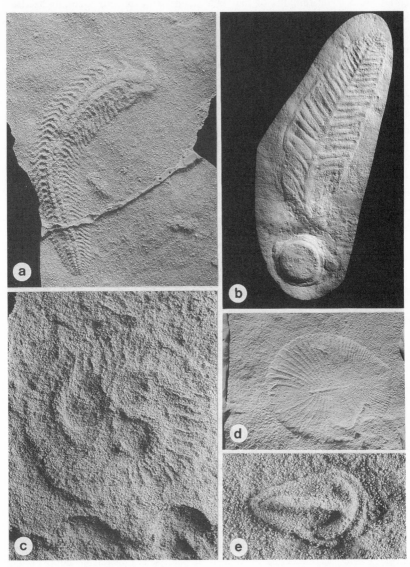

Figure 3.1. Ediacaran fossils. *(A) Spriggina,* an annelid-like form. *(B) Charniodiscus,* a frond with holdfast. *(C) Tribrachidium,* a discoid organism, proposed to be a triradial primitive echinoderm. *(D) Dickinsonia,* a planar, segmented organism. *(E) Parvancorina,* a shield-shaped organism. (*A, C, D,* and *E* reprinted from J. G. Gehling. 1991. *Geological Society of India Memoir No. 20,* pp. 181–224, by permission of the Geological Society of India. Photographs courtesy of James Gehling.)

the existence of what he called the Lipalian interval (from the Greek for "lost"). This was hypothesized to be a long period of time during which metazoans evolved, but during which no fossil-containing sediments were deposited. It was an ad hoc explanation.

A very active field of Precambrian research arose in the late 1950s that depended mainly on micropaleontology to seek traces of life. Cellular fossils preserving exquisite details could be readily studied by microscopic examination of thin sections of rocks that contained biogenic structures. As has been so well documented by Schopf and Klein and their colleagues, there is a rich record of Precambrian cellular evolution extending from almost 4 billion years ago. This record includes a record of eukaryotic cells as well, and as discussed by Knoll and by Knoll and Walter, a record of increasing diversity. There is no gap in the geologic record, no missing interval of time. Ordinary sedimentary rocks with the potential to preserve animal fossils are present in the late Precambrian. But the animals are absent. The Ediacaran fossils, which enter the fossil record at the very end of the Precambrian, helped to fill the interval with concrete remains. These fossils of exclusively soft-bodied forms also fit into the expected scenario of metazoan evolution, in which the first animals would be small, soft-bodied, and not likely to become fossilized. Animals would then grow larger, and finally acquire hard parts, such as shells, in response to some concerted selection: predation is the usual, but not the sole, hypothesis.

As significant as the Ediacaran fossils are, considerable uncertainties remain about what kinds of organisms they were. Animals had to have existed prior to their explosive appearance in the fossil record, but it is enormously difficult to determine how long before. I'll discuss the evidence for the timing of the origin of metazoans, such as it is, shortly. The second uncertainty lies in the interpretation of the fossils themselves. I'll present the Ediacaran fauna as envisioned by Glaessner, as the first flowering of the Metazoa, but will return to a baffling and even amusing quandary over just what they really were. Under the hypothesis that the Ediacaran fauna are animals, there appear to have been two bouts of early animal radiation, with a complex soft-bodied fauna emerging before the Cambrian radiation. Several hypotheses consistent with a radiation of Ediacaran metazoans are diagrammed in figure 3.2, including Glaessner's "ancestral metazoans" hypothesis.

There is a long, honorable, and richly speculative literature on metazoan origins. As Signor and Lipps have pointed out, it can be boiled down to two all-encompassing questions: when did metazoans originate, and what caused their explosive radiation? The "when" question has been answered by two kinds of hypotheses. The first is that there was a long but cryptic and unfossilizable history of the Metazoa prior to their appearance as fossils. In that view, the first fossils merely record the emergence of fossilizable body parts

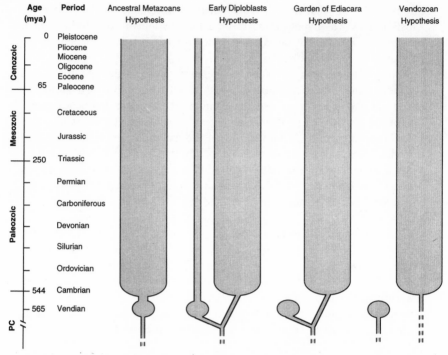

Figure 3.2.   Four hypotheses about the relationship of the Ediacaran fauna to the metazoan radiation. The determination as to which is closest to the truth remains to be made. In the "ancestral metazoans" hypothesis, the Ediacaran radiation of the Vendian is seen as directly ancestral to the metazoan phyla, which further diversify in the Cambrian radiation. This hypothesis was originally proposed by Glaessner and supported by Gehling. In the "early diploblasts" hypothesis recently suggested by Conway Morris, the Ediacaran fauna was a radiation of diploblastic metazoans ancestral to the living diploblastic phyla. The bilaterian, triploblastic animals make up the massive Cambrian radiation. In the "Garden of Ediacara" hypothesis, proposed by McMenamin and McMenamin, a now-extinct offshoot of pleated sheet animals probably living in symbiotic association with photosynthetic algae occupied the Earth before the rise of the hungry metazoans of the Cambrian radiation. The Cambrian radiation included both diploblastic and triploblastic animals. Finally, the "vendozoan" hypothesis suggests that the Ediacaran fauna is made up of creatures that were not metazoans and were displaced by the metazoan radiation. This hypothesis was put forward by Seilacher.

in long-existing phyla. The second answer to the timing question is that the record of a sudden appearance of animals more or less corresponds to what really happened. In this view, the Metazoa do not have a long Precambrian history, and the Cambrian radiation faithfully records the actual evolutionary appearance of animal phyla.

The critical events that need to be placed accurately in time are the divergence of the metazoan lineage from that of protists and fungi, the origin of multicellularity and consequent embryonic development, the origin of triplo-blastic animals, the origin of coelomate animals, and the radiation of coelo-mates into the ancestral lineages of the major phyla. Finally, we need to establish the actual phylogenetic splitting patterns.

There are two basic ideas as to why multicellular animals should have emerged and radiated at all: either the metazoan radiation was controlled by extrinsic factors, or it was controlled by intrinsic factors. Possible extrinsic factors include global environmental effects, such as the changes in atmo-spheric oxygen levels, temperature, and seawater composition discussed be-low. Possible intrinsic factors have to do with sex, genetics, or development. There is no doubt that sex eventually had to rear its head in the Precambrian algal Eden, and that "invention" of genetic and developmental systems was a necessary prerequisite for the origin of the Metazoa. The question is whether such factors are sufficient to explain the Cambrian radiation. I will defer discussion of intrinsic factors to later chapters on developmental constraints and on evolutionary stability and modifications of animal body plans. In the present chapter I'll try to encompass the extrinsic factors, timing, and pattern of the metazoan radiation.

PRECAMBRIAN LIFE AND ENVIRONMENTS

The emergence of the Ediacaran fauna rests in the context of an accelerating pattern of eukaryotic evolution in the late Precambrian. Analyses of ribosomal RNA gene sequence data by Woese and his co-workers and by Sogin show that the eukaryotic lineage is as ancient as the other major lineages of cellular life, the archaebacteria and the eubacteria. Eukaryotic cells existed for a considerable length of time before the first presumed metazoans appeared in Ediacaran times.

The reliable identification of eukaryotes in the Precambrian fossil record has been difficult because there is no criterion for distinguishing early eukary-otes from fossils of other contemporary cellular forms. As reviewed by Knoll and Walter, cells or multicellular forms that might be eukaryotes have been found in rocks that are about 2 billion years old. But these are not very secure identifications. The situation is clearer in rocks 1.6 billion to 1 billion years of age. By then, as shown by Hofmann, much larger and more elaborate cellular forms are present. Some large multicellular organisms found in these rocks, such as *Eosaccharomyces* and *Grypania,* are probably multicellular eukaryotic algae. *Grypania* was a coiled cylinder about 2 millimeters in diameter that reached half a meter in length. Between 1 billion and 540 million years ago the diversity of eukaryotes was high, and included a variety of protist and algal forms. Molecular phylogenies of eukaryotes derived from small ribosomal subunit RNA genes indicate that the divergence of lineages leading to the algae, plants, fungi, and animals was rapid. No date can be established from the molecular data per se, but Knoll and Walter suggest that a divergence time of 1.1 to 1.0 billion years ago is consistent with the fossil record. Megascopic algae were certainly present. No trace of animals is found until the Ediacaran fauna appears, but small, noncoelomic, soft-bodied metazoans would not have been detectable. Small size and lack of hard parts would have made body fossils unlikely, and lack of a substantial body cavity would have made burrowing in sediment, and thus the formation of trace fossils, impossible. A dating of the metazoan radiation might be extractable from gene sequence phylogenies if reliable molecular clock estimates could be made.

If we could walk along the shore during just about any time during the immensity of time that was the Precambrian, we would be struck by the enduring monotony. There was no life on land. In fact, there was not all that much land. Without plants to hold soil in place, erosion would have kept most of the land surface as bare rock. For 3 billion years about the only macroscopic signs of life at the edge of the sea would have been stromatolite reefs. Stromatolites are algal mats, generally rich in photosynthetic bacteria, known as cyanobacteria, but sometimes containing eukaryotic algae as well. Stromatolites often form large mounds or pillars as successive layers of sediment settle on the cyanobacterial mat, followed by layers of cells that grow over them, only to be buried beneath yet another layer of sediment. These structures were often fossilized, and they form a major part of the Precambrian fossil record. By the late Precambrian stromatolites had declined drastically. Evidence from living algal mats indicates that they are very sensitive to grazing by animals. Stromatolites today are generally found only in environments that exclude most grazing invertebrates, such as the hypersaline lagoon at Sharks Bay, Australia. Walter suggests that grazing by the earliest very small animals may well have precipitated the decline of the stromatolites that so long dominated the world's aquatic ecosystems.

Environmental changes were pronounced in the terminal Precambrian as well, but as Knoll has suggested, might have been only coincidental with the Ediacaran radiation. Nevertheless, the changes were sufficiently drastic that they deserve real consideration as the possible drivers of biological change. How to test the roles of unrepeatable and unique historical processes is another matter. The Varangian Ice Age of the early Vendian marked the last severe episode in a period of worldwide glaciation. (As documented by Hambrey and Harland, there have been several large ice ages during the history of the Earth before the most familiar Pleistocene "ice age" event that ended 10,000 years ago.) Mid-Vendian climatic warming may have produced widespread suitable environments for the first large metazoans of the Ediacaran fauna. The seawater composition changes that also occurred indicate that higher rates of erosion may have increased the fertility of the oceans.

Another significant environmental change that could well have driven the evolution of large body size was the rise in free oxygen levels. This idea, originally suggested in the 1960s by Berkner and Marshall, is that oxygen levels were low during most of the Precambrian, but began to rise sharply toward the Precambrian-Cambrian boundary. Berkner and Marshall suggested that about 1% of present oxygen levels would provide sufficient ozone to protect surface-dwelling eukaryotes from ultraviolet light. Knoll counters that this level was attained almost a billion years before the metazoan radiation. When one considers problems of oxygen diffusion and metazoan biochemistry, the evolution of large metazoans would have required oxygen levels to have passed a significantly higher threshold of several percent of modern levels. As Towe has noted, collagen, the primary connective material in animals, requires molecular oxygen for its synthesis. He has postulated that until sufficient oxygen levels were attained, animals could not have produced a sufficiently strong connective tissue to attain large sizes. Rhoads and Morse have demonstrated that contemporary animals with calcareous skeletons are excluded from anaerobic environments. Such animals cannot secrete calcareous skeletons at below 10% of present-day atmospheric free oxygen levels because acids produced by anaerobic glycolysis dissolve the deposited skeletal material. Raff and Raff have suggested that primitive metazoans probably depended on diffusion for transport of oxygen, and so must have been limited in thickness, especially if late Precambrian atmospheric oxygen levels were significantly below modern levels. They may well have been limited in organization as well by this diffusion mode of gas exchange, especially at relatively low atmospheric oxygen levels. Only later, with the rise in atmospheric oxygen to more modern levels, would more complex gas exchange systems, such as gills and a circulatory system, come into play. The large but very flat Ediacaran creatures are consistent with that view of their structural limitations. Runnegar, on the other hand, proposed that large Ediacaran forms, such as *Dickinsonia,* had a circulatory system and thus could have functioned more effectively than a diffusion animal at lower atmospheric oxygen levels.

It is a distinct possibility that we were all too easily persuaded by the Berkner and Marshall idea that free oxygen levels were low in the Vendian. The trigger for the explosive radiation of the Metazoa and their appearance in the fossil record may have been a powerful and relatively rapid rise in atmospheric oxygen before and during the emergence of the Ediacaran fauna. This new outlook has been suggested by the recent geochemical studies of Derry and co-workers and Des Marais and his colleagues. For 3 billion years or more, the Earth was a brooding planet of algal mats. Photosynthesis was fully engaged in the production of oxygen, but organic material was not buried to a sufficient extent to allow a high atmospheric level of oxygen to accumulate. The late Precambrian was marked by a rise in tectonic events that resulted in higher rates of erosion and accelerated deposition of organic material. Measurements of the carbon isotope compositions of Precambrian shales of various ages (to give the isotopic fractionation of photosynthetic carbon in buried organic matter) as well as carbonate rocks (to give the carbon isotopic composition of carbon dioxide in the seawater) were made by Des Marais and co-workers. These measurements allowed them to estimate the fraction of carbon stored as buried organic material over time. There was a large jump in the burial of organic carbon, and a concomitant release of free oxygen into the atmosphere, between 1 billion and 600 million years ago, just prior to the Ediacaran fauna. The implication of this scenario is that oxygen levels rapidly rose not to just a few percent of present levels, but to modern levels during the Vendian.

How certain can we be that the initial metazoan radiation took place during the Vendian? Three kinds of evidence support the first appearance of large animals at that time: the Ediacaran body fossils; small, enigmatic hard-part fossils; and trace fossils. Strangely, the trace fossils and tubes have so far provided less controversial evidence for metazoans per se than the Ediacaran body fossils. The few Vendian animal hard parts referred to as "small shelly fossils" cannot be assigned to any known animal, and so are called problematica by the paleontologists. However, it is certain that they are produced by animals of some sort. The first such fossils are small tubes. One of these, *Cloudina*, consists of conical, calcareous tubes. As the *Cloudina* animal grew, it moved upward in the tube and secreted another conical tube with its base inside the first. Other tubelike fossils include *Saarina*, which was an annulated organic tube similar to tubes secreted by living annelid and pogonophoran worms. *Redkinia*, on the other hand, is not a shell, but an integral part of some animal. It resembles little comblike jaws. Whether it actually represents the jaws of some enigmatic soft-bodied animal is unknown. Small shelly fossils explode in diversity at the start of the Cambrian. The presence of a few of them in the Vendian is a preview of what is to follow during the next few million years.

Trace fossils are tracks, trails, and burrows left in the sediments by animals. In most cases no trace of the maker remains. Trace fossils in Vendian and Cambrian rocks have been studied by Crimes, who reports that they showed significant diversity only in the mid- and late Vendian, and that their diversity increased with the Cambrian radiation and has remained relatively constant since. What is significant about these trace fossils, as discussed by Fedonkin and Runnegar, is that traces can be produced only by relatively complex animals that possess bilateral symmetry and a triploblastic organization of body tissues. Late Precambrian trace fossils are horizontal traces of animals moving on the sediment, but not making deep burrows. The makers of some surface traces might have been of flatworm grade, without a coelom. Clark has pointed out that efficient burrowing requires a coelom. The coelom is the mesoderm-lined body cavity in which the organs are suspended, but it is more than that. It also provides a fluid-filled hydrostatic skeleton for operation of the body wall musculature. A segmented body structure may further improve the efficiency of burrowing and locomotion. If the Vendian trace-makers possessed coeloms, then the Metazoan radiation was already well under way during the last few million years of the Precambrian. Most of the Cambrian explosion is of coelomate phyla, and deep burrowing originates at the start of the Cambrian.

ARE THE EDIACARAN FOSSILS ANIMALS AT ALL?

The Ediacaran body fossils are suggestive but unsettling. A few might represent bilaterally symmetrical segmented coelomates. *Spriggina* has been seen as the most probable such animal, although other members of the fauna have been suggested as coelomate-grade animals as well. However, most Ediacaran animals have been inferred as being of diploblastic organization. The living diploblastic organisms fall into two phyla, the Cnidaria and the Ctenophora. A number of Ediacaran animals have been placed within the Cnidaria as resembling jellyfishes, sea pens, and anemones. A selection of Ediacaran fauna creatures is shown in figure 3.1, and a pair of alternative reconstructions of *Spriggina* in figure 3.3.

Glaessner assigned the Ediacaran fossils to living classes and even families, suggesting a strong link to the subsequent radiation of animals and a strong evolutionary conservation of the Ediacaran forms (see fig. 3.2). Essentially, he conceived of that fauna as still existing today alongside a much richer modern fauna evolved from Ediacaran ancestors in the Cambrian and subsequently. Continued analysis of Ediacaran body forms by Gehling, Jenkins, Runnegar, Runnegar and Fedonkin, and Fedonkin has shown that approach to be too simplistic. Ediacaran body fossils are decidedly odd. Many are extremely large: some reach nearly a meter in length, including the ribbed *Dickinsonia,* the medusoid *Ediacaria,* and the frondlike *Charnia.* Most are flat, even the very large ones. Many are disclike or frondlike in form. These shapes could reflect an efficient solution to low oxygen concentrations, or, as suggested by McMenamin and McMenamin, a trophic strategy different from those of most living animals. No obvious mouths are preserved in Ediacaran fossils (although, as Gehling has pointed out, this doesn't mean that none were present in the living creature). We know from living examples, such as the robust 2-meter-long tube worm *Riftia pachyptila,* which lives at hydrothermal vent sites on the deep-sea floor, that lack of a mouth does not preclude large size. These extraordinary worms have tissues packed with symbiotic sulfur bacteria that metabolize hydrogen sulfide extracted from the vent water, and they live on the metabolites produced by their bacteria. Following an earlier suggestion by Fischer, McMenamin and McMenamin proposed that the Ediacaran animals harbored photosynthetic endosymbionts, with their flat construction giving their symbionts the maximal possible exposure to light. They were essentially giant kelplike animal leaves inhabiting well-lit and calm seafloors. Modern cnidarians, such as anemones and corals, are often brightly colored by their photosynthetic endosymbionts, on which they depend for a substantial part of their nutrition. But the oddness of the Ediacaran fauna goes beyond their peculiar overall morphologies.

The Vendian fauna may have been distinct from the faunas of the later world. Metazoan lineages that had an evolutionary continuity into the Cambrian evidently did exist (as in the "early diploblastic" hypothesis diagrammed in fig. 3.2), but it is not clear how many of the organisms that left body fossils also left descendants. Apparently not many, although Conway Morris has illustrated a few that may have survived into the Burgess Shale, most notably a frondlike form, *Thaumaptilon,* that resembles the Ediacaran fossil *Charniodiscus.* Most of the forms that make up the Ediacaran body fossil record do not occur in the post-Vendian record. They appear to have become extinct well before the Precambrian-Cambrian boundary. Since there is no sign of predatory damage to the Ediacaran body fossils, predatory forms may not yet have evolved. Large, flat food items that would now be rapidly eaten could have then prospered by simply lying passively on the seafloor in what McMenamin and McMenamin have called the "Garden of Ediacara" (see fig. 3.2).

A close look at a number of the Ediacaran body fossils shows that they exhibit features quite distinct from those of all later metazoans. Glaessner interpreted the "medusoids" as representing jellyfishes of various classes. However, most Ediacaran medusoids look different in structure from bona fide jellyfishes. Some have a pronounced central disc, and on their margins, rays instead of circumferential rings. Living jellyfishes have a radial central structure. They have a central mouth and stomach with four gastro-genital pouches extending radially from the stomach. Living medusas also possess circular muscles around the edge of the bell. It is the rhythmic contraction of these muscles that gives swimming jellyfishes their eerie, slowly pulsating swimming stroke. The fourfold symmetrical pattern of the digestive pouches and the circular muscle fibers of the bell margin are visible in undoubted fossil jellyfishes from younger rocks. One strong possibility is that medusoids with a central disc and radial edges are detached bases from frond-shaped creatures such as *Charniodiscus.* Another possibility is seen in restorations by Jenkins of one of the medusoids, *Ediacaria,* as a cnidarian resembling an upside-down jellyfish (mouth up) attached to the seafloor by a broad, short stalk. There is an analogous large living jellyfish, *Cassiopeia,* that has no stalk, but lives an upside-down life on quiet, well-lit tropical shallow seafloors.

Some of the best-preserved true jellyfishes occur in the Jurassic Solnhofen lithographic limestone, from which specimens up to half a meter across are known. These show the expected fourfold structure of the bell. Only a few Ediacaran medusoids *(Conomedusites)* have a fourfold symmetry preserved in their bells. Others seem to have well-preserved tentacles around their bell margins. Seilacher has disagreed, suggesting instead that the ''tentacles'' are feeding traces of a burrowing animal that made feeding excursions extending out from a central burrow opening.

The rest of the medusoids appear to be disc-shaped organisms of cnidarian grade, but have a threefold rather than a fourfold symmetry. Some of these, including *Albumares, Anfesta,* and *Tribrachidium,* have been found preserved in considerable detail in Russian faunas studied by Fedonkin. He places them in a new, extinct class of cnidarians, the Trilobozoa. It is interesting to see this progression of thought. *Tribrachidium,* an enigmatic disc with three curved ''arms'' on its surface that resembles the Greek heraldic figure, the triskelion, was traditionally considered an ancestral echinoderm. The reason for that assignment was that it resembles what an edrioasteroid (an extinct class of disc-shaped echinoderms) might look like if it had three instead of five arms. That interpretation led to considerable speculation about how echinoderms evolved pentameral symmetry. (*Tribrachidium* is not the only ancient echinoderm candidate. There is another disclike fossil, *Arkarua,* with five grooves radiating from the center, that Gehling has nominated as a primitive pentameral echinoderm.)

In sum, we have five interpretations of Ediacaran medusoids: as jellyfishes, as bowl-shaped, stalked cnidarians, as discoid three-part radial cnidarians, as detached frond bases, and as feeding traces of a burrowing animal. Some, or even all, of these interpretations could be correct because it is now recognized that the medusa-like fossils actually represent a structurally diverse group of creatures, not necessarily all related to one another or to living cnidarians.

A very substantial portion of the worldwide Ediacaran fauna consists of large frondlike forms, interpreted by Glaessner as sea pens. Living sea pens are good-sized cnidarians, colonial relatives of sea anemones. They have a bulbous base that holds them upright in the sand substrate upon which they live. Attached to the base is a thick stalk, along each side of which a series of projections is attached. These projections bear numerous feeding polyps. Some of the Ediacaran species, such as *Charniodiscus,* that resemble sea pens have a bulbous base, a stalk, and a large frond. Others, such as *Rangea,* are analogous in structure, although *Rangea* differs from the others in having several fronds to an axis. Jenkins, who has reconstructed *Rangea,* considers it to be part of an Ediacaran radiation of sea pen-like cnidarians. However, there are two difficulties with the conclusion that these forms are sea pens, or even cnidarians. First, the preservation, most often in sandstone, does not show enough detail to confirm that polyps are present. Second, Ediacaran "sea pens" are unlike living sea pens, in which the projections are separate and expose the polyps arrayed along them to water currents for filter feeding. In contrast, the Ediacaran fossils are constructed as single large fronds. Each single sheet of a frond bears a complex substructure, possibly with polyps.

There is a last peculiarity about some Ediacaran forms: many seem to be segmented. In some, such as *Spriggina* and a newly discovered "soft trilo-bite" discussed by Jenkins, the segments really do look like those of annelids or arthropods. But two other forms of segmentation appear as well. The first is exhibited by one very peculiar frondlike creature, *Pterdinium,* which can be up to a meter in length and is spindle-shaped in overall form. This creature is built with three vanes that intersect along the long axis of the spindle. Each vane is made of hollow segments that meet along the axis in alternation. This same alternation of segments is seen in a sac-shaped fossil, *Ernietta,* as well. These creatures are so peculiar that Pflug created a new phylum for them, the Petalonamae. They are not cnidarians, and they are certainly not built along a body plan seen in any other known metazoan. This kind of peculiar segmentation is seen in rather different Ediacaran fossils as well. The petalonameans have no anterior-posterior differentiation. *Pterdinium* is the same at both ends.

Other Ediacaran creatures have an apparent anterior-posterior polarity. In *Spriggina,* the head looks like the cephalic shield of a trilobite. The body is segmented and tapers to a posterior termination. In some other fossils the segments do not meet at the midline ridge. Thus, *Vendia* looks superficially like a little trilobite, but its "segments" actually have an interdigitating rather than a truly bilateral symmetry. Fedonkin calls this condition "sliding reflection." No contemporary animals are constructed in this fashion. Fedon-kin suggests that this could be a primitive form of arthropod/annelid segmen-tation that has not quite been perfected. However, such "segments" cannot be homologous to arthropod segments because the developmental rules for setting up segmental body patterns simply do not allow it: a developmental constraint exists. *Drosophila* geneticists have generated a lot of mutants in the course of their days, and exhaustive genetic screens have been done for mutations that affect the basic patterning of the fly. Some affect anterior structures, or posterior, or dorsal, or ventral, but none produce right-left disturbances. Why should that be? As reviewed by Nüsslein-Volhard, these mutations, and the detailed study of the genes they revealed, showed that two distinct systems set up the anterior-posterior and dorso-ventral axes. Segmentation follows from the initial molecular localization systems that specify the two axes. There is no system specifying right and left; that symmetry falls out of the initial conditions. That occurs because both anterior-posterior and dorso-ventral localized maternally loaded determinants (mostly mRNAs) are distributed symmetrically along their respective axes. These distributions have been visualized by use of in situ hybridization and antibodies to the protein products of the localized mRNAs. There is no asymmetry in the distribution of the localized determinants that allows a left-right symmetry break to form. The downstream gene expression systems consequently show no right-left asymmetries in patterns of expression. A genetic system that could produce sliding reflection would be different from the one that we know has been in place among arthropods, and probably annelids, since the early to mid-Cambrian. It seems likely that forms exhibiting sliding reflection either represent a lost mode of animal segmentation or were not animals at all (the "vendozoans" hypothesis diagrammed in fig. 3.2).

The strongest defense of the Ediacaran fauna as metazoans has come from Gehling, who has argued that the Ediacaran fauna is actually quite heterogeneous in body structure. It's not all big fronds and discs; there is a significant number of small taxa as well. Gehling has also argued that the flat morphology inferred for the Ediacaran fauna is based on the flattened condition of most of the fossils. These creatures may not have been foliate in life. He has noted that some of the smaller forms are preserved in relief, that some forms have an animal-like segmentation, and that signs of muscular contraction are present in some specimens of *Dickinsonia*. Gehling's arguments are important because they present the metazoan hypothesis in a testable form, and without the shackles of placing the fossils into living taxa.

In thirty years of study, we have moved away from the comfortable assurance of Glaessner's interpretation of the Ediacaran fauna as a logical step in the emergence of the modern animal phyla, simple, soft-bodied ones first. Now it is clear that the Ediacaran radiation is difficult to understand. The body plans do not all fit smoothly into the living phyla, although some may. Some may represent metazoan body plans no longer in style. In the most radical view, these creatures may not be animals at all. The issue is important because our interpretation of these fossils will influence the way we view the definitive appearance of the animal phyla in the Cambrian.

IF NOT ANIMALS, WHAT?

One of the most elegant settings for any scientific meeting is Alfred Nobel's Björkborn, his retreat deep in the Swedish countryside, where the cool white and green birch forest contrasts, perhaps a bit strangely, with the adjacent Bofors armament works. The very comfortable Bofors Hotel features both portraits of important armament works managers and bathrooms walled with a deep red marble that contains fossil stromatolites. Nobel's library in the house shows that his interests ran to explosives, history, and literature, but not to biology. Nevertheless, here the Nobel Symposia, small meetings featuring intense discussion and good food, have occasionally focused on the problems of interpreting phylogenetic histories and the record of life's course on the planet. The symposium published in 1994 as *Early Life on Earth,* edited by Stephan Bengtson, examined the planetary and biotic events leading to the Cambrian revolution.

Several researchers at this meeting discussed the nature of the Ediacaran fauna. Their interpretations were surprisingly disparate. As the final session on the metazoan radiation proceeded, I wanted to get a better idea of where each stood, so I asked a parlor game question. The most common Ediacaran fossil is *Dickinsonia.* Despite its prevalence and size, it remains an enigmatic creature. It was up to a meter in length, but no thicker than about 3 millimeters. It was segmented, and its two ends differ from each other, although nothing resembling a head is present. There are some older published interpretations of *Dickinsonia* as a cnidarian or a flatworm, but it has been quite consistently recorded in the literature as an annelid worm. In a detailed analysis of its anatomy and growth, Runnegar concluded that *Dickinsonia* was coelomic and segmented, that it had a circulatory system, and that the central axis was a gut, sometimes filled with sediment. Jenkins argued that he has located a mouth and has reconstructed a complex double intestine with branched gastric diverticulae similar to those of certain living marine worms. Given the nature of the fossils and our lack of real knowledge of what phylum they actually belong to, there is a certain danger of overinterpretation. I asked each of the seven paleontologists at the Nobel Symposium interested in the matter to consider just *Dickinsonia.* I asked each to state whether he thought it was an animal, what kind of organization it possessed, and whether it was related to the groups that took part in the metazoan radiation of the Cambrian. The answers ran an interesting gamut. All but one agreed that it was an animal. Of these, only one thought it was not related to anything in the Cambrian metazoan radiation. Of the others, one held *Dickinsonia* to be a cnidarian, one a flatworm or close to that in grade, one thought it was a low-grade bilaterian, and the other two thought it was higher in organization, possibly a stem protostome. The most radical view was that it was not an animal at all, but a "fungal grade," possibly syncytial, organism. Obviously we have not yet reached a consensus.

What of the most extreme answer, that *Dickinsonia* is not even a metazoan?

A few years ago, Seilacher introduced a profoundly contrarian approach to the problems of interpreting the Ediacaran fauna, and suggested visualizing the fauna in a radically different way. Seilacher noted that all the fossils share an unusual mode of preservation as flat impressions on sandstone. Similar younger rock facies show no such preservation of fossils. Sandstone is generally a poor preservation material for most fossils, especially soft-bodied forms. However, sand can take impressions from biological objects. Glaessner told of a seashore experiment in which he stood on a large stranded jellyfish. That treatment did no harm to the tough mesoglea, and the oral face of the jellyfish left a clear impression in the sand. I have to admit that I have done the same experiment—with the same result. Judging from the glances of passing beach walkers, it apparently does look a bit strange to be posed on top of a jellyfish. Bruton, who has done this sort of jellyfish study more systematically, found that jellyfish impressions formed readily in beach sand, but because jellyfish have the same density as water, they probably would not leave traces in submerged sand, such as would have been the case in most Ediacaran faunal settings. A major problem lies in setting up the sand around a buried soft-bodied creature to preserve the impression. Gehling, as well as Fedonkin and Runnegar, has suggested that the bacterial films that coated Precambrian seafloor surfaces might have provided a stabilizing physical and chemical medium for impressions. Allison and Briggs record other instances in the fossil record in which microbial films have preserved organic forms. One famous example is the 50-million-year-old Messel Shale of Germany, in which soft-tissue outlines of frogs and bats are preserved as films of mineralized sheets of bacteria. Thus, the Vendian seafloor may have had unusual properties that enhanced the preservation of soft-bodied animals in a way seldom duplicated in later times because of extensive algal grazing.

Seilacher suggested that the Ediacaran creatures share a "quilted" construction in which the edges between "segments" tend to stand out in the fossils as if they were stiffer than the rest of the body. That implies a sort of air-mattress body, with a stiff integument and dividers between fluid-filled compartments. The bodies were very thin. They were flexible enough to deform under stress, but too stiff to form small-scale wrinkles. Those features would preclude any of these "vendobiotans" from having a muscle-based motility. They would also mean that the "segments," say, in *Dickinsonia* or *Spriggina*, would not really be segments at all, just compartments framed by integumental dividers. Seilacher suggested that these quilted beings grew by two modes, serial addition of new units or fractal subdivision of existing units. All were basically foliate. We are used to looking at *Spriggina* as a worm preserved as if crawling on a surface, head forward. Seilacher stood it on its "head." *Spriggina*, interpreted like the other vendobiotans, becomes

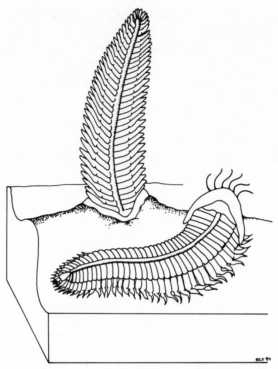

Figure 3.3. Two reconstructions of the Ediacaran fossil *Spriggina,* as an annelid and as a vendobiotan frond. The two representations illustrate the range of the interpretations that have been made of the fossils by various investigators. *Spriggina* was originally interpreted as an annelid by Glaessner, who compared it to living species. In this drawing, it is portrayed in the foreground as a chunky polychaete worm propelled by muscular parapodia. The anterior tentacles are found in living polychaetes. No such structures are preserved in *Spriggina* fossils, and so are purely poetic license. The worm *Spriggina* is shown passing by *Spriggina* portrayed as a frond. There the "head" has become an anchor binding the frond to the substrate. Seilacher interpreted *Spriggina* as a frondlike organism, but suggested that it lay flat upon on the substrate rather than standing upright. Seilacher has interpreted a number of Ediacaran fossils as recumbent or upright fronds. A recumbent frond is also in accord with the Garden of Ediacara hypothesis of McMenamin and McMenamin; that hypothesis too remains unconfirmed. There are still other interpretations. Gehling has argued that *Spriggina* is an early trilobite. However, neither legs nor antennae are preserved, so *Spriggina* has yet to be confirmed as an arthropod. (Drawing by E. C. Raff.)

an immobile frond, with its "head" reinterpreted as a holdfast or base (fig. 3.3). These quilted fronds do not look like any known animal constructional mode.

If the Vendobiota were the products of a radiation of some nonanimal group, now extinct, what were they? Seilacher, and later Buss and Seilacher, have made two quite different suggestions. Seilacher pointed out that integumental compartmentalization is a general feature of existing large syncytial organisms such as the giant parasol-shaped unicellular alga *Acetabularia* and large foraminiferans. These unicellular creatures reach large sizes in the range of a few centimeters, a good deal smaller than the meter size of the big vendobiotans. The vendobiotans could not have been uninuclear. There would simply be too much cytoplasm for a single nucleus to supply with ribosomes and messenger RNAs. However, there are large multinucleate protists and slime molds, some as large as vendobiotans. The acellular slime mold *Physarum* can contain millions of nuclei in a common cytoplasm the size of a cafeteria tray. Even more striking is the multinucleate "seaweed" *Caulerpa,* discussed by Jacobs, which grows to a meter in length and bears 10-centimeter fronds. However, these speculations have overreached the data a bit. Once outside the realm of metazoans, we might guess at a variety of other possibilities, such as brown algae. Some frond-shaped Ediacaran beings look a lot like small kelps, such as the sea palm that is common on the American West Coast.

The more recent suggestion of Buss and Seilacher is that the Ediacaran fauna constitutes a monophyletic clade, the phylum Vendobionta, which is the sister group to the Eumetazoa and of cnidarian grade (see fig. 3.2). These animals had a quilted construction and lived by symbioses with photosynthetic algae. Quilting does not in itself eliminate cnidarians. We have no living quilted forms, and simply do not know if quilting is feasible within the limits of cnidarian development and physiology. The fossils do not preserve enough detail to tell.

Perhaps we ought to look at the vendozoans in yet another way. The paleobotanist Greg Retallack has pointed out that the Ediacaran fossils are preserved in a quartzite formed under a rather great (5 kilometers) depth of sediments. Fossils are known from quartzites younger than the late Precambrian, but not of soft-bodied animals. Retallack too performed the standing-on-a-jellyfish experiment, and found them more compressible than the logs he also tested in this way. He stressed that relatively soft objects would be obliterated under the conditions of compaction under which quartzite forms. Even dense and solid remains, such as wooden logs, are flattened under these severe conditions of fossilization. He suggested that impressions in quartzite require objects with the stiffness of wood. That doesn't leave many candidates for the Ediacaran fossils. Retallack argued that only one, lichens, fits the bill. These organisms exhibit a wide variety of forms and habitats, and they often have a stiff, woody texture. Again, as large-surfaced photosynthetic symbiotic associations of algae with fungi, they could have played the role of the flowers in the Ediacaran garden envisioned by the McMenamins. Because there are few, if any, well-recognized lichen fossils in the fossil record, comparisons are scarce, and the hypothesis depends on the analogies used in its creation.

It is important to note that Gehling has questioned the vendozoan hypothesis as a universal explanation for the Ediacaran fauna on the grounds of the construction of fossils such as *Dickinsonia* and the decidedly non-flat shape of some of the smaller taxa. Conway Morris has questioned the vendozoan hypothesis by proposing a connection between the Ediacaran frond *Charniodiscus* and the Cambrian sea pen *Thaumaptilon,* which has a fused frond similar to the Ediacaran form. If one is a cnidarian animal, it is likely that the other is also. The sorting out of the cast of Ediacaran characters is guaranteed to remain a rich source of controversy, and publications. Its resolution will affect the way we view the timing and pattern of the radiation of the animal phyla.

The Ediacaran world represented a unique event that clearly had a theme of its own, not similar to the themes in metazoan evolution that have characterized the past 530-plus million years. The Ediacaran soft-bodied forms were the products of an odd and relatively short-lived experiment that may have thrived for as long as 20 million years. Very few remnants of this fauna persisted past the golden age of worldwide fronds. The radiation of mobile predatory metazoans might well, as suggested by McMenamin and McMenamin, have overgrazed them, like goats introduced to a pristine Pacific island. Most seem to have passed on almost 20 million years before the definitive Cambrian metazoan radiation began, but their absence from the record could be misleading. The enhanced recent interest in the Ediacaran fauna will, I'm sure, stimulate a lot more weary walking of upper Vendian rocks by paleontologists in search of the tombs of the last survivors.

BODIES AND BEHAVIORS ON THE CAMBRIAN BOUNDARY

If events are examined at too gross a time scale, we see great but misleading discontinuities. This is the phenomenon of seeing a distant friend's children at five-year intervals and marveling at their growth. A day-by-day acquaintance is much less dramatic. A part of the reason for the apparent suddenness of the Cambrian metazoan radiation is that for a long time the lower part of the Cambrian was not well enough explored. Nor was there (or is there yet, for that matter) a sufficient knowledge of late Precambrian metazoan life. The consequence was an exaggerated picture of the massiveness of the jump in biological events across the Precambrian-Cambrian boundary. This picture is likely to reflect both missing data and the nature of the animals themselves.

Most marine animals living today are soft-bodied. Estimates of fossilizability made by Tom Schopf ran to only about 20% of modern bottom-dwellers as good candidates for preservation as fossils. Other studies, summarized by

Allison and Briggs, are consistent with that estimate. Most modern soft-bodied animals are "worms" of various phyla and lack shells, skeletons, and teeth. We would find their bodily remains only under exceptional and rare modes of preservation. We would find trace fossils recording their activities much more commonly, but we would not know what many of the makers were. The exceptional fossils of the Burgess Shale and similar Cambrian faunas show that most Cambrian animals too were soft-bodied. There is every indication that late Precambrian animals were almost exclusively soft-bodied. Trace fossils thus offer one way of estimating the reality and extent of the metazoan radiation.

As Crimes points out, estimates made a few years ago on period-long time scales—that is, the Vendian versus the entire Cambrian—showed an enormous jump in trace fossil diversity. More detailed data are now available, and Vendian and early Cambrian faunas have been much better described. The Manykaian is the earliest stage of the Cambrian, spanning the interval of 544–530 million years ago. The Tommotian and Atdabanian stages follow. These stages are each about 2–5 million years in duration. Trace fossil diversity rises in the early Cambrian. The record really shows that there was a steady climb in diversity, not an explosion, of trace fossils over the few million years of the Precambrian-Cambrian boundary, with a doubling in the number of trace fossil "species" during the early Cambrian compared with Vendian diversity. There are about twenty-two late Vendian trace fossil forms, with most occupying shallow-water habitats. It is not clear how well numbers of trace types reflect actual diversity of animals. There could be a large difference, but estimates can be made from living faunas. To test this point, Hertweck evaluated the 268 species of macroinvertebrates of the Georgia coastal shelf in terms of their representation as traces preserved in sediments. He found that a mere forty species produced distinct traces. Of these, perhaps half would be preservable as trace fossils.

The increase in trace fossil diversity across the Cambrian boundary is significant in two ways. First, it reveals the dynamics of appearance for some phyla for which there are no body fossils. The first arthropod body fossils, trilobites and others, appear in the Atdabanian stage, but trace fossils reveal arthropod tracks from the latest Vendian and the Manykaian. Thus the earliest arthropod body fossils do not represent the earliest arthropod animals. Second, the trace fossil record reveals a distinct increase in behavioral repertoire in early Cambrian animals.

The behavioral changes are remarkable. Vendian traces are generally simple meanders, often crossing themselves. That is not a particularly efficient way for a bottom feeder to forage. By the early Cambrian, spirals and regular meandering trails that don't cross each other emerge, as do complex burrows with elaborate branches and translocations through the sediment. More complex arthropod tracks and burrows appear as well. We are probably seeing a coincidence of three important evolutionary events. The first is a direct increase in the diversity of animals. The second is an improvement in burrowing capability, which very likely represents the acquisition or perfection of a coelomic body structure and the ability to burrow by peristaltic action. Finally, there is also a record of fossilized behaviors that reflect the critical evolutionary innovations of more effective brains and nervous systems, an event Miklos and his colleagues argue persuasively "must have been subtended by gene circuit and neural circuit explosions."

Arthropod brains are evolutionarily conserved, and Osorio has suggested that the neural machinery underlying arthropod vision has changed little since the Cambrian. By as few as 10 million years into the Cambrian, Miklos and co-workers point out, there were arthropods with fully developed sensory systems and compound eyes. As these animals were preceded by the small shelly faunas, which included complex motile animals, it would seem that the arthropod brain became organized in as few as 5 million years. Vertebrates may have undergone an analogous rapid neural evolution somewhat later in the Cambrian. Miklos and co-workers conclude from the classic anatomical studies of Stensio on the brains of early fossil fishlike vertebrates that the basic vertebrate brain was established by 480 million years ago. The first primitive chordate, *Pikaia,* occurs in the mid-Cambrian, about 520 million years ago. Other evidence for primitive vertebrates, discussed by Forey and Janvier, suggests that primitive vertebrates might have evolved by the late Cambrian, 510 million years ago.

In the view of Miklos and his colleagues, the rapid definitive organization of the arthropod brain during the early Cambrian required a massive reorganization of existing genetic systems into a new neural entity. It was then rapidly locked into place, and has persisted over a very long subsequent evolutionary history. The scenario is a reasonable one, but I do not want to give the impression that I think brains just popped into being. There was a 35-million-year-long history of bilaterally symmetrical trace makers before the arthropod body fossils appeared. The trace fossils show us that behavior did grow more complex during that time. The brains and nervous systems of the most primitive living bilaterally symmetrical triploblastic animals, the flatworms, are reasonably complex. A neural system of that level of organization would have provided a good starting point for further neural evolution during the Cambrian radiation. Flatworms, as shown by Bartels and co-workers, possess the genes that encode the homeotic transcription factors that are responsible for the definition of anterior-posterior organization in the central nervous systems of higher animals, such as arthropods and vertebrates. The functions of the homeobox genes in flatworms have not been demonstrated, but the important point is that this class of genes, and other critical regulatory gene systems, would have been available for co-option in brain reorganization during the Cambrian radiation.

SMALL SHELLY FOSSILS

If trace fossils provide a continuity between the invisible animals of the late Precambrian and the great Cambrian radiation of structurally preserved animals that make up the fossil record of the 544-million-year-long Phanerozoic eon, the actual entry of those animals onto the stage is both discontinuous and peculiar. It is the overwhelming appearance of skeletons that opens the grand history of the metazoans, and the Tommotian is the age of the "small shelly fossils." It is important to note that most phyla arose during or before the start of the Tommotian, which was formerly set at the base of the Cambrian, but is now placed by Bowring and co-workers as middle early Cambrian, 530 million years ago. Skeletal elements show up suddenly in Tommotian rocks, and mark the first appearances of many still-living phyla. Thus, sponges, cnidarians, annelids, mollusks, brachiopods, and echinoderms make their skeletonized debuts during the middle early Cambrian. So do a number of weird creatures with evocative names and problematic identities. Trilobozoans, agmatans, archeocyathids, mobergellans, tommotiids, halkieriids, wiwaxiids, utahphosphids, and others take their short, enigmatic turns on the metazoan stage and vanish after a few million years. Most of these skeletal remains are in the range of a few millimeters in size, but they are complex and clearly metazoan in origins. In the Siberian, Kazakhstanian, and western Mongolian record of the Precambrian-Cambrian boundary reviewed by Rozanov and Zhuravlev, there is a rich small shelly fauna. Over 78 species, representing a wide diversity of organisms, have been recorded. Of these, only 4 extend deeply back in time to the Vendian; they are tubelike, each once housing some small, unknown animal. A few others appear just below the boundary. The rest are Cambrian appearances. Bengtson records similar events from other parts of the world, as does Jiang.

As Bengtson points out, in addition to providing fossilizable body parts, skeletons serve several functions in living animals. They provide support for body structures, a means of attachment to the substrate, leverage for muscles, enclosures for filtration feeding systems, teeth for grazers and predators, spines or shells as protection from predators, and sensory organs, and even serve as stores of calcium or phosphate that can be mobilized for metabolic purposes. Organisms secrete about sixty different minerals, but only a few are used in animal skeletons. These are mainly forms of calcium carbonate, calcium phosphate, and silica. Chitin and other organic skeletal materials are also used, sometimes in conjunction with mineralization. When skeletons appear in the fossil record, all of the commonly used skeletal mineral species show up, and they are widely spread across the phyla. The first forms to appear are small tubes, with the late Precambrian *Cloudina,* described by Grant as a laminated tube composed of a nested set of cones, holding pride of place as the earliest animal skeleton. *Cloudina* also provides a hint as to why skeletons proliferated so rapidly. Many *Cloudina* specimens have been bored into by some predator known to us now only by the traces of its lethal handiwork. Predation may well have driven skeletonization. Certainly, Vermeij has made it abundantly clear that the arms race between predators and prey has accelerated steadily during the past half billion years. Most styles of marine predator are present among the early Cambrian faunas preserved in the fossil record. So are their victims, for instance, damaged and healed trilobites. Various spicules, spines, shells, plates, and bosses appear at the start of the Tommotian. We must decide whether they represent the origins of their respective animals, or merely the invention of skeletons. It seems to be in large part the latter, although that may be too simple a conclusion.

The difficulty in finding an answer to this question that doesn't include qualifiers is that in some cases skeletons could simply be added to existing body plans without fundamental changes in the underlying biology. In other cases, the evolution of the body plan required the presence of a skeleton, and certain groups could develop their definitive forms only after a skeleton had evolved. Specific examples make the point. For clams to burrow, as Stasek and Stanley both tell us, a mineralized shell is required, both mechanically and to create a mantle cavity for the gills. If shell-less clams pose a tough proposition, shell-less brachiopods are an impossible one. According to Valentine and Erwin, brachiopods, which are so characteristic of Paleozoic marine faunas, probably provide the best-defined example of the role of a skeleton in the origin of a metazoan body plan. Cloud long ago argued that the brachiopod body plan includes its pair of shells as a basic functional component. Brachiopods are filter feeders that draw water between the shell margins and over ciliated lophophores. LaBarbera has analyzed the water flow patterns of several living brachiopod species. He has shown that the feeding currents are laminar, without lateral mixing of water within the animal. A rigid, properly configured shell is required for the coordinated action of the shell and lophophore that creates a water flow geometry suitable for efficient feeding. The very rapid radiation of brachiopods in the early Cambrian probably really does represent the success story of a newly evolved phylum.

UNEARTHING THE UNIMAGINABLE

To move beyond the hands-full-of-rocks phase of analysis of the first part of the visible metazoan radiation, it will be necessary to tie the diverse small

Figure 3.4.   An armored Cambrian lobopod. The bosses that line the sides of this animal were originally known only as small shelly fossils, called *Microdictyon*. The discovery of the intact soft-bodied animal in the Chengjiang fauna of China has revealed the nature of this enigmatic fossil. (Drawing by E. C. Raff. Redrawn by permission of *Nature*. L. Ramsköld and X. Hou. 1991. *Nature* 351:225–28. Copyright 1991, Macmillan Magazines Limited.)

shelly fossils to understandable animal groups. Many small shelly fossils are quite comprehensible. Small brachiopods, mollusks, sponges, and echinoderm parts are readily interpretable. Others allow at least good guesses. The rest have been enigmas. Their ties to real animals can come only from fossils that preserve the whole soft-bodied animal with the skeletal structures in place. With a generosity that nature seldom bestows, several Cambrian rock formations have preserved soft-bodied faunas in almost miraculous detail and have revealed the identities of some of the early skeleton bearers. The best-known of these faunas is that of the mid-Cambrian Burgess Shale. This and other Cambrian faunas from about forty localities worldwide are fossilized in such a way as to preserve soft-bodied animals as sharply defined mineral compressions. Although other soft-bodied fossil faunas are known from later geologic ages, the Burgess Shale mode of preservation is limited to the Cambrian. We don't know why that is so, but without this unique preservation, dozens of taxa of primitive soft-bodied metazoans would have disappeared without a trace.

One of the animals known first only from small shelly fossils is shown in figure 3.4. *Microdictyon*, from the early Cambrian, was long known only as a collection of peculiar hexagonal meshwork buttons, each with a central spike. A complete soft-bodied animal that bears *Microdictyon*-like plates has been described by Ramsköld and Hou from another Burgess Shale-like occurrence, the early Cambrian Chengjiang fauna of China. That *Microdictyon* is a lobopod, perhaps related to the living onychophorans, the so-called velvet worms, was a complete surprise. Similarly, *Halkieria* is described as a complete animal by Conway Morris and Peel from the early Cambrian

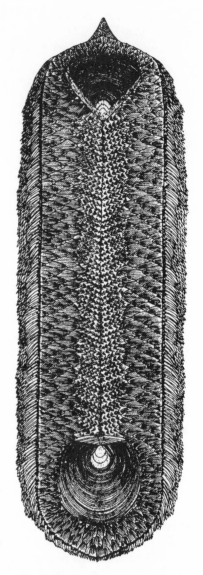

Figure 3.5. Reconstruction of an articulated hal-
kieriid from the lower Cambrian of Greenland,
viewed from the dorsal side. (From Conway Morris,
S. and J. S. Peel. 1995. *Phil. Trans R. Soc. Lond.*
B 347:305–58. With permission of The Royal Society.)

Buen Formation of Greenland (fig. 3.5.) This animal had a spicular coat
protecting a soft body and apparently traveled like a slug on a molluscan
foot. It resembles the spicule-covered Burgess Shale animal *Wiwaxia*, which
in its anatomy seems to have been close to the origins of the mollusks and
annelids. Before finding the complete creature, there would have been no
way to predict from isolated spicules how its exoskeleton was arranged. With

its spicular coat and fore-and-aft mollusklike shell, *Halkieria* is unlike any mollusk found today or in rocks younger than the Cambrian. It may give us a unique window into an early experiment in molluscan body plans. More significantly, Conway Morris and Peel have noted that its features suggest a common ancestry for annelids, mollusks, and brachiopods.

To understand the metazoan radiation, we must have a grip on timing as well as pattern of relationships. New phylogenetic tools utilizing gene sequence data may provide help with both. For example, the first sponge spicules appear in the early Cambrian. Do they mark the first sponges? It seems unlikely. Wainright and co-workers used 18S ribosomal RNA gene sequences from a variety of lower metazoans, protists, plants, and fungi to construct a phylogenetic tree and place the root of the metazoan radiation. They found that sponges are the most primitive branch of the animals, followed by cnidarians, then triploblastic animals, very much as in the classic view. Their data provide no direct time estimates for the origin of sponges, but we know from the trace fossil record that much more advanced triploblastic metazoans are already present in the Ediacaran fauna, 25 million years before the Cambrian. Sponges, or at least the sponge lineage, had to have been present too. The most plausible explanation for why sponges fail to appear as fossils before the Cambrian is that they simply had not yet invented spicular skeletons. I'll discuss gene sequence inferences in more detail in the following chapter.

The metazoan radiation began during the Vendian, and the ensuing Cambrian explosion represents the flowering of that radiation. The major phylogenetic lineages had already become established by that time. Like the sponges, many of the other phyla living today must have arisen some millions of years before we first find their remains. Early estimates of the length of the Cambrian, as discussed above, were too long. The result was a comfortable view of a prolonged period of metazoan evolution during the early Cambrian, not a rapid radiation. The early Cambrian is now seen as very short in duration. The metazoan phyla may well have begun their radiation during a 20–30-million-year interval prior to their skeletonization in the early Cambrian. Certainly, the molecular phylogenetic studies that we initiated on the metazoan radiation, reported in a paper by Field and co-workers, indicated that the time required for the splitting of phyletic lines was relatively short compared with the 530-plus million years from the divergences until now. Our results indicated that a very long period of splitting, occupying a few hundreds of millions of years, was not possible. This does not mean that metazoan multicellularity might not be quite a bit older than the Cambrian radiation, but it does mean that the coelomate phyla, at least, cannot have very long hidden histories. On the basis of the rates of evolution of globins, Runnegar has proposed that metazoan phyla might have originated as long as 800 million years ago. There is reason to be cautious about extrapolation of rates of protein evolution, and I regard these estimates as weak evidence for a long pre-Ediacaran history for anything above diploblastic animals. These kinds of uncertainties about dates show just how approximate our knowledge of the facts of the metazoan radiation is. However, there is another clue. Logan and co-workers have interpreted changes in the geochemical signal of carbon and sulfur isotope fractionation in the early Cambrian as being due to the driving of organic material to the seafloor by the world's first fecal pellets, thus dramatically changing seawater chemistry. Fecal pellets require bilaterian metazoans. That the explosion of metazoans should have left its mark in this humble but pervasive way is consistent with the hypothesis that the Cambrian fossil record approximates an actual evolutionary event.

The Cambrian fossil record is close enough in time to the initial metazoan radiation to potentially give us a good look at what transpired, although we must keep clear the distinction between phyla and phylogeny. Phyla such as brachiopods, which could not have attained their definitive body plan until a skeleton had evolved, may still have diverged as lineages from related phyletic lineages a significant amount of time before we see them morphologically. Brachiopod ancestors probably would have been small, nonskeletonized animals that fed with a lophophore, but one that did not operate as effectively as that of brachiopods. Such lophophorates still exist today. Molecular systematics can reveal the splitting of lineages, but not the evolution of form. In the best of cases, we may be able to integrate molecular and fossil sources of information to get real histories. Such histories would ideally give us the pattern of divergence of a particular phylum from its most closely related sister phylum, the time of its divergence, the time of its appearance in the morphological fossil record, and a view of the morphological features of the most primitive members of the phylum. We would then know what elements constituted the earliest versions of a body plan, how fast those features evolved, and in what directions they changed. Because we would know the patterns of phylogenetic relationship, we would also know, from the distribution of developmental features in living representatives of the phylum and its sister phylum, something about the evolution of the developmental process that produces a body plan as well. That kind of history of development is rapidly becoming available for arthropods and vertebrates, and I'll return to these in later chapters.

MORE BODY PLANS, FASTER EVOLUTION?

The famous mid-Cambrian Burgess Shale fossils have been important in three respects. First, they include soft-bodied as well as skeletonized animals and so give a balanced view of the fauna. Second, large numbers of individuals are represented, thus providing a good sample. Third, their preservation is exceptional, allowing interpretation of anatomical details. The animals, including *Anomalocaris, Waptia, Opabina, Canadaspis, Ottoia, Choia,* and most of its other denizens, were named by Charles Walcott, who discovered this fauna in 1909. Many clearly belonged to existing phyla: *Choia* is a sponge, *Ottoia* a priapulid worm, *Canadaspis* and *Waptia* are arthropods. Sponges, cnidarians, annelids, priapulids, mollusks, brachiopods, arthropods, onychophorans, echinoderms, hemichordates, and chordates are all present. The thirty-five living phyla probably all had their origins in the Cambrian; many of them occur as fossils in the Burgess Shale. Some living classes are recognizable within the Burgess Shale fauna as well. The early appearance and long evolutionary duration of many phyla are evident. That reveals a crucial aspect of the structure of the metazoan radiation: much of it involved the origins of the living phyla.

Because a great deal of the metazoan radiation included representatives of the living phyla, phylogenies erected using genes derived from living animals are relevant to the structure of the radiation, even though it included some number of other extinct phyla. Such a phylogeny should reveal patterns of branching and should bear some relationship to the timing of events. Both points are crucial to understanding the radiation. The importance of branching patterns is obvious. Information on timing is important because it affects our power to resolve the radiation by molecular means, and because knowing the timing of events would tell us something about the evolutionary rates involved. If the time between branching events was very short, as suggested by a literal reading of the fossil record, it will be difficult to resolve the deep branches of the tree by molecular means. In those circumstances, given the slowly evolving genes used to study long-ago events, very few base substitutions would have accumulated between branches. A short time between branching events would also mean that evolutionary rates had to have been rapid. If, on the other hand, the radiation involved a very long period of cryptic evolution in the Precambrian, we will be better able to resolve the branching pattern of the lineages leading to the living phyla, and slower rates of evolution will have to be posited. The finding of a long period of Precambrian evolution would severely reduce our chances of finding a fossil record of these events. It would mean that only the latter part of a long radiation, after fossilizable bodies had finally emerged, might have been recorded. The extent of diversification observed in the fossils of the 30 or so million years of the mid- to late Cambrian, as well as the patterns of gene sequence evolution discussed in the next chapter, support a relatively rapid radiation.

Walcott, in his pioneering studies of the Burgess Shale fauna, placed all of his animals in living groups. However, more detailed reexaminations of the Burgess Shale fauna by Whittington, Conway Morris, Briggs and co-workers, and Gould have shown that the picture is both more complex and more interesting. Interpretations of the structure of the animals, their relationships, and their ecology are still being shaped.

In fitting the Cambrian animals into place, we would like to know whether they include any intermediates between living phyla, and whether any extinct phyla distinct from those still in existence are present. A number of well-known Cambrian soft-bodied animals, *Opabina, Anomalocaris, Amiskwia, Odontogriphus,* and a few younger problematic animals such as the famous Tully monster, *Tullimonstrum,* from the Coal Age strip mines of northern Illinois, look very different from anything alive today. That has led to a vigorous debate among paleontologists. At one extreme, Gould proposes that the initial metazoan radiation produced many more fundamental body plans than are found today. The "losers" have been winnowed out. That is a perspective that makes a good deal of sense to me. Evolutionary radiations more recent in time, for example, that of the mammals following the demise of the dinosaurs, have followed analogous historical paths.

The problem in judging the Cambrian world is that we may be looking at unfamiliar animals that we would recognize as falling into living phyla if we understood their structures better. That is the view of other paleontologists who do not want to countenance the creation of phyla just to house animals we haven't managed to understand properly. There is something to be said for this attitude. One of the most puzzling Burgess Shale animals was *Hallucigenia,* which was restored as a truly peculiar being that walked on a row of paired stilts and had a row of long, soft processes on its back. This was a very popular animal, both for its name and for its unique and cute machinelike locomotor setup. It was a worrisome setup, because such an animal would have had poor stability and would have been blown over by even the most gentle currents. That would have guaranteed early extinction for an animal that would have had no way to right itself. What finally transpired was that all along *Hallucigenia* had been restored upside down. As more Cambrian onychophoran lobopods, such as *Microdictyon,* were discovered, it became clear that *Hallucigenia* was one of them. Ramsköld and Hou have flipped *Hallucigenia* over, thus transforming the soft processes into the legs and the stilts into lateral spines mounted along the body axis. It's still an odd animal, but one that now fits comfortably within a living phylum.

Most living phyla are represented in the Cambrian explosion, and most Cambrian animals fall into living phyla. The intriguing possibility exists that other, now extinct, body plans also evolved during the Cambrian radiation. This possibility and the apparently sudden appearance of the animal phyla suggest that rates of body plan evolution were unusually fast.

As suggested several decades ago by Simpson, there may be a pattern common to evolutionary radiations of animal groups: early in an adaptive radiation, there seem to be more lineages than later in the group's history. If this pattern holds true, many lineages are generated in the initial radiation event, but some number of these sooner or later become extinct. The surviving lineages come to dominate the diversity of the group. The reasons seem obvious. Early in the history of the radiation, lineages have not perfected their adaptations. As they become more refined and begin to expand their niches, lineages will bump into each other, and some will lose. The survivors will consolidate their control over some life zone, then radiate within it. The result is somewhat lower disparity at higher systematic levels and greater species diversity within the most successful groups. For example, within the mammals, there were several orders that evolved herbivory early on. They included the ungainly tusked pantodonts, rhino-sized animals with great canines and multiple horns, and the truly bizarre embrithopods with a massive pair of horns on their noses. These animals were varied and distinct, and fell outside the boundaries of now-existing herbivore disparity. All duly died out and left their niches to a smaller set of herbivore orders. Although several herbivore orders currently exist and current herbivore diversity is high, most of that diversity falls into one order, the artiodactyls (antelopes, pigs, deer, cattle, etc.). This is not at all an unusual story in paleontological reconstructions of the histories of animal groups, vertebrate and invertebrate.

The second approach to questions of greater early disparity comes from direct measurements of morphological differences among Cambrian and later clades in the same group. The application of this approach is just beginning, and the results are still hotly debated. One such direct measurement has been made of blastozoan echinoderms by Foote. Although prominent and diverse (there were several classes) in Paleozoic seas, the blastozoans died out in the Permian mass extinction. They were stalked echinoderms analogous to crinoids; however, instead of having arms, their feeding appendages were brachioles. Foote used 65 characters to define morphological disparity among fossil blastozoans of various ages. He also compiled their taxonomic diversity. Although both morphological and taxonomic diversity increased from the Cambrian into the Ordovician, the ratio of these two numbers showed that in the Cambrian, a few taxa occupied a large proportion of the morphospace (the range of morphologies attained by any particular body plan) occupied by the blastozoan body plan. Subsequent diversification of blastozoans merely filled up that morphospace with more taxa.

The number of echinoderm body plans that appeared and their anatomical innovations during the Cambrian are impressive. Campbell and Marshall tabulated both, although they overestimated the duration of the relevant portion of the Cambrian (Tommotian onward). During that 30 million years about eight classes arose. During the subsequent 70 million years of the

Ordovician, another ten classes arose, and none have arisen in the ensuing 430 million years. Campbell and Marshall, as well as Smith, showed that a number of important structural innovations emerged during the Cambrian. According to Smith, most involved the evolution of improved food-gathering structures and modes of attachment to the substrate via stalks. Innovations in respiratory structure and mobility appeared during the Ordovician. Smith cautioned that because of the relative sparseness of the Cambrian fossil record, evolution of Cambrian echinoderms may look more rapid than it actually was. Nevertheless, these reviews record an astonishing level of Cambrian morphological innovation and support the hypothesis that the Cambrian radiation exhibited extensive body plan evolution. As we will see in considering arthropod molecular phylogenies in chapter 4, analogous arguments swirl around that group as well.

EVOLUTION AND PROGRESS

The equation of evolution with progress has veered widely from being self-evident in the eyes of nineteenth-century evolutionists to being widely regarded as an anthropomorphic delusion in more recent times. The appearance of the living metazoan phyla in the Cambrian radiation can be taken to suggest that all the major innovations in body structure and genetic controls of development were part of that initial radiation. Although we have with us now the same phyla as in the Cambrian, we clearly don't have the same animals. There has been a massive expansion of the adaptations of animals during the past half billion years, and that expansion has been accompanied by replacements of old groups within phyla by newer ones. It has been evident since the nineteenth century that the composition of marine faunas has changed over time. For example, brachiopods were the dominant bivalved filter feeders of the Paleozoic world. Clams existed as well, but did not take over until after the Permian mass extinction. The roles of these phyla are now reversed. Brachiopods now have a low disparity and diversity and occupy cryptic niches; clams thrive in a greatly expanded set of niches. This particular example does not mean that clams necessarily have a "better" design than brachiopods. Chance factors alone may explain the long-term successes and failures of these phyla. However, a directional pattern of change exists.

In his studies of the history of diversity of marine animals during the Phanerozoic eon (Cambrian to present), Sepkoski has found that the patterns of marine family diversity through time can be explained by the existence of "three great Evolutionary Faunas." The first was the fauna of the Cambrian radiation. This fauna was dominated by trilobites, inarticulate brachiopods, and various odd and primitive mollusk and echinoderm classes. It declined after the Cambrian and was replaced by the Paleozoic fauna. The members of the Paleozoic fauna emerged during the Cambrian, but achieved their greatest diversity between the end of the Cambrian and the Permian mass extinction. The Paleozoic fauna was largely composed of attached sessile animals, including articulate brachiopods, now-extinct coral and bryozoan groups, and stalked echinoderms such as crinoids and blastoids. Motile invertebrates included nautiloid cephalopods and a trilobite fauna distinct from that of the Cambrian. This fauna was decimated by the Permian mass extinction, during which close to 60% of families and perhaps as many as 95% of marine species became extinct. The Modern fauna also had a Paleozoic origin, but expanded slowly and reached dominance only following the Permian mass extinction. The Modern fauna includes representatives of the same phyla that the earlier faunas contained, but different classes dominate. The Modern fauna has a higher biomass than the earlier faunas and features more motility among its members. For example, highly mobile echinoderms such as sea urchins and starfishes have replaced stalked echinoderms. The blastoids and cystoids are extinct. Present-day stalked crinoids are restricted to deep relict habitats.

Mass extinctions may be the result of truly catastrophic external causes. The end Permian extinction discussed by Erwin in his book *The Great Paleozoic Crisis* may have resulted from a complex of geophysical events that caused extensive regression of the Earth's oceans, drastically reducing marine habitats, as well as from global warming caused by the release of volcanic carbon dioxide. Clearly many animals that were fit to survive standard Paleozoic marine conditions were destroyed by the unusual stresses of the terminal Permian events. Nevertheless, something more than just replacement of unlucky groups by lucky ones seems to have occurred. There has been an expansion of the ecological roles of the survivors over those of the Paleozoic fauna. Bambach has analyzed the changing patterns by which animals exploit marine resources. The Cambrian fauna was very limited in feeding modes, with a few pelagic suspension feeders, some attached epifaunal suspension feeders, and a few deposit feeders, carnivores, and herbivores. Our more recent understanding of large soft-bodied animals such as *Anomalocaris* indicates that there may have more pelagic and mobile epifaunal Cambrian predators than Bambach suggested. Furthermore, the recent discovery of specialized filter plates on small Burgess Shale-like crustaceans by Butterfield suggests that arthropods evolved as specialized plankton feeders early in their history. There were also active shallow-water infaunal animals. The Paleozoic fauna expanded to include a large number of epifaunal suspension feeders as well as more epifaunal deposit feeders, herbivores, and carnivores. There were also more infaunal animals, including passive as well as active feeders. Pelagic suspension feeders and carnivores were diverse. The Modern fauna has greatly expanded over the Paleozoic modes. Post-Paleozoic faunas include diverse pelagic herbivores as well as suspension feeders and carnivores. There is a wide range of epifaunal modes, and most notably, there has been a great expansion of infaunal feeding strategies. These largely reflect the radiations of clams and sea urchins.

Other analogous expansions are evident. One is the ecological escalation between competing forms noted by Vermeij, especially between carnivores and their prey. Vermeij documents a long-term increase in adaptations that reflect the escalation process. For example, the number of marine families specialized for predation by shell breakage has increased throughout the Phanerozoic, and especially rapidly in the post-Paleozoic fauna. In true arms race fashion, various mechanical defenses have in turn evolved in mollusk shells. The invasion of the land by animals provides the second example of expansion. Terrestrial life faces a far more stressful range of environments than marine life. Only seven of the phyla have succeeded as land animals. As documented by Little, a large number of morphological and physiological adaptations have evolved among land animals. The final expansion is the neural revolution. Highly sophisticated modes of behavior have evolved among cephalopods, fishes, insects, birds, and mammals. Social insects evolved during the Mesozoic era. In mammals, the brain has retained its Mesozoic proportions in marsupials such as opossums, but has increased dramatically in size throughout the past 50 million years of the great radiation of placental mammals. It has increased even more during the past 3 million years of human evolution.

A SUMMARY: THE IMPORTANCE OF THE FOSSILS

Figure 3.6 summarizes the evidence of metazoan origins in the crucial interval of time surrounding the Precambrian-Cambrian boundary. A great deal remains to be learned, but a number of features of metazoan history can be reasonably well placed in time. The eukaryotic cell originated perhaps over 2 billion years ago. Metazoans made their first definitive appearance with trace fossils that correspond in time with the Ediacaran fauna. Molecular data have been taken to suggest a divergence of the metazoan lineage from protists and fungi a few hundred million years before that. However, just when multicellularity arose is unknown. The sponges and diploblastic metazoans, such as cnidarians, had to have arisen prior to the appearance of the first trace fossils. The Ediacaran fauna, under some interpretations, represents the radiation of diploblastic animals. Under Seilacher's and Retallack's alternative interpretations, the Ediacaran fauna, as fungal-grade vendobiota or lichens, represents a unique nonmetazoan "failed experiment," and is irrelevant to the Cambrian radiation.

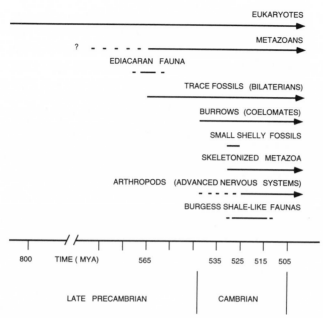

Figure 3.6. A summary of the timing of events in metazoan history near the Precambrian/ Cambrian boundary.

The nature of the trace fossils presents concrete evidence of bilaterian animals, yet still leaves us with some problems. Although the late Precambrian traces are mainly simple horizontal tracks, some were made below the sediment surface. Seilacher has taken the ability to produce such traces as evidence of a coelom. I have been more conservative here in ascribing possession of a coelom only to animals that really show signs of burrowing as competently as known coelomate animals. Thus, I have taken the order of appearance of first, simple traces and later, sediment-penetrating burrows to indicate successive achievement of first, a bilaterian triploblastic grade of organization and later, the evolution of a coelom and the ability to burrow by peristaltic motion. If Seilacher is correct, the coelom was invented at least 25–30 million years before the Cambrian explosion. When the first arthropods become visible in the record as body fossils, they possess a full set of arthropod sensory systems. Trace fossils that resemble known arthropod tracks suggest that arthropods existed in the latest Precambrian and certainly by the start of the Cambrian, but we do not as yet know when the arthropod lineage diverged from that of annelids and their relatives.

During an undetermined interval of time just before the visible Cambrian radiation, the pace of metazoan evolution seems to have quickened. It is likely that the truly defining steps in metazoan evolution occurred during that interval. Those steps included both the divergence of phyletic lineages and the evolution of morphological features and developmental processes that characterize the major animal body plans. The result was the appearance of a diverse suite of coelomate animals at the beginning of the Cambrian. Within only a few million years of the Cambrian boundary, primitive representatives of the modern phyla appeared, skeletonization became a metazoan trademark, and important innovations such as advanced sensory organs and nervous systems became established.

The fossil record of events around the Cambrian boundary is important in framing our questions about the mechanisms by which development affects patterns of macroevolution. The pattern of the metazoan radiation has left an imprint that can be interpreted and contrasted with information from genetic data. Molecular data, as discussed in the following two chapters, may supply new phylogenetic information to illuminate evolutionary relationships among metazoan groups, but only paleontology can illustrate how ancestral animals looked. The Ediacaran fauna may well be metazoan. If so, it apparently provides a picture of body plan experimentation among diploblastic animals. The fossil record of soft-bodied Cambrian animals is important in providing evidence of transitions between early metazoan body plans. There are peculiar animals in the Cambrian record that, when interpreted properly, may give us surprising portraits. One example is the halkieriids, which resemble sclerite-bearing annelids or mollusks that bear a brachiopod-like shell on each end. Do they, as renewed studies by Conway Morris and Peel suggest, represent an ancestral link between annelids and brachiopods? Finally, knowledge of the primitive features of early representatives of long-persisting body plans is necessary to understand the evolution of present morphologies, and even of present ontogenies.

4

Molecular Phylogeny: Dissecting the Metazoan Radiation

An hour passed, and his duty towards Ernestina began to outweigh his lust for echinoderms.

John Fowles, *The French Lieutenant's Woman*

INFERRING MOLECULAR PHYLOGENIES

Evolutionary biologists have luxuriated in over a century of phylogenetic deduction based on morphology and embryology. These inferences have defined most of what we know about animal relationships. Despite that remarkable effort, it has not been possible to erect a metazoan phylogeny that satisfies all investigators. Cladistic methods have helped in posing the problems more clearly, in ordering databases that go beyond idiosyncratic uses of the older evolutionary systematics, and in producing trees that can be more readily evaluated by other investigators. Nevertheless, these analyses have only exposed the problems more starkly, and they have not provided an unequivocal foundation for animal phylum relationships.

Our phylogenetic difficulties probably lie in four properties of the morphological data. First, phyla, by definition, are disjunct in their features. Phylogenetically informative shared derived features are thus rare. Second, we may have exhausted most of the usable morphological features. Perhaps additional features will come from more detailed examination of microanatomy of the invertebrate phyla, but I doubt that sufficient novel morphological features remain to be discovered to overcome a century-long problem. Third, interpretation of shared derived features is not always an unequivocal matter. Two anatomists may well interpret the same feature in diametrically different ways. The consequence is that features may be falsely homologized and incorrect inferences made about directions of change. At the phylum level, phylogenetically informative morphological features are few, so mistakes like these will have large, distorting effects on the outcome. A conflict of this type that occurred in the case of the nemertines, discussed later in this chapter, has been clearly resolved and graphically illustrates these problems. Finally, character conflicts are inevitable: one character is consistent with one phylogeny, but a second character is consistent with a different phylogeny. When this occurs, there is no a priori way to test which character is correct. One either has to depend on parsimony, that is, to consider the tree supported by the greater number of characters to be correct, or have some independent means of testing characters. There is often no way to weigh features objectively. If two features conflict, one may be more deeply integrated and thus more informative than the other. In other words, we clearly would not expect coat color to be as deeply integrated a feature as, say, organization of the coelom. However, that's a very obvious example. In most cases we do not yet have genetic data that allow us to tell whether two conflicting characters are equally deeply integrated genetically and equally informative for phylogeny. The way this problem is usually avoided is by using all plausible characters and giving them equal weightings. This strategy is an evasion, but at present it is often the only reasonable option.

Given that an average animal genome, such as ours, may contain 3 billion base pairs, there is potentially a gigantic amount of phylogenetic information in DNA. Aside from questions of the practicality of sampling so much data and deciding upon which parts of the genome will actually be informative, there is little dispute that gene sequence data offer a wealth of new phylogenetically informative characters. These new characters have the unique property of transcending the morphology of organisms. Humans, corn, and yeast share no morphological homologies beyond possession of the eukaryotic cell, but they share a vast number of homologous genes. The sequences of these genes allow the construction of phylogenetic trees of disparate organisms independently of any morphological features. The same applies with as much force across the disparate body plans of animal phyla, classes, and orders. Gene sequence data promise to let us resolve phylogenetic patterns in circumstances in which only weak inferences could be made before. The goal is to erect robust phylogenetic trees that allow us to map transitions between developmental mechanisms or between body plans. Phylogenetic inferences are pivotal to this enterprise, but as far as possible should not be based on the morphological or developmental features we hope to investigate. Genes offer the only real hope of stepping outside of morphology.

Molecular data promise to bring an immense new source of characters to bear, particularly when data from morphology or the fossil record are scarce. Even so, many of the liabilities of morphological data, including false homologies, character conflict, and even too few phylogenetically informative characters, can bedevil gene-based inferences. An honest presentation requires us to admit that molecular phylogeny is not without its difficulties. Because computer algorithms are available and easy to use, it is simple to generate gene-based trees. These trees, however, cannot be taken as literal readouts of the truth. They are hypotheses, sometimes quite weak ones, but there is a danger of being lulled into a false sense of security by trees neatly drawn by computer. At the very least, phylogenies should be drawn from more than a single gene, and those genes must yield concordant results. As we'll see below, it is still so early in this enterprise that multiple gene comparisons are only now becoming possible. Molecular biology and morphology can complement, but not replace, one another in helping us to understand evolutionary histories, and gene sequence data must be used with some sensitivity to morphological data. One way that can be done is by combining morphological and molecular datasets. However, I'll avoid that approach here, as one important objective is to be able to infer trees independent of morphological data.

Most molecular data currently used for phylogenetic analyses are gene sequences, which can be handled in ways similar to morphological data to yield cladograms and phylogenetic trees. Procedures for converting sequences into phylogenetic trees include both distance methods that consider all positions in a sequence and cladistic methods that use only shared derived characters. The best current summary of these methods has been provided by Swofford and Olsen. Readers who wish to go beyond my simple description of how molecular phylogeny is done should refer to their review for a detailed discussion and references.

In molecular phylogenetic analyses, homologous gene sequences from the species to be included are aligned to give the optimal match. No matter what approach is to be used to infer relationships between animal groups by use of DNA sequence data, the starting alignment is crucial. Despite the prevalence of computer methods for inferring trees, most alignments are done by eye. A successful alignment depends upon having a sequence context in which there is an overall high degree of identity of bases among the DNAs being compared. Bases in a well-aligned set of sequences are presumably homologous in terms of position in the sequence. Once the alignment is made, if identical bases occupy a position, no substitution is inferred. If different bases occupy the equivalent position in two sequences, a substitution has taken place. How it is scored depends on the method of analysis. A sample alignment of a segment of DNA from a gene as represented in four species is presented below. This sample alignment would represent a portion of the data to be examined in inferring a tree based on the much longer sequence of the gene.

Base position	1	2	3	4	5	6	7	8	9	10
Species 1	A	A	T	C	A	G	A	T	T	T
Species 2	A	A	T	C	A	G	—	T	T	G
Species 3	A	A	T	C	G	G	A	T	T	A
Species 4	A	A	T	C	G	C	A	T	T	C

Typically, several hundred to a few thousand base pairs are aligned for a gene. A maximal match has been achieved in this sample alignment. Achieving that maximal match required the insertion of a "gap" in the species 2 sequence, which represents a loss of a base in that sequence with respect to the others. Gaps may be validly used because insertions and deletions of bases are reasonably frequent mutational events. A stubborn refusal to use any gaps would prevent us from recognizing sequence similarity in many cases in which it is quite high. However, gaps are not a sort of fifth base that carries information. Their use must be kept to a minimum to avoid artifactually improving the match between two sequences that have many mismatched positions.

High similarity is important for successful phylogenetic inferences from sequence data because accurate alignment is dependent on base identities. If too few bases match, alignments are likely to be erroneous, and nonsystematic error will be introduced. There is a second important property of sequence divergence: the probability that parallel mutations or reversions have occurred at any site increases with divergence. Corrections for multiple mutations at a single site can be made, but inevitably, phylogenies inferred using sequences that have accumulated a high proportion of multiple substitutions at single base positions will be less satisfactory than those inferred from sequences with no or few multiple substitutions. The use of sequences that have not diverged to this point is advantageous.

Phylogenetic trees are commonly inferred using algorithms that either utilize estimated distances between sequences or are based on cladistic principles. In a distance-based analysis, all differences between the sequences are used to gauge relationship. In cladistic methods, only positions with shared changes are used. Both kinds of methods can be used to infer rooted trees. The root is the presumed point of ancestry for the whole clade represented in the tree. To find the root in the case of the four sequences in the sample, the sequence of the same gene from a fifth outgroup species known to fall outside of the clade would be needed.

The most readily understood approach to generating phylogenies from molecular data assumes that genetic distances between organisms are proportional to phylogenetic distances. Those distances can be converted into the branching patterns of a phylogenetic tree. The simplest procedure for generating a tree from genetic distance data is to build one by clustering distances. Cluster methods can be used with any metric data: commonly used data have included distances between proteins based on some quantitative measure of amount of antibody binding to homologous proteins from different species, genetic distances derived from allele frequencies, and DNA distances derived from melting curves of cross-species DNA hybrids. Although DNA sequences provide character data (each base in the alignment is a character state), sequence comparisons can be readily converted into a single metric of distance for the entire sequence (e.g., a 10-base difference between two

100-base-pair-long sequences is a 10% difference). To construct a cluster tree, one collects a matrix of distances between pairs of sequences. The two closest are used to make the first fork in the tree. Then the average distance between these and the next most distant species gives the next fork. Branch lengths are proportional to the distances. The cluster tree is a very simple kind of tree to assemble, and historically it has been commonly used. However, it can easily lead the naive tree maker astray. Cluster analyses make the assumption that all rates of molecular change are the same in all lineages, and the method depends absolutely on that property to give an accurate reconstruction. If one of the lineages in the set has evolved faster than the others, it will be misplaced, because its faster rate will translate into more distance from everything else. This misplacement will make its position in the tree look more distant from those of its true closest relatives than it really is, and if the misplacement is distant enough, may yield an incorrect branch position.

There has been an undeserved faith in molecular clocks in molecular evolution. These "clocks" depend on identifying genes that evolve at a constant rate in various evolutionary lineages, then calibrating their rates from divergence times of organisms carrying the genes derived from the fossil record. If molecular clocks were reliable, cluster methods would be the only tools necessary to construct trees. Unfortunately, rates of gene evolution are rarely constant among different lineages. Furthermore, because in cluster methods the data are used directly to build the tree, there is nothing that allows for testing for deviation from optimality. Most other methods fit the data to a galaxy of possible tree topologies, and thus criteria for the goodness of fit can be devised.

There are ways for distance methods to accommodate unequal rates if they are not too disparate. There also are ways to partially correct for another problem that obscures distances, the accumulation of multiple mutational substitutions at the same site. Additive tree methods do both. To correct for unequal rates, additive distance methods, such as the Fitch-Margoliash method or the neighbor-joining method, do not depend on the rates of evolution along the two branches connecting any pair of organisms being the same. For example, the distance between species A and B might be 10 units, but they could just as easily represent 4 units of change from the ancestor to A and 6 units of change from the ancestor to B, or any other combination summing to 10 units overall. It is not possible to tell from the distance separating a single pair what the rates are from the ancestor to the tip of each branch. However, if one adds a third species, C, there will be an A–C distance and a B–C distance to utilize as well. If the distance from A to C is greater than that from B to C, it indicates that there is a difference in the rates along the A and B arms, and how large it is. Distance methods that sum these lengths can in principle produce a unique tree that preserves all pairwise distances. These trees are nevertheless sensitive to widely different rates, and are subject to errors arising from multiple base replacements at the same site. Multiple hits occur along a branch proportionately to both greater distance and faster rates. Multiple changes at a single site can result in false homologies to other sequences in the dataset. The longer the branches, the greater the distortion. There are ways to estimate and correct for these effects, but they are far from satisfactory. The result is that a number of trees must be collected from the computer, and some method of assessing them and choosing the "best" tree must be applied. In general, the tree topology selected as best is the one that minimizes a measure of the error between pairs as measured distances versus as distances summed over the tree.

The most commonly used cladistic method is maximum parsimony. Unlike distance methods, parsimony methods use character data. Thus, the identity of each base at each position in the dataset is evaluated, and positions that qualify as "informative" are used. Informative sites are those base positions for which an unrooted tree can be drawn that requires fewer steps than the alternative trees. Thus, in a cladistic analysis, position 5 of the sample alignment given above is informative. In the shortest tree that can be drawn, its alternative derived states are shared by two pairs of species: species 1 is united with species 2 and species 3 with species 4. Only one mutational step is needed. If other species topologies are imposed, two mutational steps are required. Position 5 thus provides a molecular synapomorphy. Since parsimony implies the smallest number of mutational changes that can account for the data, either a G or an A could be primitive for this position. Which is the primitive state can be inferred only by adding an outgroup to the dataset of the sample. Position 6 would be used in a distance analysis, but would be considered an autopomorphy in a cladistic approach and therefore not used. The derived state for that position occurs only in species 4, and so does not reveal any information about the relationship of species 4 to any other species. Position 10 would be used in a distance analysis, and many substitutions have taken place in it. However, that site is not informative in a cladistic analysis because each of the three possible unrooted trees requires three mutational steps from any ancestral base. Position 5 is the only informative site for cladistic analysis in this small dataset.

Character conflicts occur in sequence data as well as in morphological data. Suppose that as we gather more sequence data to complete the sample alignment, we find that five more positions support the conclusion drawn from site 5, but that two new positions support uniting species 1 with species 3 and species 2 with species 4. If parsimony is applied, the minority sites would be considered to result from homoplasies, just as with conflicts in morphological trees. In maximum parsimony the best tree is considered to be the one that requires the least number of mutational steps. Although homoplasies cannot be eliminated, the tree with the minimum number of mutational steps will generally give the smallest number of homoplasies as well. As in distance methods, systematic errors can severely affect the outcome. The greatest problems, as indicated by Felsenstein, arise from greatly unequal rates of sequence evolution among lineages and from problems in correcting for multiple substitutions at a site.

There are other approaches to parsimony that attempt to make tree inference correspond more realistically to the evolutionary behavior of DNA. Because transitions (mutational changes from purine to purine or pyrimidine to pyrimidine) occur more rapidly than transversions (changes to or from purine to pyrimidine), they are more likely to contain homoplasies from multiple mutations at a site. Transversion parsimony methods weigh transversions more heavily because they are less likely to have accumulated multiple mutational changes. The most elaborate of these methods, devised by Lake, is called evolutionary parsimony. It employs transversions and attempts to correct for homoplasies in these. Despite advertisements to the contrary, this method appears to be as sensitive to rate variation as other methods. More sophisticated mathematical tools are being devised to deal with the classic difficulties of sequence data.

The object of any method for inferring molecular phylogenetic trees is to find the optimal tree, but for any significant number of sequences there is a large number of possible rooted trees (34 million by the time ten sequences are included; 10^{21} possible trees for twenty taxa). For situations involving relatively few taxa, up to eleven, an exhaustive search of all trees is possible. There is also an algorithm, the branch-and-bound method, that provides exact solutions for trees containing over twenty taxa by use of combinatorial optimization. For cases in which the number of sequences has grown too large, there are heuristic approaches that seek optimum trees by approximate methods. These methods reduce computing time, but may provide a local solution rather than the true global one. One generally asks the computer for several of the shortest trees because they may be close to one another or even equivalent in length, and the most conservative solution is sometimes a consensus of the equivalent or shortest trees rather than the single shortest tree alone. That choice depends on an assessment of just how robust the shortest tree appears to be.

The significance of the results can be examined by several approaches that test to see whether there is a more phylogenetically informative signal in the data than would be expected for random sequences (such tests have been described by Archie as well as by Hillis and Huelsenbeck). The strength of individual nodes on the tree is commonly tested by a process that Felsenstein calls bootstrapping. To do a bootstrap test, positions are deleted from and replaced at random by other positions in the alignment, and trees are inferred.

Then a different set of positions is deleted and replaced, and again trees are inferred. Thus some sequence positions are lost in each replication, whereas others are present more than once. After 100 or so replications of this procedure, the number of replications in which branches in the "best" tree have emerged is scored. Branches that recur close to 100% of the time are considered strong. These positions locate the informative branches in the tree, whereas inconsistent ones become multichotomies that show that for some parts of the tree no informative distinctions can be made. The problem with bootstrapping is that it indicates which of the conflicting signals in a dataset are the strongest, not which necessarily correspond to phylogenetic history.

There is another useful way to test robustness, called "jackknifing." In this method, described by Lanyon, one or more taxa are eliminated from the dataset and the best tree is again found. If the topology holds, it can be judged robust. If, as commonly occurs, removing taxa affects tree topology, one can look to see which relationships persist. Often the shifting of tree topology in jackknifing is due to a systematic effect, such as the deleted taxon having a long branch length compared with the others in the tree. Bootstrapping and jackknifing are useful analyses for estimating the solidness of a result. These methods can be applied to distance trees as well as to parsimony trees. They often show that simply feeding sequence data into a computer and taking the "best" result is naive and potentially misleading.

In parsimony methods, the major evolutionary hypothesis used is that the tree with the smallest number of steps will be closer to the correct pathway of evolution than one with more steps. By working backward from the sequences at the branches of the tree, one can predict the ancestral sequences at the nodes that join the branches. That too has been a popular pastime, although some caution is necessary because multiple base substitutions at the same site can produce false homologies here as well, and thus both incorrect signals and wrongly predicted ancestral sequences. In another class of methods, those exploiting maximum likelihood, more complete models of evolution are used to work backward from the branch tips to predict ancestral sequences and branch lengths. Maximum likelihood utilizes models that take into account preferred kinds of base substitutions, such as transitions versus transversions, to estimate the likelihood that a change will take place. In finding the optimum parsimony tree, the single best solution is found as the shortest tree. Finding the optimal likelihood tree requires finding the likelihood of all ancestral nucleotides that are consistent with the observed branch patterns and lengths. Not surprisingly, the likelihood methods are far more difficult computationally than other methods, and thus are more severely limited in the number of taxa that can be included in a tree than parsimony or distance methods.

When data are relatively well behaved—that is, when the genes of the included lineages evolve at a rate commensurate with the time of separation of the branches to be defined, and when the rates of evolution of the genes in each lineage do not deviate greatly from one another—all methods give similar answers. As the real data deviate from the ideal, the reliability of all methods falls. Thus far, although cladistic methods are preferred on theoretical grounds, they do not necessarily perform better than distance methods. Analyses are often performed with more than one method, and various tests of the robustness of the results are applied. I'll discuss some other applications, pitfalls, and concerns of molecular phylogenies in chapter 5. The reader should keep in mind that gene sequence data have not replaced morphological data in assessing phylogenetic relationships. Molecular phylogenetics, however, has provided some strong tests of phylogenetic hypotheses, and has allowed us to test relationships among organisms for which few meaningful morphological homologies exist.

OUTLINING A MOLECULAR PHYLOGENY OF THE PHYLA

The phylogeny of the thirty-plus metazoan phyla still remains an unsolved problem, and one that has become more pressing. I began serious work on it with my colleagues (Field and co-workers) a few years ago, when bacterial phylogenists Carl Woese and Norman Pace developed RNA sequencing methods that made it possible to generate large amounts of sequence data by direct sequencing of the small ribosomal subunit RNA. Pace and his co-workers presented an explicit program for the application of this technique to microbial phylogeny in 1985. The 18S ribosomal RNA molecule was particularly suitable for deep phylogenetic studies. The gene is present in all organisms; it occurs in only one version in animals; it contains a significant number of bases (about 1,800); it evolves slowly; and it appears not to be subject to horizontal gene transfer. In addition, the ribosomal RNAs are highly prevalent in cells, so direct sequencing techniques could be readily performed. DNA sequencing has since superseded RNA sequencing in accuracy and rapidity with the introduction of polymerase chain reaction (PCR) techniques. However, direct RNA sequencing remains important historically in that it made large-scale phylogenetic sequencing possible.

   In our initial study, we determined ribosomal RNA sequences from twenty classes in ten phyla. As we began to produce trees with the data, we suffered all the pangs of anticipation of children on Christmas Eve. We thought that our substantial metazoan database would finally let us settle the old phylogenetic problems. It didn't, and we have since grown more modest in what we expect from molecular phylogenies. The results did, however, show us a number of important things. As expected from their relatively simple grade of morphological organization, the deepest animals in the tree were the cnidarians. With the distance method that we used at that time, they plunged

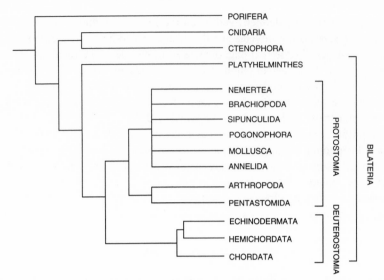

Figure 4.1. A summary molecular phylogenetic tree of relationships among major metazoan phyla as inferred from 18S ribosomal DNA sequences. The tree represents no single analysis, but is a consensus based on the results of several investigators.

unexpectedly deep to join the fungi. This result was in tune with an early origin of diplozoan-grade animals, but embarrassing in indicating a polyphyletic origin for animals. Our later analyses using more complete data, as well as Wainright and co-workers' recent analyses, show that Cnidaria branch deeply, but are part of a monophyletic metazoan clade. They had been placed deeply in our initial tree because their 18S rRNA sequences share primitive features with those of fungi. The second major conclusion of our early study was that platyhelminths were the sister group of the coelomate phyla. That was very much in accord with morphological expectations.

The results we obtained for the coelomate phyla resolved some issues, but proved to be of lower resolution than we had hoped. We discerned four major coelomate animal radiations—arthropods, coelomate protostomes, chordates, and echinoderms—but could not resolve the branching pattern among those groups. Our initial analyses used a distance method. Reanalysis of our data by other methods, by Patterson using parsimony and by Lake using evolutionary parsimony, revealed more structure, although Lake's analysis introduced some serious new ambiguities. His analysis, for instance, split the arthropods and made them paraphyletic. The collection of further sequences, especially complete sequences, and continued analysis has helped to produce a more coherent 18S ribosomal DNA sequence tree for the metazoan phyla. A consensus 18S rDNA gene tree, which I've sketched on the basis of the studies discussed in this chapter, is shown in figure 4.1. This tree has several important features that reflect on the metazoan radiation. Sponges are the basal lineage. Cnidarians branch next, consistent with the early origin of diploblastic animals. The noncoelomate bilaterian platyhelminths form the next major branch. The coelomates are monophyletic, with distinct deuterostome and protostome clades. These results are consistent with Hyman's view of metazoan phylogeny, and with a massive radiation of coelomate animals in a relatively short interval of time immediately prior to the Cambrian. The striking feature of our initial 18S rRNA trees, and those of others, is that they reveal a distinct protostome coelomate clade of closely related phyla. The phyla that we have found so far to belong to this group are the annelids, mollusks, pogonophorans, sipunculids, nemertines, and brachiopods. I'll discuss this grouping further below.

THE BASE OF THE METAZOAN RADIATION

Metazoans have traditionally been shown as arising out of an amorphous group of unicellular eukaryotes, the protists. Now, through the application of both 18S rDNA sequences by Sogin's group and protein sequences by Hasegawa and co-workers, the protists are seen as actually comprising a large number of disparate phyla. With this knowledge, the determination of the sister group of the animals has become possible as well. The first problem has been to resolve the relationship of the metazoans to the other two great multicellular eukaryotic kingdoms, the fungi and the plants. Because the divergences of these three groups were apparently close in time, earlier gene sequence studies had given conflicting results. Vossbrinck and co-workers and Douglas and co-workers using ribosomal RNA sequences, as well as Gouy and Li using protein and RNA sequences, inferred that plants and animals were sister groups. Hasegawa and co-workers, Herzog and Maroteaux, and Sogin and co-workers settled on a close fungal-animal linkage. With larger databases and improved analyses, the most recent reassessments (by Hasegawa and co-workers using elongation factor-1α, von Ossowski and co-workers using catalase, Wainright and co-workers using 18S rDNA sequences, and Baldauf and Palmer using multiple protein sequences) are consistent with fungi and animals as sister groups. I don't know whether this will be the final word, but the congruence of results from three different genes is impressive. As far as identification of the immediate sister group to the metazoans, Wainright and co-workers have given us the first study of the relationships of the taxa that have been classically proposed to constitute the base of the metazoan tree: placozoans, sponges, and choanoflagellates.

The sponges have a cellular grade of organization, with no discrete tissues or organs. They possess only a few cell types. It has never been certain whether they are really metazoans or possibly "parazoans," with an ancestry

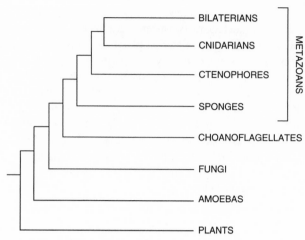

Figure 4.2. An 18S ribosomal DNA molecular phylogenetic tree revealing the base of the metazoan radiation. This tree, which summarizes the results of Wainright and co-workers, supports the monophyly of metazoans, the close relationship of metazoans to fungi, and the root of the Metazoa in the choanoflagellate protists.

within the protists separate from that of true metazoans. Sponges draw water into themselves through pores. Water circulation results from the action of a specialized collared flagellated cell type, the choanocyte. There is a group of colonial protists, known as the choanoflagellates, that possess a similar collared flagellated cell. The similarity has been noted by zoologists, but a definitive test of the hypothesis of a relationship has been unavailable. In their 18S rDNA maximum likelihood tree, Wainright and co-workers found that sponges occupied the base of a monophyletic metazoan tree. They further inferred that choanoflagellates were the sister group of the metazoans including the sponges, and fungi the sister group of the choanoflagellates plus metazoans (fig. 4.2). As pointed out by Bergquist, the possession of choanocytes is not a feature that defines a sponge plus choanoflagellate clade. Choanocytes are present in true metazoans as well.

If sponges represented an independent origin of multicellularity, we would not expect them to share basic features of metazoan cell-cell interactions in development. Morris has presented evidence that the complex extracellular matrix of metazoans, composed of collagen, proteoglycans, adhesive glycoproteins, fibronectin, and the integrin link to the intracellular cytoskeleton, is present in sponges and is a homology that unites sponges with other metazoans as a monophyletic group. Erwin, in his recent review of the origin of metazoan development, argues that the case for extracellular materials as metazoan synapomorphies still needs to be solidified. Erwin also makes the very interesting suggestion that metazoan origins involve what he calls a "correlated progression model." Sponges would have progressed in the metazoan direction by the addition of several differentiated cell types and the ability to produce an epithelial sheet of cells. The diploblastic-grade cnidarians would have added the ability to produce a basement membrane and to organize tissues.

Our later 18S rRNA metazoan trees (Field and co-workers in 1990) place the cnidarians within the Metazoa, and Wainright's tree places cnidarians and ctenophores as branching next in the tree above the sponges. These results are consistent with cladistic analyses of morphological features. These diploblastic phyla form the sister group to the triploblastic bilaterians. There is another interesting result as well. Wainright and co-workers also included in their study the enigmatic and very simply organized animal *Trichoplax adherens,* which has been placed in its own phylum, the Placozoa. *Trichoplax* has only a few thousand cells organized into a dorsal sheet of flagellated cells, a more columnar ventral sheet of flagellated cells, and a few mesenchymal cells in the extracellular gel in the space between the cell sheets. *Trichoplax* lacks discrete muscle and nerve cells, but has contractile cells similar to the epithelio-muscular cells present in diploblastic animals. Grell has suggested that *Trichoplax* is a simple diploblastic animal, and that the dorsal and ventral cell sheets are homologous to the ectoderm and endoderm of cnidarians. The 18S rDNA sequence tree bears him out, with *Trichoplax* falling into the tree with cnidarians and ctenophores. Simplicity cuts two ways, so it's not clear whether the simplicity of *Trichoplax* is primitive or is due to loss of features as a consequence of a history of miniaturization in its lineage.

The happy consensus of morphology and 18S rDNA sequence data that the Metazoa are monophyletic was questioned by Christen and co-workers on the basis of partial sequences of 28S ribosomal DNA. Adoutte and Philippe have reconsidered that possibility in a reanalysis of 18S rDNA trees. Their distance trees supported a separate ancestry for diploblastic animals. Their parsimony trees yielded either metazoan monophyly or diphyly depending on the species used in the analyses. Results like these show that gene sequence phylogenies can be shaky, but the preponderance of data, including a set of shared molecular features compiled by Shenk and Steele, do support a single origin for the animal phyla.

THE BASE OF THE RADIATION OF THE BILATERIA

The origin of the bilaterian, triploblastic metazoans provided an innovation in metazoan body organization that allowed the evolution of the majority of animal body plans. The fossil evidence described in chapter 3 shows that this radical departure from the diploblastic animals occurred prior to the Cambrian radiation. A few phyla have been proposed as lacking coelomic cavities and thus representing the basal grade in bilaterian organization. The placement of these phyla, the platyhelminths, nemertines, and mollusks, has been controversial, and even monophyly of the acoelomate condition has been disputed.

The platyhelminths have been the most crucial group for phylogenetic speculations on bilaterian origins. Cladistic analysis of morphological features of the flatworms by Ehlers indicates that they are monophyletic (although the strength of this conclusion has been questioned on morphological grounds by Smith and co-workers). Analyses of 18S rRNA sequence data by Riutort and co-workers and by Baverstock and co-workers also support the monophyly of the platyhelminths. Thus, the diversity of free-living and parasitic flatworms apparently constitutes a coherent platyhelminth radiation, not a polyphyletic assemblage of superficially similar, morphologically simple acoelomate bilaterian lineages. Because of their position as the sister group of all other Bilateria, the flatworms provide a view of the primitive bilaterian body plan. Investigations, such as those of Bartels and co-workers, of homeobox-containing genes of flatworms are just beginning, and offer an important opening into understanding their roles in the establishment of the body axis in bilaterian animals. One other phylum of very small "worms," the gnathostomulids, has been tied to the flatworms on the basis of a cladistic analysis of morphological features by Ax. Sterrer and co-workers have questioned the link, and no molecular data are available.

A recent, and if correct, spectacular reassignment of a phylum on the grounds of gene sequences has been suggested for the myxozoans. Myxozoans are peculiar and tiny obligate parasites. Although they exhibit a degree of multicellularity and cell differentiation, they have been traditionally considered to be protists. On the basis of 18S rDNA sequences, Smothers and co-workers have treed them within the Metazoa as a sister group to the Bilateria.

The phylogenetic placement of the few phyla that have been considered to be acoelomic triploblastic animals has been controversial on both morphological and molecular grounds. One such controversy involves the nemertines, or ribbon worms. Nemertines belong to an obscure phylum of about 900 species of animals that to the insensitive eye resemble slimy shoelaces. Most are marine and small, but they do hold the record for longest "worm": 30 meters. Nemertines are bilaterally symmetrical, triploblastic animals. Nemertine embryos exhibit spiral cleavage and form a larval mouth in the protostomous manner. The traditional textbook view is that nemertines are the sister group of flatworms, and not much more complex in body plan. Unlike flatworms, which have no body cavities other than a gut, nemertines have two kinds of body cavities: a rhynchocoel, which houses the long, evertable proboscis with which they capture prey, and a system of cell-lined cavities traditionally considered to be a blood vascular system. Much hangs on the interpretation of these cavities, particularly the blood cavities. If they are circulatory, they may be homologous to invertebrate blood spaces. However, if they are homologous to the coelom of eucoelomates, then the phylogenetic position of the nemertines and the origins of their specialized body cavities must be interpreted very differently. The minority view, which for the most part has been published in German and thus has been easy for English speakers to ignore, is that the body cavities of nemertines are coelomic. More recently, Turbeville investigated the ultrastructural anatomy of these spaces in order to improve the quality of the morphological database. He argued from his observations that the nemertine spaces were coelomic homologues on the basis of the classic criteria of homology: their position (lateral), their properties (cell-lined as in coeloms rather than lined with extracellular matrix as in invertebrate blood vessels), and their origin (from mesoderm). Because some other morphological features have been interpreted as supporting nemertines and platyhelminths as sister groups, there has been no strong impetus among zoologists to revise the traditional interpretation of their body cavities on morphological grounds. There is no consensus on interpretation of features, and so these animals pose an ideal test case for molecular phylogenetic study derived from data independent of the morphological data underlying the classic debate.

We (Turbeville, Field, and Raff) determined a nemertine 18S rRNA sequence to address this controversy. Once analyzed, the molecular data fell strongly on the side of nemertines as members of the eucoelomate protostomes rather than a sister group of the platyhelminths. Thus, the nemertine body cavities appear to be coelomic spaces. The nemertine 18S rRNA sequence is highly similar to those of coelomate protostomes, and not to those of flatworms.

The phylogenetic relationships of mollusks have been similarly controversial. Mollusks develop very similarly to annelids, but the protostome mode of development does not allow an unequivocal decision because flatworms also exhibit spiral cleavage. Adult anatomy too has failed to decide the question of whether mollusks possess a coelom. The molluscan pericardial cavity has been held to be a coelomic homologue because it is mesodermal and is associated with the gonads. Based on the lack of a definitive coelom and the movement of some primitive molluscan groups on a slime trail produced by the flat foot, Clark has concluded that they are part of an acoelomic radiation made up of a platyhelminth-nemertine-mollusk clade. In a recent analysis of morphological, embryological, and molecular features, Scheltema has come to the opposite conclusion, that is, that mollusks are the sister group of the unsegmented sipunculans. She further places this clade as the sister group of the annelids, within the coelomate protostomes. The 18S rRNA sequence results of Field and co-workers also put mollusks strongly into the protostome coelomates, again conforming to one of the two favored phylogenetic hypotheses drawn from morphology. This result has been confirmed by Kojima and co-workers, who used elongation factor-1α sequences to infer the relationships among several animal phyla. Annelids, pogonophorans, and mollusks formed a eucoelomate protostome sister group to the arthropods consistent with our 18S rRNA sequence trees. The tight link of mollusks with annelids and other coelomate protostomes means that the molluscan pericardial cavity is probably derived from a coelom. Annelidan segmentation may have been acquired after the divergence of the annelidan lineage from the molluscan lineage, or primitive molluscan ancestors may have exhibited body segmentation and subsequently lost it with evolution of the definitive molluscan body plan. The determination of whether or not segmentation is primitive to this group is a crucial issue. I suspect it will be solved only when sufficient data on the genetic regulation of annelid segmentation is available to decide whether segmentation in the annelids is homologous to segmentation in the arthropods, the outgroup to the entire coelomate protostome clade. If that is the case, segmentation has been lost in the ancestors of the nemertines, sipunculids, and other nonsegmented protostome coelomate phyla.

The removal of nemertines and mollusks from the ranks of the acoelomate animals leaves only the platyhelminths, the flatworms, as a living group with that grade of body organization. Perhaps the flatworms are the last remnant of a great unrecorded radiation of acoelomate phyla early in metazoan history. Alternatively, their body plan may have limited the acoelomates to a low disparity. Perhaps it was the "invention" of a coelom that opened most metazoan niches and triggered the evolution of the animal body plans that appear in the Cambrian radiation. If this scenario is correct, the flatworms present us with a living fossil that has continued the primitive bilaterian body plan to the present. These animals deserve more attention in revealing how the bilaterian body axis evolved.

STRANGE TERRITORY: THE PSEUDOCOELOMATES

The pseudocoelomates, which are also called aschelminths, are a bizarre collection of ten phyla, representing almost a third of all living body plans: rotifers, gnathostomulids, priapulids, gastrotrichs, kinorhynchs, nematodes, nematomorphs, entoprocts, acanthocephalans, and loriciferans. They pose phylogenetic ambiguities even worse than those that plague the establishment of relationships among the major metazoan phyla. The classic criterion for inclusion in this group has been the presence of a coelomic cavity, but one lacking a peritoneal lining. The pseudocoelom has been traditionally considered to be an embryonic blastocoel that persists in the adult. Most pseudocoelomate phyla are made up of very small, often parasitic, animals. The disparities of millimeter-sized animals are less immediately obvious than those of larger animals. The largest pseudocoelomates are the priapulids, which attain sizes of up to 20 centimeters in length. Although there are only fifteen living species, priapulids were an important component of the Cambrian Burgess Shale fauna, and thus are an important group for reconstructing the events of the Cambrian radiation. Of the other pseudocoelomates, the nematodes are the most diverse, and make up a sizable proportion of the world's animal biomass. Many are parasitic on plants and animals. Several have unpleasant associations with humans, including hookworms, pinworms, ascarids, and the filarids that cause elephantiasis. The small free-living soil nematode *Caenorhabditis elegans* is now one of the major model systems for studies in developmental genetics. Because it is used so extensively by developmental biologists as a model for animal development, its phylogenetic position relative to the major metazoan phyla has clear importance aside from any other phylogenetic questions about pseudocoelomates.

Although they are united in the pseudocoelomate condition, the anatomy of the various pseudocoelomates gives few solid features that allow them to be comfortably linked with one another as a monophyletic group. Ruppert has shown that the body cavities of pseudocoelomates do not all arise in the same way and are not persistent blastocoels, and therefore that the pseudocoelom is not a homologous feature uniting these phyla. Lorenzen has suggested that the properties often seen in pseudocoelomates are the result of extreme size reductions in larger eucoelomic ancestors. For example, eutely, the extreme consistency of position and numbers of cells seen in pseudocoelomates, is a correlate of low cell numbers. Copulation likewise is required of very small animals that simply cannot produce enough gametes to broadcast spawn as do most larger invertebrates. The trend to copulation as a consequence of small size has occurred in small members of many nonpseudocoelomate phyla as well. Lorenzen has carried out a cladistic analysis of the pseudocoelomates and suggests that some groups can be united. Thus, he links the rather similar nematodes and nematomorphs, and links rotifers with acanthocephalans on the basis of a very few apparently shared structures. To show what a difficult mess this is, a cladistic study by Schram has concluded that the pseudocoelomates are monophyletic! Nevertheless, the most likely conclusion is that the pseudocoelomate condition was attained independently by a number of lineages, and that pseudocoelomates are thus a polyphyletic assemblage. Our analyses of 18S rDNA sequences obtained from some pseudocoelomate phyla also fail to support monophyly, but suggest that pseudocoelomates arose early in metazoan history (see Raff, Marshall, and Turbeville for a more complete discussion of pseudocoelomate phylogeny).

The ancestors of the pseudocoelomates were probably eucoelomates. The hallmarks of extreme size reduction are present in the pseudocoelom. It is variable in its features, with some groups having almost no cavity and others a large pseudocoelomic cavity that contains mesenchyme cells. As suggested by Brusca and Brusca, the pseudocoelom is probably a convergently acquired developmental modification that abbreviated coelomic development to leave a blastocoel-derived cavity in small adults. Most pseudocoelomate phyla have entire guts, including an anus, and other complex anatomical features that are consistent with ancestors more anatomically elaborate than flatworms.

ORIGIN OF THE COELOM

The enormous disparity of coelomic animals and their great success in most niches marks the origin of the coelom as one of the key innovations in animal evolution. The picture as sketched out thus far is that metazoans had a monophyletic origin in the late Precambrian. Choanoflagellates are the most closely related protist group. Sponges diverged next, and form the sister group of all other metazoans. Diploblastic animals form the next branch of the tree and form the sister group of the bilaterian animals. Last, the flatworms, which are acoelomate triploblastic bilaterians, branch off as the sister group of all coelomate animal groups. The time of origin of the coelom is not directly available either from the fossil record or from molecular phylogenies. Since coelomate phyla are present by the Tommotian stage of the middle Cambrian, about 530 million years ago, the coelom must have arisen before the Cambrian radiation. The existence of bilaterian trace fossils dates to about 565 million years ago. The proliferation of burrows in rocks deposited just prior to the Cambrian indicates that a coelom allowing efficient burrowing had appeared by that time.

As shown by the cladistic analyses of morphological features carried out by Eernisse and collaborators and by the trees of other workers that they review, there is no consistent support for coelomate monophyly. The 18S rDNA trees have been more consistent with coelomate monophyly, but they are sensitive to which outgroup taxa are used. The result is that a major problem remains in discovering the origin(s) of one of the defining features of animal body plans. Coelomate animals have traditionally been divided into two superphyla, the deuterostomes and the protostomes, on the basis of embryological features. Gene sequence trees have been generally (but not universally) consistent with this division, and I will use these superphyla as reasonably solid clades.

COELOMATE PROTOSTOMES

When Field and co-workers introduced new data in the form of 18S rRNA sequences, we had as a null hypothesis the expectation of confirmation of most of the traditional conclusions about metazoan relationships. Many were confirmed; some were not. However, sometimes the molding of the landscape of our ideas is more subtle and more profound than mere confirmation or refutation of earlier hypotheses. One unexpected and strong signal that has emerged from analyses of the 18S rRNA sequence data is that there was an early divergence of the protostomes into arthropods and their relatives and the other protostomes, which we call the coelomate protostomes. An early radiation of this coelomate protostome lineage gave rise to a number of major phyla: annelids, mollusks, sipunculids, pogonophorans, nemertines, and brachiopods. This grouping suggests some unexpected features of the metazoan radiation. The coelomate protostomes represent a distinct and major radiation of one of the coelomate clades after the initial coelomate radiation. The fossil record of the metazoan radiation shows that the coelomate protostome radiation cannot have followed long after the origin of coelomates, but was distant enough in timing to have left a strong signal in the 18S rDNA gene. It also produced a high degree of disparity, as recognized by the phylum status of its products. Of the coelomate protostomes, the annelids, mollusks, sipunculids, and pogonophorans share common features in their development, most notably a trochophore larva. The nemertines and brachiopods don't produce this larval form, and traditionally, nemertines, mollusks, and brachiopods have not been considered part of such a clade. The inclusion of these phyla in the protostome coelomates may appear surprising. However, in each of these cases, conflicting morphology-based hypotheses are on the books. The molecular data in all three cases fall on one of the contending morphological sides, not just out on some morphologically insupportable branch. If the 18S ribosomal gene sequence data are correct, then molecular data have revealed a major evolutionary event in metazoan evolution that was not recorded in the fossil record.

Additional molecular support for a coelomate protostome clade has come from the distribution and sequence of the oxygen transport protein hemerythrin. As noted by Runnegar and Curry, this protein is found only in priapulids, sipunculans, lingulid brachiopods, and some annelids, a puzzling distribution under most classic phylogenies. The sequences of brachiopod, priapulid, and sipunculan hemerythrins are readily aligned. The distribution pattern for this protein supports the placement of brachiopods with the coelomate protostomes by 18S rRNA sequence analysis, and also suggests that the priapulids, which are pseudocoelomates, may belong to this group as well. Kojima and co-workers have also found that the sequence of elongation factor-1α confirms an annelid-pogonophoran link.

Important recent fossil evidence also supports an annelid-mollusk-brachiopod link. As noted in chapter 3, the early Cambrian halkieriids, described by Conway Morris and Peel, bore brachiopod-like shells on each end of a body armored in scalelike sclerites. Halkieriids evidently glided over the seafloor on a broad foot like that of living snails and slugs. The sclerites were similar to those of another Cambrian animal, *Wiwaxia,* which also had mollusklike features and has been recognized as a sister group to the annelids. Conway Morris and Peel suggested that brachiopods might have evolved from a larval halkeiriid. These long-departed animals furnish a remarkable example of correspondence between the fossil record and gene sequence data.

A major consequence of the discovery of the coelomate protostome radiation is that arthropods and annelids cannot be each other's sister group to the exclusion of other groups, as would be expected if the Articulata hypothesis were correct. Arthropods are the sister group of all coelomate protostomes. The relationship is deeper than the hypothesized split of the articulates. That requires either that segmentation is primitive in the ancestor of the arthropod-plus-protostome-coelomate clade, or that annelids and arthropods evolved segmentation independently. These two evolutionary scenarios suggest quite different consequences for the pattern of the Cambrian radiation and for those seeking genetic and developmental mechanisms for segmentation. We don't know which is correct based on current data.

LOPHOPHORATES

Lophophorates (phoronids, brachiopods, and bryozoans) are often considered to form a third coelomate superphylum. These phyla are united by a shared structure, the lophophore, a hollow ring of tentacles surrounding the mouth. The anus lies outside of the ring. Hyman suggested that the lophophorates are related to the protostomes. Others (see Ax as well as Brusca and Brusca) have argued that lophophorates are deuterostomes or a sister group to the deuterostomes. The morphological features that have been suggested as indicating membership in the deuterostomes are the shared tripartite coelom, radial embryonic cleavage, and the origin of coelomic pouches in development from the sides of the archenteron. Some lophophorates form the larval mouth in a deuterostome manner, other in the protostome manner from the blastopore. However, the embryological features of lophophorates are overall consistent with a deuterostome relationship for the group.

The placing of the brachiopods in the coelomate protostomes by Field and co-workers on the basis of 18S rRNA sequences was surprising in view of their developmental features. The brachiopod sequence showed a very high similarity to those of coelomate protostomes, but not to those of deuterostomes. Halanych and co-workers have sequenced the 18S rDNAs of other lophophorate phyla, the bryozoans and the phoronids, and have found that they too are related to annelids and mollusks. They have applied the fairly colorful term ''lophotrochozoans'' to the coelomate protostomes. When morphological and gene sequence results are in as apparently wide disagreement as seen in the case of brachiopod 18S rRNA, we have to consider the possibilities. One is that brachiopods really are deuterostomes, but that during their history a horizontal transfer of ribosomal genes from a protostome occurred, and the deuterostome sequence was lost. This is an unlikely scenario, but one that could be readily tested by determining the relationships of other gene sequences. Even if a horizontal transfer of ribosomal genes had occurred, other genes would reveal the deuterostome genetic background. It is more probable that brachiopods are protostomes by ancestry, as indicated by the 18S rRNA sequence, and that their deuterostome-like developmental features are the result of convergent evolution. Caution about the phylogenetic "value" of developmental features is necessary, and early development can exhibit a very high evolutionary flexibility.

ARTHROPOD MONOPHYLY VERSUS POLYPHYLY

As the most diverse animal phylum, the arthropods, not surprisingly, present some of the most complex and interesting phylogenetic problems of any animals. Not the least of these problems has been whether the arthropods really belong to a single phylum at all. Arthropods are characterized by their "arthropodization"; that is, unlike annelids and other coelomates, they are covered with a hard cuticle. Arthropods, however, are not merely annelids dressed up in suits of armor. Their hard covering has resulted in a profound reorganization of the body plan relative to that of annelids or, indeed, any of the coelomate protostomes. The motility of annelids resides in their hydrostatic skeleton, in which the body wall musculature works on the fluid-filled coelom. This machinery allows the animal to utilize its segments to expand or contract parts of the body, providing a powerful ability to burrow in sediments. Arthropods are covered with a stiff exoskeleton, so they cannot alter segment sizes. Thus, arthropods have lost their presumptive ancestral hydrostatic skeleton and motor abilities, and have evolved jointed appendages and the associated muscles. In the more primitive arthropods, the segments and appendages are more or less alike. One of the major trends in arthropod evolution has been tagmosis, the morphological differentiation of appendages and segments for specialized functions. Knowledge of the genetic basis of tagmosis, so well studied in *Drosophila,* has opened an immense window into understanding how genes control body plan and pattern in development. The phylogeny of arthropods and the historical tracing of tagmosis patterns is of crucial importance to exploring the genetic basis of their development.

Arthropods share not only the primary feature of arthropodization but also a suite of other major features, including a plan of external and internal body segmentation. With the exception of the advanced long germ band insects

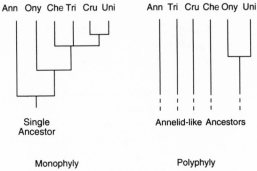

Figure 4.3.   Two hypotheses for arthropod origins. Ann = annelids; Ony = onychophorans; Che = chelicerates; Tri = trilobites; Cru = crustaceans; Uni = unirames.

discussed in later chapters, they add segments in a manner similar to that of annelids. That is, the body is first laid out with only a few head and possibly thorax segments. Additional segments are added via the growth and patterning of a posterior growth zone, the teloblast. The arthropod nervous system is similar in layout to that of annelids. There is a dorsal brain with connections surrounding the esophagus. The paired main nerve trunk runs down the ventral side of the body, linked by lateral connectives. The coelom itself has been drastically reduced and replaced by a hemocoel (blood cavity). The dorsal heart characteristic of protostomes has been retained, but instead of circulating through a system of closed blood vessels, the blood of arthropods enters the hemocoel, which fills the body cavities in each segment and constitutes an open circulatory system.

Weygoldt has catalogued the shared anatomical features of arthropods (α-chitin cuticle, mixocoel, dorsal blood vessel with paired ostia, pericardial sinus, complex brain, nephridia with sacculi, appendages with extrinsic and intrinsic muscles and terminal claws, and centrolecithal eggs). These shared features form the obvious basis for considering arthropods to be monophyletic. However, the interpretations of morphological details have not been unanimous, and there has been an interminable debate over whether arthropods are monophyletic or polyphyletic. The two basic phylogenetic views of arthropods are shown in figure 4.3. The monophyletic tree is based on cladistic analyses of morphology by Ax and by Weygoldt. The crucial conclusions are that arthropods are monophyletic, that onychophorans form the sister group of the euarthropods, and that branching patterns can be established between groups within the arthropods. Thus, crustaceans (shrimps, lobsters, etc.) and unirames (insects, millipedes, etc.) form a larger clade, the mandibulates, and the extinct trilobites are the sister group of the chelicerates (horseshoe crabs, scorpions, spiders, ticks, etc.). In the polyphyletic hypothesis of

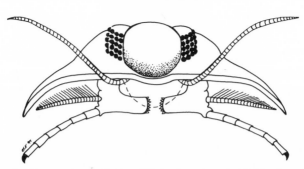

Figure 4.4. The biramous leg of trilobites. This head-on view of a trilobite shows the structural features of this primitive biramous limb. The thick base of the leg is the coxa. The tooth-bearing part of the coxa forms the gnathobase used for chewing food. The walking leg is called the telopod, and the gills are borne on the filamentous branch. (Drawn by E. C. Raff.)

Manton and Anderson, the major arthropod clades have separate origins in various hypothetical extinct annelid-like ancestors. The onychophorans are the sister group of the unirames, but otherwise the major clades all have separate origins.

As recounted by Tiegs and Manton in an influential 1958 review of arthropod evolution, the onychophorans posed a real problem in how one should account for arthropodization. Onychophorans were recognized by nineteenth-century zoologists as sharing features of both arthropods and annelids. We would now regard the annelid-like features as plesiomorphic, but they were classically regarded as showing that onychophorans formed a link between the two phyla. On the basis of such shared features as trachea, Haeckel more specifically envisioned the onychophorans as direct ancestors of unirames. He regarded the crustaceans as arising from a separate ancestor. Tiegs and Manton pointed out that no matter how arthropods arose, some convergent evolution had to have occurred, and they suggested that arthropodization itself may have been a convergent feature.

Manton, from her perspective as an arthropod functional morphologist, went on to develop a detailed view of arthropods as polyphyletic in origin based on the very distinct locomotory structures and functions of the major living arthropod groups, the chelicerates, crustaceans, and unirames. Fossils of some trilobites are well enough preserved to show that the legs of trilobites were biramous, with an internal walking leg and an external gill (fig. 4.4). The living crustaceans possess multiramous or biramous legs analogous to those of trilobites. The primitive crustacean leg consists of an inner walking leg attached to two basal limb segments, the coxa and the basis. Small processes called endites extend from the basis, and serially arranged exites are also present on leg segments. The crustacean limb bears a prominent exite that functions in swimming and possibly in gas exchange (like the gills of the trilobite leg). The largest exite is emphasized in considering the limb to be "biramous." On some legs, the side of the coxa facing the body can be modified to form a food-grinding surface called a gnathobase. Whether these structures are homologous to the biramous limbs of trilobites is moot, but the trilobite limb clearly exemplifies the biramous arthropod leg. Living chelicerates and unirames have walking limbs without gills or other processes. Manton could not envisage how one type of locomotory system could have given rise to the other. She was also troubled by the distinct differences between the mandibles of crustaceans and of unirames. Manton held that these feeding structures were functionally as distinct and nonhomologous as jaws. The working surface of the arthropod mandible is a gnathobase. Manton concluded that the jaws of unirames are whole-limb mandibles in which the tip of the modified limb forms the food-working surface. It is interesting that there are now genetic data that contradict the concept of the insect mandible as a whole-limb homologue. Panganiban and co-workers have found that unlike whole limbs, the mandible does not express the gene *Distal-less* in its development (see chapter 12 for a fuller discussion of insect appendages).

The most recent case for arthropod polyphyly was presented in 1990 by Willmer, who accepts the arguments of Manton and Anderson and who also stresses the autopomorphic features unique to each group rather than the synapomorphic features that unite them. What is most interesting about Willmer's arguments is the contention that if annelid-like worms were to evolve exoskeletons, the features that we regard as characterizing arthropodization would arise inevitably given the mechanical demands. Therefore, convergence and polyphyly would be inescapable. These morphological hypotheses that treat common arthropod features equally well as homologies or convergences require testing on independent, molecular grounds.

As shown in table 4.1, the tagmosis patterns of onychophorans and of the major arthropod groups are distinct. Onychophorans have a much simpler tagmosis pattern, with only three head segments. Among the bona fide arthropods there are five or six head segments behind the anterior terminal segment, the acron. No acron can be detected in onychophorans. Each arthropod group has a highly distinct tagmosis pattern. Thus, chelicerates all have six postoral appendages, but no preoral appendages. Crustaceans have two preoral and three postoral head appendages, and unirames one preoral and three postoral appendages. The nature of the appendages is also characteristic of each group. These patterns exhibit an immensely long evolutionary conservation through hundreds of millions of years. Yet, as Manton and Anderson, as well as Gould, have noted, the tagmosis patterns of the Burgess Shale arthropods are in most cases quite different from those of the long-persisting major lineages (fig. 4.5). Some trilobites are present, as are a few crustaceans and a possible chelicerate, but most Burgess Shale arthropods have their own,

TABLE 4.1. Identities of Arthropod Segments and Appendages

Segment No.	Onychophora	Trilobita	Chelicerata	Uniramia	Crustacea
1	Antennal	Acron	Acron	Acron	Acron
2	Mouth, jaw	Antenna	Chelicerae[a]	Antennae	1st antennae
3	Slime palp	Mouth, 1st pair legs	Mouth, pedipalps	Mouth, premandibular[b]	Mouth, 2d antennae
4	Trunk limbs	2d pair legs	2d pair legs	Mandibles	Mandibles
5	"	3d pair legs	3d pair legs	Maxillae	1st maxillae
6	"	Trunk limbs	4th pair legs	Maxillae[c]	2d maxillae
7	"	"	5th pair legs	1st thorax app.[d]	1st thorax app.
8	"	"	No appendages	2d thorax app.[e]	2d thorax app.
9	"	"	"	3d thorax app.	3d thorax app.
10	"	"	"	Abdominal app.[f]	Abdominal app.
11	"	"	"	"	"
12	"	"	"	"	"
"	"	"	"	"	"
"	"	"	"	"	"
n	"	"	"	"	"

Source: Data summarized from Anderson, Brusca and Brusca, and Raff and Kaufman.
[a]Chelicerae arise postorally in development, migrate to a preoral position
[b]Embryonic only
[c]Labium in insects, lost in millipedes
[d]First pair of legs in insects, maxillipedes in centipedes, and collum with no legs in millipedes
[e]Second pair of legs in insects, and first pair of wings
[f]Abdominal appendages reduced or lost in insects, trunk legs in millipedes and centipedes

often bizarre, tagmosis patterns. To Manton and Anderson, these highly divergent tagmosis patterns so early in the history of the arthropods indicate a hidden evolution of convergent "arthropods" from non-arthropod ancestors. In this scenario, there is no common descent relating the various arthropod groups and their tagmosis patterns, and the developmental rules for each group are fixed early in their history. This is a decidedly different view from that of Gould, who envisions a common descent with a low level of developmental constraint among early arthropods. Only later would tagmosis patterns become fixed by rigid developmental rules. The consequences of these two modes of thought for interpreting the evolution of development in arthropods are very different.

Weygoldt, in reanalyzing the case for arthropod polyphyly versus monophyly, argued that Manton's arguments for polyphyly were weakened by her failure to consider synapomorphies and divergent evolution after splitting of lineages bearing a homologous structure. Certainly, the fully evolved gnathobasic mandible of crustaceans and the whole-limb mandible of unirames might not be derivable from one another. However, both could have descended divergently from a primitive arthropod head appendage that served in feeding and had gnathobases as well as an endite. In the crustacean lineage, evolution of a gnathobasic mandible was selectively favored, whereas in

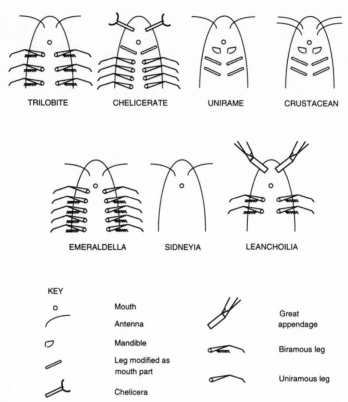

Figure 4.5. Patterns of head appendages in living and Cambrian arthropods. The patterns of four major post-Cambrian clades are shown in the top row; three Cambrian forms from the Burgess Shale are shown in the bottom row.

unirames, the limb tips became the primary food-handling element of the jaw. As far as the differences in walking limbs between a biramous and a uniramous state, Weygoldt has suggested that the primitive unirames had biramous appendages. Unirames and chelicerates, as well as the few terrestrial crustaceans, have lost the outer parts of these appendages in favor of evolving efficient terrestrial walking legs. There are no known Cambrian unirames, but millipede-like marine arthropods have been described from lower Silurian rocks by Mikulic and his co-workers. The limbs of these animals appear to be uniramian, as in terrestrial millipedes. The implication is that the uniramian limb may have originated before unirames took up a terrestrial lifestyle, but other evidence suggests otherwise.

The importance of considering fossils in unraveling the history of the unirames has been made abundantly clear by Kukalová-Peck's immensely detailed investigation of well-preserved Paleozoic fossil insects. Her studies show that primitive insects had multiramous limbs. The uniramous limb of modern insects was derived from the more primitive biramous or even multiramous leg after insects were successfully established on land. As I'll discuss in chapter 12, the multiramous condition of the primitive insect limb was portentous for the evolution of important novel features of insects. Knowledge of primitive insect anatomy derived from the fossil record has significance both for systematics and for understanding the evolution of regulatory gene systems involved in the patterning of the insect body plan during development.

Other complex organ systems have also been interpreted as more consistent with monophyly than with convergent evolution. One such system is the compound eye, which is present in living and fossil arthropod groups, but not in any non-arthropod phylum. Compound eyes are thus part of the suite of features that constitute arthropodization. On the basis of the anatomy of the compound eyes of insects and crustaceans, Paulus has concluded that both eyes and arthropods are monophyletic. The eyes of Cambrian fossil trilobites discussed by Clarkson and by Ready underwent the same mode of accretive growth as seen in development of the *Drosophila* eye. Homology is thus supported by fossil evidence for a long-conserved developmental mechanism. As so well portrayed by Peter Lawrence in his book *The Making of a Fly,* the details of the regulatory gene expression patterns that establish the complex arrangement of cells in the insect eye during development are now becoming well known in the fruit fly *Drosophila*. The morphogenetic steps at the cellular level and the underlying genetic and cellular signaling processes are complex. There are eight photoreceptor cells, or rhabdomeres, in the compound eye. Several genes have been characterized that govern the differentiation of various of these cells. The *sevenless* gene encodes a membrane-bound receptor with an intracellular tyrosine kinase activity present on the R7 rhabdomere. That receptor responds to signal molecules produced by the gene *bride-of-sevenless* expressed by the R8 rhabdomere. Without proper expression of this receptor-ligand system, the R7 cell never differentiates. The expression of other genes encoding transcription factors, such as *glass, seven-up, sina,* and *rough,* is, as discussed by Moses, required for the differentiation of other specific groups of rhabdomeres. To fully establish the homology of insect and crustacean compound eyes, the crustacean versions of these genes will have to be cloned and their expression patterns determined. It might be objected that, as de Beer has observed, structures that are clearly homologous do not necessarily arise through the same developmental pathway. If that were true for the compound eyes of insects and crustaceans, our gene-level comparisons would at best be of no help. The reason for predicting that this will not be the case comes from the few evolutionary comparisons of the actions of pattern-forming genes in development. These patterns appear to be conservative. If the same gene complex is involved in compound eye formation in both subphyla, powerful evidence for homology of the structure and a common phylogenetic heritage for arthropodization will be established.

The development of the central nervous systems of insects and crustaceans has been shown to have high levels of similarity by Thomas and co-workers and by Whitington and co-workers. Whitington and co-workers presented a detailed evaluation of similarities and differences between insect and crustacean nervous system development, and concluded that although significant differences exist, a conclusion that these nervous systems are homologous is warranted. Descent from a common mandibulate ancestor followed by modification appears to be well supported for insects and crustaceans.

The second line of evidence for a polyphyletic origin of arthropods was drawn by Anderson from a detailed consideration of the embryology of annelids, onychophorans, and the major arthropod groups. Anderson took the approach of using fate maps and developmental features of the embryos as indicators of the uniqueness or relatedness of arthropod groups to one another as well as to various annelid groups. His conclusion was that arthropod development is too disparate for arthropods to be monophyletic. Anderson united the onychophorans with myriapods, insects, and the other minor groups that make up the unirames. All exhibit a syncytial mode of cleavage with subsequent formation of a cellular blastoderm. The conceptual fate maps of their blastoderms, which he inferred on the basis of embryological data, are similar, and development of midgut, mesoderm, and ectoderm is similar, as is formation of somites. Of all arthropods, Anderson considers that unirames and onychophorans have the greatest similarity in development to annelids. Polychaete annelids produce a feeding larva, the trochophore, and much of their fate map reflects the development of this larva. However, the oligochaetes and leeches have abandoned the trochophore for a direct mode of development. They have retained the same pattern of spiral cleavage as polychaetes and a similar fate map, although with a modified allocation of cell mass. Onychophoran and unirame fate maps are consistent with an ancestry among these annelids.

Crustacean development differs from unirame development. Crustaceans with very large eggs cleave syncytially and form a cellular blastoderm. Other crustaceans with smaller eggs exhibit complete cleavage, which Anderson has concluded represents the primitive mode of early development in crustaceans. The cleavage of barnacles is spiral, and Anderson has attempted to homologize it with the classic spiral cleavage of annelids. Unfortunately, the pattern of cleavage is sufficiently different that it is impossible to substantiate this interpretation. Anderson has argued that the most significant attribute of crustacean

development is the crustacean fate map, which is quite different from the
unirame/annelid fate map. Crustaceans produce a highly specialized nauplius
larva. Regional fates in development presage the patterning of the nauplius.
Finally, chelicerate development is different from that of either annelids or
other arthropods. Anderson has concluded that nothing in the chelicerate cleavage pattern or fate map indicates a close tie to any of the other groups.

Anderson's enormously detailed and profound analysis of arthropod development has provided a classic interpretation of embryological data in an
evolutionary mode. His main tool was the establishment of apparent homologies of embryological structures by deriving fate maps from embryological
data. This was a heroic task because fate maps have not been determined
experimentally for most of the animals he considered. It was also a risky
procedure because the fates may not correspond to the actual behavior of cells
in the embryos. A critique of this analysis is important because it provides us
with an improved methodological perspective for utilizing embryological data
in phylogeny. The first difficulty is that Anderson's analysis was typological.
Each group of arthropods was considered as a class possessing a set of
concrete features defining it. Emphasis was put on the unique features of
each group rather than on features shared with other groups. Because any
definable group will have some number of these autopomorphies, this approach put undue stress on differences. The consequence was that gaps were
magnified. Even within such groups as the unirames, differences in developmental features were interpreted as indicating separate ancestries for clades
such as insects versus centipedes, millipedes, and other myriapod groups.
This interpretation produced the unwanted problem of multiplying unknown
hypothetical ancestors. The more informative approach is to stress shared
derived features. When that is done, the observed divergent states are seen
as the products of modification of a common ancestral state. One can then
ask whether appropriate tests for homology can be devised, and whether
polarity of change can be determined and ancestral properties assigned.

Polyphyly arguments based on differences in embryology depend on estimates of how much evolutionary change is possible. For instance, if two
groups exhibit some difference in fate map, then an argument that they cannot
have a common ancestor means that transitions between fate maps organized
in particular ways are constrained in such a way that a transition is impossible.
The difficulty with such statements is that they are generally unsupported by
empirical data on the occurrence of particular evolutionary changes. The fact
that a particular developmental feature has been long conserved in evolution
says nothing per se about whether it can be modified. It has been a long-held
assumption that early development is highly constrained and resistant to evolutionary change. Until recently, there has been little reason to suppose otherwise. Studies of closely related species that differ in developmental mode,

such as those discussed in chapter 7, show that quite dramatic alterations can occur in early development over short evolutionary times. Changes in egg size, cleavage pattern, spatial localization, fate map, and gastrulation processes that are equivalent to those observed between classes or phyla can occur between two species within the same genus.

Finally, the differences observed between classic aspects of development may not be an adequate representation of the mechanisms that govern pattern formation and morphogenesis. For instance, the new molecular genetic descriptions of insect development present a very different picture of how the insect body plan is specified and executed than do traditional descriptions. Direct molecular genetic comparisons between insects and other arthropods are just beginning. However, some of them do show that genes whose expression controls segmentation in *Drosophila* operate similarly in other insects and in crustaceans. These studies show that despite differences in some of the mechanisms of early pattern formation, a considerable commonality exists. It may well be that the arthropods share an extensive common genetic underpinning upon which the visually large variations in their development have been played out during evolution.

ARTHROPOD MOLECULAR PHYLOGENIES

Several recent attempts have been made to obtain arthropod molecular phylogenies. The maximum parsimony trees resulting from these studies are summarized in figure 4.6. The trees of Turbeville and co-workers and of Abele and co-workers are based on partial 18S rRNA sequences. The tree of Van de Peer and co-workers is based on complete 18S rDNA sequences and utilizes a new method for correction for different rates of evolution in various regions of the 18S rDNA. The tree of Ballard and co-workers is based on 12S mitochondrial ribosomal DNA sequences. The questions addressed by these gene sequence studies are whether arthropods are monophyletic, what their sister group is, the relationships among the major arthropod groups, the relationships within the major groups, and the relationships of some minor phyla thought to be related to the arthropods. The five trees shown in figure 4.6 let us suggest some conclusions. First, all support arthropod monophyly. Second, in some, the arthropods emerge as related to the coelomate protostomes, but they do not emerge as any closer to annelids than to mollusks; thus these trees provide no support for the classic annelid-arthropod clade, the Articulata. The final tree, by Wheeler and co-workers, based on 18S rDNA and ubiquitin, differs from the others on this point in inferring an annelid-arthropod clade. Recall that on the basis of morphology, Eernisse and co-workers argued against the validity of the Articulata. No molecular dataset (including that of Wheeler and co-workers) supports the Articulata,

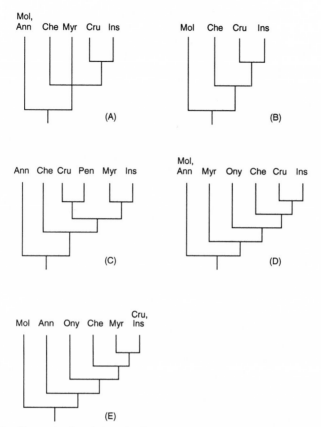

Figure 4.6. Five molecular phylogenies for the arthropods, inferred by *(A)* Turbeville and co-workers, *(B)* Van de Peer and co-workers, *(C)* Abele and co-workers, *(D)* Ballard and co-workers, and *(E)* Wheeler and co-workers. Ann = annelids; Mol = mollusks; Ony = onychophorans; Che = chelicerates; Cru = crustaceans; Ins = insects; Myr = myriapods; Pen = pentastomids.

nor does the preponderance of all data. All trees indicate that the chelicerates stand distinct from the crustaceans and the unirames, but gene sequence trees do support a mandibulate clade composed of insects (as representative unirames) and crustaceans. The myriapods are positioned close to the insects in some trees, but are found deep in the other trees, breaking up the unirames. It is not clear at this point which result is correct. The difficulty may lie in inadequate sampling. Only a single myriapod 18S rRNA, from a millipede, has been sequenced, and it exhibits a high rate of sequence evolution. The resulting long branch length may have pulled the millipede too deeply into the tree. The counter to this hypothesis is that when three myriapod 12S ribosomal DNAs were sampled, myriapods again did not fall into a unirame clade with insects. The strong morphological features uniting unirames suggest that the sampling of too few genes may be giving an artifactual result.

However, it is also possible that the unirames are a polyphyletic assemblage and the morphological features supporting them are convergences. This possibility is strengthened by the comparison of centipede central nervous system development with that of insects made by Whitington and co-workers. They found that, in contrast to the profound similarities observed between insect and crustacean nervous system development, centipedes differ in dramatic ways from insects in nervous system ontogeny. This observation either indicates a lack of constraint in nervous system development among closely related arthropods or suggests that the unirames are not a monophyletic group. If they are not, myriapods would have to be recognized as a subphylum equivalent to the crustaceans, chelicerates, and hexapods (insects and their closest relatives). It may be that an important and unsuspected problem in arthropod phylogeny has been revealed by a discordance between molecules and morphology.

Two of the molecular phylogeny studies of arthropods included small phyla thought on anatomical or embryological grounds to be related to arthropods. The pentastomids, a small phylum of parasitic "worms" that live in the nasal passages of tetrapod vertebrates, have been argued to be related to crustaceans on the basis of sperm morphology and nauplius-like larvae (Brusca and Brusca provide a brief summary of these creatures, and Walossek and Müller a critical evaluation of proposed relationships). The 18S rRNA sequence study of Abele and co-workers supported a link to crustaceans. Larvae (less than a millimeter long) of these animals have, remarkably, been shown by Walossek and Müller to be present in the late Cambrian fossil record. Pentastomids apparently had a marine origin. Some primitive morphological features that link them to arthropods are present in these fossils, but none that link them to any specific arthropod group. Intriguingly, they seem to have evolved a parasitic lifestyle before their vertebrate hosts had evolved. The onychophora, which play such a central role in arguments about arthropod origins, fall into the arthropods in the 12S mitochondrial rDNA tree. If correct, this is an important result in that onychophorans are no longer a suitable outgroup for arthropods. A tree with onychophorans within the arthropods either is consistent with loss of arthropodization by the onychophoran lineage or requires that myriapods became arthropodized independently of chelicerates, crustaceans, and insects. In the combined dataset tree of Wheeler and co-workers, the onychophorans emerge as the sister group of the arthropods, more consistent with the classic view of their relationships. A third phylum of tiny arthropod-like animals, the tardigrades, has not yet been considered by molecular phylogeny. However, these animals have been discovered by Walossek and co-workers to be exquisitely preserved in middle Cambrian rocks.

In a different molecular approach to metazoan evolution, Bartnik and Weber examined the distribution of intermediate filaments in twenty-eight invertebrate phyla by means of electron microscopy and immunostaining with a broadly reactive antibody prepared to vertebrate intermediate filaments. Intermediate filaments were observed in nearly all of the phyla examined, from cnidarians, acoelomates, and pseudocoelomates to lophophorates, deuterostomes, and protostomes. However, intermediate filaments were absent in all arthropods, in which they are apparently replaced functionally by microtubules. Onychophorans had peculiar filamentous structures and so could not be placed with the arthropods by this criterion. Pentastomids lacked intermediate filaments, consistent with their placement as an arthropod offshoot. Admittedly the loss of a feature is not a very strong candidate for a synapomorphy, but the loss of intermediate filaments and their functional replacement by a different class of intracellular fibers in all arthropod groups is consistent with other molecular data indicating arthropod monophyly.

So, now we have come full circle with the available morphological and molecular data. These, as discussed thus far, weigh in substantially on the side of arthropod monophyly. But paleontology is a restless science, and the Cambrian still has a few surprises in store. A well-preserved and entirely new lobopod animal has been discovered by Budd in rocks of the lower Cambrian from Greenland. This animal, *Kerygmachela kierkegaardi,* is extraordinary in that it possesses huge frontal appendages and dorsal gills arrayed along the length of the body. Budd suggests that the possession of these appendages and the arrangement of the gills are consistent with a relationship to some other problematic Burgess Shale animals, notably *Opabina* and *Anomalocaris,* that also have large frontal appendages and bear similar dorsal gill-like appendages. He also suggests that this animal bears on the question of arthropod monophyly, with the Cambrian lobopods actually being a paraphyletic assemblage. In this view, one descendant clade of gilled lobopods contains the ancestor of biramous arthropods, and a second lineage of uniramous lobopods contains the ancestor of modern onychophorans and unirames (at least the myriapods). Arthropodization in this model is convergent in biramous and uniramous lineages. I think it more likely that arthropods are monophyletic, but that the lobopods had a tremendous and littleknown radiation of morphologies in the Cambrian. Most, like the fantastic *Kerygmachela,* probably departed without leaving descendants.

THE DISPARITY OF CAMBRIAN ARTHROPODS

As the animals of the Cambrian radiation are becoming better defined by paleontologists, it is ironic that it may be growing more difficult to relate the three arthropod subphyla living today with the arthropods of the Cambrian.

Gould has looked at the Burgess Shale arthropods and has noted that their arrangements of head appendages in most cases do not fall into those characteristic of the extinct trilobites, nor of the living arthropod groups, the chelicerates, crustaceans, and unirames. Since these higher categories are characterized by their patterns of head appendages, the presence of different patterns in Cambrian arthropods would suggest that they belong to major lineages that failed to prosper over the long run. If any of them had, their modern descendants would stand as distinct from other major groups as crustaceans do versus chelicerates. These unique taxa represent a substantial gang of Burgess Shale arthropods, about twenty genera. If we had to elevate them to ranks equivalent to the four recognized subphyla, that would represent an enormous disparity just within the Cambrian arthropods.

Briggs and co-workers and Wills and co-workers dismiss this reliance on appendage arrangement patterns as lacking objectivity, and argue for a different conclusion. They agree that Cambrian arthropods don't fall readily into the major post-Cambrian arthropod clades, but they question whether the Cambrian designs are equivalent in rank to trilobites, crustaceans, unirames, and chelicerates. As they note, some features of modern arthropods, such as insect wings, are as unusual as some individual features of extinct Cambrian forms. These individual idiosyncrasies do not prevent us from recognizing their bearers as members of the major clades, so peculiar structures are not enough in themselves to declare new body plans. Briggs and co-workers and Wills and co-workers carried out a multivariate analysis of Cambrian and modern arthropod morphologies using a large number of characters, without giving the arrangement of head appendages any special weight. The result was that the Cambrian and modern arthropods did not deviate from each other appreciably in degree of disparity. A cladogram based on the same characters revealed the three living subphyla, but no others composed exclusively of Cambrian forms. If the three living subphyla are truly distinct in their features from the Cambrian groups, we would expect them to occupy distinct branches on the tree. Although the degree of homoplasy was extremely high, the Cambrian arthropods branched with the living groups in a strict consensus tree.

Two very different outlooks on biology are represented in the debate about the distinctness of Cambrian arthropod body plans. The approach of Briggs and co-workers and Wills and co-workers is purely one of systematics, considering features rather than considering organization. If the features that unite all members of a phylum or class are merely colored balls in a teacup, incidentally associated in some particular combination, then there really isn't a body plan held together by any particular constraint of genes or developmental integration. Given this scenario, the use of these features as discrete characters would be justified. It is not justified if these features are fundamental to underlying organization.

Briggs and co-workers and Wills and co-workers assumed that the features of Cambrian and modern arthropods were discrete characters to be scored in multivariate or cladistic analyses. Difficulties are evident in the various characters used in their analyses. To give the nature of the cuticular surface (smooth versus tuberculate) equal weight with the presence or absence of tracheae or compound eyes fails to take into account the complexities of genetic information required to produce each structure. However, even the inclusion of various complex features, such as compound eyes, may not reflect the genetic and developmental controls that produce the organization of a particular body form. The number of segments included in the cephalon and the form of appendage on each segment were among the characters used in these studies. The features that define major post-Cambrian clades constituted 9 of 59 total features scored by Wills and co-workers. We don't know how characters such as these should be weighted. We do know that the arrangement and number of appendages is under the control of a complex set of homeotic genes. The control exercised by those genes seems very fundamental. We don't know how complex the gene controls are that govern the ontogeny of the other dozens of morphological features used in these analyses. My own suspicion is that the arrangement of appendages does reflect fundamental features of the body plan organization of arthropods, and that the Cambrian radiation did include a greater disparity of arthropod and other designs than existed later. Because most of these lineages died out early, they never became very diverse, nor did they have time to refine autapomorphies that would have separated them more clearly from other clades.

DEUTEROSTOMES

The second major traditional branch of the eucoelomate animals is the deuterostomes. This is a much smaller superphylum than either the arthropods or the coelomate protostomes, but it has the virtue of containing among its member phyla the vertebrates and other chordates. The body plans of adult deuterostomes are so disparate from one another that, as discussed in chapter 2, the group is constituted for the most part on the basis of shared embryological features. The currently proposed members of the deuterostomes, based on embryological considerations, are urochordates, cephalochordates, vertebrates, hemichordates, echinoderms, chaetognaths, and the lophophorate phyla. The inclusion of chaetognaths and lophophorates is in doubt. Furthermore, hypotheses on the relationships among phyla within the deuterostomes have the feel of medieval bestiaries. The application of molecular data is of crucial importance in answering the most basic questions of membership and relationship as well as of which of the invertebrate deuterostome taxa is the sister group of the chordates.

These questions have been addressed using 18S rDNA sequence data by Field and co-workers, by Telford and Holland, and by Turbeville and co-workers. The results of all these studies supported a monophyletic deutero-stome clade. However, not all taxa proposed to be deuterostomes were confirmed as members when their 18S rDNA sequences were examined. As I've noted above, the results of Field and co-workers and Halanych and co-workers placed the lophophorates elsewhere in the metazoan tree, among the protostomes. Telford and Holland tested the placement of the chaetognaths, or arrow worms. These animals are planktonic carnivores shaped like the darts thrown in British pubs. They range in size from a few millimeters to 12 centimeters in length and are voracious predators on planktonic animals, including fish fry. Brusca and Brusca justified the placement of chaetognaths among the deuterostomes on the basis of several basic deuterostome traits: their mesoderm arises from the archenteron, their coelom is tripartite, and their subepidermal muscles arise from mesoderm derived from the archenteron. However, as summarized by Willmer, a number of other supposedly deuterostome traits are not so unambiguously clear for chaetognaths. Their nervous systems are predominantly ventral and share protostome features. Furthermore, they possess chitin, a probable protostome trait. Chaetognaths also share some pseudocoelomic features, such as lack of a peritoneum.

The 18S rDNA study of Telford and Holland agreed that chaetognaths are not deuterostomes. They instead fall deep into the tree with flatworms. The difficulty with accepting this result at face value is that only one chaetognath species was used, and it exhibits a high rate of sequence evolution, potentially causing long branch length effects. Thus, an artifactually deep placement is possible, and we are still not certain of what the relationships of chaetognaths really are.

In our study (Turbeville and co-workers), we focused on the three major unequivocal deuterostome groups, the hemichordates, chordates, and echinoderms. The relationships among the three groups are controversial. A cladistic analysis of the morphological data indicates that a number of characters support each of three possible deuterostome phylogenetic trees, as shown in figure 4.7. The first is the orthodox hypothesis found in many textbooks, which places hemichordates as the sister group of the chordates. The major shared features are the possession of gill slits and a dorsal hollow nerve cord. The second hypothesis places hemichordates as the sister group of the echinoderms. The major shared feature supporting this pairing is the heart/glomerular complex. The third hypothesis places the echinoderms as the sister

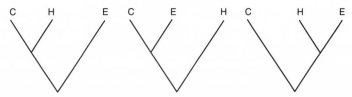

Figure 4.7. Three phylogenies for the major deuterostome phyla: the chordates (C), echinoderms (E), and hemichordates (H). The first tree has been traditionally favored. The second corresponds to Jefferies's hypothesis of an echinoderm ancestry for chordates. A cladistic analysis (by Turbeville and co-workers) indicates that all three trees are equally well supported when only morphological features are considered. A combined morphological and 18S rDNA analysis indicates that the third tree, in which hemichordates are the sister group of the echinoderms, is most likely. Both morphological and molecular data suggest a rapid early divergence of the three phyla from each other.

group of the chordates. This is the view propounded by paleontologist Richard Jefferies, who has interpreted the features of an extinct lower Paleozoic echinoderm group, the carpoids, as reflecting their position as ancestral chordates, which he has called calcichordates. The carpoids were unusual bilaterally symmetrical echinoderms covered with large calcite plates, and some, the stylophorans, bore a single arm. Jefferies has interpreted this arm as homologous to the chordate tail, complete with notochord. This is a controversial suggestion. The calcite plates are typical of echinoderm skeletons, and the evidence supports the "tail" as an echinoderm feeding arm. The carpoids themselves are long extinct, but the hypothesis that echinoderms are the sister group of the chordates can be tested. The first attempt was made by Holland and co-workers using relatively short 18S rDNA sequences. They found weak support for hemichordates as the sister group of the chordates.

Turbeville and co-workers used complete 18S rDNA sequences and a variety of inference methods. The problems that arose with this analysis illustrate some of the difficulties that still plague molecular systematics. Although neighbor-joining, Fitch-Margoliash, and maximum likelihood inferences supported chordate monophyly, the maximum parsimony method did not—that is, the urochordates (ascidians) grouped with the hemichordates. Wada and Satoh found weak support for an echinoderm-hemichordate clade, but they too found that the ascidians fell out of the chordates. The morphological features (such as a notochord, dorsal nerve cord, equivalent fate maps, and somites, as well as the possession of vertebrate-like muscle actin genes) that link ascidians with chordates are so persuasive that this 18S rDNA-based inference is unacceptable without strong independent support. The problem may well lie with the high rate of 18S rDNA evolution in the ascidian lineage. The phylogenetic signal was weak with 18S rDNA, and Turbeville and co-workers concluded that the most robust result was obtained by combining morphological and molecular datasets. When that was done, not only were the chordates solidly monophyletic, but a solid case could be made for the hemichordates as the sister group of the echinoderms. Another approach is to shift genes. When we (Kissinger and Raff) used muscle actin genes, we found that there is a very strong signal supporting the linking of ascidians with vertebrates.

If the result of an echinoderm-hemichordate clade is confirmed, it will be important in understanding the evolution of deuterostome features and gene controls in deuterostome development. It will also provide a strong set of clues as to how a bilaterally symmetrical ancestral deuterostome body plan was transformed into the highly modified vertebrate and echinoderm body plans that were derived from it. The ancestral deuterostome would have been bilaterally symmetrical and would have possessed a dorsal hollow nerve cord and gill clefts. While the chordate lineage added a notochord and somites to produce a mobility system that allowed swimming by side-to-side motion, echinoderms underwent a transformation to a unique water vascular motility system and pentameral symmetry, which resulted in a ring-shaped central nervous system with radial projections into the arms of the animal.

MOLECULAR BIOLOGY AND THE METAZOAN RADIATION

At this time the bulk of gene sequence data used for phylum-level phylogenies comes from 18S ribosomal RNAs and DNAs. As the sequences of other suitable genes enter the database, they will become important in testing the initial molecular results. Gene sequences have resolved some controversies in phylum-level relationships, but not all results have been robust or consistent. However, it is notable that most phylogenetic inferences drawn from molecular data show a reasonable concordance with morphological data. Conway Morris has rightly contended that paleontology is crucial to molecular biology. Molecular phylogenies and the amazing detail now emerging from studies of Cambrian soft-bodied animals provide complementary views of the radiation of metazoan body plans. The gene sequence inferences made thus far are consistent with all phyla having their origins deep within the tree. These data are also consistent with a rapid divergence of the phyla relative to their durations since the early Cambrian. Molecular data thus support the great conclusion of paleontology that the Cambrian radiation was the event that generated all of the basic animal body plans. Despite the astonishing evolutionary modifications that have occurred within body plans since the Cambrian, no new phyla have originated. An increase in phylum-level disparity has not been a part of progress in the post-Cambrian metazoan world.

The better extraction of a meaningful phylogenetic signal from gene sequence data will require sequencing more genes. Ribosomal RNA genes are just a beginning. Better inference methods may improve tree construction, and should provide better statistical tests of the robustness of inferences. We probably now understand better how to retrieve phylogenetic information from gene sequences than from morphological data. Nevertheless, we are hardly ready to discard morphological data in favor of molecular data, nor should we do so if we want to achieve a full understanding of metazoan evolution.

5

Recovering Data from the Past

"And once the computer has analyzed the DNA, how do you know what animal it encodes?"

"We have two procedures," Wu said. "The first is phylogenetic mapping. DNA evolves over time, like everything else in an organism—hands or feet or any physical attribute. So we can take an unknown piece of DNA and determine roughly, by computer, where it fits in the evolutionary sequence. It's time-consuming, but it can be done."

"And the other way?"

Wu shrugged. "Just grow it and find out what it is," he said. "That's what we usually do."

Michael Crichton, *Jurassic Park*

THE CRYSTAL PALACE DINOSAURS

Within the heavy urban tangle, the traffic noise, and the exhaust smell, deep in the south London district of Sydenham, there is a park that contains in one of its obscure corners the remnants of one of the great mid-nineteenth-century intellectual leaps in our interpretation of the history of life on Earth.

In 1854, the spectacular Crystal Palace, a giant pavilion of cast iron and glass a third of a mile long, which had been constructed for the Great Exhibition of 1851, was moved from Hyde Park and reassembled in the open space of Sydenham's Crystal Palace Park. The Great Exhibition was a symbol of the divinely ordained progress so dear to Victorian England, and provided a handy structure for historians (such as James Morris in his engaging trilogy of the times). Queen Victoria dedicated the new enterprise, and thousands of Londoners took weekend excursions out to visit it. The Crystal Palace burned in 1936, leaving only a pilastered concrete base, a few derelict sphinxes, some broken statues and urns, and a great grassy expanse bounded by wide stairways that lead to other grassy patches of forgotten purpose. The park can be entered from the Crystal Palace side by crossing a weedy lot and passing the Crystal Palace Museum in its red brick urban camouflage. The museum is open only from 2 to 5 P.M. on Sundays; in the best tradition of missed opportunities, I was there on a Friday. The long, grassy Crystal Palace site itself lies below the sphinx-flanked stair. The sphinxes were erected by

Joseph Bonami, an early excavator in Egypt, and reflect a mid-nineteenth-century mania for things ancient Egyptian. Stretching beyond the site is the huge National Sports Centre. Between the Crystal Palace site and the sports complex is a Victorian-style red-painted iron signpost that points in various directions to the amenities: "toilets," "museum," "maze." One panel reads "monsters." The sign that points to monsters does so in the unpromising direction of the Sports Centre. Inquiry of any of the local children on bicycles elicits patient replies. It's evident that anyone ought to know that all you have to do is go through the Sports Centre to the petting zoo and the stagnant ponds with the paddleboat rentals, and the monsters are very obviously there. But of course the kids refer to them as dinosaurs, and indeed they are.

In 1822, Gideon Mantell, a physician practicing in the town of Lewes in Kent, discovered the teeth and bones of a gigantic and enigmatic animal from the Cretaceous rocks of the Wealdian region of southern England. Mantell had a passion for fossils that transcended all other concerns, including his family and his medical practice. His wife finally felt herself squeezed out by his collections and left him. Despite initially discouraging suggestions from experts that the fossil teeth represented nothing more than those of a relatively recent extinct rhinoceros, Mantell persisted, and eventually showed that the Wealdian teeth resemble those of a large present-day lizard, the iguana, but on a very much larger scale: thus, the *Iguanodon*. It was the dean of Victorian anatomists and eventual opponent of Darwinian evolution, Richard Owen, who first conceived of a wholly novel group of animals, the Dinosauria, based on the remains of the *Iguanodon* as well as other partial remains of very large fossil reptiles. *Iguanodon* was betrayed by its teeth as a herbivore. *Megalosaurus,* discovered at about the same time in the same strata as *Iguanodon,* was clearly shown by its daggerlike serrated teeth to be a very large carnivore. What Owen saw was that these gigantic animals combined the features of reptiles with those of modern very large mammals such as elephants and rhinos. Elephantine lizards that share some features with mammals do not exist today, but Owen reconciled the unexpected mosaic of features of these novel animals and visualized them as the dominant life forms of the Mesozoic era.

Owen's job of reconstruction was made more than a little problematic by the very incomplete nature of the dinosaur remains known at the time. The discoveries of complete *Iguanodon* skeletons in a Belgian coal mine and their brilliant reconstruction by Louis Dollo were still a generation in the future. Owen attempted to give dinosaurs life by contracting in 1854 with the enterprising, technically capable, and oddly named sculptor Waterhouse Hawkins to build full-sized 30-ton reinforced concrete replicas to be set into the shrubbery at Sydenham. These replicas were designed with flair, and perhaps with some sense of fun. As the models neared completion, Hawkins served a banquet in his studio to celebrate their progress. The head table, with

Hawkins, Owen, and, it is said, nineteen other guests, was inserted into the unfinished lower half of the body of the larger *Iguanodon*. The replica is big, but it must have been cramped and hot inside for so many worthies in Victorian frock coats. Illustrations of that evening suggest that the distinguished guests suffered from no shortage of libations, however crowded they might have been. These shenanigans have continued to be a joyous part of the paleontological tradition so fondly described by Edwin Colbert and by Martin Rudwick.

The models were set up in Sydenham in an evocative natural setting of ponds and conifers. Two iguanodons and a threatening *Megalosaurus* form the dramatic center of the tableaux. The *Megalosaurus* gazes hungrily at the iguanodons and at a fourth monster, the *Hylaeosaurus,* an impressively spiky if overimaginatively restored dinosaur, which waddles off from center stage. The scene is framed by some truly demonic-looking pterosaurs and, in the water, a suite of ludicrous big-eyed ichthyosaurs, sinister snake-necked plesiosaurs, and weirdly froglike labyrinthodonts. To be true to proper Victorian notions of higher educational purpose, the models were placed as a panorama in geologic order from the Coal Age labyrinthodonts through the Jurassic marine reptiles to the big Cretaceous dinosaurs. They are all still there in good repair and fresh paint, lurking on the edge of the petting zoo, kept safe from the vandal's hand by moats filled with well-fertilized murky green algae and patrolled by flotillas of hungry ducks. The giant *Iguanodon*s are vibrant and lively even now, but they predate the large body of knowledge that enlightens our much different appreciation of dinosaurs.

Incidentally, the Crystal Palace display still colors our presentation of imagined worlds and life forms to the public. Just up the road from the mysterious Urquhart Castle, which broods over the north shore of Loch Ness, there is an "official" Loch Ness monster exhibition center in the town of Dromnadrochit. There, in the best Crystal Palace style, is a full-sized "Nessie" posed on a rock in a pond. She is reified as a giant dark green plesiosaur, with a small reptilian head, long serpentine neck, plump rounded body, four big flippers, and a short tail. Her setting amid Scottish sheep fields and busloads of camera-laden tourists comes full circle to the spirit of the Victorian Great Exhibition.

For all the quaintness of the Crystal Palace setting and the datedness of his models, the core of Owen's genius as an anatomist shows through in his restored dinosaurs. The Sydenham monsters very powerfully transcend just being big lizards. Owen got it right that dinosaurs represented something new. But the details are wrong, and in some cases wrong in ways that prevent full appreciation of just what dinosaurs were as living animals. Some errors are inconsequential. The *Iguanodon* is restored with a nose horn like a rhino's. The "horn core" was later discovered to be the enlarged spikelike thumb. The bodies are covered by huge scales in accord with the best lizard stereotype. Much later fossil finds of mummified dinosaurs complete with preserved skin impressions revealed that large dinosaurs were not covered with scales, but with leathery skin similar to that of elephants. Posture poses a more substantial problem. One Crystal Palace *Iguanodon* has a very reptilian sprawl; the other stands in a much more mammalian pose. As befits an active and aggressive predator, the *Megalosaurus* walks upright on great graviportal limbs tucked under the body mammal-style. There is an ambiguity evident in Owen's thought: big, sprawling cold-blooded lizard vies with upstanding warm-blooded rhino. The traditional concept of a reptile with belly slung low to the ground was hard to shake. These animals are reptiles, yet they are something more.

The representation of what Rudwick calls *Scenes from Deep Time,* the restoration of extinct animals in settings appropriate to their time, was an invention of the nineteenth century. It drew its inspiration from paleontology and comparative anatomy as well as from an older tradition of the painting of Biblically recorded events. Since so much about restored extinct animals is inferred, the reconstructions inevitably present the restorer's hypotheses about former creatures and the former world in a powerful visual way. Adrian Desmond, in his book *Archetypes and Ancestors,* suggests that Owen's recon-struction of dinosaurs as mammalian analogues served a conscious ulterior motive of helping to suppress the dangerous idea of "transmutation." That term was applied to the still poorly formed ideas about evolution that troubled the complacency of a number of biologists prior to Darwin. Transformational-ist ideas were fueled by the Lamarckian view of organisms striving for im-provement, as well as by the outlook of transcendental biology. That view insisted on a fundamental unity of life in a chain of animals extending from lower to higher, despite the differences in anatomical plans that characterize the various groups. Evolutionary coloring came from attempts at the time to link dissimilar animals on the basis of resemblances, some quite fanciful, in embryonic development.

Owen considered that if only degenerate lizards exist today as descendants of the once glorious reptiles of the Mesozoic world, then there can be no pattern of progressive change in the history of life. Charles Lyell, the father of British geology, agreed. Lyell's newly published *Principles of Geology,* which provided the first synthesis of that discipline, accompanied Darwin on the *Beagle.* Although Lyell influenced Darwin, and they later became close friends and scientific correspondents, Lyell was to have a difficult time in coming to terms with Darwinian evolution. Lyell read the fossil record as antiprogressional in terms of both geologic processes and the complexity of multicellular organisms. It is in this context that Lyell, as early as 1835, commented on dinosaurs and noted that "it appears also that some of these

ancient saurians approximated more nearly in their organization to the type
of living mammalia than do any of our existing reptiles.'' In 1844 the Scottish
editor Robert Chambers anonymously published his *Vestiges of the Natural
History of Creation*. Chambers, who evidently had no taste for martyrdom,
showed considerable good sense in publishing the book anonymously. *Vestiges* was subjected to a storm of furious criticism. Ironically, the fuss over
Vestiges ultimately reduced the adverse reactions to Darwin's later *Origin
of Species*. *Vestiges* was explicit in propounding evolution and in seeking
mechanisms for its workings, including a heavy leaning on embryonic development. Both because of its heterodoxy and because it contained some substantial factual errors, the book infuriated the scientific establishment. Chambers saw the history of life as both evolutionary and progressive. Owen's
suggestion of ''degradation'' as a major component of that history provided
useful ammunition for the orthodox to use against *Vestiges*. The Crystal
Palace dinosaurs concretely showed that there had not been an improvement
in reptiles since the Mesozoic era; instead, living reptiles were poor stuff by
contrast with the old days. There were thus no good grounds to suggest a
progressive evolution of life forms.

It took further discoveries to show that both *Iguanodon* and *Megalosaurus*
were bipedal animals that stood well clear of the ground balanced on long
hindlimbs. It took an even longer time for us to appreciate that the phylogenetic tree is bushy. Lizards are not the degenerate descendants of dinosaurs,
but represent yet a different ancient lineage. Carnivorous dinosaurs are the
sister group of birds. Some of them are extremely birdlike, and again pose
difficult problems in interpreting creatures not quite like any living today.
An 1858 book by Edward Hitchcock about the fossil trackways of the Connecticut River Valley is especially suggestive of the tortuous way in which
science can work. These three- or four-toed tracks are from the early Jurassic,
and we now know that most of their makers were bipedal dinosaurs. Because
no dinosaurs had yet been restored as bipedal, the best model available to
Hitchcock for what the track makers might have been was large birds. He
calculated that the biggest of his birds weighed up to 800 pounds and stood
12 feet high. But Hitchcock was a good enough observer to realize that the
ancient world he glimpsed might have contained animals unlike any now
living. He says of one of his larger birdlike tracks, ''When I first saw its
track, although it had a small fourth toe, I thought it a bird; but when I found
soon after that it had left the distinct trace of a tail, that opinion must be
abandoned; for the trace could neither be explained by referring it to the
dragging of the feet, nor to the large tail feathers of a bird. Yet, if a biped,
its body must have had somewhat the form of a bird, in order to keep it
properly balanced. The tail, although evidently rather stout, was not enough
so to help prop up the body. And how very strange must have been the

appearance of a lizard, or batrachian, with feet and body like those of a bird, yet dragging a veritable tail!'' He also realized from the trackways that although these animals walked primarily on their two hindlimbs, they occasionally lowered their forelimbs to the ground. These are not bad descriptions of dinosaurs and their behaviors, but they had not yet been integrated with Owen's dinosaurs.

Because Owen's models were so evocative, they serve as a useful caution to us in our attempts to reconstruct the history of metazoan life. His reconstructions were innovative and fit the data at hand. The mammalian analogy was in the right direction, but without further fossil evidence there was no way to infer the true anatomy of the animals. Even now, big questions remain. Dodson has estimated that we have discovered about a fourth of an estimated 1,200 once-living dinosaur genera. Some newly discovered forms, like *Mononychus*, with its tiny one-clawed arm, are remarkably queer. This animal, reported by Altangerel and co-workers, is a birdlike dinosaur or dinosaur-like bird (it's not clear which is a better description). The ambiguity and unexpectedness of the discovery show that big surprises remain in dinosaur anatomy and physiology. A decade of concentrated effort and flamboyant public debate has not resolved the issue of whether or not the mammalian and avian aspects of dinosaur anatomy prove that these animals were warm-blooded (for the determined reader, the arguments of equal and opposite experts can be found in *A Cold Look at the Warm-Blooded Dinosaurs*, edited by Thomas and Olson). As in Owen's time, a good bit of the battle over warm-blooded dinosaurs is waged by means of reconstructions of living animals whose poses and activities assume a warm-blooded physiology. It's not so clear that we can establish the boundary for the transition from cold-blooded to warm-blooded even within the mammal-like reptile to mammal sequence. This uncertainty remains even though a phenomenally complete fossil record exists for this lineage, a record so good that it documents fully the complex transition from the reptilian jaw and ear to the mammalian condition. The uncertainty has led one author to propose in three different papers published over a ten-year period that some advanced mammal-like reptiles, the therapsids, were warm-blooded, cold-blooded, and had a metabolic status between those two conditions. Bennett and Rubin, who have recently contemplated the problem, have wryly suggested that this indicates a certain "elusiveness of the subject."

As we push our inquiries into the earliest traces of the animal fossil record and to the limits of gene sequence data, our best new discoveries force us to draw inferences based on incomplete and imperfect data, just as the fragmentary remains of dinosaurs from southern England did Owen. We may ultimately get definitive evidence—our Belgian iguanodons—but we have no guarantees on that score.

LOSS AND RECOVERY OF DATA FROM THE PAST

At this juncture of our deep immersion in the marketplace of phylogenetic controversy and our concern for the big picture of the Cambrian radiation, it's important to remind ourselves about the operational use of phylogeny for understanding the evolution of body plans. The following chapters will in large part explore the role of developmental processes underlying the evolution of body plans, because it is the modification of ontogeny that transforms morphology. The phylogenetic background presented in the earlier chapters was intended to provide a view of our current understanding of body plan origins, the uncertainties of our phylogenetic understanding, and the methods currently available. A meaningful study of the evolution of developmental processes requires that data on developmental mechanisms be interpreted within a phylogenetic framework. A knowledge of how phylogenetic inferences are generated is necessary, as is an appreciation of the sources and limitations of phylogenetic information. The conclusions that we draw about evolution by combining experimental data about development with phylogenetic inferences will be sensitive to the quality of those inferences. There is a strong tendency for those who don't do phylogeny themselves to accept nice, neat published phylogenetic trees as unequivocal sources of reliable information. A modicum of informed caution is desirable.

The present chapter has two goals. As the moral of my dinosaurian tale, we should consider that all data from the past become degraded with the passage of time, and that as a result, our reconstructions of past events, organisms, and phylogenetic relationships are all made through filters that remove information and distort our vision. A part of the present chapter explores this loss of information, particularly from gene sequence data (morphological information is also subject to information loss from a variety of causes). I discuss approaches to the process of constructing phylogenies that are aimed at recognizing information losses and ambiguities and ameliorating their effects. I will also discuss the search for sequence information from ancient DNA preserved in special fossil environments. Just as conventional morphological fossils supply information about evolution unavailable from living organisms, these molecular fossils may supplement phylogenetic inferences drawn from gene sequences of living organisms.

INTERPRETING LOST BODY PLANS

The oddest evolutionary episode recorded in the fossil record may well be the appearance of the Ediacara fauna discussed in chapter 3. The first interpretations, made by Glaessner, placed all of these fossils within living animal phyla and classes, even to the family level. Glaessner's approach was not so bad. He had to have some organizing principle, a guiding hypothesis of relationships, on which to build a picture of the Ediacara fauna. But as more of their anatomy has become apparent, it has become clear that attempts to homologize them with modern animals founder on the unusual features of their structure and the lack of crucial detail in the preservation of most of them. We are finally left trying to decide whether the Ediacaran body fossils are best interpreted as a short-lived, aberrant radiation of a primitive diploblastic metazoan group or, as suggested by Seilacher and by Retallack, as fungal-grade syncytial organisms or woody lichenlike symbioses. If we have to concede that the Ediacaran fauna represents something other than animals, we not only have the problem of accounting for a mysterious radiation of multicellular life, but must also retreat from having a convenient soft-bodied animal fauna that predates the Cambrian metazoan radiation. We also have an illustration of the difficulties of projecting modern biological analogues back into the remote past.

The Cambrian radiation, although it clearly involved real animals with living relatives, provides more than its share of mysteries. Our difficulties in interpreting the animals of the Cambrian radiation come from both the loss of data inherent in fossilization and the lack of exact relatives or analogues among living animals. Ultimately, to understand fossil beings, we absolutely depend on comparisons and analogies with living species. As we find ourselves dealing with creatures for which living relatives or analogues are not obvious, we become extremely limited in our confidence to interpret what we think we see. Those uncertainties extend to the taxonomic status of the organisms, to interpretations of ecological role, to morphology, and to inferences about physiology, biochemistry, and development.

THE NEANDERTHAL'S MISSING VOICE AND DNA'S FORGOTTEN BASES

There is a breathtaking quality in the preservation of some fossils, but as in all attempts to interpret the past, information loss is inevitable, and what is preserved cannot give us unequivocal answers. For example, our extinct cousins, the Neanderthals, are represented by an excellent fossil record, with several nearly complete skeletons and perhaps 500 individuals represented altogether in the present sample. Nevertheless, as Trinkaus and Shipman show, over a century of study has not placed them comfortably within humanity. Thus, "Neanderthals have been cast in virtually every imaginable relationship to ourselves. They have been subsumed under our own modern species by some and thrust far out on the most remote branch of our family tree by others." Older restorations envisioned them as shuffling lowbrows destined for replacement by improved moderns, our own superior ancestors. More recent restorations show instead a vigorous people, well adapted in anatomy to cold glacial conditions and in body robustness to strenuous big-game hunting. Neanderthal brains are as big as ours, but perhaps differently proportioned because their skulls are longer and lower than modern human skulls. Their faces are expanded in the nasal area, apparently to allow more effective warming of cold air in breathing. It is hard to extract some crucial aspects of biological organization and behavior from the fossil record. Could Neanderthals speak? This deceptively simple question has led to a conflictual and fluid debate that illustrates the ultimate problem of massive loss of data as organisms move from the biosphere into the geologic record. In the case of the Neanderthals, there are some ambiguities in reconstructions of the skull, and even the recent discovery of a Neanderthal hyoid bone identical to that of modern humans has not resolved the issue of whether the Neanderthal throat was built suitably for speech like ours. The behaviors that emerged from the billions of connections in the soft tissue of their brains of course died with them. What we infer about behavior from the Neanderthal archaeological record is often negative. Their lack of art or personal adornment is documented. What that means about their capabilities we may never know.

Can we at least place the Neanderthals phylogenetically with respect to modern humans? Despite superb evidence on Neanderthal anatomy, and even culture through their tools and archaeological remains, the place of the Neanderthals remains obscured and hotly debated. Two books published within a few months of each other during 1993 were devoted entirely to the Neanderthals. They are well argued, persuasive, and present diametrically different conclusions. The school represented by Trinkaus and Shipman leans toward accepting the Neanderthals as members of our species, albeit highly adapted in anatomy to the severe cold and dryness of Ice Age Europe. Under this scenario, our modern lineage branched from theirs relatively recently, and modern Eurasians arose in situ from Neanderthal ancestors. The other view, articulated by Stringer and Gamble, is that Neanderthals and modern humans share a common ancestry well back in the Pleistocene. The Neanderthals were eventually displaced by modern humans who radiated from Africa late in the Pleistocene. The Neanderthals are thus a different species from our own, with a distinctive set of adaptations differing from those of modern humans. This "out of Africa" scenario has been tested by molecular means.

In 1987 Cann and co-workers took a magnificent swipe at the Gordian knot of modern human origins. They produced a human phylogenetic tree based on the mitochondrial DNAs of a number of human races. The object was to see whether the sequences of this molecule present among living races could be traced back to a relatively recent ancestor, and if so, whether the continent of origin could be determined. The feasibility of such a study lies in two properties of mitochondrial DNA. The first is its rapid rate of evolution. The second is its mode of inheritance: Human mitochondrial DNA is transmitted maternally and does not undergo recombination. The first study was based on analysis of restriction sites rather than sequencing. The result was that all human mitochondrial variants could be traced back to a single female ancestor who lived in Africa about 200,000 years ago. This ancestor was dubbed "mitochondrial Eve." Some aspects of the analysis, such as its lack of adequate rooting and statistical justification for the result, raised eyebrows. Thus, somewhat later, the same group (Vigilant and co-workers) redid the analysis, this time using sequence data from a rapidly evolving domain of the mitochondrial DNA and an expanded database of individuals sampled. The reported result was the same.

However, despite a database of 135 different sequences, the enterprise has failed to prove its point. It is instructive to see why. Although a large number of different sequences were used in the parsimony analysis, there were only 119 informative sites. That works out to about one site per node on the tree and, of course, indicates weak support for each branch. Given the large number of sequences, there is an enormous number of equally parsimonious trees to the one they presented as their best. Templeton, as well as Maddison and co-workers, has experimented further with parsimony analyses of the data. They found that trees with an African origin are no more favored among most parsimonious trees than those with a non-African origin. They also found that the use of chimpanzee sequences to root the tree is unreliable as well. That is probably due to the mutational saturation of sites between species in this fast-evolving DNA. So far, then, phylogenetic inferences from gene data have actually been less successful than paleontological data in resolving the origins of modern humans. However, as we'll see in a moment, a direct approach to Neanderthal DNA may be possible.

Reconstructions of the past by any method will always remain problematic. If we are to use molecular tools effectively, we have to realize that the ambiguities and information losses that plague the fossil record also plague the genome with its constant mutational remodeling. It is difficult to accept, but much information about the evolutionary past is irretrievably lost. In the case of mitochondrial Eve, too much faith was put in simply reading out the answer from a parsimony analysis because it matched expectations. The data simply may not contain a sufficient signal to resolve the branching among modern human populations and their geographic site(s) or time(s) of origin. The failure of the "mitochondrial Eve" approach does not mean that the hypothesis of an African origin of modern human populations is not correct. It may be. Data on gene polymorphisms presented by Cavalli-Sforza is consistent with an early split between African and non-African populations. The overall arguments based on paleontology, archaeology, population genetics, and molecular data reviewed by Lewin also support an African origin.

CAN OUR METHODS RECOVER PHYLOGENIES FROM GENES?

In principle, the enormous amount of sequence information recorded in each animal genome, on the order of a billion or more base pairs, should be sufficient to give us unique evolutionary histories. Each species is genetically distinct, but the Linnaean hierarchy shows that species share morphological features with their relatives that record evolutionary history. The genes that encode those features should be even more informative. There is, of course, the trivial problem of the expense of reading billions of bases worth of sequence data from a large number of organisms. But that's not the real issue. The question is whether we could always infer unequivocal phylogenies if we had the sequences in hand. It seems most likely that we could not. Molecular evolution is no more Haeckelian than is morphological evolution. New sequence elements do not simply get grafted onto old ones. Sequences evolve by substitution of bases; by deletion of bases; by recombination, domain shuffling, and gene conversion events that recopy sequences to produce hybrid genes; and by duplication and divergence events that produce families of related genes. The same base position thus incurs multiple substitutions as time increases. As a consequence, phylogenetic information is lost through randomization of bases at individual sites along the sequence. Evolution thus erases information as it creates new sequences by descent with modification. In inferring trees, it is common to assume that if one arrives at a most parsimonious tree, it has resulted from an informative dataset. However, as shown by Hillis and Huelsenbeck, one of the surprising outcomes of tree inference using parsimony methods is that one finds a most parsimonious tree even if the dataset is random and thus actually contains no phylogenetic signal.

One way of directly approaching the problem is to create known phylogenies and test the ability of the inference methods to recover them. That has commonly been done by generating artificial phylogenies on paper or in computers. The difficulty with these artificial phylogenies is that they are of necessity built from our preconceived ideas of how sequences evolve. Models of evolutionary change, as shown by Felsenstein and by Hendy and Penny, can reveal cases in which inference methods will not be able to recover the correct topology. These cases include instances in which homoplasies exceed informative changes, and cases in which rates of evolution differ greatly between evolving lineages included in the analysis. Penny and his co-workers have evaluated the approaches used for inferring gene sequence trees, and have suggested attributes that molecular phylogenetic inference procedures should have: they should be efficient, powerful, consistent, robust, and falsifiable.

Efficiency refers to the speed with which the overall optimum tree can be found. For anyone seeking trees with large numbers of members, it seems that there is bad news. Trees with large numbers of species pose what mathematicians call nondeterministic polynomial-complete problems. It appears unlikely that any general mathematical solution of such problems will be found. As a result, when large numbers of species are being treed, heuristic methods are used that are fast and can find good solutions, but not necessarily the best solution.

The second desirable criterion is power. That term refers to how long a sequence is needed before the solution converges on a single tree. There is a noticeable practical consequence in terms of time and money spent in the laboratory if 100,000 bases are needed for each species instead of only 1,000. There does not appear to be a simple relationship between the amount of sequence data available and the information actually used by inference methods. The various methods each ignores certain kinds of data, and there are systematic losses of information as well. For example, distance methods use more information than parsimony methods when the dataset is small, but much less of it when large numbers of species are used. Parsimony methods omit "noninformative" sites that are used in distance methods.

The third criterion is consistency. Consistency refers to convergence of trees on the correct model. This is a tricky issue. Inference methods are more likely to produce good results if they make realistic assumptions about the mechanisms by which evolutionary changes have occurred in the sequences being considered. We do not necessarily know that the models proposed are good descriptions of nature, but clearly there can be bad ones. Penny and co-workers, for example, suggest that a model that relates tree topology to the alphabetical order of the names of the species in the tree would not be likely to correspond to any real evolutionary mechanism. In some inference methods only minimum assumptions are made (for instance, that evolution is described by a tree at all). Other methods, such as maximum likelihood, make substantial use of a model in the inference process. Studies of such standard methods as parsimony and distance have shown that under some circumstances these methods give answers inconsistent with standard models. A related issue arises from the criterion of robustness. A method may produce consistent answers when ideal conditions are met, but fail when real circumstances deviate in some way from the model. Penny and his co-workers feel that there is little knowledge of how well the methods perform with real data.

Finally, Popperian falsifiability has become an inevitable touchstone for judging whether hypotheses are scientifically valid. Falsifiability posits that hypotheses are not scientifically useful unless they predict consequences that can be tested and potentially proved false. Phylogenetic trees are hypotheses. Ideally, the results of phylogenetic inferences should be testable in such a way that they could be falsified if incorrect. Unlike discussions of other kinds of molecular biology studies, discussions of molecular systematics are often inconclusive and come down to sometimes emotional arguments on first principles (furniture has been thrown on at least one occasion). I suspect that much of the heat generated in discussions of inference methods and their results comes from our inability to effectively falsify molecular (or indeed any other) phylogenies. An example of the kind of problem we accordingly face is the fact that inference methods will produce a tree even if the version of a gene being sequenced for one of the included species was introduced to its genome by horizontal transfer from another species. That sequence would violate the model of a treelike evolutionary process for sequence evolution, but the tree-building procedure would incorporate it into a tree without blinking. Any other genetic process that results in the creation of a sequence by mechanisms outside of the model would have similar results. At present, we may be limited to comparing the results we obtain with those obtained from other databases. Morphology thus far has offered the most comprehensive database. Some phylogenetic conclusions are so well supported by morphological features that it is unlikely that a contrary inference from a gene sequence tree would overturn them. For example, the monophyly of vertebrates is supported by numerous shared morphological features. A molecular tree result that found otherwise would be highly suspect. One that found vertebrate monophyly would be likely to receive strong support. A variety of genes may eventually provide sequence datasets to augment the ribosomal RNA gene datasets available now.

In a more subtle way, a single database can be sampled to reveal the signal its components contribute in order to assess its internal consistency and to assess whether the database has a significant phylogenetic signal. Data points of the sequence can be randomly sampled by bootstrapping to see how well they support the favored topology. Analogously, by jackknifing, one or more lineages can be omitted to see how sensitive the topology is to included sequences. Finally, Hillis and Huelsenbeck have recently observed that even random sequence data will yield a "best" tree with parsimony. They have shown that the distribution of tree lengths for a random dataset will be bell-shaped, whereas an informative dataset will produce a highly skewed distribution. The skewness for a real dataset can be determined, and it thus becomes possible to assess the informativeness of a particular database.

The phylogenies to which inference methods are applied do not have to be unknown. Hillis and his co-workers have produced a direct test of phylogenetic inference methods by generating a phylogeny in a very rapidly evolving lineage, bacteriophage T7. In this study, phage were grown in the presence of a mutagen. A phylogeny was generated by taking stocks from a phage culture and creating bifurcating lineages. After a period of time, bifurcating lineages were again produced. In the end, the investigators obtained a set of terminal branches analogous to the terminal taxa existing in a natural phylogeny. These differed from a natural phylogeny in that a true phylogeny was known for all terminal branches. The investigators also knew the number of mutations acquired along each branch and the character state for each ancestral node. These are all vital pieces of information not available in real phylogenetic studies.

In the study by Hillis and co-workers, a symmetrical topology was generated with an equal time distance between nodes. Their tree contained nine lineages, and thus 135,135 possible topologies. They inferred trees from the experimentally generated restriction maps for each lineage by standard cluster, parsimony, and distance methods. All methods found the correct branch topology. However, none predicted the actual branch lengths for all branches. Parsimony did best at this, and cluster analysis, which is known to be very sensitive to branch length variation, did poorest. Parsimony is the only method that reconstructs the ancestral states at the nodes on the tree. Since the restriction maps at the nodes were known, the ability of the parsimony method to correctly reconstruct these was also tested. In this experimental case parsimony inferred 97% of the ancestral states correctly.

The utility of this program of research has been questioned by Sober. He has argued that a model of how the evolutionary process works is necessary for the laboratory simulation to be meaningful. An atheoretical approach suffers from the weakness of providing no criteria for judging whether the laboratory phylogeny accurately reflects natural processes. Sober concluded that experimental simulations will be useful in testing the kinds of natural evolutionary processes that affect the success of inference methods in retrieving a phylogeny. Hillis and his co-workers have responded that the experimental phylogeny they generated does not represent all of nature, but that it falls within the realm of known natural processes. They also noted that the amount of homoplasy in their tree was very similar to homoplasy levels found in analyses of natural taxa. The argument remains unresolved, but it should be possible to set up experimental phylogenies that mimic various potential real phylogenies. The performances of the various methods can be evaluated in this heuristic fashion to give at least an estimate of their reliability under various circumstances.

GENE TREES VERSUS SPECIES TREES

Animal genomes contain thousands of potentially informative gene sequences for phylogenetic analysis. However, the adequacy of methods is not the only problem facing us in seeking molecular phylogenies. To be useful for this purpose, sequences have to evolve at a rate commensurate with the rate of organismal divergence. Thus, ribosomal DNAs are useful for inferring deep splits such as those encountered between kingdoms or phyla, but are useless for analyzing recently diverged lineages. The control region of mitochondrial DNA, because of its very rapid rate of evolution, has been used for recent divergences, such as between species in a genus. Other genes have been used for phylogenies with time frames between those extremes. Thus, hemoglobin genes and mitochondrial 12S ribosomal DNA have been used to analyze relationships among mammalian orders. The choice of appropriate genes for phylogenetic sequencing also depends critically on establishing that truly homologous sequences are being compared.

DNA sequences have the potential to supply phylogenetic information about the organisms carrying them only if the splitting pattern for the gene sequences corresponds to the splitting pattern for the organismal lineages. Imagine that an ancestral species carries a single gene encoding protein A. Upon speciation, each descendant species carries a single *A* gene. As the two gene *A* sequences evolve, the sequence changes reflect the organismal splitting event and thus have phylogenetic content. Such sequences are known as orthologous. The large and small ribosomal subunit RNA genes have been found to be orthologous in all animals examined thus far. However, most genes in any animal are members of small families of related genes. The duplication events that produce members of gene families do not correspond to cladistic events in the lineage of the organisms that carry those genes. Thus, if gene *A* in an organism duplicates to produce gene *A'*, descendants of the species in which the gene duplication occurred will possess both genes *A* and *A'*. These genes are called paralogs. Their subsequent evolutionary histories are distinct and do not necessarily match the splitting patterns of organismal lineages. In a series of related species, some might contain only gene *A*, whereas others contain both *A* and *A'* members of the gene family. There can be further paralogous splitting of gene families in some lineages, or even losses of family members. The result can be that not all species in a lineage contain the same family members, only paralogs. Thus, one would go very wrong in comparing the sequence of gene *A* from one species with that of gene *A'* from another in trying to adduce an organismal phylogeny. In many instances, such as in the globin genes, there is no difficulty in sorting out the orthologous family members, but that is not always the case.

As we go beyond the few genes presently in use for molecular phylogenetic studies, paralogy will become a major practical problem. Most animal genes are members of gene families. Genes that are members of multigene families will have to be used, but the orthologs will have to be identified in each species being compared. In some cases that may prove impossible because some gene families are evolutionarily fluid, and there may be no strict ortholog.

SETTING THE MOLECULAR "CLOCK"

To estimate rates of molecular evolution, we depend on paleontological estimates. The so-called molecular clock is often advertised as a distinct way of timing evolutionary events by rates of molecular evolution, but any estimate of these rates ultimately rests on time estimates from the fossil record. Typically, some measure of distance is made for a gene or set of genes from a number of organisms; for example, the number of amino acid differences among β-chains of hemoglobins could be used. If the times of evolutionary divergence between the ancestors of the organisms for which the gene distance comparisons have been made are known, the number of amino acid differences per million years of separation can be estimated. Such estimates can then be applied to other organisms for which molecular data, but no fossil record, are available. That was done in a very spectacular way by Sarich and Wilson in 1967 to estimate the divergence of human and ape lineages. Their estimate of 5–10 million years ago shocked anthropologists, who had estimated a date of 15–20 million years ago on the basis of the poorly known fossil record of the time. More recent hominid fossil finds support the molecular clock estimate. Despite its successful application, the molecular clock cannot be accepted as a universal timekeeper, and must be used with caution. Indeed, various genes evolve at different rates, with genes that produce highly functionally constrained proteins being the slowest. Other effects are also important. The same gene may evolve at different rates in different evolving lineages, perhaps as a result of different selective pressures or because of different efficiencies of DNA repair. Rate variations are so common that mathematical inference methods for finding phylogenetic trees, such as those reviewed by Swofford and Olsen, now almost universally take them into account. The methods used to compensate for variable rates help, but cannot entirely remove errors caused by rate variation.

Since the triumph of molecular biology in predicting the timing of the human-ape split, molecular biologists have trusted in textbook paleontological dates for molecular rate estimates. That trust can be seriously misplaced for two reasons. Evolving lineages may be imperfectly understood, or classified improperly. But there is a more subtle source of error that is generally not appreciated. Because the fossil record is incomplete, stratigraphic ranges (durations of fossil lineages as estimated from the rock layers in which they occur) must always be underestimates of the real durations. This problem arises from several causes. Not all populations of a fossil species are sampled. In the case of small populations, geographic sampling may be incomplete, or samples may not be large enough to include a relatively rare form. Even with species that had large population sizes, all populations may not be represented because they may not have died in suitable areas of deposition, or the sediments that buried them may have been chemically inhospitable to fossil preservation, or they may have been fossilized but lost through subsequent processes such as erosion. Deficient sampling will be most likely for organisms with small populations or limited geographic ranges. Those are just the situations one might expect for new and not yet established groups, or for relict forms. The effect, of course, is that we really do not ever know when a lineage actually appeared or disappeared, even when absolute dates for the known fossils are accurate and phylogenetic relationships between lineages are understood. These are the same problems that bedeviled Darwin in his attempt to account for the lack of intermediates in the fossil record. He devoted a chapter of *The Origin of Species,* entitled "On the imperfection of the geological record," to discussing the causes of that incompleteness. It is not so much a matter for despair as for recognizing that single textbook divergence times cannot be used without an evaluation of the uncertainties, which can be very large.

In a particularly imaginative approach to this very inconvenient sampling problem, Charles Marshall has shown that the completeness of a particular lineage's known fossil record can be used to make a statistical estimate of confidence intervals (millions of years above or below the known range in which we might have a 95% or 99% chance of finding the true end of the range). Thus, if a lineage is rich in occurrences, the chance that the first occurrence in the fossil record will be near the actual appearance is good. For organisms with poor or spotty fossil records (which describes some of the creatures we might find most interesting: for example, humans, platypuses, birds, insects, and soft-bodied Cambrian animals all pose this problem), confidence intervals can be very large, even tens of millions of years. Marshall provides a striking example of just how strongly confidence intervals can affect studies and interpretations of molecular clocks. Some estimates suggest that birds have slower rates of mitochondrial DNA evolution than other vertebrates. To test that hypothesis, Shields and Wilson estimated rates of mitochondrial DNA evolution for the goose genera *Anser* and *Branta.* They used an estimated divergence time for the two genera of 4–5 million years ago, which is the age of the earliest known fossils for each genus. Unfortunately, the goose fossil record is very sparse: four bones from three localities for *Branta* and a somewhat better fourteen bones from five localities for *Anser.* Marshall has plotted the confidence intervals for the origins of each genus based on the existing record. What his plot shows is that both genera may have significantly older origins than their first appearances indicate. What that does to the Shields and Wilson estimates is to make them compatible with either a rate equal to that of other vertebrates, as they conclude, or a slower rate. One cannot tell which to accept, because the time estimate is too poor to allow a convincing rate estimate to be made from existing data.

EXTINCT LINEAGES AFFECT MOLECULAR PHYLOGENIES

The promise of molecular phylogeny lies in the vast number of extant organisms from which DNA can be extracted and genes sequenced. The thirty-five living animal phyla are formally divided into seventy-five living classes. Several phyla are not sufficiently diverse to be divided into classes. In those cases order or family is the largest division made. It is hard to know what to make of animals that are not very disparate in morphology in terms of evolutionary distance. For example, the two orders of onychophorans are probably as old and distinct as classes in more varied groups, but they are not very distinct on the grounds of observable morphology. Among those phyla that are divided into classes, the classes are further divided into much larger numbers of orders and families. Some of these divisions pose important problems in systematics, such as the relationships among the 20 living mammal orders, that are impossible to resolve solely on morphological grounds. About half of the living phyla and many classes have long histories in the fossil record. The rest do not, and are prime candidates for phylogenetic reconstruction from gene sequence data. The power of this approach is amply shown by the success of such reconstructions for groups such as bacteria and protists for which morphological data have been inadequate to reveal relationships.

The reconstruction of phylogenetic histories by use of gene sequence data is limited in a peculiar way by the extent to which a clade contains extinct lineages. Living representatives of old lineages provide gene sequence data from which phylogenies can be constructed, but old groups contain many extinct lineages, known only from the fossil record. The problem this poses for understanding the evolution of major groups is not a trivial one. If the group being examined has been pruned too much by extinction, the consequences of not being able to sample extinct members can be confounding, or at least limiting to the depth of interpretation. The problem is well illustrated by the recent debates over the relationships between lungfishes, coelacanths, and the land vertebrate classes—amphibians, reptiles, mammals, and birds—that constitute the tetrapods. This is a major issue, because understanding the crucial steps of the vertebrate invasion of the land requires that we identify the true branching pattern for the ancestors of the tetrapods. Gene sequence data would seem an ideal means of doing this.

The issue of tetrapod origins is often presented in a somewhat peculiar let's-startle-the-natives way. Lungfishes, in overall body form, look like fishes, but in this style of storytelling, it is revealed to the multitudes that among the animals lungfish, trout, and human, the correct phylogeny is lungfish plus human, with trout a more distant branch in the bigger vertebrate tree. The illustration is intended to show that although lungfishes are more similar to other fishes in primitive features (plesiomorphies), they share derived features (synapomorphies) with tetrapods that put them closer to tetrapods in phylogeny if not in appearance. This has to raise the question of whether the adaptations of lungfishes to air breathing that resemble those of tetrapods are homologues representing a lungfish-tetrapod clade, or parallelisms between two clades that independently faced similar environmental challenges.

The existence of living fossils should, in principle, be very helpful in evaluating relationships. The discovery of the coelacanth presented us with a living animal that originated in the mid-Paleozoic and was thought to have been extinct for 80 million years. The uses made of it in phylogenetic studies illustrate that the use of living fossils is not straightforward. As Forey has shown, comparison of the living coelacanth with fossil coelacanths has helped us in better understanding the fossils, but even the availability of living animals has not solved the phylogenetic problems. Coelacanths have been proposed as the sister group of Chondrichthyes (sharks, etc.), of lungfishes plus tetrapods, of lungfishes alone, and of tetrapods alone. They also have some unique specializations. Once again morphology has left us with a confusing picture of its meaning that begs for a molecular solution. Several attempts to resolve the origins of tetrapods by using gene sequence phylogenies have been made. These studies have utilized genes from various tetrapods, ray-finned fishes, lungfishes, and the only other potential living fossil representative of the rhipidistians, or lobe-finned fishes, from which tetrapods arose, the coelacanth. Analyzing the sequence data has been difficult enough, but a deeper problem arises from the fact that in the Paleozoic the rhipidistians were very diverse, but most of those lineages are extinct and not accessible for gene sampling. That in turn profoundly affects the inferences we can draw, even if we have a firm molecular phylogeny of the living taxa in hand.

A variety of molecules and methods of phylogenetic inference have been used to assess tetrapod origins. Gorr and co-workers, for instance, used hemoglobins, and concluded that the coelacanth was the closest relative of the tetrapods. Their methods drew considerable fire from other investigators for two reasons (see Meyer and Wilson, Sharp and co-workers, and Stock and Swofford). The distance analysis of Gorr and co-workers utilized a cluster analysis method sensitive to branch length variation, and the idiosyncratic parsimony method that they used produces artifacts. When Sharp and coworkers redid the distance analysis, they used neighbor joining, which revealed the branch length differences and a new topology. Gorr and coworkers inferred a coelacanth-tetrapod clade on the basis of a very unusual similarity. They used both adult frog and tadpole β-globins, and found that the coelacanth mapped closest to the larval globins. Sharp and co-workers showed that adult frogs have long branches, whereas tadpoles and coelacanths have short branches. The affinity is an artifact of short branch lengths producing closer nodes in a cluster analysis. It also reveals something else: the dangers of using paralogous genes.

A rather different approach to the phylogeny of the coelacanth has been taken by Litman and his colleagues (reported in a paper by Amemiya and co-workers). This group has focused on the evolution of the organization of the immunoglobulin genes among vertebrate groups. Immunoglobulin genes have not been detected in cyclostomes; thus, the most primitive immunoglobulin gene arrangement found thus far has been characterized from sharks. Sharks and other chondrichthyans have a clustered arrangement of the elements of the immunoglobulin heavy chain gene. This is very different from the extended multiple unit arrangement characterized in mammals and found as well among ray-finned fishes and tetrapods. The coelacanth gene possesses an organization combining features of both shark and bony fish genes. Amemiya and co-workers interpreted the features coelacanths share with sharks as plesiomorphic, and the features they share with ray-finned fishes and tetrapods as synapomorphies. They concluded that the coelacanth thus shares a sister group relationship with a tetrapod-plus-ray-finned clade that is intermediate between cartilaginous and bony fishes. This suggestion might remove coelacanths from contention as a tetrapod sister group, but does not solve the problem of tetrapod origins. This work is also interesting in showing that information on genomic organization distinct from sequence data also exists and can yield phylogenetic information.

Slowly evolving mitochondrial gene sequences were used by Meyer and Wilson to sort out the ray-finned-coelacanth-lungfish-tetrapod branching problem. They used partial sequences from both the 12S ribosomal RNA gene and cytochrome *b* genes. Their results supported a lungfish-tetrapod clade, with coelacanths as the next branch out, and ray-fins even more distant. Meyer and Dolvin, as well as Normark and co-workers, evaluated the results with full-length sequences of these genes, and found less solid support for this tree than did Meyer and Wilson. Yokobori and co-workers, using mitochondrial cytochrome oxidase I sequences, suggest equal support for a coelacanth-lungfish clade as for a lungfish-tetrapod clade. So, the molecular phylogeny is not yet out of the woods. The most telling analyses come when we assume that Meyer and Wilson's result is correct and see where it leads when the fossil record and morphology are also considered.

Meyer and Wilson looked at the morphological features of the living lungfishes, tetrapods, coelacanths, and ray-finned fishes to identify true shared derived features as well as parallelisms and convergences. They inferred,

based on their phylogeny, that the features that modern lungfishes and tetrapods share relating to air breathing, circulatory systems, and locomotor structures are all homologous features derived from a common ancestry. However,
what the fossil record shows is that the living lungfishes, coelacanth, and
tetrapods represent only the few surviving clades of the rhipidistians, which
in the Devonian were a diverse group. Marshall and Schultze assumed that
Meyer and Wilson's molecular phylogeny was correct, and evaluated it in
light of the "exquisite fossil record of the earliest lungfish." They came to
three remarkable conclusions. The first is that the features present in modern
lungfishes were not present in the earliest lungfishes. Modern lungfishes are
air breathers; the earliest lungfishes had gills and were deep-sea forms. Second, the morphology of the earliest lungfishes is relatively unimportant in
assessing the vertebrate water-to-land transition. Third, very few of the morphological features modern lungfishes share with tetrapods are really shared
derived features: rather, they are convergences. An especially striking demonstration of that conclusion is that the earliest known tetrapod, *Acanthostega*,
from the upper Devonian of Greenland has been shown by Coates and Clack
to have had functional internal gills. It probably also possessed lungs, which
were a primitive feature shared by bony fishes. Tetrapods have lost their gills
in becoming more terrestrial. The first tetrapods thus convergently resembled
modern lungfishes more than they resembled the earliest lungfishes.

The reason that the pairing of lineages by molecular data provides less
insight into the events of the past if many taxa are extinct is shown in
figure 5.1. All of these phylogenies relate the living lungfish, coelacanth,
and tetrapod lineages to extinct rhipidistians. The trees are all consistent with
the finding that modern lungfishes are the sister group of tetrapods. But note
how different the trees are. If tree A or B is correct, lungfishes are the sister
group of tetrapods, and the features in which lungfishes are similar to tetrapods are potentially derived from a common ancestry. If tree C is more
correct, as appears to be the case, the features of modern lungfishes are
irrelevant to inferring anything about the tetrapod conquest of the land. The
real sister groups are extinct, and the few survivors are outliers. Placing the
living groups in phylogenetic trees using molecular data might produce a
correct topology for them, but would be unhelpful in understanding the actual
phylogeny of the group's radiation. Analogous difficulties may occur in attempts to determine molecular phylogenies of echinoderms. There were
twenty-five lower Paleozoic classes of echinoderms, but there are only six
living ones. The living classes did not appear until late in the Cambrian or
in the succeeding lower Ordovician. Most Cambrian forms belong to unusual
extinct classes. The same problem may apply to arthropods, which the Burgess Shale shows to have contained far more lineages than the three living
subphyla. We do not know how deeply this problem applies to the metazoan

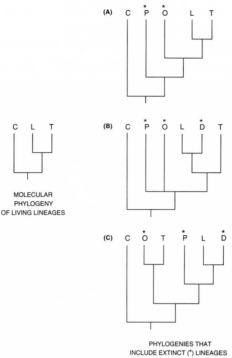

Figure 5.1. The effect of the pruning of stem lineages by extinction. The phylogeny inferred for living tetrapods, lungfishes, and the coelacanth on molecular grounds by Meyer and Wilson is compared with three morphological phylogenies that include extinct as well as living lineages. The third phylogeny is the preferred one. The molecular phylogeny is consistent with all three more complete phylogenies when the extinct forms are removed. Thus, even if a molecular phylogeny for living members of a group is correct, it may not be informative about ancient divergence patterns when most of the ancient lineages are extinct. C = coelacanth; D = *Diadolepis;* L = lungfishes; O = osteolepiforms; P = porolepiforms; T = tetrapods. (After Marshall and Schultze 1992.)

phyla. If most of the radiation involved phyla with living representatives, we are on solid ground. If there was a large radiation of extinct phyla, we will have the obvious trouble in sorting out their relationships and understanding the patterns of evolution of the features of the primitive metazoan groups and the origins of living phyla from among them.

FOSSIL GENES

In Crichton's novel *Jurassic Park,* the most outrageous dreams of molecular evolutionists are funded by a billionaire who wants live dinosaurs for his Mesozoic theme park. He gets them. DNA is recovered from dinosaur blood cells preserved in the bodies of blood-sucking insects trapped in amber. The sequences of these fossil genes are read, and missing bases are filled in from homologous genes from frogs or other related animals. Finally the DNA is implanted into artificial eggs containing cellular machinery derived from crocodile eggs. With a little plausible hand waving it all works beautifully. Well, almost. Many of the protagonists of *Jurassic Park* come to deeply, if briefly, regret their success in retrieving carnivorous dinosaurs. This is another Faustian science fiction story in which scientists are led astray by inappropriate temptations and pay the ultimate price for the bargain. However, few of us will face the penalty of being eaten by revived extinct animals. In most cases, the remains of extinct organisms, although they may provide a great deal of morphological information, have lost most or all of the information directly encoded in their DNA. Nonetheless, the dream of restoring dinosaurs to life from fossil DNA, as achieved in the fictional world of *Jurassic Park,* reflects a real interface between paleontology and molecular biology whereby some gene sequence information can be recovered from ancient organisms. Encouragement comes from the remarkable instances in which DNA has been extracted from extinct forms and has yielded phylogenetically meaningful sequence data. However, the DNA recovered from amber and other ancient sources is only a minuscule part of the content of a complete genome. It is unlikely that enough sequence data will ever be recovered from fossils to understand the logic of development of extinct species. That any significant genetic information can be recovered at all is remarkable enough. We may be able to recover some Cenozoic and Mesozoic DNAs (including dinosaur sequences), but DNA is progressively lost with age. The genes of even older forms—trilobites, eurypterids, all the wonderful Cambrian oddities—are presumed to be permanently lost to us. Fossil DNA is unlikely to help us unravel the origin of phyla, although it may supply information about subsequent events in body plan evolution.

The recovery and sequencing of DNA from paleontological sources has meant reaching for the edge of technical feasibility, so it is wise to pay attention to skeptics. Lindahl, in considering the processes that degrade DNA, has suggested that it is unlikely for DNA to last significant (geologic) lengths of time without being dehydrated and protected from oxygen and bacteria. Amber potentially provides these conditions, but it is unlikely that DNA in fossils in hydrated sediments, such as in ordinary sedimentary rocks, meets these requirements. Lindahl also suggests that given the extreme sensitivity of the polymerase chain reaction method to contamination by minute amounts of modern DNA, the DNA sequences obtained from amber or other fossils are not unequivocally ancient. Nevertheless, in the short time that I have been working on this chapter, reports of the isolation of fossil DNAs have expanded dramatically. By the time the book is read, ancient DNA studies will be a considerable enterprise, and people will be bemoaning the destruction of valuable amber specimens. Clearly, good controls will have to be devised to guard against contamination from modern sources, and replication will have to be done by more than a single laboratory.

The pioneering attempts to recover genetic information from fossils were aimed at proteins. Jerold Lowenstein and his collaborators developed sensitive assay methods that used radioactively labeled antibodies to detect specific proteins in the remains of extinct mammals. Their method could be used quantitatively to measure relatedness between proteins. With these methods, Lowenstein and co-workers were able to show that albumin recovered from frozen baby mammoth tissue was equally similar to that of living African and Asian elephants. Shoshani and co-workers showed that elephant-like albumin was present in fossil mastodon bones. Rainey and co-workers showed that the extinct Steller's sea cow was closely related to the living dugong. Sea cows and elephants are members of an old clade within the mammals. The availability of these ancient proteins allowed the generation of an immunological phylogeny of the elephants, sea cows, and their recently extinct relatives. The extraction of phylogenetic information from protein residues is no longer the approach of choice, but has been displaced by the recovery of DNA from fossil sources. Nevertheless, Loy has reported that protein residues remaining on ancient stone tools can be probed with antibodies to reveal what parts of the phylogenetic tree our ancestors were carving up. The adequacy of immunological methods and even the survivability of immunologically recognizable proteins on stone tools has been experimentally questioned by Eisele and found wanting.

The first demonstration that DNA could be recovered from extinct creatures came in 1984 with the cloning by Higuchi and co-workers of short pieces of mitochondrial DNA from the hide of an extinct zebra, the quagga, preserved in a museum since the nineteenth century. Sequence data from the recovered quagga DNA fragments show similarity to the living Burchell's zebra. Similarly, DNA sequences recovered by Thomas and co-workers from museum specimens of the marsupial "wolf," the now probably extinct thylacine, show this animal to be related to other Australian marsupial carnivores.

DNA has been isolated from ancient mummified human remains recovered from archaeological contexts (a paper that contains a "materials and methods" section featuring mummified ancient Egyptian priest liver DNA is really hard to top). In these early studies, DNA was also recovered from recently extinct animals, including a ground sloth represented by a 13,000-year-old skin preserved in a cave and a 40,000-year-old frozen mammoth. Unfortunately, as demonstrated by Pääbo and his colleagues, DNA even a few thousands of years old is usually not in prime condition. Old DNA has undergone considerable chemical modification from such processes as hydrolysis or oxidation, which results in breaks in the DNA strand, modification or loss of some of the bases or sugars, and cross-links between strands. Such DNA is very difficult to clone directly because damaged DNA is not replicated well by the bacteria used in cloning, and, in addition, damaged DNA that is replicated becomes subject to repair processes in the bacteria that can change bases in the sequence.

The best successes in recovering ancient DNA have come from the use of the polymerase chain reaction technique. PCR employs short DNA primers that can bind to particular DNA sequences. Once the primer has bound, a DNA polymerase can use the primer as a site at which to initiate the copying of the strand to which it has bound. The use of two primers directed in opposite directions allows both strands of the sequence lying between the primers to be copied. What makes this technique so powerful is that repeated rounds of these reactions can be run by melting the complementary strands, reannealing the primers, and doing another round of DNA polymerization. Repeat cycles are made possible by the use of thermostable DNA polymerases derived from thermophilic bacteria, which live at temperatures approaching the boiling point of water; thus, the heating required to melt complementary strands between rounds of DNA replication does not denature the enzymes. This repetition allows an enormous amplification of the desired DNA sequence. Mitochondrial DNAs are used in most studies of tissues derived from extinct animals because these DNAs are present in high copy number in tissues, and because the primer sequences giving access to the DNA sequences can be readily predicted. The damage suffered by ancient DNAs results in only short fragments, on the order of a few hundred bases, being amplified. The presence of overlapping damaged DNA templates results in jumps by the polymerase between damaged DNAs during the polymerase chain reaction and yields a longer mosaic product. That in itself should pose no problem if only homologous template strands are present. However, jumping is also accompanied by insertion of adenosine in place of lost pyrimidine bases, and can thus yield sequence artifacts. The use of primers that give products in the 150-base-pair range should avoid jumping. However, such short segments are a far cry from the long DNAs of *Jurassic Park*. As reviewed by Pääbo and by Soltis and Soltis, there are a variety of problems in amplifying DNA from ancient materials, and controls to guard against contamination by modern DNAs also need to be addressed to make ancient DNA reports reliable.

The use of PCR techniques has blended the forensic with the phylogenetic, and indeed the problems are much the same. This point is nicely made by the use of PCR by Gill and co-workers to identify the bodies of the Russian royal family, murdered by the Bolsheviks in 1918 and hidden in a shallow grave. Pääbo and his co-workers have determined DNA sequences from

7,000-year-old human brain tissue, and more recently, Handt and co-workers analyzed the DNA relationships of the Tyrolean ice man. This roughly 5,000-year-old late Neolithic man, with his clothing and equipment, was found frozen in an ice field in the Alps in 1991, and became an instant celebrity. Handt and co-workers carried out a PCR analysis of the variable control region of his mitochondrial DNA, which revealed four significant points. First, the amount of DNA recovered was six orders of magnitude lower than for fresh tissue; therefore, only mitochondrial DNA present in about 500 copies per cell was likely to have survived. Second, the DNA was largely degraded to about 150-base-pair lengths. Third, there was significant contamination by modern human DNA, so that sampling and analysis had to circumvent the contamination. Finally, the ice man was most closely related to modern Europeans.

Even more significant for phylogenetic purposes, the polymerase chain reaction technique has been successfully applied to fossil bones. Janczewski and co-workers have obtained DNA sequences from both mitochondrial and nuclear genes from 14,000-year-old saber-toothed cat fossils preserved in the Rancho La Brea tar pits of Los Angeles. The 10,000–38,000-year-old fossils of the tar pits are not very ancient by geologic standards, but the result is significant for two reasons. First, the animals from which the bones were obtained underwent postmortem decay, but the osteocytes deep in the bone were apparently protected from bacterial action until they were finally dehydrated and permeated by organic compounds in the tar. This discovery suggests the possibility that DNA has survived in the fabric of other ancient bones. The second important thing about this result is that the tar pit fauna is very diverse, and includes 465 animal and 159 plant species. Of the 59 mammals, 23 are extinct species, including some of the Pleistocene giants such as dire wolves, saber-toothed cats, American lions, mammoths and mastodons, and giant ground sloths. The tar pit fossils offer a remarkable window into the Pleistocene, and may let us sample some of the genetic features of its recently extinct fauna. These samples may provide us not only with phylogenetic information, but possibly also with information on rates of gene evolution, population structure, and even on viral evolution and infection if appropriate probes for retroviruses can be devised. The saber-toothed cat, incidentally, turns out to be a close relative of the recent large pantherid cats. Its phylogenetic placement is of obvious interest. Its teeth and body structure indicate a hunting style very different from that of any living cat. These powerful animals were probably slow ambush hunters of very large mammals. Their sabers enabled them to stab deeply to produce profusely bleeding wounds. The saber-tooth style has evolved three times independently, in extinct South American marsupial predators, in the extinct catlike nimravids, and in the true cats.

Studies of other recently extinct animals are possible because subfossil bones are readily available from caves and other protected sites. Methods for extracting amplifiable DNA from ancient bone are becoming more reliable. For example, Cooper and co-workers have investigated the potential relationship of a living New Zealand flightless bird, the kiwi, to the extinct moas by using mitochondrial 12S ribosomal RNA genes from moa bones and tissues. Surprisingly, the kiwi seems to be more closely related to Australian flightless birds than to the moas. Höss and Pääbo have obtained mitochondrial DNA sequences from a 25,000-year-old extinct Alaskan Pleistocene horse, *Equus hemionus,* that show it to be related to, but not identical to, the living domestic horse. Happily, DNA sequences have finally been obtained from that touchstone of the Pleistocene, the mammoth, by Höss and co-workers and by Hagelberg and co-workers. The sequences from several Siberian mammoths ranging in age from 10,000 to 50,000 years are similar to those of both living elephant species. The sequences of 40,000-year-old cave bear DNA determined by Hänni and co-workers place this species as a sister group of the European brown bear.

DNA may survive in other surprising venues. Bonnichsen and co-workers have shown that sequenceable DNA survives in buried hair from archaeological sites. Loy suggested that DNA might survive on ancient tools, and was the first to extract DNA from a 2,000-year-old tool from northern Canada. It would seem highly probable that DNA on an exposed stone surface would in short order become free lunch for any passing soil microbe. Yet the surfaces of tools in some cases preserve protein residues. It may be that blood proteins in association with microscopic cavities on the tool surface denature to form an anoxic and hydrophobic microenvironment that can resist the ravages of time and life. My colleague Bruce Hardy is studying 35,000-year-old Neanderthal-made tools from France for DNA. He has recovered mitochondrial DNA sequences consistent with the identified animal bones found at the site. Pork seems to have been a favorite.

Incidentally, while on the subject of frozen mammoths, there are those persistent stories of their scientific excavators having mammoth steak banquets. The accounts of fossil mammoth hunters such as E. W. Pfizenmayer, one of the excavators of the famous Beresovka mammoth, make it clear that a great deal of bacterial degradation occurred during the freezing of these big carcasses, making them very unappetizing. The only claim by a scientist to have eaten an ancient carcass that I've come across is Björn Kurtén's account of a 36,000-year-old Alaskan frozen bison stew. He says the meat was "fresh with an unmistakable beef aroma."

Only a few really old DNA sequences have been claimed to date. The first was a chloroplast sequence reported by Golenberg and co-workers from 17–20-million-year-old organically preserved magnolia leaves from the Miocene Clarkia beds of Idaho. Although exciting, the result is still moot, and the recovery of DNA from organically preserved fossils in ordinary geologic settings has still to be unequivocally demonstrated. Another research group, Sidow and co-workers, used the polymerase chain reaction method to recover DNAs from leaves preserved in the Clarkia bed, but recovered only bacterial sequences. The plot has thickened further with a report from a third laboratory, by Soltis and co-workers, that a chloroplast DNA sequence was obtained from a fossil bald cypress from the same formation. The reported sequence is similar but not identical to that of living bald cypress.

Although preserved DNA potentially exists in a variety of fossil settings, perhaps the most extraordinary preservation modes involve ambers of various ages. Amber forms from tree resins, which quickly surround their small, generally arthropod, victims. The terpenes in the resin have a bactericidal action, and during polymerization the tissues are dehydrated. An entomologist, George Poiner, was set to looking at amber insects by finding a piece of Baltic amber while strolling on a Danish beach. It had been generally thought that only an external mold of the preserved insects remained, but as shown by Poiner and Hess, extraordinary morphological detail is preserved in ambers that extends to cellular structures as well. The isolation of dinosaur DNA from the stomachs of mosquitoes preserved in amber was the basis for the resurrection of dinosaurs in *Jurassic Park*. With the recovery and sequencing of DNA from an insect preserved in amber, DeSalle and his co-workers showed that ambers, with their rich preserved faunas of small animals, contain a potentially diverse library of ancient DNAs. DeSalle and his co-workers recovered DNA from a large termite, *Mastotermes electrodominicus,* preserved in 25–30-million-year-old amber from the Dominican Republic. This evocatively named termite is closely akin to the living *Mastotermes dawiniensis,* now found only in Australia, an evidently relict distribution of a formerly widespread termite family. The recovered sequences were short spans of both mitochondrial and, significantly, nuclear ribosomal DNA. These were adequate to provide analyzable sequences, which showed that the fossil *Mastotermes* is distinct from its living congener, but more closely related to it than any other termite.

In the study of the amber-preserved termite, negative control reactions indicated no contamination from sources in the laboratory environment. Some flylike sequences were recovered from the DNA prepared from the fossil termite. These were regarded as contaminants and disregarded. However, the study was done in the American Museum of Natural History, an old building with a potential for contamination by bits of old insects. The fossil sequence was similar to that of the living *Mastotermes dawiniensis,* but differed by 17 bases in a 100-base sequence. That is strong evidence that the result is correct, although it does not wholly eliminate the contamination issue. The contaminant DNA might have come from some other termite species. Only duplication of studies like this under stringent conditions can assure investigators that they are dealing with old DNA and not modern contaminants.

There are many ambers available, including Mesozoic ones, that contain well-preserved fossils. The potential for recovery of DNA will no doubt generate a spectacular industry for molecular paleontologists over the next few years. Some of the older ambers, from Mesozoic times, offer the potential of yielding sequence data from forms ancestral to living groups as well as species belonging to living clades. DNA apparently can be recovered from these old ambers. Within only a few months of the report by DeSalle and co-workers on DNA sequences recovered from the amber-preserved termite, Cano and co-workers reported the isolation and sequencing of DNA from a 120–135-million-year-old weevil preserved in a Mesozoic amber, the oldest amber known to contain insects. The early Cretaceous weevil belongs to a primitive family, and the sequence recovered from it matches closely that of a living member of the same family of weevils. Sequence data from very old ambers could be highly significant in testing predictions of ancestral sequences inferred from sequences of living descendant species. The sequences of very old DNAs have the potential to be more useful in recovering phylogenies than working backward from sequences from taxa at the tips of the branches of the tree. The fossil termite sequence, when added to the tree constructed from living species, allowed a test of the hypothesis that termites are most closely related to roaches. Use of sequences from a living termite did not give an unequivocal answer. The fossil termite sequence pulled termites away from roaches, and suggested that the traditional tree with its roaches-gave-rise-to-termites scenario is wrong.

The limits of information recovery from fossil DNA have yet to be explored. Even in the ideal case of ambers, the amount of information that can be recovered will attenuate rapidly with age. Some ambers with fossil algal and fungal cells are known from Triassic rocks, but the oldest known animal fossils in amber are of Jurassic age. Most of the amber fauna is limited to insects. If DNA can be recovered from cells embedded in fossil bone or other hard tissues, ancient vertebrate and marine invertebrate groups might also yield gene sequence data. DNA is currently being sought in dinosaur bones.

The amount of sequence recoverable is not known. Thus far, only repeat sequence ribosomal and mitochondrial genes have been exploited. These represent only a tiny fraction of the theoretically recoverable information in a genome. The most severe limits to information recovery from fossil DNA, however, will reside not in the simple loss of variable portions of the DNA, but in the destruction or mutation of the information in recovered sequences. Individual base pairs can be deleted or mutated and thus may yield incorrect

sequences. Such damaged DNA may also increase the error rate in the copying of DNA by the polymerase chain reaction method.

SUMMING UP: PHYLOGENY AND THE EVOLUTION
OF DEVELOPMENT

Phylogenetic information is vital in three major respects. First, a knowledge
of phylogenetic relationships among distantly related taxa enables us to have
an overview with which we can map relationships among organisms. Such
a phylogeny allows us to map the evolution of developmental features or
particular regulatory genes onto the major branching patterns. It also allows
us to judge how closely related our experimental model systems are to other
organisms to which we may want to extrapolate genetic or developmental
features. It is our knowledge of phylogeny that allows us to decide which
features are likely to be homologous and which have been acquired convergently. The distinction is crucial.

The second major use of phylogenetic information is that, combined with
the appropriate paleontological data, it allows us to infer rates of change of
genes or morphological features. These estimates can have significant bearing
on evolutionary hypotheses. Finally, and most significantly, as I'll discuss
more thoroughly in chapter 7, phylogenetic knowledge is basic to any meaningful investigation of the evolution of developmental processes or features.
Without such knowledge one can only guess at the polarity of change of any
feature, whether a gene or a bit of morphology. Polarity lets us infer the
primitive state versus the derived state, and it allows us to know whether a
feature has been lost in one of the lineages. These are all such basic parts of
our study that no investigation in the evolution of development should be
undertaken without establishing a phylogenetic history. Cladograms and trees
need to become as familiar to us as embryonic model systems or Hox genes
in our discussions of the evolution of development. However, phylogenies
of any kind must be treated in an informed manner. Uncritical acceptance of
phylogenies can lead the unwary to preposterous conclusions.

Finally, although this chapter is intended to provide caution about the reach
of molecular data in phylogenetic studies, it is also intended to look forward
to new applications and new sources of molecular data. My discussion of
ancient DNA studies betrays my enthusiasm, but I consider them important
because they represent a wholly novel source of phylogenetic information.
That any DNA should survive in the geologic record and that polymerase
chain reaction methods would make rare bits of DNA accessible for sequence
analysis was completely unexpected a decade ago. That ancient DNAs can

answer questions about evolutionary events of an antiquity sufficient to tell us about the origins of phyla or even of many classes still appears improbable. However, DNAs even as old as can be obtained from Mesozoic sources can provide important data on the origins of some important major living groups, such as the social insects. Ancient DNAs might also directly provide sequences that can be used to test predictions drawn from sequences obtained from living animal groups. Further development of techniques for analyzing DNA and for phylogenetic inference may only be poising us to effectively read evolutionary information from animal genomes.

6

The Developmental Basis of Body Plans

As Gregor Samsa awoke one morning from uneasy dreams he found himself transformed in his bed into a gigantic insect.

Franz Kafka, *The Metamorphosis*

BODY PLANS AND DEVELOPMENTAL BIOLOGY

To this point, animal body plans have occupied us as a historical problem. The tracing of their history is critical because it presents us with necessary information about rates and patterns of evolution. Although much about the origin of animals remains obscure, we know that the radiation of phyla was rapid, explosive in geologic terms, and that it was qualitatively unique. All of the known animal body plans seem to have appeared in the Cambrian radiation. Although our tracing of the phylogenetic relationships among phyla is still inadequate, we feel that we are closing in on the history of body plans. Now we can approach the central evolutionary problems they pose. These are paradoxical problems of opposites. On one hand we celebrate the rapid and almost protean changes produced by half a billion years of animal evolution that have populated the planet with such a rich diversity of forms, such elaborate neural systems and behavior, such improbable convergences, and so many novel structures. But we face an equally great stability of underlying body plans. Profound change within a conserved body plan points obscurely, but powerfully, to a richer role for the interplay between development and evolution than we have suspected.

The recognized metazoan body plans were rapidly and firmly integrated relatively early in the history of animals. But they cannot have been firmly set until the metazoan radiation was already under way. We know that the phylogeny of the animals is a branching one, a bush spreading from a single ancestor. Thus, now-disparate phyla once shared common ancestors. Those ancestors had body plans of some sort, and those ancestral designs had to have been transformed to yield the descendant body plans of the Cambrian phyla. Why should that have occurred only as the phyla became established in the Cambrian, and not since? It is possible that until the final establishment of definitive phylum body plans, developmental integration was poor. Because the ancestors of the phyla did not possess tightly integrated body plans, they were free to evolve in fundamental ways. A great deal of evolution of animal morphology has occurred since the Cambrian. Most have involved evolution of novel features within recognizable body plans, but some have involved modification or addition of elements. Phylum-level stasis suggests that perhaps later evolution did not reach as deeply into the basic constructional rules of development as did pre-Cambrian evolution.

Body plans are emblematic of the entire problem of how morphology has evolved. To solve this problem we must understand the integration of the evolutionary mechanisms of variation and selection with the developmental mechanisms that translate genetic information into body form in each generation. We can begin to understand body plans from the perspective of developmental biology, and we can attempt to pick apart some striking apparent paradoxes.

This chapter has two purposes. The first is to consider the paradoxical evolutionary history of phylum-level body plans since the Cambrian radiation. The paradox arises because there would seem to have been opportunities for the appearance of new phyla resulting from episodes of mass extinction and from the conquest of the land by animals. Nonetheless, all phyla are old, and their basic body plans have been retained since their Cambrian appearances. Extensive evolutionary changes have occurred within phyla, but the underlying patterns have been conserved. The second purpose of the chapter is to consider the developmental phenomena associated with the conservation of body plans and to evaluate the developmental mechanisms that would appear to make for both conservation and change.

WHY HAVE NO NEW PHYLA APPEARED
SINCE THE CAMBRIAN?

The metazoan radiation occurred over half a billion years ago. One of its most extraordinary features is that all metazoan body plans seem to have originated in the few million years of the early Cambrian. The Cambrian also produced the largest number of problematic forms interpretable as extinct phyla. Only two such problematic phyla are known from fossils from younger periods, and their times of origin are unknown. They too may have had an as yet undiscovered Cambrian ancestry. The lack of origination of new phyla since the Cambrian has a very curious corollary. As Gould, as well as Jacobs, has observed, nothing equivalent to the Cambrian radiation happened at any other time, even following mass extinctions such as the terminal Permian event.

We have to ask whether the exclusively Cambrian origin of metazoan phyla is real or artifactual. The major skeletonized phyla have fossil records that start in the Cambrian. Until recently, statements of Cambrian origins for the many other phyla without such records have been heavily suffused with poetic license. Half of the living phyla (i.e., mesozoans, platyhelminths, rotifers, gnathostomulids, gastrotrichs, acanthocephalans, tardigrades, kinorhynchs, nematodes, loriciferans, urochordates, phoronids, chaetognaths, echiurids, pogonophorans, sipunculans, and entoprocts) either lack a significant or reliable fossil record or (as noted by Erwin and his co-workers and in an unpublished compilation provided to me by Jan Bergström) are first recorded from times long past the Cambrian. These phyla might have arisen in the Cambrian radiation, but no current fossil evidence excludes a later origin. More detailed contemporary studies of the Cambrian radiation are revealing unexpected fossils that are shortening this list; four phyla have been recently removed. Walossek and co-workers have found Cambrian tardigrades, Walossek and Müller have discovered Cambrian pentastomids, Hou and Bergström have reported Cambrian nematomorphs, and a possible Cambrian ctenophore is known from the Burgess Shale (see Briggs, Erwin, and Collier). Gene sequence phylogenies favor an early origin for all the phyla sampled so far. Those data, combined with the unequivocal evidence from paleontology that the phyla with good fossil records had an early Cambrian origin, make a strong case for early origins for most phyla.

The reason for the single radiation of phyla is not intuitively obvious. Ten times as much time has elapsed since the Cambrian radiation than the most generous estimate of the duration of the radiation, and opportunities for the appearance of new phyla have again occurred.

MASS EXTINCTIONS AND BIG OPPORTUNITIES

Animals have faced more than just the gradual attrition of background extinction. There have been numerous metazoan mass extinctions in the past half billion years. The terminal Cretaceous mass extinction, justly celebrated for its efficient execution of the dinosaurs, appears to have involved an extraterrestrial event, an impact with a comet on the order of 10 kilometers in diameter. The collision left a huge crater (about 300 kilometers in diameter, as estimated by Sharpton and co-workers) buried beneath the Yucatan Peninsula of Mexico. The impact caused massive worldwide forest fires and oceanic pollution, and the dust it raised produced a freezing darkness that may have persisted for weeks. Over half of the living species became extinct as a result. That this catastrophe was caused by an impact with a large extraterrestrial object has led to suggestions that a cyclic pattern of mass extinctions could be the result of some process that periodically affects the orbits of comets in the Oort Cloud, where comets spend their quiet moments, and sends them into the inner solar system like cosmic bullets. Such collisions have occurred throughout geologic time. I have visited one giant impact site, the evocative Siljan Ring in Sweden, a 55-kilometer crater with a rim of upturned limestone largely filled by a semicircular lake. That impact occurred coincidentally with a Late Devonian mass extinction. Whether that particular smoking gun caused the mass extinction is still moot. Raup and Sepkoski have suggested that mass extinctions occur with about a 26-million-year periodicity, consistent with an extraterrestrial cause. Whether such a cause exists is still debated, and mass extinctions may have resulted from idiosyncratic events. Not all mass extinctions have obvious cosmic connections. The Permian mass extinction seems to have had terrestrial causes, arising from climate and sea level changes resulting from the formation of the supercontinent Pangaea. Benton, using a new database in analyzing the history of diversity of life, has concluded that no mass extinction periodicity exists.

Mass extinctions are not obviously the result of inferior body plans. Raup has suggested three kinds of mechanisms that might operate during mass extinctions. First is the "field of bullets" hypothesis, which suggests that extinction is random with respect to fitness. Species are simply mowed down; duration is a matter of luck. Second is the "fair game model," in which extinction is selective in a Darwinian sense, with the most fit species surviving. This model fits most closely with our ideas about what happens in background extinction, but suggests that selection is acting with a higher intensity in mass extinctions. Third is the "wanton extinction model," in which extinction is selective, but not for features that lead to better adaptation to the organism's normal environment.

Raup has argued that if the field of bullets hypothesis were correct, large groups would weather bad luck far better than smaller ones. However, that is often not the case. Selectivity of some kind must operate even in catastrophic extinctions. If fitness in the normal environment were advantageous, successful groups would be unlikely to die out. Yet dominant groups are often decimated. The notion of wanton extinction, in which selection operates, but not on features that make for successful adaptation under normal circumstances, is supported by several examples. One of the hallmarks of the terminal Cretaceous extinction is that small land animals survived whereas all the big ones died out. In the marine realm, the small organisms, the phytoplankton, were decimated, and some hitherto highly successful metazoan groups died out. Others survived with little apparent loss. In the end Permian extinctions summarized by Erwin, the highly diverse brachiopod families were reduced by 90%, but the clams were unaffected. Nevertheless, there was nothing magical about being a mollusk. Of the cephalopods, almost all ammonites died out, whereas the related nautiloids were unaffected. Soon after the extinction event the ammonites underwent a great radiation to recover their dominance. The brachiopods did not. Finally, the sea urchins nearly perished, with only one or two genera surviving. Nevertheless, they went on to become the dominant echinoderms of modern oceans. The crinoids, which had been far more diverse than the echinoids during the Paleozoic, never fully recovered. The survivors of mass extinctions are not necessarily superior to the victims.

The Permian mass extinction, with its spectacular decimation of marine animals to about 5% of their earlier diversity, set the stage for a reradiation of metazoan life. It was the most profound extinction event known, and it reduced marine animal diversity to levels not seen since the early Cambrian. Consequently, it provided the best chance for a second experiment in phylum origins. With the enormous reduction in diversity, whole suites of niches were vacated. The survivors were not necessarily the forms best adapted to occupy those niches. Life rebounded rapidly in the Triassic, but although the Triassic was as long in duration as the Cambrian, and numerous new families and even orders appeared, there were no new phyla.

INVASION OF THE LAND

Mass extinctions offered vacated niches for experiments in body plan origins. The expansion of life into wholly new habitats offered other opportunities. The greatest of these was the movement of life onto the land. This invasion was well under way by the early Silurian, 430 million years ago. The structure of the invasion is being actively investigated, and good discussions of this unique early world are presented by Gray and Shear, Jeram and co-workers, Labandeira and co-workers, Shear, and Shear and Kukalová-Peck. The expansion of life on land involved both higher plants and animals. However, its evolutionary consequences were instructively different in these two kingdoms.

The earliest land animals were small arthropods, including centipedes and the spiderlike trigonotarbids, revealed as body parts that can be dissolved with acid out of terrestrial shales of that age. Wingless insects, true spiders, scorpions, mites, and other arthropods followed not long after. The ecological setting of these animals may have resembled the forest floor litter communities of today. There are no herbivorous forms, just detritus feeders and carnivores. Vertebrates appeared on land late in the Devonian, about 365 million years ago. By the end of the following Carboniferous, about 280 million years ago, more complex ecosystems included herbivorous and carnivorous vertebrates, 2-meter-long millipede-like arthropleurids, and the first winged insects. Animals thus took 150 million years to evolve fully terrestrial ecosystems. During the same interval, land plants appeared and evolved a fully terrestrial vegetation.

It is remarkable that, as tabulated by Little, only seven animal phyla

evolved adaptations allowing them fully terrestrial lives above the soil surface. These conquerors of the land are the platyhelminthes, nemertines, annelids, mollusks, onychophorans, arthropods, and vertebrates. Several other
phyla of small animals that live in soil in essentially aquatic environments
include rotifers, nematodes, nematomorphs, and tardigrades. Despite spectacular morphological innovations (notably by insects and vertebrates), no new
phylum-level body plans arose to accompany the movement of animals into
an ecospace as empty as the marine realm at the end of the Precambrian. Most
Paleozoic terrestrial arthropods and vertebrates were recognizably similar to
their marine relatives. Scorpions are all terrestrial now, but most Paleozoic
scorpions were aquatic. (We're lucky these days that we have only mosquitoes to contend with on visits to the ol' swimmin' hole.) Uniramous arthropods—millipedes, centipedes, and insects and other hexapods—are another
story. Students of arthropod phylogeny, including Whittington, Manton and
Anderson, and Kristensen, have pointed out the obscurity of unirame origins.
Unirames appeared in early terrestrial faunas. The Cambrian lobopods, like
Aysheaia, are thought to be onychophorans, and ancestral to unirames by
some hypotheses. No genuine marine unirames are known. The origin of
unirames, particularly the hugely diverse hexapods, may be the closest event
we have to the origin of a new animal body plan as a consequence of expansion onto the land. However, unirames are still very much arthropods.

Although terrestrial animals have close and structurally similar marine
relatives, there is nothing like the land plants among marine plants (aside
from those land plants that secondarily took up life as seaweeds later in their
careers).The closest primitively marine relatives of land plants are the green
algae. The adaptations of land plants set them apart from any marine forms,
and the body plans of land plants are unique. The major groups of plants are
called divisions rather than phyla. Apparently the number of divisions is in
flux at the moment in the minds of botanists. However—and I have consulted
with my paleobotanist friend David Dilcher on this—there are probably about
seven well-supported disparate body plans, and thus divisions, among land
plants. All of these body plans are of a much more complex anatomical grade
than those of algae. Plant divisions include bryophytes, the mosses and their
kin; lycopodophytes, the club mosses that provided huge Coal Age trees;
rhiniophytes and trimerophytes, two extinct groups; sphenophytes, the
horsetails; pteridophytes, the ferns; and spermatophytes, the seed plants that
dominate today. Multicellular animal body plans were already established
when animals moved onto the land. Multicellular plants invented theirs in
the process of conquering the land. It is true that two very disparate kingdoms
with different evolutionary dynamics and quite different modes of development are being compared. However, the point is that at least some organisms
evolved new body plans in conjunction with the major expansion of life onto

the land. Animals did not, despite an equally great ecological transition that both opened empty adaptive zones and required extensive physiological and morphological innovations.

HYPOTHESES ON THE STABILITY OF BODY PLANS

Three kinds of explanations have been advanced for the notable lack of evolutionary innovation among animal body plans following the mass extinctions in the post-Cambrian world. It's not clear that they apply with equal force to the invasion of the land. The first is an ecological one, suggested by Erwin and co-workers. Based on an estimate of ecospace utilization during the Triassic, Erwin and co-workers argue that after the Permian mass extinction, the surviving body plans remained, although at low diversity, across the adaptive zones. Bottjer and co-workers argue that ecospace utilization was even lower immediately following the Permian mass extinction than estimated by Erwin and co-workers for the entire Triassic. Their estimate is that ecospace utilization at that time approximated that of the late Cambrian to early Ordovician. However, this is still higher than ecospace utilization during the Cambrian explosion. Ecospace utilization does appear to have been important. Jablonski and Bottjer observed that introductions of higher-level taxa, such as families and orders, were more frequent following mass extinctions. If preemptive competitive exclusion at the phylum level was strong enough following the Cambrian radiation, new players could arise and diversity could increase, but new phyla could not emerge. Surviving representatives of existing phyla effectively filled any vacant niches suitable for particular body plans far more effectively than less well adapted new phyla could. This argument seems to have real force when the survivors of mass extinctions are assessed. Although reduced in diversity, well-integrated and well-adapted animals remained. In the Cambrian, all competitors would have been amateurs. In viewing the invasion of the land, it is true that well-integrated body plans were represented in the shore parties, but as the new adaptations required for terrestrial life were so far removed from those of marine animals, one might imagine that new body plans would have had the advantage.

The second explanation is a probabilistic one offered by Stuart Kauffman. He has suggested that the world can be thought of as an imaginary landscape of adaptive peaks surrounded by nonadaptive valleys. A body plan is a device for occupying an adaptive peak. Once a newly originated body plan has landed on the slopes of the peak, selection favors a climb higher on the peak; that is, change within that body plan. Once the nearby peaks were occupied in the Cambrian radiation, longer jumps would be required to get to a better adaptive peak. The probability of successful long jumps decreases more rapidly than that of short jumps, but the peaks within range of short jumps would already have been filled. The Permian mass extinction thinned out the inhabitants on the peaks, but left each peak in the hands of some survivor, because no phylum went extinct.

The third explanation, offered by Jacobs and by Gould, is that a body plan cannot be transformed into any other. The developmental and genetic programs underlying each body plan are so tightly integrated that significant change is severely constrained. This is a different argument from the ecological one. The ecological argument contends that new body plans are unable to compete as well as established ones. In Kauffman's metaphor, the incumbents are near the top of the adaptive peak, and the new challengers are trying to climb up. The developmental constraints argument implies that once an integrated body plan is assembled, it can't be reintegrated without fatally disrupting ontogeny. This proposal implies the very interesting idea that the integration of genetic and developmental controls during the evolution of a body plan is irreversible. A thought experiment illustrates the point. Suppose that through some incredibly bad luck, all animal phyla became extinct except echinoderms. Could the survivors eventually evolve into new phyla? They would start from within a unique and unpromising body plan with tube feet, a water vascular system, a circular "brain," and pentameral symmetry. Could they evolve bilateral symmetry and various features that we associate with the other phyla, or would sea urchins go on munching algal mats until the Sun swells into a red giant and parboils the entire lot? There is a lot of variability within the echinoderms, and they have done some remarkable things. One group of sea urchins, the heart urchins, has evolved a secondary bilateral symmetry. Who in Paleozoic times would have predicted that one group of crinoids, the comatulids or feather stars, would lose their stems and become highly motile, swimming or walking with their arms and clinging to their perches with cirri that resemble multiple articulated appendages? I don't know the answer to this question, but I'm sure that in the right circles a lot of beer and peanuts could be consumed while it was being debated.

A *Tyrannosaurus* doesn't much resemble an amphioxus or a primitive jawless vertebrate like a lamprey. It's obvious that considerable evolution goes on within body plans, and that not all phases of development or all processes can have been constrained since the Cambrian radiation. Yet aspects of development are obviously constrained in the evolutionary retention of body plans. The suggestion of constraints on development limiting body plan evolution also includes the less easily testable proposition that developmental controls were relatively loose in the early Cambrian. This looseness would have allowed a great deal of experimentation with body plans and thus the rapid radiation of the highly disparate Cambrian fauna. Then things settled down, and the existing body plans became fixed by internal constraints. As

Gould put it in 1991, "The key to disparity is not Cambrian lability, but later and active stabilization." This idea may be right, but there are substantial problems to consider.

We can have no direct knowledge of how precisely or loosely organized Cambrian developmental/genetic systems were, and thus we have no practicable test of this hypothesis. The increased constraint as body plans became fully wired in is only an inference from the observed morphological stability of post-Cambrian groups. Is that stability positive data, or is it merely a lack of change because selection has not elicited any modification of body plans? Evolutionary stasis in itself does not demonstrate nonevolvability.

On logical grounds, there are two reasons for not translating the observation of historically stable body plans into statements about ability to change. First, body plans are by definition distinct. Suppose that a transformation to a different body plan had occurred. Unless the transitional forms were preserved as fossils, the transformation would not be traceable in the morphology of the descendants. They too would appear as distinct body plans. The fossil record is unlikely to be very helpful. Because the body plans of the fossilizable animal phyla were already established by the early Cambrian radiation, any transitions connecting them had to have occurred in the late Precambrian. So far, a fossil record for pre-radiation metazoans has eluded us, and indeed, may not exist. It also follows that if some of the nonfossilized phyla arose from existing phyla in post-Cambrian times, no record of those transformations would be likely to exist either. Second, there is an empirical reason to question the assumption that stasis demonstrates inability to evolve. Although larval body plans are generally conserved, as shown in chapter 7, they can change radically in a short time.

PATTERNS OF DEVELOPMENT IN THE METAZOAN RADIATION

Whittington, as well as Gould, suggests that if we look at the Cambrian world through the window of the Burgess Shale and the other well-preserved soft-bodied faunas, it appears that the early radiation of the Metazoa produced a higher disparity of body plans, although a much lower diversity of species, than is characteristic of present-day animal life. The apparent lack of origination of new phyla since the early Paleozoic despite dramatic modifications within body plans raises three critical questions for developmental biology.

The first is the basic question of the polarity of change: What is the ancestral state, and how was it modified in the descendant lineages? Evolutionary changes in development must be mapped on a reliable phylogeny. Our ability to order additions to and changes in developmental and anatomical features in this way is still crude, but we are already getting powerful lessons of this sort from existing data. For example, as discussed by McGinnis and Krumlauf, the homeotic genes of arthropods have homologues arranged similarly in mammalian chromosomes. Despite the gulf of time separating vertebrates from arthropods, and the differences between them in body plan and developmental patterns, homeobox genes are expressed in the same order along the body axis and play a role in central nervous system determination in both phyla. We can map this commonality on a phylogenetic tree and show that these regulatory genes and functions predate the radiation of coelomate phyla. We can predict that mollusks, hemichordates, and a host of other unstudied protostomes and deuterostomes will possess the homologues of these genes, and that they will use them in some aspects of body axis specification during development.

The second question to ask is, what are the developmental mechanisms that underlie the extensive modifications of animal body plans that have occurred during 500 million years of Phanerozoic history? After all, even though zebras and zebrafish share a common vertebrate body plan, they are plainly very different animals, and not just in body shape. The physiology, nervous systems, and behavioral repertoires of these two animals are rather disparate as well. The descriptive embryologies of mammals and zebrafish are well known, and molecular studies are showing that the basic genetic rules involved in generating the phylotypic stage are much the same. However, finding the features in development that account for the evolutionary modifications that lie between them is no simple matter. Part of the problem is the wide phylogenetic gulf separating these two animals. Looking at close relatives is crucial to lower the evolutionary noise and to isolate those parts of the genome that produce developmental or morphological differences.

Finally, we face the confounding question of what defines a body plan, and whether one can be transformed into another. We don't yet know the history of events that produced the body plans that appeared in the Cambrian. We don't know whether there is something inevitable about them. Gould has suggested that contingency has played a dominant role in determining which animal groups survived the lottery of extinction, and he has suggested that if the Phanerozoic were run over again, completely different groups might have survived to populate the present Earth. But we really don't know whether a series of contingent accidents has shaped the zoology of our world, or whether some body plans are essentially inevitable given the structural rules basic to metazoans. As an example of a pattern that is inevitable under those rules, Newman has suggested segmentation, on the basis of a simple "generic" hypothesis for the formation of segments along a body axis. Newman claims that if a cell adhesion molecule is expressed in a periodic fashion, segmentation will result. Once that has occurred, other regulatory genes could be recruited to stabilize the generic pattern. It may well be that the body plans we have before us today are the result of contingencies, selection, and internal constraints.

The directions of evolutionary change for both morphological features and underlying genetic controls can be inferred by considering metazoan phylogenetic trees. For example, since both segmented and nonsegmented phyla are included among the coelomate protostomes, the question arises whether segmentation arose independently in the protostome clade or was derived from a common ancestor shared with the segmented arthropods. In an initial study of this question, Patel and co-workers investigated the expression of the *engrailed* gene. This gene is one of several segment polarity genes involved in the interpretation of positional information along the body axis of the insect embryo. As gene action establishes the positions of the segments in insects, expression of the polarity genes determines anterior and posterior within a segment. The data of Patel and co-workers suggested that *engrailed* did not play an equivalent role in annelids. However, their negative result appears to have been due to antibody specificity. Wedeen and Weisblat showed with a probe specific to the leech homologue of this gene that *engrailed* does play a role in annelid segmentation similar to that in arthropods. What is important about these results is that use of a different genetic program would suggest that the two similar body plans are not homologous. That is why the suggestion of Patel and co-workers that annelids were not using *engrailed* as arthropods do was so striking. The contrary demonstration by Wedeen and Weisblat was important because it kept us from a premature decision on the homology of segmentation among protostomes. It is beginning to appear (according to Weisblat and co-authors) that the expression of Hox genes as well as *engrailed* in annelids resembles that in arthropods. A segmented ancestor utilizing similar genetic mechanisms to generate body segmentation is implied.

Figure 6.1 shows the mapping of some important regulatory innovations on a consensus phylogenetic tree inferred from 18S ribosomal DNA sequences. Once phylogenetic mapping is possible, inferences can be made about regulatory gene evolution in as yet unstudied phyla. Such inferences are amenable in turn to experimental testing.

Defining phylogenetic relationships among developmental regulatory genes is crucial to determining the times and phylogenetic branch points at which particular regulatory gene systems evolved. The features of these genes among living lineages allow us to infer the course of evolution of regulatory gene systems. For example, as inferred by Holland as well as by Kenyon and Wang, the Antennapedia and Bithorax clusters of homeotic genes that regulate the patterns of segmental identity in *Drosophila* evolved from a primitive bilaterian homeotic gene cluster now shared by insects, vertebrates,

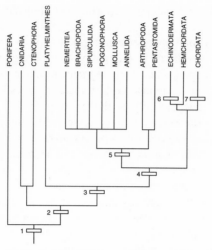

Figure 6.1.   Some important events in regulatory evolution mapped onto a metazoan phylogeny. Significant regulatory innovations include: 1, multicellularity; 2, tissues; 3, anterior-posterior axis and central nervous system; 4, Hox gene expression in patterning of the protostome and deuterostome body axes and central nervous systems; 5, metameric segmentation (lost in some phyla); 6, pentameral symmetry; 7, neural crest and amplification of Hox clusters. The tree is based on 18S ribosomal DNA results from several laboratories.

nematodes, and certainly many other phyla. The evolutionary relationship among Hox clusters is shown in figure 6.2. The presence of shared regulatory genes in distantly related organisms does not guarantee that those genes do the same things in development. We can attempt to link regulatory genes to body plans, but the mapping of genes and body plans will not be straightforward. A substantial number of major morphogenetic regulatory genes are widely shared among coelomates, and possibly among all Bilateria. Their roles are functionally conserved in some cases, but in others homologous genes have been recruited to different functions.

The range of novel functions acquired by conserved genes is an extremely important point, and it can be briefly demonstrated by a dramatic example. As discussed by Westerfield and co-workers, the developing vertebrate eye and ear employ several homeobox-containing genes in their patterning. As reviewed by Kessel and Gruss and by McGinnis and Krumlauf, many of the developmental control genes identified in mice and other vertebrates are homologous to homeobox-containing genes characterized in *Drosophila*. These genes encode transcription factors that act as decision makers during the differentiation of tissues and structures in development. Of the genes involved in eye development, *Pax-6* is a member of a family of vertebrate genes homologous to the *Drosophila* gene *eyeless*. Another gene, *Dlx-1*, is

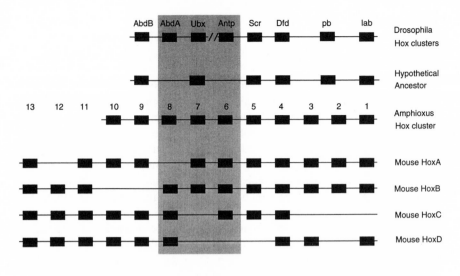

Figure 6.2. Evolutionary relationships among Hox gene clusters. In *Drosophila* (but not in all insects) the cluster is split (indicated by line breaks); however, the split cluster can be aligned with Hox clusters in other phyla. A hypothetical ancestral Hox cluster is posited to have given rise to the Hox clusters of coelomate metazoans. *Amphioxus,* a primitive chordate, has a single cluster. Vertebrates have duplicated the cluster. Mice and other mammals have four Hox clusters, HoxA through HoxD. The Hox clusters contain three groups of genes: those with anterior boundaries of expression at the anterior end of the body axis (3′ end of the cluster), those with anterior boundaries in the middle, and finally, those with the most posterior boundaries (5′ end of the cluster). The genes of the anterior group fall into distinct *lab, pb, Dfd,* and *Scr* groups. It is unclear how many middle group genes were present in the ancestor; a single gene is posited. The posterior group genes of vertebrates are all members of the *AbdB* group. Some genes have been lost from individual mammalian clusters. (After Garcia-Fernàndez and Holland 1994.)

a homologue of the *Drosophila* gene *Distal-less,* which is essential for the development of appendages. Finally, the *Hox-7* gene contains a homeobox domain homologous to the *Drosophila* muscle-specific *msh* gene. Nothing morphologically similar to the vertebrate eye or ear exists in insects. Thus, these regulatory genes play distinct and novel roles in animals with disparate body plans. These are only a few of a growing number of examples of genes shared by insects and mammals that carry out apparently similar regulatory roles in morphologically distinct bodies (Manak and Scott have provided a thorough review of these genes).

HOX GENES AND BODY PLANS

Because metazoans are monophyletic, all animal body plans arose as a result of transformations of a single ancestral body plan. Further, because all bilaterian animals share a common ancestor, they arose from a common bilaterian body plan. The interesting question then becomes whether there is a set of genetic rules that bilaterian animals share. If there is, the diverse body plans of bilaterian phyla have been built upon shared developmental genetic themes, which might constitute a conserved genetic body plan. Alternatively, the evolution of different body plans might have been based upon different sets of genetic rules. The evidence strongly favors the former scenario. Slack and co-workers have called this hypothetical Hox gene–centered genetic body plan for most animal phyla the "zootype." Whether their particular vision of a primitive animal body plan based on extrapolation from the Hox gene expression patterns of living animals will prove adequate remains to be seen.

The concept of the zootype rests on the conserved function of the homologous genes of the Hox gene cluster in determining the anterior-posterior body axes of animals as diverse as cnidarians, nematodes, annelids, arthropods, and vertebrates. The genes in the cluster have several interesting properties. They were initially defined genetically in *Drosophila* in the laboratories of E. B. Lewis and T. C. Kaufman. Mutations in these genes were observed to cause homeotic transformations among segments of the *Drosophila* body. The transformation of one segmental identity to another was first noted in 1894 by William Bateson, who called this phenomenon "homeosis." Homeotic mutations of the Hox cluster were discovered in flies early in the twentieth century. Homeotic mutations transform segmental identities, and produce dramatic effects on body organization. For example, mutations in genes of the Bithorax complex studied by Lewis caused transformations in thoracic and abdominal segments, such as the famous example of conversion of the third thoracic segment and its halteres into the identity of a second thoracic segment, accompanied by conversion of the halteres to a second pair of wings. The halteres are evolutionarily derived from a second pair of wings present in the dipteran ancestor. Homeotic genes thus became understood as selector switches that make choices among potential developmental fates. Once it was realized that the homeotic genes encode transcription factors, it became evident that they acted by controlling the transcription of genes downstream from them in the pathways that execute segmental developmental fates. The Antennapedia complex defined by Kaufman and his co-workers (see papers by Wakimoto and co-workers and by Kaufman and co-workers) determines the identities of head and thoracic segments. Mutations in genes of the Antennapedia complex affect the identities of anterior segments. For example, the dominant *Antennapedia* mutation converts the antennae of the fly into legs through the expression of the gene in the head as well as in the thorax, its correct site of expression. Recessive mutations of *Antennapedia* in which the gene fails to be expressed in the thorax as it would be normally cause conversions of the legs to antennae. We'll return to the roles of Hox genes in the evolution of insect body plans in chapter 12.

The genes of these two complexes occupy two sections of the third chromosome of *Drosophila,* but are united into a single contiguous Hox cluster in most other animals. A single Hox cluster appears to be the primitive state. The molecular structure of the Bithorax complex of *Drosophila* was defined by Bender and co-workers in 1983. As discussed by Peifer and co-workers and by Beachy, the cluster encodes only three transcription units, that is, expressed genes; however, it contains a number of complex regulatory sequences. The transcription units make up only about 10% of the 300,000 base pairs of the complex. The Antennapedia complex was described in molecular terms by Kaufman and his co-workers, and was shown to be similar in size and transcriptional complexity to the Bithorax complex.

The most surprising observations to arise from the molecular characterization of these genes were the finding by Scott and Weiner and by McGinnis and co-workers that the genes shared a structural motif, the homeobox of 180 base pairs, and the observation by McGinnis and co-workers and by Holland and Hogan that the motif was present in the genomes of members of other phyla, including vertebrates. When the homeoboxes were first discovered, it was suggested that they somehow controlled segmentation. The reaction to this idea was mixed. Their similarity to nuclear tags or DNA binding domains led us (Raff and Raff) in 1985 to suggest that their function lay in basic aspects of cellular function rather than regulation of segmentation per se. Both we and Brown and Greenfield, in back-to-back letters to the journal *Nature,* also questioned whether vertebrates should be thought of as segmented in any way. The information that subsequently emerged on homeoboxes and the proteins of which they are a part has forced some interesting reevaluations of basic ideas about both development and evolution. Homeobox-containing proteins do indeed function solely as transcription factors. Many of the genes whose transcription they regulate have no roles in segmentation. The discovery that homeobox-containing genes occur even in nonsegmental phyla such as nematodes and echinoderms caused their developmental roles to be further assessed in the more general context of axial determination. Nonetheless, the patterns of Hox gene expression in rhombomeres, vertebrae, and somites and the generation of experimental homeotic mutations in mice have shown that a basic segmental organization is present in chordates, and that it has co-opted Hox genes in its developmental execution.

The homeodomain encodes a 60-amino-acid domain that folds into a helix-turn-helix structural motif (determined by the physical studies of Qian and co-workers, Kissinger and co-workers, and Otting and co-workers, and nicely reviewed by Treisman and co-authors). The third helix of this motif contacts the major groove of the DNA. The specificity of binding to promoters of genes regulated by homeobox genes is sensitive to the amino acid sequence in the DNA-binding region. For example, Treisman and her co-workers showed that a single amino acid change in the homeodomain could change which promoters the homeodomain protein would bind. Recognition of the correct binding sites is vital for correct activation of genes downstream from the homeotic genes.

Not all homeobox-containing genes are in the Hox cluster. There are a number of other homeobox-containing transcription factors. These "orphan" genes play important roles in the development of individual organ systems. Some contain other DNA-binding domains as well, such as the POU domain. The POU domain family, as discussed by Herr and co-workers, includes the Pit-1, Oct-1, and Oct-2 transcription factors of mammals and the *unc-86* gene of the nematode *C. elegans*. These genes regulate the differentiation of specific tissues. For example, Pit-1 is specific to the pituitary, where it controls synthesis of growth hormones and prolactin. Like the Hox genes, the orphan homeobox-containing genes act as major switches for the transcriptional regulation of downstream genes.

The Hox genes themselves are regulated by genes whose products determine the rough axial specification of eggs and early embryos. There is in *Drosophila* a thoroughly described cascade of genes that determines the ultimate patterns of Hox gene expression. This cascade is well discussed in reviews by Nüsslein-Volhard, by Gilbert, and by Lawrence. The cascade begins with the protein products of the maternally expressed genes *bicoid* and *nanos,* which designate the anterior and posterior poles of the embryo. These genes activate or repress the transcription of the gap gene *hunchback. Hunchback* levels regulate the expression of other gap genes, and the interactions among gap genes regulate each gene's domain of expression. The gap genes encode transcription factors, and they regulate the expression patterns of the pair-rule genes. The bands of activity of pair-rule genes establish parasegment boundaries in the embryo and determine the patterns of expression of the segment polarity genes *engrailed* and *wingless*. These genes are expressed in adjacent one-cell-wide bands. Communication between these cells stabilizes the expression of *engrailed* and *wingless* in adjacent cells at the boundary and provides the information required to establish the anterior and posterior domains of the segments. Once segmental boundaries are established, the homeotic genes come into play to determine segmental identities. The features of gene cascades establishing ever more finely subdivided modules within the embryo appear to be general in development and have profound meaning for the evolution of development. We'll return to this issue in chapter 10.

The abrupt transformations caused by homeotic mutations in *Drosophila* led Richard Goldschmidt, early in the twentieth century, to propose that homeotic mutations could generate "hopeful monsters" as potential ancestors for rapid macroevolutionary steps. We now realize that homeotic mutations, so prominently observed in insects, rarely provide viable raw material for evolution. Instead, they indicate the nature of the genetic controls exploited by insects in the evolution of the specialized tagmosis patterns characteristic of their various orders. Nonetheless, homeotic transformations may have played a role in the evolution of insect ordinal body plans. Whiting and Wheeler have reinvestigated the relationship between the Diptera and a small order known as the Strepsiptera. The two appear to be sister groups, and they differ from each other in a very odd way. Both have a single pair of wings and a pair of halteres. In dipterans, the wings are on the second thoracic segment and the halteres on the third. In strepsipterans, the situation is reversed: the wings are on the third thoracic segment and the halteres on the second. Whiting and Wheeler have made the interesting suggestion that the *Ubx* gene, which is expressed in the *Drosophila* third thoracic segment, might be expressed instead in the second thoracic segment of strepsipterans. This suggestion is testable by molecular means, and a successful test would make a search for other homeotic transformations between higher taxa worthwhile.

The remarkable and evolutionarily conserved role of the Hox gene cluster is the determination of identities of body parts along the anterior-posterior axis. Homeotic mutations were first described in insects, and have been defined in flies, moths, and beetles. Spontaneous homeotic mutations have not been observed in mammals. It was suspected that this was because vertebrates were not built according to a segmental body plan. It is now evident that although vertebrates are not built of segments in the same sense as insects, segmentation is present in the vertebrate body, and particularly plainly so at the phylotypic stage. Vertebrae, somites, and the rhombomeres that give rise to the hindbrain are all obviously serially repeated. Homeotic mutations can be experimentally generated in mouse embryos through gene knockout techniques. Thus, knockouts of genes *Hox A-2* by Gendron-Maguire and co-workers and by Rijli and co-workers, *Hox C-8* by Le Mouellic and co-workers, *Hox B-4* by Ramirez-Solis and co-workers, and *Hox A-5* by Jeannotte and co-workers and by Ramirez-Solis and co-workers all produced homeotic transformations in axial skeletal elements. Several of these caused a transformation of the identity of a vertebra into that of the next most anterior one, although knockout of *Hox A-5* caused a vertebral transformation into a more posterior identity. Other Hox gene knockouts caused defects in specific regions of expression along the body axis, but not homeotic transformations. For example, the knockout of *Hox A-3* by Chisaka and Capecchi resulted in absent or severely deficient thymus, thyroid, and parathyroid glands along with heart and blood vessel malformations. Like the classic homeotic genes of *Drosophila,* the Hox genes of other phyla work as switch genes in deciding between alternate developmental fates in localized regions of the embryo.

In the days when the homeotic genes were known only from insects, a reasonable evolutionary hypothesis for the elaboration of Hox genes was that the control of segmental identity evolved progressively from head to body as annelids gave rise to arthropods and primitive arthropods gave rise to more highly specialized arthropods. In such a scenario, the Antennapedia complex genes would have arisen first, as annelids have specialized head segments but body segments that are identical. The Bithorax complex would have evolved by duplication and divergence of Antennapedia complex genes to regulate body segment specialization. This scheme cannot be entertained any longer. The Hox gene cluster is more ancient than the origin of arthropods, and the sister group relationship of annelids and arthropods is moot. What is likely is that this gene cluster originally functioned in primitive metazoans to regulate axial regionalization. There is an elegance, at once elusive and concrete, in the co-option of Hox genes for generating quite different phylotypic stages among the phyla. In arthropods, these genes regulate segmental tagmosis patterns, and in vertebrates, the identities of neural crest derivatives. Both display axial information, but very differently.

EVOLUTIONARY STABILITY OF EARLY DEVELOPMENT

A traditional feature of metazoan phylogenetics ever since Haeckel has been a reliance on evolutionary conservation of larval forms. For example, mollusks and annelids are distinct in the fossil record from the early to mid-Cambrian. Consultation of a text on comparative embryology such as that by Kumé and Dan shows that both phyla retain similar cleavage, cell determination, and larval forms. There is no doubt that these developmental features have been retained for at least 540 million years from the ancestor of these two phyla. Although most of what we know about the evolution of larval features has to be inferred in this way, some surprisingly good fossil evidence also supports long-term conservation. As shown in figure 6.3, fossil larvae illustrated by Müller and Walossek and by Walossek from the upper Cambrian (just over 500 million years ago) bear a striking resemblance to nauplii of living crustaceans. These extraordinary fossils preserve the entire developmental series of some species, and it has been possible to show that the fossil nauplii are truly the larvae of a primitive crustacean. Walossek and Müller have also found fossil larvae of parasitic pentastomids from the upper Cambrian. These are remarkably similar to the larvae of living pentastomids. Fossil mayfly larvae are known from the lower Permian (270 million years

(A)   **(B)**

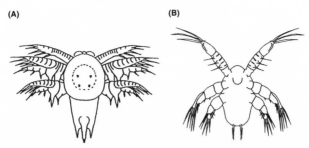

Figure 6.3.   Morphological conservation of a larval body plan. *(A)* Dorsal aspect of the first nauplius of *Rehbachiella,* a Cambrian stem-lineage crustacean. *(B)* Ventral aspect of the first nauplius of the living crustacean *Eurytemora.* Note the similarity of the unsegmented body and the small number of appendages in both living and ancient nauplii. Additional segments and appendages are added in subsequent growth in both. (*A* after Walossek 1993. *B* reproduced with permission of the Royal Society of Edinburgh and Müller and Walossek from *Transactions of the Royal Society of Edinburgh: Earth Sciences* 77 part 2 [1986], pp. 157–79.)

ago) and, as illustrated by Kukalová, resemble modern mayfly larvae, but have a more primitive pattern of attachment of larval wings. Fossil pipid frog tadpoles from the early Cretaceous (120 million years ago) have been shown by Estes and co-workers to have persisted with little change from early in the history of this group. There are a few other cases. For instance, well-preserved fossil sea urchin pluteus skeletal rods from the upper Jurassic discovered by Deflande-Rigaud provide direct evidence for the existence of the pluteus larval form 160 million years ago. Apparent trilobite embryos preserved in mineralized form from the middle Cambrian of China have been shown by Zhang and Pratt to have cleavage patterns similar to those of living horseshoe crabs.

The sea urchins represent a relatively well-defined radiation for which we have an excellent fossil record, with living representatives of most clades, and embryological data from all major clades. Thus, as Wray has shown, the evolution of the sea urchin pluteus provides an exceptional opportunity to evaluate the extent of evolution within a larval body plan. The echinopluteus larva retains many primitive features shared with the larvae of both other deuterostomes and other echinoderms, but also possesses several derived features shared among all sea urchin plutei. Although there are modifications in arm numbers, skeletal morphology, and other structures, the basic pluteus architecture is remarkably conserved in all lineages. The pluteus thus represents a larval body plan conserved for 250 million years. This larval body plan is completely distinct from the adult body plan generated in the juvenile rudiment and released at metamorphosis.

It has been suggested especially cogently by Arthur that evolutionary conservation of early development can be expected on mechanistic grounds, because all subsequent steps of development depend on the correct completion of the initial processes. This idea suggests an upwardly expanding and branching tree whose initial starting point is the fertilized egg. Each branch point in the tree represents a new state that gives rise to subsequent new cell states in development until terminally differentiated states are reached. With each cell division, state changes occur and complexity increases. If any lower point is affected by mutation, the whole structure above it should be affected and possibly even collapse. The study of mutations in genes that govern individual developmental processes shows that when action by a succession of genes is involved, this expectation is true. Is it true for ontogeny as a whole? The observation of evolutionary stability in larvae, and the theoretical view that development is hierarchical and dependent on early stages, suggest strongly that early development must be refractory to substantial evolutionary change. Surprisingly, this view, as attractive as it is, turns out to be wrong. Although it is incorrect, it has nonetheless had a substantial impact upon our ideas of body plan stability.

RADICAL EVOLUTIONARY CHANGES
IN EARLY DEVELOPMENT

The hypothesis of a hierarchical ontogeny of expanding complexity requires that it be difficult, if not impossible, to make substantial modifications in early development. One conception of this kind of constraint on evolution is Wimsatt and Schank's notion of generative entrenchment. They define generative entrenchment of a feature or trait as "some measure of the number of other traits affected by changes in that trait." They argue that features appearing in early development should be more conserved in evolution than features appearing later. The concept of generative entrenchment may be sound, but I think that it may not apply so much to early development as to later in the phylotypic stage. In many ways, the experience of developmental geneticists bears out the idea that early development ought to be entrenched. Early development within a species seems to be precisely regulated with gene cascades that exhibit increasing complexity and absolute dependency on the action of the preceding genes in the cascade. Whether such a view of early development is meaningful in an evolutionary context can be determined only by looking at the evolution of early development. What is observed is that despite the constancy of early development within a species, dramatic evolutionary changes have occurred among closely related species in several animal groups.

Members of many taxa have modified their early larval development in favor of direct development to the adult without going through the ancestral feeding larval stage. The best known examples are among ascidians, frogs, and sea urchins. Significant changes in early development have occurred a number of times among closely related species in these groups. Direct-developing ascidians have modified the localization of maternal information and have truncated the differentiation of cell lineages. As a result they have lost the notochord and somites of the larval tail. Direct-developing frogs have greatly modified cleavage, gastrulation, and the relative timing of developmental events. Development proceeds from egg to frog without a tadpole. Direct-developing sea urchins have extensively remodeled development, with differences in genome size, egg size, sperm head shape, cleavage patterns, cell fate maps, blastulation, gastrulation, and timing of aspects of development. Larval body plans that were stable for hundreds of millions of years were readily abandoned in short spans of time within each of these groups. Such radical shifts presumably could occur because selective pressure for a different mode of development was sufficient to overcome whatever internal constraints might have maintained the integrity of larval body plans. There is no denying that cascades of gene action are profoundly important in early development, as has been so well documented for *Drosophila*. However, old cascades can be eliminated and new early cascades substituted for them in evolution. Early development can evolve without necessarily affecting later stages or the adult.

Evolutionary changes in early development do not arise only from changes in modes of larval development. Early development is also subject to other changes in reproductive mode, as well as to changes in egg size and embryonic feeding strategies. A striking example will suffice to make the point here. Vertebrate embryos, as shown in figure 6.4, were noted in the nineteenth century to all pass through a typical vertebrate phylotypic stage, although their later development diverges. This is the famous diagram by Ernst Haeckel, using the comparisons of vertebrate development made earlier by von Baer. Haeckel employed it to illustrate that "early" stages of development were essentially the same, and that became the dominant view of the evolution of early development. There are some significant difficulties with this traditional view. First, as discussed by Richardson, Haeckel doctored his drawings to exaggerate similarities. Second, the early stages shown at the top of the diagram really occur well into development. Third, as figure 6.5 shows, the early stages of development of vertebrate classes are actually very different from one another. Perhaps this observation is unsurprising as the representative animals are far apart phylogenetically. However, close cousins can also show a great disparity of early development. Monotreme mammals, whose living representatives include only the nearly legendary platypus and echidnas of Australia, are relatively primitive mammals. Although they are warm-blooded, have hair, and produce milk, they also lay eggs of a very reptilian style. Their embryology was investigated in the 1930s by Flynn and

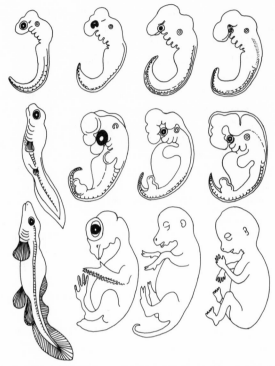

Figure 6.4.   A part of Haeckel's diagram of progressive stages of development of representative vertebrates. The earliest stages shown are at the top of the diagram, and progressively later stages below. The animals shown are (from left to right) a bony fish, chick, calf, and human. Haeckel, who was adducing evidence for his recapitulation theory of evolution, used this diagram to show that developing animals recapitulate earlier stages in the phylogeny of their lineages. He argued that the chick, calf, and human embryos shown passed through the same early stage as the fish. He exaggerated the resemblances of the early stages to one another to emphasize the point. (From Raff and Kaufman 1983; drawn by E. C. Raff. Reprinted with permission of Indiana University Press.)

Hill. (A thorough set of comparisons of vertebrate developmental stages, including those of monotremes and reptiles, is available in Nelsen's *Comparative Embryology of the Vertebrates*.) Cleavage in monotremes is highly meroblastic, with cleavage confined to a small region of the surface. These embryos go on to form a blastoderm like those of reptiles and birds, but not like that of placental mammals with their secondarily small eggs.

BODY PLANS AND HOW THEY DEVELOP

Three ideas are entangled in the concept of a body plan as it is currently
used. The first is the concept of the adult body plan. This idea stems from
Richard Owen's archetype, which was, as pointed out by Desmond, later

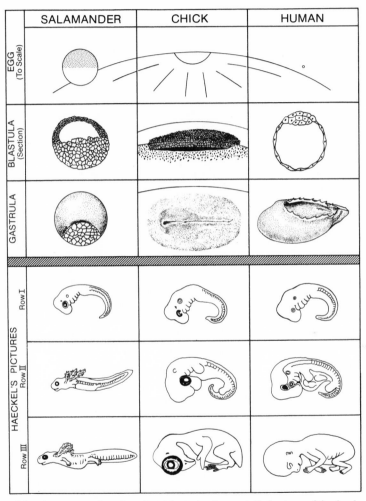

Figure 6.5.   A more complete view of vertebrate development. Early stages of the development
of salamander, chick, and human are as varied as the postphylotypic stages, but the phylotypic
stages (row I) are similar. (From R. Elinson in R. A. Raff and E. C. Raff, eds. *Development
as an Evolutionary Process.* Copyright Wiley-Liss, Inc. 1987. Reprinted by permission of John
Wiley & Sons, Inc.)

placed in an evolutionary context. The term *Bauplan* (Anglicized as body plan) was introduced by Joseph Henry Woodger in 1945 to express the idea of a homologous structural plan underlying evolutionary transformations within a taxonomic group. (Woodger's peculiar entry into the body plan enterprise and the history of the idea are well told by Hall in his book *Evolutionary Developmental Biology.*) Homologous elements in anatomy are recognized by spatial and compositional criteria, and when available, transitional criteria from the fossil record. The possession of a suite of such features by a set of animals constitutes a similarity of organization. Commonality of organization exists because of shared ancestry. This concept does not require that the body plan arise from a shared developmental pattern. What matters is that the adult anatomies correspond.

A second concept of body plan refers to early developmental stages. Many animal groups develop via a feeding larval stage and then metamorphose into the adult. The larval form can be dramatically and decidedly distinct from the adult form. Larval forms are often highly conserved through long periods of evolutionary time, and they frequently are very different in their organization from both the corresponding adult body plan and other larval forms. They can thus be considered to represent larval body plans, as real as the adult body plans. Their development may be quite distinct from the processes that lead to the adult, suggesting that larvae are products of constructional rules distinct from those that produce the adult.

A last body plan concept relates not to the archetype-derived concept of an adult architecture, which is independent of any developmental component, but to the actual developmental generation of the adult body plan. This developmental body plan rests upon a commonality of developmental patterns leading to a finished adult structure. Despite having vastly different developmental environments, animals such as fishes and mammals share the vertebrate phylotypic stage. Selection would seem to have little role in maintaining phylotypic stages under such disparate circumstances. Instead, constraints operating on development itself are often proposed to underlie this commonality of organization: the phylotypic stage is thus constant because the internal mechanics of development allow no other way to build a vertebrate.

Species that exhibit similar developmental body plans do not necessarily produce similar adults. For example, all chordates, including vertebrates, pass through a pharyngula stage in which a dorsal nerve chord, notochord, and somites are integral components. Ascidians share this developmental body plan, but develop into very nonchordate adults. The developmental body plan reflects the processes of ontogeny that produce the "phylotypic" stage that foreshadows the features elaborated in the adult. The phylotypic stage occurs after organogenesis has begun, and is shared by all members of a phylum before the divergence of their later stages. Adult structures derived from further development of the phylotypic stage are indeed homologues in having historical continuity in evolution. However, the phylotypic stage, as I show below, has the troublesome characteristic of being the most evolutionarily conserved stage of development, but also being attainable through nonconserved developmental processes. A telling way of illustrating this concept of body plan and its evolutionary peculiarities was devised by Elinson, and is shown in figure 6.5. Early development in vertebrates is divergent, and so is late development. But the phylotypic stage is conserved.

We are faced with a paradox. Body plans are clearly stable over long evolutionary spans. If they were not, we would not recognize persistent higher taxonomic categories such as phyla and classes. Yet basic elements of body plans are attained by different developmental pathways. They have been extensively modified by additions, and in the evolution of direct-developing species, losses and remodeling of larval body plans. Adult body plans have also been remodeled. Since shared body plans can be attained by different pathways, why should they be so stable and channel development through evolutionarily conserved phylotypic stages?

FIDDLING WITH THE RULES

Because the vertebrate body plan is so well known, it is worth taking a look at the wide variation among vertebrates in the construction of their body plan elements. The vertebrate phylotypic stage is the pharyngula. As can be seen in Haeckel's famous diagrams illustrating von Baer's laws (see fig. 6.4), this is the stage with gill pouches, limb buds, and a tail through which vertebrates from fishes to philosophers pass before they diverge toward their adult forms. A notochord, a dorsal nerve tube, and paired body muscle bands (somites) are all present. It should be noted that von Baer's laws provide an incomplete description of development in that von Baer considered the phylotypic stage to represent early development. In fact, he was dealing only with the later half of ontogeny. Although all vertebrates pass through a very similar pharyngula stage, early development, including egg size and yolk content, cleavage mode, gastrulation, and cellular movements, differs dramatically between classes. The vertebrate developmental body plan arose by accretion to earlier elements: the dorsal neural tube, notochord, and paired tail muscles that are present in the basic chordate body plan seen in its most rudimentary form in ascidians and more fully developed in cephalochordates (the classic *Amphioxus*, now called *Branchiostoma* by taxonomists). As pointed out by Gans and Northcutt, development of the vertebrate head includes additional features, including the neural crest cells and sensory placodes, such as the optic placode that produces the eye, the otic placode that forms the ear, and the nasal placode (see Webb and Noden for a review of the processes that produce these structures). The neural crest cells contribute to an amazing array of neuronal, glandular, and pigment cells and to the major skeletal and connective tissues of the head.

In the vertebrates, structural additions have been accompanied by genetic additions. Structural gene duplications have accompanied body plan elaboration, as has been demonstrated by Vandekerckhove and Weber in the duplications and divergences of muscle actin genes. The single muscle actin gene of ascidians and cephalochordates was duplicated in bony fishes to two genes, and to four in amniotes, with distinct tissue-specific expression. More interestingly, regulatory gene duplications—that is, increases in number and diversity among regulatory genes—have accompanied the evolution of basic features of the vertebrate body plan, as discussed by Holland and by Ruddle and co-workers. Thus, the branchial region of vertebrates is also present in the cephalochordate *Amphioxus*, where it lies at the anterior end. Hunt and co-workers have shown that the anterior parts of the domain of expression of the homeotic genes lie in this region in the mouse. These genes have apparently been co-opted into a new role in patterning the neural crest of vertebrates. One homeobox gene (*Hox 7.1*, a homologue of the *Drosophila msh* gene) has been shown by Hill and co-workers and Benoit and co-workers to be expressed in the neural crest, limb buds, and visceral arches. As shown by Holland, this *msh* homologue is represented as a single gene in ascidians, but as three genes in vertebrates. The *msh* class homeobox genes are not part of the Hox cluster, but the parallel in the duplication and divergence of these regulatory genes in the evolution of the vertebrate body plan is evident. Garcia-Fernàndez and Holland have shown that there is a single Hox cluster in the cephalochordate *Amphioxus* that is the prototype of the four parallel Hox clusters of vertebrates (see fig. 6.2).

The anterior portion of the vertebrate head and brain has been proposed by Gans and Northcutt and by Hunt and co-workers to represent a new addition to the older chordate body plan. For this to have occurred, co-option and reintegration of regulatory genes would have been required. Hunt and his co-workers have suggested that in accord with this idea, the new anterior head required the evolution of new genetic patterning mechanisms not based on the genes in the homeotic cluster. This prediction has been borne out by the remarkable discovery by Shawlot and Behringer that a homeobox-containing gene, *Lim1*, is the regulator of the primary head organizer in vertebrates. When this gene is knocked out in mice, the embryo develops normally, but lacks the anterior part of the head. However, the determination of the anterior head by a regulatory gene not in the Hox cluster does not really require the anterior head to have been added late in chordate evolution. A different light has been cast on the origin of the forepart of the vertebrate head by Holland and co-workers, who used Hox gene expression patterns to ask whether a region homologous to the vertebrate forehead exists in a primitive nonvertebrate chordate. *Amphioxus* has a chordate body plan but lacks the craniofacial region present in vertebrates and lacks a subdivided brain as well. But the "headlessness" of *Amphioxus* isn't all that clear-cut. Lacalli and co-workers have suggested that the frontal eye of *Amphioxus* may be homologous to the paired eyes of vertebrates. If so, its presence supports the argument that the anterior region of the *Amphioxus* central nervous system is homologous to the vertebrate brain.

It is to a basic chordate resembling *Amphioxus* that the anterior portions of the head should have been added in the origin of the vertebrate body plan. Holland and co-workers found that *Amphioxus* possesses Hox genes homologous to those of vertebrates. They examined the expression of an anterior-class Hox gene (homologous to mouse *Hox B-3*) in *Amphioxus* development, and found that its anterior boundary of expression was well back from the front end of the animal (in a position corresponding to between rhombomeres 4 and 5 in the vertebrate hindbrain), just as in vertebrates. The addition hypothesis would have predicted that the gene would instead have its anterior boundary of expression well forward in the *Amphioxus* "head." This result is powerfully suggestive, but since homology cannot be defined only by position in development, this observation alone cannot be conclusive.

Other genetic observations also reduce the necessity of suggesting that a new vertebrate forebrain was added to a cephalochordate grade of organization. Rather, it looks as if preexisting anterior portions of the ancestral bilaterian animal were present in the ancestral chordate and were co-opted in the evolution of vertebrate head and brain structures. Simeone and co-workers have shown that two homeobox-containing genes, *Emx* and *Otx*, respectively the homologues of the *empty spiracle* and *orthodenticle* genes of *Drosophila*, are expressed in the developing mouse forebrain. *Empty spiracle* and *orthodenticle* are involved in establishing head segmentation in *Drosophila*, but their exact roles are not yet known. Once again (as discussed by Holland and co-workers), we are presented with the extraordinary observation that, over large phylogenetic spans, homologues of regulatory switch genes are expressed in conserved spatial domains. The roles of these genes have surely diverged in vertebrate brain development, but the retention of the regionalized pattern of expression is consistent with an early evolved role for these genes in metazoan head specification. The patterns of expression are also consistent with an early origin of the anterior part of the vertebrate brain in chordates, rather than its later addition in vertebrates. Definition of the sites of expression in *Amphioxus* of genes characteristic of the vertebrate anterior head will be an important test of ideas on the origin of the vertebrate head. It should be noted that even if the addition hypothesis for the anterior portion of the brain is incorrect, the neural crest cells, and possibly the sensory placodes, do represent novel additions to the chordate body plan, and thus pose interesting evolutionary problems.

Holland and co-workers have also made the interesting suggestion that other homeobox-containing genes may have long-conserved functions in the specification of other organs in metazoans. For example, the *Drosophila* gene *cad* and its homologues, the vertebrate *Cdx* genes, were shown by Frumkin and co-workers to share an expression site in the gut. These observations again imply that although metazoan body plans vary, they are based on an ancestral set of instructions that has been maintained beneath the disparate phylotypic stages of the various phyla. Additions to these instructions parallel the substantial modifications within body plans characteristic of evolution within the phyla.

The vertebrate developmental body plan has been subjected to extensive modification. Thus, although salamanders and frogs gastrulate similarly, there are major differences in cellular behaviors between them. The precursors of the mesodermal cells lie beneath the surface in some frogs, but, as discussed by Keller and by Hanken, in salamanders these precursors lie on the surface in the early gastrula. The somites, which give rise to the segmental muscle bundles of the body wall, are present in all chordates, and are one of the distinguishing features of the chordate phylotypic stage. Nevertheless, as shown by Radice and co-workers and by Malacinski and co-workers, the pattern of somite formation from mesoderm differs between salamanders and frogs, and even among frogs. Analogous differences among vertebrate embryos extend to other basic elements. In amphibians, the notochord, another characteristic element of the chordate phylotypic stage, arises from cells that have involuted as a sheet into the interior of the embryo. This is the primitive pattern in tetrapods. The notochord arises in quite a different way in the chick, in which it is formed from an aggregation of mesodermal cells that have ingressed individually into the interior. Analogous changes have evolved independently in other great branches of the vertebrates. In sharks, the notochord arises from a cell sheet, whereas in advanced bony fishes, the teleosts, it arises from an aggregation of mesenchymal cells.

Other ''difficult'' modifications to elements of the vertebrate body plan have occurred in tetrapods. Because they are familiar animals, we take turtles for granted. However, the formation of the unique turtle trunk represents an astounding modification of vertebrate anatomy. If you feel your upper back, you will notice that your shoulder blades, which are a part of the shoulder girdle, are external to your rib cage. That's the standard tetrapod arrangement. In turtles, the rib cage lies over the shoulder girdle instead of under it. The shoulder girdle has been shifted to become internal to the rib cage, a novel topology. This unique modification of the vertebrate body plan has been shown by Burke to have been achieved in development by a change in the trajectories of movement of the cells that give rise to the ribs and the carapace: instead of moving down along the body wall with the body wall muscle precursor cells, these cells extend over the limb buds. She proposed that a novel set of interactions arose between a continuous epidermal placode and cells deep in the body wall, which gave rise to an outward-growing "carapacial ridge" that influences the expansion of the somitic mesoderm and the condensation of the ribs. The origin of this new interaction lies in the co-option of standard mechanisms of epithelio-mesenchymal interaction such as those that operate in the formation of more usual projections from the body, such as the limb buds. If turtles didn't exist, we might have predicted that the placement of the shoulder girdle is internally constrained and impossible to change. The mere prevalence and long-term conservation of a particular body plan element does not indicate that it is impossible to make a transition to a different element.

Extensive modifications to the developmental pathways that build a phylotypic stage are not restricted to vertebrates. The phylotypic stage of insects is the stage at which the segments have formed and distinct head, thorax, and abdominal regions are present. As discussed by Anderson, this basic segmental body plan is produced in two different ways, and insects thus fall into two developmental classes. The more primitive short germ band insects produce abdominal segments progressively from a terminal growth zone by a process reminiscent of the process of segment formation in other arthropods and in annelids. In the more advanced long germ band insects, all segments are formed simultaneously. Both groups produce the same number of similarly arranged segments. The big question, as has been briefly reviewed by French, is whether morphogenetic and determinative processes differ between the two groups. The basic patterning of the long germ band insect *Drosophila* occurs very early in development, before the nuclei of the embryo become separated into discrete cells by cell membranes. A small set of maternally loaded and localized messenger RNAs encodes transcription factors that activate another small set of genes, the gap genes. The remarkable generation of pattern by the formation of bands of expression of gap genes within the cytoplasm of the noncellularized *Drosophila* embryo has been summarized by Hülskamp and Tautz. Briefly, these genes set up broad and exclusive domains of expression. They encode transcription factors in the syncytial cytoplasm of the early embryo, and these transcription factors subsequently activate the pair-rule genes that are expressed in stripes in the still syncytial embryo. The embryo becomes cellularized by formation of cell membranes between the nuclei, and the boundaries of expression of the pair-rule genes establish the expression of the gene *engrailed*. This gene is expressed simultaneously in the cells that form a narrow band in what will be the posterior margin of each of the fourteen segments of the body. The *engrailed* gene is thus a segment polarity gene. This cascade of gene action proceeds to the segment-specific activation of the homeotic genes that determine the identity of each segment. This tale of genes generating pattern is now becoming well understood for *Drosophila,* and is nicely summarized in Peter Lawrence's *The Making of a Fly.* Sommer and Tautz have shown that it applies as well to the nasty housefly *Musca domestica,* which, although separated from it by perhaps as much as 100 million years, is a long germ band dipteran similar to *Drosophila* in development. There appears to be a strong evolutionary conservation of patterning in dipterans. Langeland and Carroll, for example, showed that the complex controls exercised by the products of the gap genes over the pattern of expression of the pair-rule gene *hairy* have been conserved over the 60 million years that separate *Drosophila melanogaster* from *D. virilis.* This very brief summary does not do justice to the elegant details, but it will do to illustrate the problem posed for gene regulation by the existence of alternative developmental ways of generating the insect phylotypic stage.

The evolution of short germ band insects is significant because the cascade of gene action that sets up the patterning of segmentation takes place in a different cellular environment than in long germ band insects. As in long germ band insects, the maternal patterning elements can act in an acellular environment. In the flour beetle *Tribolium,* a short germ band insect studied by Sommer and Tautz, the gap gene *Krüppel* is indeed activated in a syncytial environment. However, the subsequent activation of pair-rule genes appears to occur after cells have formed. Thus, the pair-rule gene *hairy* may be under the control of the gap gene *Krüppel* in a cellular environment. It is not yet clear how the segment polarity gene *engrailed* is controlled in this insect. It is activated progressively (not simultaneously as in the long germ band insect *Drosophila*) in a cellularized environment as the growth zone elongates and produces segments. Sommer and Tautz suggest that activation of one gene by the products of another could occur by the same mechanisms as seen in *Drosophila* if the cells are coupled by cytoplasmic connections.

The studies of Patel and his co-workers on the grasshopper *Schistocerca* suggest that some aspects of short germ band segmentation may involve different genetic controls than in long germ band insects. In *Drosophila,* the pair-rule gene *even-skipped* is intimately involved in establishing the proper pattern of *engrailed* expression. Patel and co-workers have found that in the grasshopper, *even-skipped* is expressed, but never in a pattern of stripes. The expression of *engrailed,* however, appears as a band at the posterior of each segment as it appears. *Even-skipped* seems to play a conserved role in the development of the grasshopper central nervous system, but not in the establishment of segments. Possibly it was co-opted for that role during the evolution of long germ band insects.

Once segments have been established in short germ band insects, segmental identities are determined by homeotic genes. As shown by Beeman and by Stuart and co-workers, these genes in the short germ band beetle *Tribolium* are organized similarly to the classic homeotic clusters of *Drosophila*. Thus the homeotic genes set up the insect body plan, but the events leading to their actions have evolved with the shift from the short to the long germ band pathway.

STABILITY OF PHYLOTYPIC STAGES

Gould, as well as Jacobs, has pressed the intriguing argument that post-Cambrian arthropods have retained a set of tagmosis patterns that is much more restricted than those of Cambrian arthropods because of internal constraints on development. That is, the surviving classes have stabilized the genetic and developmental mechanisms underlying their segmentation and tagmosis patterns, and so lack the developmental flexibility of the Cambrian arthropods. It's a little hard to assess Cambrian developmental mechanisms from a half billion year distance. However, this suggestion offers some possibilities, and these may have explorable consequences. The first possibility is that the argument is true. The second is that the argument is irrelevant because Cambrian developmental and genetic systems were just as tightly constrained as those in living arthropods, and the survival of only a few groups has given the illusion of a deterministic stability. The final possibility is that the argument is false in a different way, and living classes could change tagmosis patterns just as readily as Cambrian arthropods given suitable selective pressures. These three hypotheses have the dubious virtue of being equally difficult to evaluate. A consideration of the processes involved in development of the phylotypic stage suggests the existence of features that should allow evolutionary change and features that should restrict it. Some general mechanistic considerations are important.

Evolutionary changes are facilitated by the duplication of genes, including control genes. The duplicate genes are similar to, but distinct from, the ancestral gene, and can be co-opted to carry out related but different functions. An example of a gene family resulting from such a process is the steroid receptor superfamily. It encodes many different proteins, which function similarly to one another in binding steroids and related compounds. Once the specific hormone has bound, these proteins act as transcription factors for quite distinct genes. They play diverse and dramatic roles in morphogenesis. In some systems, their misactivation, as shown by Mohanty-Hejmadi and co-workers and by Maden, causes homeotic transformations in which one structure replaces another.

Regulatory genes are conserved and recombined into novel control path-

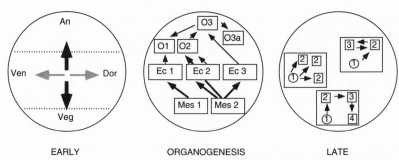

EARLY   ORGANOGENESIS   LATE

Figure 6.6.   Qualitative changes in modularity over early, middle, and late stages of development. The model shown here is based on general features of deuterostome development, particularly of echinoderms and amphibians. Events of early development are dominated by axial information systems (animal-vegetal and dorso-ventral), which establish initial patterning processes and the first localized gene action as cells divide. In these embryos, these axes are global, and allow considerable developmental flexibility. Late development also shows considerable developmental flexibility, which arises because the body is highly modularized by division into separate organ primordia. Signaling events (small circles and boxes with arrows) within the primordia (shown as large boxes) are little influenced by events in other primordia. Mid-development, however, exhibits a high interconnectivity between elements that will later come to represent separate modules. In this diagram, mesodermal tissues signal ectodermal tissues, which in turn interact with sensory placodes. Clearly this represents only a few of the inductive processes involved in the formation of the phylotypic stage. (From R. A. Raff, in *Early Life on Earth,* edited by S. Bengtson. Copyright 1994 by Columbia University Press. Reprinted with permission of the publisher.)

ways containing such elements as receptor-ligand systems, second messengers, transcription factors, and cis-acting domains of responder genes. New regulatory pathways can be composed of existing regulatory elements rearranged to act through novel connections. Finally, many parts of development are modular, and can thus be dissociated readily to produce heterochronies and other changes. This means that genes or developmental features can be retained in the evolution of new ontogenies, with their expression shifted in time or relocated in site of expression.

The mechanisms to which I've just alluded should facilitate evolutionary changes in development. Yet in the phylotypic stage, we face extreme evolutionary stability. The stabilization of phylotypic stages may well arise because, as I suggested in a paper with Greg Wray and Jon Henry in 1991, embryos are modular, and their modularity changes qualitatively as development progresses. The schematic diagram in figure 6.6 illustrates this hypothesis. Early embryos contain few modules, and are built from a few informational elements organized along axes established in the cytoplasm of the egg or by reorganization of the cytoskeleton caused by fertilization or other external signals. On the other hand, animals at late stages of development contain many modules, complex within themselves but more or less independent from one another. The developmental programs required to give rise to limb buds, or lungs, or salivary glands are all individually complex and involve local cell-cell interactions and gene cascades. However, they do not need to speak to one another. The phylotypic stage really represents a midpoint in development, at which the early embryo has blocked out the primary germ layers and the modules characteristic of late development are just beginning to appear. It is at this developmental midpoint that ultimately widely separated and distinct modules interact with one another in complex and pervasive ways. What could be more starkly illustrative of this than the observation of Jacobson that heart mesoderm helps to induce the vertebrate eye?

Conversations among modules do evolve, and knowing how will be important in testing the role of modularity and integration in the evolutionary stability of body plans. In a striking study of this kind, Sommer and his colleagues have investigated the evolution of the cell lineages that produce the vulva in development of soil nematodes. This structure is a part of the definitive nematode body plan, and in *Caenorhabditis elegans,* arises as the result of an inductive signal from a gonadal cell called the anchor cell. The vulva is produced in soil nematodes of other families, but a number of substantial evolutionary changes have occurred in the way it is generated. In all genera examined, homologous vulva equivalence group cells arise. Some of these cells are induced by the anchor cell to produce primary or secondary vulval fates, and the remainder assume nonvulval fates. However, the number of these cells varies, as does the number of vulval precursor cells with a primary vulval fate, the number of progeny cells, and the migration of progeny cells before morphogenesis. Formation of a two-armed gonad with a central vulva appears to be primitive. Some species produce a single-armed gonad and a posterior vulva. In some of these, the vulval precursor cells are still induced by the anchor cell, despite their posterior migration. In others, no induction occurs, and the vulval precursor cells differentiate autonomously. Thus both induction and the competence of cells to respond can evolve. Sommer and his co-workers made the interesting observation that because nematodes that form a posterior vulva use the same equivalence group cells as species that form a more central vulva, rather than more posterior cell precursors, there may be a constraint in the formation of the vulval precursor cells. As with the generation of the vertebrate phylotypic stage, conserved structures are generated by varying processes.

EVOLUTION AFTER THE PHYLOTYPIC STAGE

The developmental processes that predominate during the early, middle, and late stages of development profoundly affect the kinds of evolutionary changes that can occur in different stages. Early development has a relatively simple modularity built around a global axial organization, and evolutionary changes in early development are frequent and pervasive. Changes in the steps in early development that produce the larval stage can be readily modified without affecting the development of the phylotypic stage. Mid-development is highly constrained because of complex interactions among what will eventually be discrete and noncontiguous modules. The modules established in the phylotypic stage continue their differentiation in late development, but because they no longer interact decisively with one another, they can often undergo localized modifications. Such modifications can produce striking changes in the anatomy of appendages or other body parts while still maintaining the overall body plan. Limb buds, organ primordia, and segments are examples of developing features that will become separate and isolated domains after the phylotypic stage. The modularity of late development also accounts for the widely recognized phenomenon of mosaic evolution of adult features within a lineage. Some examples are worth a brief look.

Salamanders are about the most primitive of living tetrapods, and possess a body plan not far removed from that of late Paleozoic amphibians. Nevertheless, they exhibit remarkable departures from the primitive in some specialized body structures. Larval salamanders possess a branchial arch apparatus that supports the gills and opens the buccal cavity during suction feeding. In primitive salamanders, the branchial skeleton of the adult acts as a buccal pump to fill the lungs, and also acts as a support for the tongue. Since plethodontid salamanders are lungless, the branchial skeleton has been released from the constraints of acting as a buccal pump. Lombard and Wake have shown that in plethodontids, the branchial skeleton is highly modified to provide an elongated support for the highly projectile tongue that these salamanders use to catch prey. A conserved body plan is thus coupled with a specialized and highly derived projectile tongue apparatus. Alberch has investigated the developmental origins of the adult branchial skeleton. In primitive salamanders, the anatomical changes in this structure associated with metamorphosis occur through remodeling of the larval structure. However, in plethodontids, the complexity of the change seems to have been constrained by the limits of remodeling a differentiated structure, and so two developmental compartments have evolved. One produces the larval cartilage; the other produces the adult branchial skeleton upon degeneration of the larval structure at metamorphosis. The temporal modularity of larval and adult branchial cartilage development has allowed a novel structure to evolve without disrupting either aquatic larval specializations or conserved adult structures.

An entirely different set of modifications to the vertebrate body plan has evolved in snakes. These animals have an elongate body and have substituted walking on their ribs for walking on legs. Snakes have up to 450 vertebrae, all of which are serial homologues. As illustrated by Lillie for the chick, vertebrae are derived from groups of sclerotome cells that condense between the somites. This means that the great increase in number of vertebrae in snakes has resulted from a regulatory change in number of somites. A number of developmental events have had to follow in consequence. For example, neural crest cells (as discussed by Weston, Thiery, and co-workers, as well as LeDouarin and co-workers) migrate along specific pathways over and between somites to form serially repeated nerve ganglia and pigment cells. With an increase in the number of somites, these cell migratory pathways were repeated as units corresponding to the repeated somites. The basic rules governing unit structures and interactions have remained; the number of modules has changed.

Other changes have occurred in snake anatomy. Snakes have lost their external and inner ears. Their eyes have become simplified, and have evolved a glassy covering that has replaced the conventional movable eyelid. Loss of the eyelid means that snakes sleep with their eyes "open." Some (the pit vipers such as rattlesnakes) have added a fully innervated pair of facial heat sensors. All traces of forelimbs and forelimb girdles have been lost. This makes it possible for snakes to expand their body diameter freely when swallowing their prey whole. The loss of forelimbs in snakes extends to the absence of the forelimb bud in the embryo. The bones of the skull also have been modified to facilitate the swallowing of large objects. This developmental loss and gain of body elements is mosaic, and results from events grafted onto a pattern of early development that is similar to that of lizards. Each of the separately evolved changes in snakes has involved distinct modules in postphylotypic development. The only modification of the phylotypic stage itself has been the loss of limb buds and the repetition of the basic somitic units. These are changes in number rather than in composition of homologous features, and they are not changes in the basic relationships of developing structures within the embryo.

Change in number of units is a common event in evolution. It is apparently an "easy" change because the counting of rudiments for repeated structures is under relatively simple control. This was indicated long ago by the work of Sewall Wright on control of toe number in guinea pigs. The forefeet of guinea pigs normally lack digit 1, and the hind feet digits 1 and 5. Earlier, W. E. Castle had selectively bred lines of guinea pigs that produced an additional normal small toe, in which bones, muscles, and toenail were consistently present. Wright analyzed some of Castle's guinea pigs by making crosses between four-toed and three-toed lines. His backcross experiments with various F_2 lines showed that alleles representing as few as four genes control whether a guinea pig has three or four toes. The observation that a separate set of genes controls the number of toe rudiments is important. Once

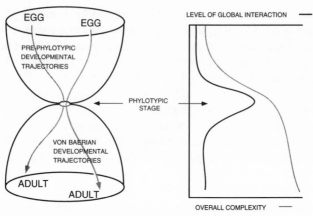

Figure 6.7. The developmental hourglass. The graph at the right plots the rise in overall complexity of an individual embryo moving through the hourglass. It also plots the level of interaction among modules. Interaction rises as the embryo approaches the phylotypic stage, and then declines as modules become more autonomous in later development.

a rudiment is established, the complex pattern of its development can be played out no matter how many genes are involved in the actual execution of the structure. The implications for morphological evolution are obvious.

In bats, an unexceptional mammalian body is paired with a unique pair of wings. The wings are highly modified forelimbs in which the joints of the fingers are extremely elongated and act as struts to support the wing membrane. Despite these extraordinary hands, the bat forelimb arises from a perfectly ordinary limb bud. The phylotypic stage of bats resembles that of other vertebrates. The pattern of ossification of the hand resembles that of other mammals, but subsequent differential growth results in elongated digits. This is a common theme in mammalian evolution. A look into a book such as Carroll's *Vertebrate Paleontology and Evolution,* which illustrates the long history and great diversity of vertebrate skeletons, quickly shows how standard are the parts from which tetrapods, including mammals, are built. Many of the same bones are present in the skull, dentition patterns are all modifications of a standard ancestral formula, and limbs are modifications of an ancient pattern of limb structure extending back to the first tetrapods.

THE DEVELOPMENTAL HOURGLASS

Both early and late stages of development evolve relatively freely compared with stages in between that we have recognized as "phylotypic." The developmental hourglass (fig. 6.7) expresses that observation. In this hourglass, development, not sand, flows from top to bottom. The arrows represent variant developmental trajectories that pass through the hourglass, converging on the phylotypic stage, then diverging again. The volume of the hourglass represents probability space, with the width of the hourglass at any level representing the probability that a change can be successfully incorporated into a developmental pathway at that level. In accordance with the schematic model of development presented in figure 6.6, early and late stages with their low linkage between component modules can accommodate changes more readily than the highly interlinked middle stage. The hourglass does not tell us what kinds of changes might be absorbed. In the evolution of sea urchin early development, which I discuss in the next chapter, it appears that changes may involve replacement of one set of mechanisms by another. That is, the system has not simply modified the existing pattern, but novel patterns (for example, in cell lineages and in the mechanics of gastrulation) have replaced ancestral ones. The new patterns arise out of existing mechanisms, but suggest that developmental changes can involve qualitative thresholds.

The developmental hourglass is a metaphor, and a metaphor may have more than a single meaning. Duboule has also arrived at the hourglass as a metaphor for conservation of the vertebrate phylotypic stage. He calls his version the "phylotypic egg-timer." In Duboule's view, the constrained phylotypic stage of vertebrates is a consequence of the linearly organized Hox genes. He has hypothesized that the action of these genes during the phylotypic stage is constrained by their genomic organization. This temporally and spatially constrained gene expression is crucial to the establishment of the anterior-posterior axis of the animal. As establishment of the body axis is crucial for organizing subsequent morphogenesis, and as the axis is organized during the phylotypic stage, the phylotypic stage has to be conserved. I think that this explanation suffers from being too narrowly tied to a much-loved set of genes, and not sufficiently cognizant of the full suite of cooperative actions required to lay out a body plan in development. The action of the Hox genes can be only a part of that suite.

Although the developmental hourglass is a good diagram for vertebrates and echinoderms, and probably other phyla as well, we should not take it as necessarily universal. There are about thirty-five living phyla. That they present phylotypic stages like that defined for vertebrates remains to be determined. For example, the phylotypic stage of the highly mosaic soil nematodes (discussed in the next chapter) seems to represent a different proposition than those of vertebrates or echinoderms with their highly regulative and interactive ontogenies. Soil nematodes have rigid early developmental patterns, and thus embryonic cell lineage patterns might be suggested as an integral part of their phylotypic stage. However, the marine nematodes discussed by Malakhov are not mosaic in development, and so nematodes, like other phyla, seem to have quite distinct early developmental trajectories converging on a shared body plan. Nematodes, seen from a broader perspective than soil nematodes, may conform to the developmental hourglass. Perhaps we should regard the phylotypic stage, in those phyla in which it can be clearly defined, as providing the best definition of the body plan for a phylum because it contains the most conservative association of constructional elements. The body plan concept will itself continue to evolve.

The existence of the phylotypic stage, and the apparent complexity of the interactions involved in its generation, would argue for limits to body plan evolvability arising from developmental constraints. Yet, if so, one would also predict that the processes that produce the constrained phylotypic stage should also be constrained. Instead, we find that early development varies widely, and that even processes that contribute more proximally to the phylotypic stage vary substantially. In sum, current information on mid-development implies that constraints on the evolution of developmental pathways arise as a result of complex interactions between modules. This statement then implies that evolutionary conservation of body plans has a component determined by internal processes in ontogeny, and that natural selection is thereby constrained in its results. The developmental constraints that define phylotypic stages could have resulted from divergent evolution and modification of the body plans of the early metazoan ancestors of the phyla that arose in the metazoan radiation. We probably see evidence for those earlier body plans in developmental controls such as the Hox gene cluster, with its deep integration into the development of basic axial features of living animals. However, it remains possible that no developmental feature really prevents the transformation of the body plan of any phylum into another. Instead, ecological constraints arising from competition in Phanerozoic environments might not have allowed animals the luxury of experimenting with new body plans that would have been competitively disadvantaged in a world of existing well-integrated anatomies. This idea suggests that experimentation was possible in the early Metazoa because no body plan was yet well adapted ecologically. The true explanation probably lies in a complex mix of developmental and ecological constraints.

7
Building Similar Animals in Different Ways

They stood there alone on a hill on prehistoric Earth and stared each other resolutely in the face . . .
"Arthur," said Ford.
"Hello? Yes?" said Arthur.
"Just believe everything I tell you, and it will be very, very simple."

Douglas Adams, *Life, the Universe and Everything*

THE EVOLUTIONARY SIGNIFICANCE OF EARLY DEVELOPMENT

The developmental hourglass poses a significant problem for evolutionary developmental biology. The neck of the hourglass presents us with a region of low probability of evolutionary change: the phylotypic stage. Prior to that stage, development is far less constrained to evolutionary change. That leads to an extraordinary phenomenon. The unconstrained upper portion of the hourglass allows a number of early developmental trajectories to reach a particular phylotypic stage. The evolutionary freedom of early ontogenetic stages is significant in providing novel developmental patterns and life histories. These stages also contain an unexplored potential for the discovery of mechanisms of evolution of the developmental features composing animal body plans.

This chapter will focus on the evolution of early development for theoretical and experimental reasons. Early development contains those processes that produce the phylotypic stage, and thus the body plan. An evolutionary understanding of early development is needed to understand those processes and their permitted variation. Early development is readily approached experimentally, and a vast body of knowledge already exists. Finally, closely related species exist that exhibit diverse modes of early development that nonetheless lead to the same phylotypic stage. These species offer a great, although little-investigated, resource for experimental studies of the mechanisms that underlie evolutionary differences in form.

The processes of early development block out the embryo. Despite this crucial role in the ontogeny of the individual, early development is evolutionarily flexible in many phyla, and thus offers an accessible way to study the evolution of development experimentally. Early embryos are simple in organization, contain few cell types, and are readily investigated. Because developmental variants occur among closely related species in many groups, comparative studies can be made in a meaningful manner with a large number of potentially interesting developmental systems. Early development is thus one of the best windows on the interface between development and evolution. As this is still a new enterprise, I'll discuss the rules for carrying out this inquiry before considering the results of current investigations. The results demonstrate the wide range of developmental processes engaged by evolution, as well as the possibilities for investigation of the role of development in all aspects of the evolution of body plans.

DICHOTOMIES AND MODEL SYSTEMS

Biology has its share of deep divisions arising from long-standing philosophical differences about how to approach biological problems. To Ernst Mayr, the most critical gap yawns between populational thinking and essentialism. Populational thought recognizes the variation among individuals of a class as real, whereas essentialistic thought seeks the typical and sees variation as noise. To Garland Allen, the chief divide lies in the naturalist-experimentalist dichotomy. The naturalist tradition continues the comparative and descriptive aspects of biology that yielded such rich insights to Darwin and his contemporaries. The naturalist tradition focuses on organism-level problems such as phylogeny, diversity, adaptation, and ecology. Its approaches are generally observational and descriptive. The experimentalist tradition intends to make biology more like the physical sciences in methodology and questions. It often takes a reductionist approach to functional problems. Rather than simply observing, it proceeds by experimentally perturbing the system under study and observing the response. A strong reductionism was not always part of this program. Embryology, which became an experimental discipline in the 1890s and has remained so since, did not become heavily reductionist until the 1960s, when studies of gene expression became a major component of developmental biology. It is significant that a joining of developmental biology with genetics was not possible until developmental biologists began framing their questions at this level. Dichotomies are visible at both ends of the twentieth century. At the beginning of the century the experimentalist-naturalist dichotomy was translated into the falling away of embryology from phylogeny. At the end of the century we see an uneasy coexistence of organismal biology and ecology with molecular biology, genetics, and developmental biology in academic biology departments. The schisms of department politics rooted in these disciplines testify to a continuing lack of intellectual integration of reductionist and essentialistic outlooks with whole-organism and populational outlooks in biology.

Dichotomies of perspective have retarded the study of evolutionary developmental biology. Embryology switched from being an observational science to an experimental one in the 1890s with the acceptance of Wilhelm Roux's program for experimentally dissecting embryos to reveal the roles of their constituent cells and processes. At that point, evolution became irrelevant because it was not seen as necessary or even helpful in the working out of mechanistic problems in development. However, a nonevolutionary approach has limited any appreciation of diversity in embryology. If one looks at most major textbooks of developmental biology, the development of each important group of animals considered is presented in its "typical" form. No doubt there is a heuristic reason for this in not confusing students with more descriptions than necessary, but there are underlying prejudices as well. Perhaps the strongest ones are the tendency toward essentialism (a fly is a fly, after all) and the seeking of common mechanisms. In these terms, diversity becomes merely an inconvenience. That disdain is unfortunate. Most nineteenth-century biologists were typologists who regarded the ideal specimen of a species as the typical one, with variants as less perfect. Darwin saw otherwise, and recognized variation as a critical part of evolutionary change. Diversity has much to teach us about development.

In the hunt for general mechanisms, the use of model systems, that is, favored organisms for the study of particular processes, has become increasingly the norm in developmental biology. Most work is now being done on the fly *Drosophila melanogaster,* the nematode *Caenorhabditis elegans,* the house mouse, *Mus musculus,* and the frog *Xenopus laevis,* and there is a new emphasis on the zebrafish, *Brachydanio rerio.* Analogously, studies of plant development are now mainly concentrated on the small weedy flowering plant *Arabidopsis thaliana.* The historically prominent marine model organisms, a few species of sea urchins, ascidians, and mollusks, have fallen out of favor. The major advantages of most of the currently favored model organisms are their amenability to study by genetic approaches and our ability to maintain them as standardized, genetically homogeneous laboratory stocks.

There has been little evident phylogenetic thought in the choice of the new model organisms, and there is a naive outlook that, although not often stated, pervades the model systems approach. As elegantly pointed out by Hanken, this pervasive linear phylogenetic thinking runs from the bacterial model *E. coli* through various animal model species and so on up to mammals (a version of this kind of implicit phylogenetic thinking is shown in the left-hand diagram of fig. 7.1). Discussions of model systems usually ignore the branching order of phylogenetic relationships (as shown in the right-hand diagram of fig. 7.1). The zebrafish is a case in point. Teleost fishes represent the crown of a whole vertebrate branch separate from the tetrapods, one that separated over 400 million years ago. In addition, teleosts have a highly

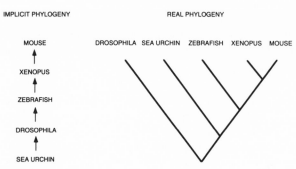

Figure 7.1.   Implicit and real phylogenies of some animals used as model systems in developmental biology.

specialized mode of development. Studies of the few living primitive bony fishes (such as the sturgeon) show that their development differs substantially from that of the zebrafish. The zebrafish may well come to be an important vertebrate model system, but not for its phylogenetic closeness to humans or its implied primitive mode of vertebrate development. Amphibians illustrate much better the primitive pattern from which all tetrapod development arose. Devoting attention to a frog genetic model system would have made more sense on phylogenetic grounds. For practical experimental reasons the zebrafish is probably the better choice, but people should be clear as to what it represents.

THE LIMITS OF MODEL SYSTEMS

If model organisms are not chosen for phylogenetic reasons, they must be chosen for some other of their properties. As my colleague Jessica Bolker has pointed out, species used as model systems may be atypical members of their groups. Some model systems have been selected accidentally, but most share crucial features that have made them particularly convenient for handling in laboratory situations. This generally means that they have short generation times, are small in size, and that their development is robust and highly canalized, and thus resistant to environmental perturbation. Model animals also often have streamlined genomes. They thus convergently share atypical attributes that make them favorable as laboratory animals. They may be more highly determined by maternal factors than typical members of their groups, and thus present a skewed set of early developmental processes. Because they are so canalized, environmental and ecological effects on development that may be quite prevalent in nature may be rendered invisible in the laboratory.

Another negative consequence of the focus on model organisms is that developmental biologists become ever more remote from an appreciation of, or even knowledge of, the diversity of developmental modes in nature and what they offer. Although textbooks feature the standard mode of development for the most intensively studied animals, there is a wide diversity in development even among quite closely related species. In part, this diversity has been ignored because of geographic as well as disciplinary chauvinism. Our texts have been written from the perspective of European and North American embryologists and species. For example, as I discuss later in this chapter, temperate-zone frogs exhibit only one life history mode and a single mode of development, the traditional one we all learned as children and subsequently through any course we took in embryology. However, Duellman and Trueb note that tropical frogs exhibit almost thirty life history modes and numerous modes of development, some very bizarre indeed. It is not that alternative developmental modes are unknown to biologists. What has happened is that developmental biologists have maintained little professional connection with the biologists who study those other species. Developmental biology as a discipline has not included the study of life history, which falls generally into the realm of natural history, ecology, or population biology. For example, much of the important work on fertilization has been done with sea urchin eggs. There are hundreds of published studies on the subject spanning a century, but none, until Pennington's 1985 study, attempted to learn how sea urchin eggs are fertilized in nature (see also Babcock and co-workers and Levitan and co-workers). Laboratory developmental biology is largely oblivious to how selection might interact with development, as well as to the diversity of developmental modes in nature.

There are other blocks to the practice of a broader developmental biology. The first is the operating demands of the discipline. Sophisticated studies on selected organisms amenable to genetic analysis are more likely to yield immediately significant insights into developmental mechanisms (and therefore more likely to be funded) than are preliminary explorations of poorly known species. The second has been the failure of developmental biologists to realize that evolutionary approaches and the exploration of developmental diversity offer a powerful probe into developmental mechanisms and the architecture of ontogeny, in a distinct and perhaps complementary manner to that provided by developmental genetics.

There can be little argument with the enormous fruitfulness of the developmental genetics program. It has yielded the most effective means of studying ontogeny ever devised, and because of its integration with molecular biology, has provided the basis for a powerful connection between evolution and development. Without the concepts, information about genes, and tools that these disciplines have produced, a meaningful investigation of nonstandard species would be very difficult, perhaps impossible. Molecular genetics, through its intensive focus on a few organisms, has in effect opened up the ontogeny of almost any organism to investigation. Evolutionary raw material in the form of different modes or pathways of development among similar organisms can thus be readily studied. The discovery of the sharing of homeotic and other regulatory genes over vast phylogenetic distances has already begun the process of bringing such nonstandard animals as crayfishes and centipedes into the experimental mainstream.

RULES FOR EVOLUTIONARY DEVELOPMENTAL BIOLOGY

I would like to briefly sketch a conceptual framework for an experimental science of evolutionary developmental biology. The examples that I will discuss are drawn from studies of early development because that is where my own work lies. However, the principles apply as well to studies of processes and features characteristic of later stages of development. An integration of data from three sources is necessary, and is immanent in the studies that I will be discussing. The first source is the comparative approach to variation in homologous features or processes in related organisms. The second is the use of a phylogenetic context that allows direction of change to be inferred. The third is the detailed study of the function of the features and processes being compared. Developmental genetic and experimental embryological studies provide the core of functional data.

There are in effect two distinct approaches to connecting developmental and evolutionary biology. A third, older approach, that of using embryological data to derive phylogenies, is nearly extinct. The most prominent current approach is to make comparisons between model systems that are phylogenetically widely separated. These comparisons have shown that major developmental regulatory genes are often highly conserved. These conserved genes may play similar or divergent roles. The homeotic genes of the Hox cluster, for example, that mediate axial polarity in flies also do so in vertebrates, although in importantly different ways. The gene *hedgehog,* which produces a signaling protein involved in determining segment polarity in *Drosophila,* also generates polarities in the development of the vertebrate central nervous system and limbs. These and other analogous observations have electrified developmental biology (see discussions of the role of *hedgehog* in *Drosophila* by Ingham and Martinez-Arias, and in vertebrates by Echelard and co-workers, Krauss and co-workers, and Riddle and co-workers). Comparisons between gene sequences have also revealed conserved domains in regions upstream of the coding sequences. These have often revealed motifs important in the binding of conserved transcription factors. Evolutionary conservation has thus become the major way of discovering universal controls in development.

I will focus on the second approach, which is the attempt to find out what mechanisms underlie evolutionary changes in development between closely related animals. This approach represents a focus on what makes ontogenies different, as opposed to an emphasis on what makes them the same. Ultimately, integration of the results of looking for conserved regulatory genes with the results of seeking proximate mechanisms of change will give us a fuller view of how body plans arose and evolved.

In developmental genetics, mutations are used to disrupt the functions of developmentally important genes. The resulting phenotypes are compared with the wild type. The defects are analyzed to define the role of the mutated gene. In evolutionary studies of development, an analogous method of comparison is used: that is, ontogenies of related species are compared. In such comparisons, the primitive mode of development must be inferred. It serves the same role as the wild type in a genetic comparison. The differences between the approaches come from the fact that in the evolutionary case, (1) a viable modified ontogeny serves in place of the defective mutant of developmental genetics; (2) the genetic differences that lie between the primitive and derived modes of development are not defined by the experimenter; and (3) usually little experimental manipulation of the genotype is possible.

To carry out a meaningful study of how development evolves, it is important to use related organisms. Close relationship minimizes genetic divergence, which can be substantial even among related species. In any detailed comparison of the ontogenies of two species, we are seeking to define specific genes and processes important in producing the differences in development. In some cases, as in sex determination and caste formation, discussed in chapter 11, or in the development of the small marine polychaete worm studied by Levin and Bridges, two distinct developmental forms occur within a single species. The studies of heterochronies in the evolution of flower development carried out by Lord and Hill exploited the alternate developmental modes that occur in reproductive parts of the same plant. An analogous pattern of modification may occur in colonial animals with multiple body units. As will be shown below, even when closely related species are compared, the differences between their developmental processes can be numerous and complex.

A second necessary part of any evolutionary investigation of development is that phylogenetic relationships must be inferred. If that is not done, all that can be determined is that the ontogenies of two organisms differ. Once organisms are placed into a phylogenetic tree, primitive and derived states of ontogenetic features can be ordered. This information is of crucial importance in going beyond scenario-making, a practice that characterizes much of the older literature. For example, the traditional hypothesis of chordate origins, discussed by de Beer in *Embryos and Ancestors,* is that the chordate condition of a dorsal neural tube and underlying notochord arose from the tadpole larva of ascidians by neoteny, a sexual maturation of the larval form. However, the gene sequence phylogeny of deuterostomes discussed in chapter 4 does not specifically support a larval origin hypothesis for chordates. The ascidians are close to the base of the chordate clade, and are thus perhaps the most primitive living chordates. However, their sedentary adults probably evolved from a motile ancestor. The immediate ancestor to ascidians may have been quite chordatelike both as larva and as adult, with a sessile adult stage evolving later. We simply don't know yet how the chordate features originated, but we are better off if we stand clear of scenarios that make us think that we know more than we do.

In principle, a cladistic analysis of features other than developmental ones (to avoid circularity) should be carried out, and developmental features mapped onto it. Molecular features offer the best approach to finding informative characters in many cases. In addition, molecular analyses allow inference of minimum rates of change by allowing estimates of times of divergence of closely related lineages. If a robust phylogeny can be found, the direction, or polarity, of evolutionary change can be readily determined. There are complex situations in which a group contains several species with derived modes of development. A phylogenetic analysis can show whether these modes were derived from a common ancestor, or whether they were derived in parallel. Parallel or convergent changes are data, not noise. They reveal preferred evolutionary pathways to particular developmental ends. As I will show later in this chapter, convergences are common in the evolution of early development.

No evolutionary comparison can be made without establishing the homologies of features being compared. Using closely related species allows homologies between developmental features to be established more readily than they can be between distant forms. The general criteria discussed in chapter 2 for establishing any homology can be applied to embryonic and larval features: position within a complex of features, similarity in structure and composition, and historical transition. All of these, with suitable adjustment for the nature of the material, can be applied in comparisons between cells and structures within embryos.

The application of the homology criteria to real cases requires sufficiently detailed data from both the primitive and derived embryos. Determination of the origins and fates of cells requires the mapping of cell lineages. Mapping is generally done by labeling precursor cells early in development by injection of a fluorescent dye or other marker and following the fates of the marked cells in some later stage. Both cell lineage data and cell fate maps for the embryo can be obtained through this procedure. Simple observation is not enough for tracing where cells arise or end up. Without specific tracing experiments it is very easy to misinterpret what is going on, because cells and even whole sheets of cells can change position without a dramatic change in the overall shape of the embryo. Even in cases in which it is obvious that movement has occurred, visualizing the paths of movement is generally not possible by inspection alone. Cell tracing experiments can yield data that fulfill the positional criterion of homology. They can also provide data to which a kind of criterion of historical transition can be applied, but in a different way than that is normally done. Historical transition is usually determined by tracing the shifts in a structure through evolutionary time by use of fossils to determine intermediate states—not a likely course with embryos. Determination of properties of cells to fulfill the criterion of structural and compositional similarity comes from cell biological and molecular studies of cellular structure, behavior, and gene expression.

It must be noted, as discussed by de Beer, by Roth, and by Shubin, that because developmental features themselves evolve, developmental criteria can be misleading and must be cautiously applied. Elements of cell behavior or gene expression can be lost, truncated, or shifted in timing. More seriously, structures inferred as homologous on the basis of anatomical similarity and historical transitions read from the fossil record have been observed to arise from different places or precursors in the development of extant species. The consequence may be a less than perfect inference of cellular homologies. Nonetheless, without at least a provisional inference of homology, evolutionary comparisons cannot be made. Again, the use of closely related species reduces the chance of incorrectly inferring homologies. I will return to this issue in discussions of the specific examples presented below.

The judicious choice of experimental organisms, the determination of phylogenies, and the mapping of features to yield solid inferences of homologies can in themselves answer significant questions. More important, they allow experimental studies to be solidly grounded in a matrix of information required for evolutionarily meaningful comparisons. Related species for which these substantial starting positions have been established are few.

LIFE HISTORY AND DEVELOPMENTAL STRATEGIES

The odd consequences of life history are all around us. One of the favorites of television nature programming is the spadefoot toad of the American Southwestern deserts. This toad sleeps away below ground for most of the year and emerges with the brief spring rains for feeding and reproducing. Its tadpoles develop in temporary pools. Their development has responded to this situation by becoming very rapid, and they have evolved another adaptation to increase their chances of survival. Although most of the tadpoles are omnivorous, some spadefoot tadpoles in a pool become carnivores, feeding on fairy shrimp and even cannibalistically. Their heads and intestinal tracts are remodeled accordingly. A revealing analysis of the consequences of this outrageous developmental behavior has been made by Pfennig. After all, if the carnivores get to be on top of the food chain, how is it that the omnivores haven't been selected away? It turns out that there is a balance between these two developmental modes within a single species. If the pools last only a short time, the carnivores develop faster and produce toadlets. However, if the pools last longer, the omnivores survive much better, and more of their genes get passed on. The genetic equilibrium is stable, and the species can use these two developmental modes to cover all prospects in an unpredictable environment.

There are many analogous situations in which some environmental feature changes ontogeny. Barnacles, which, as described by Lively, develop as truncated cones in the absence of carnivorous snails, grow bent to one side in the presence of the snails. This is an example of a developmental phenomenon known as phenotypic plasticity, which, as reviewed by Dodson, by Stearns, and by West-Eberhard, is well known to ecologists and evolutionary biologists, but is unstudied by developmental biologists. Creatures such as *Daphnia* and many other small animals develop spines or modified shapes in response to the presence of predators. In still other animals, food supply may determine developmental pathways. Tannin levels have been shown by Greene to determine developmental features of caterpillars of the moth *Nemora arizonaria*. In the spring, when caterpillars feed on oak catkins low in tannin, they mimic catkins in form. Summer broods of caterpillars feed on oak leaves with a high tannin content, and come to mimic twigs. Phenotypic responses analogous to these are widespread, and define the phenotypic responses possible to a particular genotype. These responses provide organisms with discrete variations that are adaptive under particular environmental conditions. They provide developmental biologists with a poorly exploited opportunity to explore the links between environmental cues, receptors for those cues, internal signaling and hormonal systems, and ultimately, the regulation of gene expression underlying alternative developmental programs within a single genotype.

The comparisons that I discuss in this chapter involve closely related species that differ in mode of early development in a genetically fixed manner. I'll compare species that exhibit the primitive mode of development via a feeding larval stage with those that exhibit the derived mode involving an alternative path of development from egg to adult. There is some considerable diversity of alternative developmental pathways in many phyla. Frogs that develop without tadpoles are an obvious example. I will focus on the developmental aspects of these embryos, but must mention first that the ecological context affects the evolution of development. In turn, the mode of development influences the population genetics of a species. Palumbi has shown that sea urchin species with a planktonic mode of development share genes over very wide geographic ranges. However, as McMillan, Palumbi, and I have found, species with a nonfeeding larva that remains in the water column for only two or three days show genetic differentiation over relatively small geographic ranges. Local populations become genetically differentiated because sea urchins are relatively sessile and are usually dispersed as larvae. In direct developers, the larvae tend to stay home. As Jablonski has shown for fossil mollusks, developmental mode also influences the susceptibility of species to extinction. Direct developers give rise to more species with a small geographic range, which have a higher probability of extinction than planktotrophic species.

A large number of animals exhibit indirect development via a feeding larval stage and metamorphosis to the juvenile adult. In many groups, feeding larvae have been lost, and development has become more or less direct. McEdward and Janies have suggested that style of development and ecological role of developmental stages are often conflated. Thus, among marine invertebrates, indirect development is generally associated with planktotrophy, or larval feeding in a pelagic environment. Upon metamorphosis, these larvae give rise to benthic adults. Forms in which some degree of direct development has evolved can produce pelagic nonfeeding larvae (lecithotrophs), which derive their food from stored nutrients. Finally, some of the most extreme direct developers (i.e., with loss of most larval features) are brooded by the mother. Thus, embryological, habitat, and nutritional features are all involved in patterns of development.

Developmental features, as discussed by Jablonski, by Roughgarden, and by Strathmann and Strathmann, are the consequence of selection on aspects of life history. Direct development and brooding are strongly associated with small adult body sizes because small invertebrates cannot produce a sufficient number of gametes to broadcast the large numbers of eggs needed to succeed in the plankton. It has been long known (see the reviews of Thorson and of Mileikovsky) that there is also a strong geographic correlation with the larval mode of development of marine invertebrates. Nonfeeding pelagic marine invertebrate larvae are more common in polar than in low latitudes, probably because of more limited periods of plankton productivity. Direct developers are also common in deep-sea environments. A particular developmental mode may serve more than one ecological purpose. In a case studied by Cameron and co-workers, a deep-sea urchin with a large floating egg may actually have found an improved means for long-range dispersal. The buoyancy of the egg allows the larva to float up into currents near the sea surface, which carry it far away from its parents before it sinks to the bottom to metamorphose. Direct-developing frogs are common in tropical environments because

TABLE 7.1. Feeding Larval versus Direct Development: A Summary of Costs and Benefits

Feature	Planktotrophic	Non-planktotrophic
Parent's reproductive effort	High	Low
Egg size	Small	Large
Number of eggs produced	Large	Small
Success per embryo	Low probability	High probability
Larval dispersibility	High	Low
Settlement	Unpredictable; high mortality	Predictable
Distribution	Wide range	Limited range
Genetic differentiation between populations	Low	High
Speciation	Low rate	High rate
Extinction	Low probability	High probability
Success in extreme environments	Low probability	High probability
Adult body size	Large to moderate	Small adults possible

tropical waters harbor a higher predation pressure than do temperate ponds. The message (as recognized in 1922 by Garstang) is that development is subject to selection, and developmental mode is a part of an organism's adaptation to its environment. Selection may not be operating directly on development at all, but rather, on the need to meet ecological demands, such as for a small adult to produce fewer but larger eggs, for an embryo to develop rapidly, or for larvae not to have to feed in the plankton.

I've summarized the trade-offs between alternative modes of development in table 7.1. Most of them are broadly applicable, but individual species can vary from these general rules without consulting biologists. Thus, not all trade-offs are necessarily true for all cases. For example, Rouse and Fitzhugh have observed that among some polychaete annelids there is a strong correlation between body volume and the total volume of eggs produced. They also found a correlation between body size and reproductive mode, with brooders being small and broadcast spawners large. However, there was no correlation between body volume and egg size.

CONSERVATION AND CHANGE IN EARLY DEVELOPMENT

As the earliest metazoans were, by all indications, small animals, they would have been direct-developers. With the advent of large animals in the Cambrian radiation, production of large numbers of planktotrophic larvae became possible. The feeding larval forms conserved over 530 million years likely evolved as part of the Cambrian radiation. The long conservation of larval forms implies that early development is prevented from evolving because subsequent steps of development are mechanistically constrained by early development, as suggested by Arthur. However, early development in various animal groups has undergone radical evolutionary change, sometimes in short spans of evolutionary time. There is a vast diversity of evolutionary changes in early development, occurring both in the Cambrian radiation and subsequently in well-established phyla.

DEVELOPMENTAL MODES IN SEA URCHINS

There are about 1,000 living species of sea urchins. Many species live in shallow waters, and their eggs are easily fertilized in laboratory situations. For these reasons, sea urchins rapidly became one of the favorite animals of the experimental embryology that grew up at the end of the nineteenth century. The majority of species develop in a highly stereotypic manner, as diagrammed in figure 7.2. As noted by Wray, a number of features of echinoid development have been conserved for 250 million years. These include the maternal specification of the animal-vegetal axis, radial cleavage, and an unequal fourth cleavage to form a three-tiered 16-cell embryo. The mesomeres give rise to the larval ectoderm; the macromeres to the gut, secondary mesenchyme cells, and some of the coelom; and the micromeres to the primary mesenchyme cells that secrete the skeleton and to another cell lineage that produces coelomic cells. The classic work on sea urchins has been reviewed in detail by Sven Hörstadius in his *Experimental Embryology of Echinoderms.* A much more precise cell lineage diagram and fate map for one sea urchin species, *Strongylocentrotus purpuratus,* has been assembled by Cameron and Davidson and their colleagues.

Once the hollow blastula is reached at the 256-cell stage, primary mesenchyme cells ingress into the blastocoel, and invagination of the archenteron, or larval gut, begins. Extension of the archenteron has been shown by Etten-

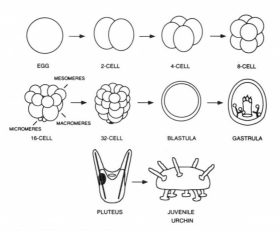

Figure 7.2.   Indirect development of a typical sea urchin.

sohn, by Burke and colleagues, and by Hardin and colleagues to involve two processes. The first is an initial invagination of a short, wide archenteron, driven by cells surrounding the site of invagination. This initial invagination is followed by extensive cell rearrangements that produce a long, narrow archenteron. This second phase of gastrulation is rather like rearranging bricks in a chimney to make it taller and narrower without using any more bricks. Filopodia produced by secondary mesenchyme cells at the tip of the archenteron were found by Hardin and McClay to guide it to the roof of the blastocoel and help to establish the site of the larval mouth. The skeleton, made of calcareous spicular rods, is secreted by the primary mesenchyme cells. These cells pattern the skeleton and produce spicules that support the extension of the initial four arms of the pluteus. A ciliary band used for swimming and food capture forms along the arm surfaces, and a larval nervous system appears at the base of the arms. The dorsal-side ectodermal cells form an epithelium. The tip of the archenteron fuses with the oral ectoderm to form the larval mouth, and the archenteron differentiates to form a functioning gut.

The pluteus feeds and grows for several weeks. It ultimately develops eight arms, and within itself begins to form the echinus rudiment, the precursor of the definitive juvenile adult. The echinus rudiment combines cells from two sources. One is an ectodermal invagination, called the vestibule, that forms on the left side of the larva. The vestibule interacts with a coelomic outgrowth, the hydrocoel, that arises from the left coelomic cavity. Together these cell sheets produce the structures that will organize and produce the juvenile sea urchin upon metamorphosis.

The pluteus mode of development has at least a 250-million-year history, and the basic pluteus design is conserved in most orders. A detailed examination by Wray showed that evolutionary modifications of the pluteus have been limited. Long conservation of a larval design can serve as an argument for the view that sea urchin development is refractory to change because of intrinsic constraints on early developmental processes. However, mechanistic conclusions based on evolutionary conservation, while providing a useful clue to function, must be drawn with sensible caution. Despite the conservation of the structure of the pluteus in the species that possess it, about 20% of living species of sea urchins lack a feeding pluteus and develop more or less directly from gastrula to juvenile.

Strathmann has considered evolution from feeding to nonfeeding larval forms as a general phenomenon in the evolution of larval stages in marine animals. Loss of feeding larvae apparently has occurred much more frequently than their gain. That inference depends upon how polarity of evolution is interpreted, given the lack of a fossil record for most larval forms. Among the diverse modes of development observed in sea urchins, the indirect mode via a pluteus is inferred to be the primitive one. There is a broad similarity among the feeding larvae of all echinoderm classes, and more specifically, all feeding sea urchin larvae are plutei. Nonfeeding sea urchin larval forms are more diverse: some exhibit some pluteus features, others essentially none. It does not appear impossible for feeding larvae to evolve from nonfeeding forms. However, it is not likely that diverse nonfeeding larvae would evolve convergently into the same feeding larval form unless there are strong constraints that limit what will work as a feeding larva to a single solution.

There are functional constraints on larvae that feed in the plankton, but a number of different mechanical solutions to larval feeding are found among marine animal phyla. Quite distinct feeding larvae exist not only among the phyla, but even within one class, the polychaete annelids. Although the trochophore larva appears to be the primitive type of feeding larva among spiralian phyla, Strathmann concluded that in a few spiralian phyla, it has been lost and a different feeding larval stage reacquired. If nonfeeding forms were evolving into feeding larvae with any significant frequency, we ought to expect a situation in sea urchins such as Strathmann hypothesized for annelids, a diversity of larval feeding structures. However, sea urchins have either nonfeeding larvae or plutei, but no other styles of feeding larvae. There is another argument that applies to the direction of evolution in sea urchin larvae: the individual sea urchin species that possess nonfeeding larvae are embedded among species that possess plutei. A phylogenetic tree for the living echinoids shows that direct-developing sea urchins are rooted among species with plutei (fig. 7.3). The topology of the tree shown in figure 7.8 likewise forces the conclusion that in this phylogeny, a direct-developing species has as its sister species an indirect developer, and that its ancestor had a pluteus larva. I'll return to the issue of evolutionary reversibility of larval features in chapter 11.

Direct-developing sea urchins are rare among Northern Hemisphere inter-tidal sea urchins, but as the summary of Emlet and co-workers shows, are prevalent in deep-sea faunas and in the shallow-water faunas of Australia and Antarctica. As argued by Emlet and co-workers, by Raff, and by Wray and Raff, the shift to direct development has been made independently among members of almost all orders of sea urchins. These deviations from ancestral planktotrophic forms show a progression. Indirect-developing sea urchins produce eggs of 65–320 μm in diameter, with most about 100 μm. Direct developers produce eggs ranging from 300 to 2,000 μm in diameter. Although egg size alone does not determine developmental style, the correlation between developmental mode and egg size is strong, and exhibits a transition point. A very few species produce eggs of about 300 μm that develop via a feeding pluteus. Of these species, at least the one that has been tested by

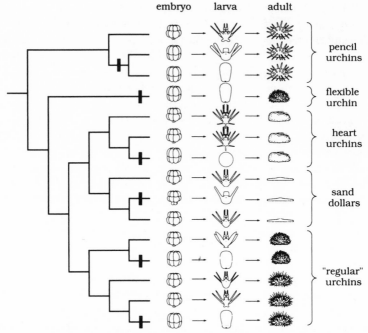

Figure 7.3. Evolution of developmental mode in sea urchins. Adult and larval forms are mapped on a phylogeny containing the major living sea urchin orders. Adult morphology is conservative in most lineages, but sand dollars and heart urchins have greatly modified adults. The pluteus larva is maintained by most lineages, and is the primitive mode of development. Direct development has arisen independently in over 20 lineages of sea urchins. (After Wray 1994; drawn by G. A. Wray.)

Emlet, *Clypeaster rosaceus,* which has a 280 μm egg, develops facultatively. It has a typical feeding pluteus, but can develop to metamorphosis if experimentally starved. *Brisaster latifrons* has 330 μm eggs, but again, produces a feeding pluteus. The eggs of *Peronella japonica* are 300 μm in diameter. They produce a two-armed partial pluteus that cannot feed and is a direct developer.

Species with large eggs produce floating lecithotrophic larvae. These larvae demonstrate that mosaic evolution occurs in features of early development as it does in features of late development. As shown in the case of *Asthenosoma ijimai* by Amemiya and Emlet, these larvae can still secrete tiny larval skeletal elements in reduced arms. Or, as shown by Emlet in the case of *Heliocidaris erythrogramma,* which retains no external trace of arms, they can still produce a tiny pair of reduced pluteus arm rods. Some direct developers retain development of other larval features, such as a ciliary band. The largest known eggs, such as the 1.3 mm egg of the brooding Antarctic heart urchin *Abatus cordatus* studied by Schatt, exhibit no larval features at all. Direct-developing larvae spend a short time in the water column and then undergo metamorphosis to a juvenile urchin.

This overlap in egg size and mode of development in some sea urchin species provides an important clue to the key innovation that opened up the direct-developing mode to so many species. A few sea urchin species produce a large egg that develops into a pluteus capable of feeding. This condition opens the interesting evolutionary possibility that once an egg large enough to produce a facultatively feeding larva exists, feeding can be eliminated, larval feeding structures lost, and development of the adult accelerated. Facultative feeders thus lie on an unstable cusp. In achieving direct development from such an ancestor, selection need not have operated on development per se, but on the ecological consequences of the change in developmental mode. West-Eberhard has emphasized that through the facultative expression of important features (such as ability to feed), new adaptive zones can be entered without passing through a nonadaptive "valley" between distinct adaptive peaks.

Developmental modes are diverse among sea urchins, even among closely related species (fig. 7.4). The embryology of two of the big primitive Australian slate pencil urchins of the genus *Phyllacanthus* has been described by Parks and co-workers and by Olsen and co-workers. Both species are direct developers, but they are quite different from each other. *P. imperialis* has a partial nonfeeding four-armed pluteus, whereas *P. parvispinus* has a football-shaped larva with no pluteus-like features. Even more surprising is the case of the most common southeastern Australian shallow-water sea urchins, *Heliocidaris tuberculata* and *H. erythrogramma*. The first of these has a small egg of 90 μm and produces a typical pluteus. The other, *H. erythrogramma*, has a 430 μm egg, about 100 times the volume of the *H. tuberculata* egg, and has gone along a very different ontogenetic track (fig. 7.5). This species was shown by Williams and Anderson to undergo direct development from egg to juvenile in about three and a half days. When I first learned that there were direct-developing sea urchins, it struck me that *H. erythrogramma* offered an embryo that could be used in the experimental dissection of the processes involved in evolutionary modifications of early development. I have been accused of making regular trips to Sydney ever since then to escape the cold Midwestern winters. In truth, our travels are driven by the pulse of the animal's reproductive cycle.

H. tuberculata was placed in the same genus as *H. erythrogramma* by taxonomists on the basis of the morphology of the adults, but I was skeptical that so large a difference in early development could occur between species of the same genus. We determined genetic distances between the species by single-copy DNA hybridization (Smith and co-workers) and by mitochondrial

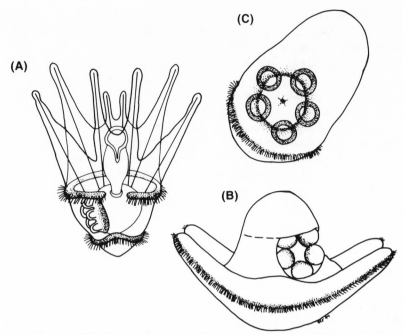

Figure 7.4.   Three styles of larvae in sea urchins. *(A)* A typical feeding pluteus of *Strongylocen-trotus droebachiensis*. The rudiment that will give rise to the juvenile sea urchin and its five primary podia is visible through the largely transparent body, adjacent to the stomach. *(B)* The nonfeeding partial pluteus of *Phyllacanthus imperialis,* with the juvenile rudiment lying in the oral region between the reduced pluteus arms. Several pluteus features are retained, including four arms, an internal larval skeleton, and a ciliary band running along the arms. *(C)* The direct-developing larva of *Heliocidaris erythrogramma*. A few pluteus features are retained: rudimentary larval skeletal elements, a larval nervous system arrayed along a relict larval oral-aboral axis, and a reduced ciliary band.

DNA restriction site differences (McMillan and co-workers). The single-copy DNA distances were consistent with a 10–13-million-year divergence between the species, and the mitochondrial distances suggested a 5–8-million-year divergence, as expected for urchins within the same genus. The separation of the lineages is also consistent with the shift to direct development in the *H. erythrogramma* lineage having been driven by the cooling of the southern oceans about 5–10 million years ago, at the end of the Miocene epoch. The shift in developmental pattern has not taken the full 10 million years or so since divergence, because the west and east coast populations of *H. erythrogramma* have been separated for about 2 million years, but develop in the same way. My suspicion is that the transition to direct development took place relatively soon after the separation of the lineages, because there

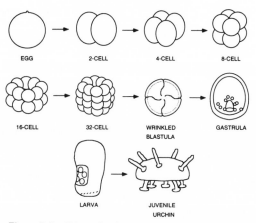

Figure 7.5.   Direct development of *H. erythrogramma*.

would have been strong selection for an effective ontogeny. Additional modifications may have occurred more gradually following the initial shift.

RADICALLY REORGANIZING DEVELOPMENT

One might consider the evolution of direct development as merely a form of degenerative evolution, analogous to the loss of eyes in cave fish, in which dispensable larval features are reduced or lost altogether. Not so. Simple loss of features is an inadequate description of the modifications that have occurred. Other simple models are also inadequate. My first hypothesis about *H. erythrogramma* was that the evolution of direct development involved a suite of heterochronic changes in developmental timing. Timing changes are evident in the abbreviation of development of the archenteron and larval skeleton, and in the early onset and acceleration of development of coelomic cavities and the adult skeleton. However, these heterochronies appear to be the results of more fundamental underlying changes that do not necessarily involve timing per se. The entire program of morphogenesis in *H. erythrogramma* has been radically remodeled. The striking degree of change became obvious after Greg Wray and I began to determine the fate map of *H. erythrogramma* by injecting fluorescent tracer dyes into individual embryonic cells and tracing the labeled descendant cells into larval stages. A comparison of the 32-cell fate map of *H. erythrogramma* with that of an indirect-developing sea urchin is shown in figure 7.6.

Modifications of both internal and external cell fates in *H. erythrogramma* are extensive and show that no unitary mechanism or single-gene change underlies the switch in developmental mode. In both modes of development, all animal-tier cells, along with some derived from the vegetal cells, give

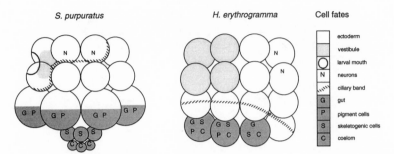

Figure 7.6. The larval cell fate map of the 32-cell *H. erythrogramma* embryo compared with that of an indirect-developing sea urchin 32-cell embryo. Larval cell types are indicated in the key. (After Wray and Raff 1990.)

rise to larval ectoderm. However, at least four topological differences in the origins of particular ectodermal cells have evolved in *H. erythrogramma*. First, there has been an assumption of ectodermal fates by the most vegetal cells on the dorsal side. In indirect development, the fate map shows no strong dorso-ventral difference. Experiments by Henry and Raff, described below, indicate that the maternal organization of the egg cytoplasm has changed in *H. erythrogramma* to substantially modify its dorso-ventral symmetry. Second, cell labeling reveals that much of the ventral-side ectoderm becomes internalized in *H. erythrogramma*. That has resulted in different patterns of ectodermal shape change as the larva forms. Moreover, it has provided a basis for the evolution of a novel gastrulation mechanism. Third, the ciliary band has changed in origin from primarily animal-tier cells to entirely vegetal-tier cells. Is it homologous to the ciliary band of the indirect developer despite its change in cellular origin? Recent studies on the cellular origins of the pluteus ciliated band by Cameron and his colleagues in an indirect-developing species provide an important clue. Cell tracing experiments show that the position of the ciliated band is not fixed by cell lineage, but that cell-cell communication must be involved. The signal and its source are not yet defined, but a change in the site of the signaling cells may have repositioned the ciliated band. Both the band and the developmental processes would be homologous, although shifted in space. This model of evolutionary change in inductive signal source remains to be tested. Finally, the larval mouth has been lost. We don't know what cell fate changes occurred in that loss, or if autonomous or cell-cell signaling mechanisms characteristic of mouth formation in the pluteus were abandoned.

There also have been large-scale shifts in the allocation of ectodermal cells. In the 32-cell indirect-developing embryo, 16 cells produce only ectoderm; thus a large structure in the juvenile can begin with a small investment of embryonic cells. Structures grow as the larva feeds. In *H. erythrogramma*,

26 of 32 cells have an ectodermal fate. In large part, the difference arises because the direct-developing larva cannot grow. Making a juvenile sea urchin directly requires reapportionment of cells in the embryo, and this has happened in the evolution of *H. erythrogramma* development. In the indirect developer, only parts of 2 left-side cells of the 32-cell embryo are fated to produce the vestibule later in larval development. In *H. erythrogramma,* 8 cells, a full quarter of the embryo, are fated to produce the vestibule early in larval development. The recruitment of a large fraction of the ectoderm to a vestibular fate makes morphogenetic sense in that the vestibule has to form early and at its final size.

The internal cell fates of *H. erythrogramma* have also been modified extensively, and cell types homologous to those of indirect developers have different precursors. Embryonic cleavage patterns are different. For example, all of the early cells in *H. erythrogramma* are the same size. Thus, the unequal cleavages typical of indirect-developing sea urchins have been eliminated. However, the modifications in processes that give rise to internal cell types run deeper than simple cleavage geometry. At least three kinds of changes have occurred. The first are alterations in timing. Some cell type precursors arise later than in indirect development, and others earlier. The fact that heterochronies occur at the cellular level suggests that some timing changes in basic regulatory processes may have occurred. The heterogeneity of the timing changes suggests that no single mechanism is involved. A second alteration mirrors the early dorso-ventral determination of ectodermal fates and larval symmetry. In *H. erythrogramma,* the most dorsal vegetal cells have lost their primitive internal fates. Third, there are differences in the order of founder cell origins. The simple cell lineage comparison shown in figure 7.7 makes this point. Indirect developers contain a cell type, the micromere, that gives rise to two founder cells in its next division. In subsequent divisions, these founder cells give rise respectively to coelomic cells and skeleton-secreting cells. No cell type equivalent to the micromere exists in *H. erythrogramma.* A change in cell lineage organization has occurred, and it would appear to involve a change in maternal information. As I'll show below, a change in cell lineage organization is unusual among the suite of evolutionary changes in early development, and implies a fundamental reorganization of early development in *H. erythrogramma.*

We are only beginning to understand the mechanisms that have changed the early development of *H. erythrogramma* so radically. Maternally encoded information plays a large role in the organization of early embryos. Messenger RNAs and proteins required for early development are important components of eggs, and it appears that at least one of the primary axes of the embryo is rigidly determined in the eggs of all animals. The animal-vegetal axis of sea urchin eggs is firmly established in the unfertilized eggs of both

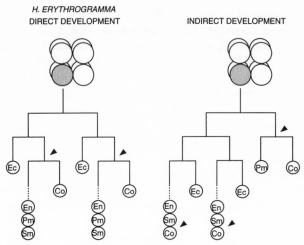

Figure 7.7. Embryonic cell lineages in *H. erythrogramma* are highly modified in comparison with those of indirect-developing sea urchins. Eight-cell embryos of *H. erythrogramma* and a related indirect-developing sea urchin are shown. The cell lineage of a single vegetal pole blastomere is traced for both. The four vegetal blastomeres of the indirect developer all give rise to the same set of lineages. In *H. erythrogramma*, ventral blastomeres differ from dorsal blastomeres in lineages produced (not shown). The arrows indicate the founder cells for coelomic fates. No lineage in *H. erythrogramma* undergoes the typical division of the micromeres to produce a coelomic founder and a primary mesenchyme cell founder. Co = coelomic cell founder; Ec = ectodermal; En = endodermal; Pm = primary mesenchyme (skeleton-secreting cells); Sm = secondary mesenchyme. (After Wray and Raff 1990.)

indirect and direct developers and, as noted by Wray, is an invariant feature of cell determination in echinoderm eggs. Our cell fate studies strongly imply that the dorso-ventral axis is set maternally as well. Henry, Wray, and I have found that neither the site of sperm entry nor the plane of the first cleavage sets up the dorso-ventral axis in *H. erythrogramma*.

The dorso-ventral axis appears from the experiments of Hörstadius to be resident in the eggs of indirect-developing sea urchins, but not as rigidly set as in *H. erythrogramma*. The evolutionary changes in *H. erythrogramma* show that features that are co-localized in the eggs of indirect developers are mechanistically distinct and dissociable. We showed by culturing separated parts of *H. erythrogramma* embryos that the capacity to form certain cell types varies with position. Thus, both halves of an embryo separated at the 2-cell stage produce a half-sized larva. The dorsal half, however, has little ability to form mesenchyme cells, in contrast to the embryo derived from the ventral half. This result is consistent both with the fate map and with the hypothesis that maternal informational elements are segregated toward the ventral side of the egg. In indirect-developing eggs, the determinants that specify mesenchyme cell differentiation are located vegetally, and have no ventral tendency. There has been a relocation, and a revealing one. The vegetal pole of the animal-vegetal axis and the site of primary mesenchyme cell specification are coincident in the indirect developer, but not in *H. erythrogramma.* Further, the distinct mode of vegetal pole cell unequal cleavage characteristic of indirect developers has been obviated (compare figs. 7.2 and 7.5).

Although the dorso-ventral pattern is specified maternally in *H. erythrogramma,* its expression requires embryonic gene expression and cell-cell communication. Henry and Raff showed this very simply by separating dorsal and ventral halves at successively later times in development. There was a progressive decline in the ability of dorsal halves to produce ventral structures, consistent with a repressive signal propagated along the dorso-ventral axis. The use of maternal information to block out pattern, followed by increasingly complex patterns of cell-cell interaction (classically called induction by embryologists) and transcriptional specialization of groups of cells within the embryo, is a very general strategy in early development. An evolutionary look reveals that these mechanisms are strikingly flexible.

CHANGES IN MORPHOGENESIS

Morphogenesis is distinct in *H. erythrogramma* from cleavage onward. As Henry and I have shown, a wrinkled blastula is formed, which is not an unusual event for large-egged echinoderms. The wrinkles then disappear through a process of cell elongation and tighter packing, and finally a typical-looking blastula is formed. Williams and Anderson, as well as Parks and co-workers, showed that a very large number of mesenchyme cells, about 2,000, enter the blastocoel, and the archenteron invaginates.

If any process should be conserved in evolution, it is gastrulation. After all, it's in the gastrula that the morphogenetic movements that set up the basic tissue layers take place. Yet profound changes in gastrulation have occurred in *H. erythrogramma.* The first phase of gastrulation resembles that of indirect developers, but as Wray and I found, the second phase is entirely novel. In indirect developers, archenteron elongation results from cell rearrangement, but in *H. erythrogramma,* a sheet of cells involutes from the ventral side of the embryo. As the embryo cannot grow, it must use an alternative means of producing a large sheet of cells from which it can construct a coelom. This new means of elongating the archenteron has evolved not to produce a larval gut, but to provide raw material for subsequent accelerated adult morphogenesis. This novel process cannot have evolved by modification of the ancestral cell rearrangement, but had to replace the ancestral mechanism. The evolutionary precursor to the new mechanism is not known.

SIMILAR GENES, DIFFERENT EMBRYOS

Current ideas of how development works are largely founded on the principle that differential gene expression underlies differences in morphological development. The very distinct morphologies that we have observed in indirect- and direct-developing sea urchin embryos suggest that their patterns of gene expression should be quite distinct. *H. erythrogramma* was observed by Raff and co-workers to have a 30% larger genome than *H. tuberculata*, and Smith and co-workers showed that much of that had to be single-copy DNA (potentially genes). Such differences could underlie quite profound genetic differences between the species. There are two kinds of hypotheses we might use to connect the differences in morphogenesis with differences in gene expression. In one, we could, as is classically done, regard the direct-developing embryo as having lost many features and thus as being reduced or simplified. In this case, one might expect loss of expression of genes involved in constructing the larva. Alternatively, the direct developer might have evolved novel developmental features. In this case, new genes, as well as changes in the timing of gene expression, may have evolved to regulate the accelerated development by a novel path to the phylotypic stage.

Because structural genes are the ultimate targets of regulatory genes, changes in their expression may reveal patterns of regulatory gene evolution. The changes in gene expression in *H. erythrogramma* indicate that a wealth of subtle modifications may have occurred. For example, there is a substantial delay in secretion of the skeleton relative to indirect-developing sea urchins, and the expression of a mesenchyme-specific protein (msp130) involved in skeleton secretion is concomitantly delayed in its expression. Studies of the control of expression of the *msp130* gene by Klueg, Harkey, and I show that transcriptional regulatory sites upstream of the gene have not changed in function. The delay in expression is apparently due to changes in transcription factor activity. It would seem that modifying the promoter of the *msp130* gene would be a more probable and parsimonious change than modifying the transcriptional regulation of the cell. Indeed, this kind of single-gene cis-regulatory change has occurred in the recruitment of genes to new sites of expression (as discussed in chapter 11). However, skeletogenesis requires many genes along with *msp130,* and so the timing of the whole transcriptional program has been modified.

Structural genes and their expression patterns have evolved in the *Heliocidaris* species in ways that are not fully understood, but which indicate a surprising fluidity of gene evolution. The evolutionary behavior of the actin gene family is illustrated in figure 7.8. *Strongylocentrotus purpuratus* and *Lytechinus pictus* are outgroups closely related to the two *Heliocidaris* species, but have different numbers of cytoplasmic actin genes. Despite the

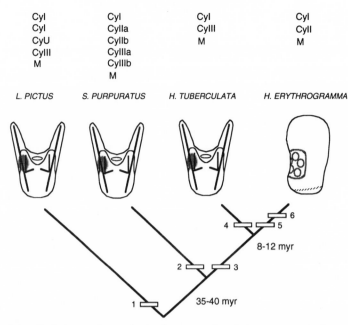

CyI	CyI	CyI	CyI
CyI	CyIIa	CyIII	CyII
CyU	CyIIb	M	M
CyIII	CyIIIa		
M	CyIIIb		
	M		

| *L. PICTUS* | *S. PURPURATUS* | *H. TUBERCULATA* | *H. ERYTHROGRAMMA* |

Figure 7.8. Evolutionary changes in development and evolutionary events in the actin gene family mapped on the phylogeny of some camarodont sea urchins. Cladisic events: (1) novel CyU gene, (2) duplication of CyII and CyIII genes, (3) reduced number of actin genes expressed embryonically, (4) loss of CyII embryonic expression, (5) loss of CyIII gene, (6) switch to direct development. (Data from Fang and Brandhorst, Lee and co-workers, and Kissinger, Hahn, and Raff.)

closeness of relationship and the similarity of development of *S. purpuratus* and *H. tuberculata,* we (Kissinger and co-workers) found that both *Heliocidaris* species express only two cytoplasmic actin genes. These are a subset of the *S. purpuratus* genes. Actins play important roles in cellular behavior during development, and one would presume that nearly identical patterns of development would have similar molecular requirements. Like *S. purpuratus, H. tuberculata* develops a pluteus larva. Nevertheless, both the number of actin genes and their tissue-specific patterns of expression have changed. These alterations in gene expression correlate with evolutionary time, not with mode of development. *H. erythrogramma* embryos also express only two cytoplasmic actin genes, again a subset of the *S. purpuratus* genes. One actin is shared with *H. tuberculata;* the other is not. *H. erythrogramma* has only a pseudogene for CyIII actin instead of an active gene, a change that does correlate with change in development. *H. erythrogramma* does not make the squamous epithelial cells that express CyIII actin in the pluteus. The pattern of evolution of actin genes in sea urchins suggests that both the passage of evolutionary time (neutral evolution) and changes in developmental mode (adaptive evolution) are occurring. How it is that genes that are so highly regulated in sites of expression can substitute so freely for one another among related species remains unexplained.

Detection of regulatory genes known from model systems in other organisms is currently a favored research approach. It offers known genes and defined probes for cloning or sequences from which PCR primers can be prepared. For evolutionary developmental biology, known regulators supply the investigator who is using a nonmodel organism with genes whose expression and function can be compared with a better-known system. Not surprisingly, known regulatory genes are present in the *Heliocidaris* species, and we have begun to compare their expression patterns across developmental modes. As an example, Ferkowicz and Raff have isolated clones of members of the *wnt* gene family, which play roles in cell-cell signaling in a number of developmental settings (see Parr and McMahon). The *Heliocidaris wnt-5A* gene is a homologue of one of the genes that Takada and co-workers found to be expressed in the mouse primitive streak. This gene has a complex set of expression patterns in vertebrates, and has been implicated in morphogenetic movements. In situ hybridization reveals that in the gastrula embryos of both *Heliocidaris* species it is expressed similarly, in the vegetal plate. Somewhat later in *H. erythrogramma* gastrulation, its major expression is in the coelomic pouch, which is destined to contribute the mesodermal component of the juvenile. In later embryos, the *wnt-5A* expression pattern resolves into a pentameral array. It appears that there is an acceleration of mesodermal patterning in the direct developer.

An alternative way of seeking posited genetic differences between *H. erythrogramma* and *H. tuberculata* is by making a global screen for gene expression differences. A comparison of mRNAs of late gastrula through early prism stage *H. tuberculata* with mRNAs of late gastrula through early larval stage *H. erythrogramma* is being undertaken by Haag and Raff. The morphologies of the two embryos already differ at these stages, and morphogenetic pathways diverge widely from these stages on. *H. tuberculata* is differentiating a skeleton and a functioning gut. *H. erythrogramma* is not yet secreting a skeleton, but is forming a large coelomic outpouch. The coelomic pouches of *H. tuberculata* are still many hours off. Yet the vast majority of prevalent mRNAs are shared. Despite the nascent differences in morphogenesis, the transcript inventory is conserved. Although the transcript catalogs of *H. tuberculata* and *H. erythrogramma* are similar, we have detected some gene expression heterochronies and some *H. erythrogramma*-specific and *H. tuberculata*-specific sequences, but the significance of these transcript differences is still unknown. The similarities seen in a raw transcript screen no doubt conceal many subtleties, but the structural building blocks of these two morphologically distinct embryos are very similar. The differences, as in the case of apes and humans, which also exhibit high structural gene similarity, lie in regulatory genes. Differences in morphogenesis may lie in a few key selector genes, in spatial differences in the expression of shared genes, in different patterns of processing of transcripts, or most likely in some combination of these.

PARALLEL AND DIVERGENT MECHANISMS

I have stressed *H. erythrogramma* both because one loves one's own children and because it has been so amenable to comparisons of developmental processes at several levels. Comparisons between *H. erythrogramma* and related indirect-developing sea urchins demonstrate that the evolution of direct development was a rapid, but not a simple, event. Direct development arose from an indirect-developing ancestor. Features of the pluteus were lost or reduced, and features of adult development appeared early. These shifts in timing provide only a superficial description of what has occurred. A deeper examination of the differences in developmental pattern between *H. erythrogramma* and related indirect developers has revealed that ontogeny has been radically remodeled. The most striking aspects of change are in cell lineages and cell fates. The *H. erythrogramma* embryo gives rise to the typical echinoderm phylotypic stage, but by a pathway quite different from that of indirect-developing sea urchins.

Another important question is illuminated by investigation of direct-developing sea urchins: whether the same or distinct mechanisms underlie the independent origins of direct development in various lineages. If the same mechanisms occur in parallel, that would indicate that the course of evolution of early development reflects internal constraints. Some features of gametes that Raff and co-workers investigated in 1990 suggest strong parallel trends. Direct developers have larger genome sizes than sister taxa with indirect development. Parallelism is also suggested by the fact that all direct developers have large eggs, and most have elongate sperm heads. The adaptive role of large eggs is apparent, and elongate sperm heads are needed to penetrate the thicker jelly coat of the larger eggs. An adaptive role for a larger genome size is not apparent. A real test of mechanistic diversity will require a detailed examination of the cell lineages and embryology of other direct developers. Few data are as yet available, but observations on *Holopneustes purpurescens* by Morris suggest that other mechanisms could govern early pattern formation. Unlike that of *H. erythrogramma,* the dorso-ventral axis of *Holopneustes* appears to be established before first cleavage by cytoplasmic movements relative to the axis of sperm entry.

EXPERIMENTAL APPROACHES TO THE EVOLUTION
OF EARLY DEVELOPMENT

During the past few years, the pace of experimental studies on the evolution of development has accelerated, and they have begun to engage a great phylogenetic span of animals. I won't pretend that what follows is a complete review of these studies, but it's a sampling of important current work that includes a range of taxa and developmental processes. Most model systems have not been chosen with an eye to their phylogenetic positions, but they are scattered among several phyla: nematodes, arthropods, mollusks, echinoderms, fishes, and tetrapod vertebrates. Results from model species related to species with divergent developmental patterns can be readily applied to evolutionary comparisons involving the divergent species. The use of a range of taxa allows us to seek commonalties of mechanism. In some cases, similar evolutionary alterations in development may have evolved independently in disparate taxa. Evolution by convergent means may indicate powerful constraints on the architecture of developing systems or on the processes available.

Nematodes

Soil nematodes have emerged as useful organisms for studying evolutionary developmental biology because a number of close relatives of *Caenorhabditis elegans* vary in anatomy and development. *C. elegans* has become a major model organism for developmental genetics. Although it is a complex metazoan, the adult hermaphrodite has only 959 somatic cells. *C. elegans* exhibits a highly stereotypic pattern of cell lineages and developmental events. Much of its development is cell-autonomous, which has simplified the analysis of developmental controls. The cell lineages of other soil nematodes can be readily compared with *C. elegans,* with its well-defined cell lineage and thoroughly studied developmental genetics. Ambros and Fixsen have noted that three general kinds of cell fate transformations are observed in mutants of *C. elegans:* temporal transformations, in which cells express fates normally occurring at another time in development; spatial transformations, in which cells adopt a fate normally expressed at another site; and sexual transformations, in which cells express fates normally expressed in the other sex. They argued that cell lineage differences among nematode species can be interpreted as corresponding to these mutational changes within *C. elegans.*

An example of alteration in cell fate is shown by the development of the gonads of *Panagrellus redivivus* compared with those of *C. elegans.* These two soil nematodes belong to related families. They differ in that whereas *C. elegans* has two ovaries opening into a common vulva, *P. redivivus* has a single gonad. Sternberg and Horvitz found that the cell lineages of the precursor cells to the gonads are similar, but differ in one crucial respect. In

C. elegans, these cells produce a cell called the distal tip cell at the end of each ovary. This cell is vital for growth of the gonad arm and for mitosis of germ cells. In *P. redivivus,* one distal tip fate has been replaced by a programmed cell death fate, itself under genetic control.

Single-gene mutations that cause switches in cell fate are well known. Most such changes don't produce novel fates, but rather, shifts from one existing fate to another or production of the same set of cell fates but in new locations. In the evolution of features such as the number of gonad arms, existing processes are co-opted. In a formal sense, cell fates are mediated either by autonomous regulation of differentiation or by directions provided by other cells. What is required for evolution is that these processes have a genetic basis and be accessible to evolutionary modification under selection.

Studies of *C. elegans* have been especially effective in revealing mechanisms of cell fate determination. Kemphues and co-workers in 1988 first obtained maternal effect genes, called *par-1* through *par-4,* that affect the differentiation of cells during early development. In consequence, *par-1* embryos produce extra pharyngeal cells. These mutations apparently interfere with the localization of maternally encoded cytoplasmic determinants. Control of early cell lineage patterns is also governed by maternal effect genes. Schnabel and Schnabel showed that maternal expression of *cib-1* is required to specify a set of stem cell-like blastomeres. Lack of proper expression of *cib-1* causes a set of early stem cells that normally give rise to both germ line and somatic cell fates to fail to produce the germ line lineage. Instead, only somatic cell lineages are produced. Another set of maternally expressed genes, *skn-1, pie-1,* and *mex-1,* have been shown by Mello and co-workers and Bowerman and co-workers to be required for proper spatial differentiation of pharyngeal cells. The *skn-1* gene encodes a presumptive transcription factor that is localized in nuclei. This protein accumulates to a higher level in cells that are destined to produce pharyngeal cells autonomously than it does in sister cells in the embryo that do not do so. It may well be a master switch that determines gene expression and cell fate in these cells. The *pie-1* and *mex-1* genes regulate the localization of the *skn-1* gene product to particular cells. Mutations in these genes cause a mislocalization, and they also cause particular embryonic cells to follow cell lineage patterns characteristic of different embryonic cells. They are thus cell lineage homeotic genes, and exert their effects on cell lineages through maternally encoded products. These cell fate transformations suggest that modifications of analogous maternally encoded localization and cell fate determinants might play a role in the evolution of early cell lineages, as observed in the direct-developing sea urchin *H. erythrogramma.*

Skiba and Schierenberg have compared early cell lineages of several soil nematodes with those of *C. elegans.* The early cell lineage patterns of species of *Rhabditis* and *Panagrolaimus* were identical to those of *C. elegans,* but showed heterochronic changes in the germ line lineage. The *Cephalobus* species examined showed heterochronic changes in early cell lineages as well as an inversion in the spatial arrangement of cell products in the germ line cell lineage. In studies of postembryonic cell lineages among nematode species, Ambros and Fixsen have found both cell lineage and heterochronic alterations. These show that cell lineage transformations are not prohibited among related soil nematodes despite an overall conservative mode of development.

Evolutionary conclusions reflecting dependence on model organisms can be limiting or even misleading. The nematodes powerfully illustrate this point. Until very recently, developmental studies have concentrated on soil nematodes. *C. elegans* development is highly mosaic. Most differentiation is cell-autonomous, although some important cell-cell interactions occur. Priess and co-workers found that the embryonic cells that eventually produce the pharynx depend on an inductive interaction early in development with a neighboring cell. The gene *glp-1,* which encodes a membrane receptor, is required. This same gene, according to Kimble, appears to be required to determine other major processes of axial polarity in the embryo as well. Nonetheless, the mosaic development of *C. elegans* has been extrapolated to nematodes as a whole. The extension of developmental studies to a marine nematode by Malakhov and his co-workers shows that some nematodes are not like soil nematodes, but can be very regulative. At this point few experimental data are available, but the study of more diverse nematodes is likely to yield a range of early developmental trajectories that may rival those of other phyla despite a highly conserved adult body plan. The rigid mosaic development of soil nematodes may be a consequence of rapid development.

A Molluscan Diversion

The spiralian groups, mollusks, annelids, and other protostomes, are very similar in development. Spiral cleavage and cell lineage patterns are highly conserved and have been used as phylogenetic features linking these phyla. (A generalized view of spiral cleavage is presented in figure 2.3.) However, even within these conserved patterns, there have been remarkable changes in mechanisms. The most important is the way in which the D quadrant, which plays a vital role in establishing differentiation patterns in the spiralian embryo, is determined. It has traditionally been thought of as being rigidly maternally determined. However, Freeman and Lundelius have shown that among a wide spectrum of mollusks, there are two modes of D quadrant determination, the classic maternal mode and a second mode by cell-cell interaction in the embryo. These examples show that radical changes in cell determination mechanisms occur in groups other than sea urchins.

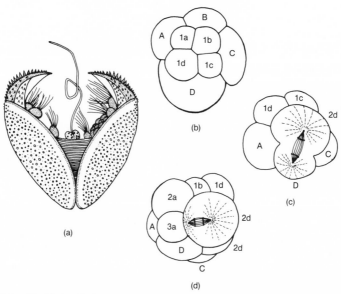

Figure 7.9. Modified cell cleavage patterns and the development of the glochidium larva of the freshwater clam *Unio*. *(A)* The glochidium larva, with its bear-trap valves and sensory trigger hairs. *(B)* An eight-cell embryo with typical spiral cleavage pattern. *(C)* Modification of cleavage to produce a large 2d micromere, which produces the larval shell gland. The shell gland secretes the huge larval shell. In marine mollusks this micromere is primitively smaller than its sister cell, the 2D macromere. *(D)* Continued cleavage. Note the large 2a micromere, which gives rise to the larval adductor muscle. (From Raff and Kaufman 1983; drawn by E. C. Raff. Reprinted with permission of Indiana University Press.)

Morphologically striking modifications of early development have evolved in mollusks and annelids. The first study ever made of an evolutionarily changed pattern of early development was carried out by F. R. Lillie in the 1890s using the freshwater clam *Unio*. Unionid clams live in freshwater streams and so face a serious problem of dispersal. The adults are effectively sessile, and planktonic larvae would be swept downstream and lost. These clams produce large numbers of embryos, which the mother broods until they reach the larval glochidium stage. As shown in figure 7.9, the glochidium resembles a tiny bear trap. This larva is a novel form. It does not feed planktonically, nor is it a direct developer. To develop, it must attach to the gills or fins of a fish. There it assumes the life of a parasite until it finally drops to the stream floor as a little clam. Two problems, dispersal and larval feeding, are solved in one snap of the trap. In some species, the glochidia are simply launched into the water to wait on the bottom for their ride to stop by. In other, more sophisticated species, the female clam plays a mean trick. She possesses a mantle with its margin modified to resemble the body,

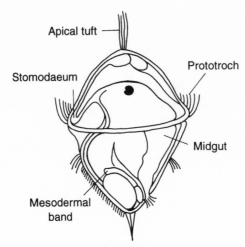

Figure 7.10.   Trochophore larva. (From Raff and Kaufman 1983; drawn by E. C. Raff. Reprinted with permission of Indiana University Press.)

fins, and eyes of a minnow. When the female is ready to release her larvae, she undulates her mantle margin. If a passing fish investigates this Trojan minnow, it gets bombarded by a cloud of glochidia. The glochidium has sensitive sensory hairs, and the slightest disturbance triggers the powerful valves with their large hooks to snap shut—with luck, on soft tissue.

Lillie showed that although *Unio* has modified its early development to produce a novel larva, it has retained the basic spiralian cell lineage pattern. In most mollusks and annelids, the micromeres of the first tier give rise to the ciliated apical region and the ciliary band (prototroch) of the trochophore, the primitive larval form in these phyla (figure 7.10). These structures are not produced in *Unio*. Accordingly, the rate of cell division of the first tier of micromeres in *Unio* is retarded relative to the second tier. Material has been reallocated to meet altered constructional needs. The first-tier cells have undergone two evolutionary modifications, a slower division rate and a loss of primitive cell fates. Fewer cells are needed because ancestral features produced by these cells have been deleted. The second tier of micromeres gives rise to most of the larval structures and mass.

Certain embryonic structures of the glochidium are enlarged relative to other parts of the larva. The shell gland that produces the massive larval shell is derived from a single second-tier micromere, 2d. Cleavage has been modified such that the 2d cell is larger than its sister, the 2D macromere, and is the largest cell in the embryo. In primitive spiralian development, the macromeres are the largest cells and the micromeres are smaller and relatively uniform in size. The eight-cell *Unio* embryo shown in figure 7.9 exhibits the typical spiralian arrangement of macromeres and micromeres. As the embryo continues to divide, the D macromere produces the large 2d micromere. A second large cell of the second tier of micromeres is the 2a cell, which gives rise to the larval ectomesoderm that is destined to form the large adductor muscle of the glochidium larval shell. The daughters of the 2a and 2d cells divide more rapidly than the other second-tier micromeres. The 4d micromere of *Unio* appears early relative to its appearance in the idealized spiralian. The fates of the cells are unchanged, but size and timing changes in early cell lineages have allowed a morphological remodeling of the larva.

Lillie recognized that several factors are involved in the evolutionary modifications he observed. Cell lineages are not altered, and cell fates are conserved in the general sense of which cells give rise to the basic ectodermal, mesodermal, and endodermal cell types of the larva. There are fate changes, or at least losses, in the first-tier micromeres. The fates of the other micromeres are conserved even though cell sizes and cell division schedules are changed. The changes in relative cell sizes result from alterations in the placement of the mitotic spindle and actin ring of the cleavage furrow. These presumably reflect changes in the cytoskeleton.

As early as 1904, Wilson was able to show that in another mollusk, *Patella,* four cells of the first tier of micromeres are already determined as precursors of the ciliated prototroch. As cleavage continues, each of these cells is destined to divide twice more and then, at a set time, grow cilia arranged in transverse rows. Wilson isolated and cultured individual precursor cells and showed that these isolated cells carry the same program of two cell divisions and subsequent assembly of cilia. Thus, both the differentiative capacity and the control of rate and timing are autonomous properties of the cells and are probably the result of maternal gene action. The evolutionary modifications of early development in *Unio* are probably a consequence of genetic changes affecting maternal determination of embryonic cell behavior. Genes with analogous functions observed in *C. elegans* provide important clues to the genetic basis of evolution of early development in organisms such as mollusks and echinoderms for which genetic studies are unavailable.

Evolutionary elimination of spiralian development has occurred, apparently very anciently, in the origins of some major groups. The cephalopods are indisputably mollusks, but show no trace of spiral cleavage. Arthropods too apparently had spiralian ancestors but retain little trace of that heritage in their early embryology. What this means is that although a pattern of early development, such as in the Spiralia, can be conserved for a very long time in the face of substantial modifications of larval development, conservation is not absolute. The whole complex has been discarded and early development reorganized in at least two instances. Conservation does not imply that change cannot happen. In a very rough way, evolutionary modification of early

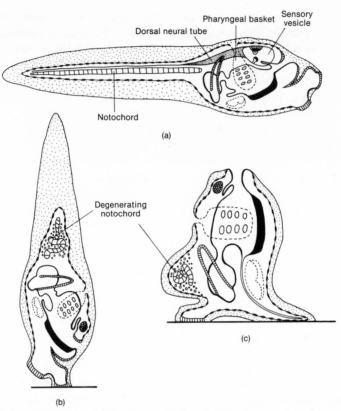

Figure 7.11. The ascidian tadpole larva and its metamorphosis into a filter-feeding adult. *(A)* Tadpole larva. *(B)* Larva attached to substrate and undergoing metamorphosis, with loss of motile and sensory structures. *(C)* Adult. (From Raff and Kaufman 1983. Drawn by E. C. Raff, after Korschelt and Heider. Reprinted with permission of Indiana University Press.)

development seems to occur in two ways: modification of a common pattern *(Unio)* or a discontinuous switch to a novel pattern (cephalopods). Both are interesting, but as always, novel features are more difficult to explain.

Evolution of Development in Primitive Chordates: Ascidians

Ascidians are the most primitive chordates. Their adults are saclike filter feeders that have an elaborate siphon and filter system, but no trace of chordate anatomy. They were thought to be mollusks until the discovery of their larvae by Kovalevsky in the 1860s, which showed that they are the most humble of the chordates. Ascidians develop via a tadpole larva that possesses a notochord, sensory organs, brain and dorsal nerve cord, and somites. Upon settling, they abandon the racy life of the plankton, resorb their tails, and metamorphose into water-pumping adults (fig. 7.11).

Berrill, in his pioneering work on the evolution of development in ascidians, found that their egg sizes range from 80 μm to about a millimeter. Ascidians do not parallel sea urchins in showing a correlation between egg size and developmental mode. Perhaps this is because ascidian larvae don't feed. They develop very rapidly through the larval stage, and metamorphosis occurs as early as twelve hours after fertilization. The tails and sense organs are involved in finding suitable settlement sites, not in feeding. Large-egged ascidians produce typical tadpoles, and gastrulation is not modified in response to a larger egg volume. Evolutionary modifications to early development in ascidians, as described by Berrill as well as by Jeffery and Swalla, include loss of the tadpole tail; adultation, in which larval and adult programs are superimposed by concomitant retardation of larval features and acceleration of adult features; and caudalization, in which extra rounds of cell division produce extra tail cells without adding other larval cells.

Tailless larval development has arisen several times among ascidians. Swalla and Jeffery have focused their studies on two closely related ascidians, *Molgula oculata*, which develops a tailed larva, and *M. occulta*, which forms a tailless larva. The larva of *M. occulta* is analogous to the direct-developing larvae of sea urchins in its abbreviation of larval structures. The eggs of the two ascidian species are similar in size. Their cleavage patterns are the same, and the same cell lineages appear to be established, but as development proceeds, cell lineages associated with tail formation fail to fully differentiate in *M. occulta*. Although some notochord cells form, they are reduced in number and fail to intercalate, swell, and extend. Tail muscle cells also fail to appear. Although the muscle cell lineage has been shown to be present in *M. occulta* (see Whittaker; Jeffery and Swalla; and Satoh and Jeffery), this tailless species has lost the capacity to differentiate tail muscle cells. The cells that would be the precursors of the muscle cells are placed properly in the embryo, and they synthesize the enzyme acetyl cholinesterase, a specific biochemical marker of muscle cell differentiation. However, instead of the normal thirty-eight cells, only about twenty cells express the enzyme. Thus, a fate change has occurred in some of the primary muscle cells. Neither muscle actin nor myosin heavy chains characteristic of muscle cells are synthesized in *M. occulta* embryos.

Cross-species hybrids in which eggs of the tailless species *M. occulta* were fertilized by sperm of the tailed species *M. oculata* regained some of the properties of the tailed larva. A small number of notochord cells organized to form a short tail, the brain pigment cell was restored, and more presumptive muscle cells synthesized acetyl cholinesterase. However, the ability of presumptive tail muscle cells to express actin and myosin was not restored. Swalla and Jeffery concluded that multiple mechanisms control the shift to tailless development. The features that were restored by the portion of the hybrid genome brought in by the sperm had to result from embryonic gene action. The features not restored were likely part of the maternal determination system. As reviewed by Venuti and Jeffery, differentiation of some ascidian embryo cells is autonomous. Maternal information is present and sufficient to determine the fates of these cells with no dependence on contact with any other cells in the embryo. Other cells depend upon inductive signals from neighboring cells. This mix of mechanisms is common to all animal embryos.

When ascidian eggs are fertilized, they exhibit an amazing behavior. Several colored cytoplasmic domains sort themselves out and move to defined locations in the egg. The colors are due to particulates such as mitochondria that in themselves have no direct role in differentiation. However, these colored particles fortuitously mark the localization of determinants of autonomous cell differentiation, and map the sites of origin of particular cell types. The determinants are maternal in origin, with muscle, endoderm, epidermis, and notochord all showing autonomous differentiation (Nishikata and co-workers, Crowther and Whittaker, Whittaker, Whittaker and Meedle, Meedle, and Swalla and Jeffery). In addition, the anterior-posterior axis has been found by Nishida to be maternally determined, and to lie in a cytoplasmic constituent located in the posterior-vegetal part of the fertilized egg.

The region from which primary muscle cells arise is marked by the cytoplasmic domain of the myoplasm. Experiments by Whittaker and by Deno and Satoh that involved moving myoplasm into neighboring cells showed that the maternally loaded myoplasm determines muscle cell fate. The myoplasm was shown by Jeffery and Meier to consist in part of a microfilament-like cytoskeleton. Swalla and co-workers have found that a particular microfilament protein is present in the eggs of both tailed and tailless species, but localized to the myoplasm only in tailed species. The failure of tail muscles to differentiate in *M. occulta* is a maternal effect, and could have resulted from a failure of the tailless species to properly organize its myoplasm.

Since the tailless state has resulted from losses of genetic function, losses of maternally expressed genes are expected. Swalla and co-workers screened for messenger RNAs expressed in eggs of the tailed *M. oculata* and absent in those of the tailless *M. occulta*. Two of the three identified genes encode strictly maternally expressed messenger RNAs. One gene encodes a nonreceptor tyrosine kinase, and the other encodes a novel protein with domains characteristic of DNA-binding proteins. The third and best-characterized gene, an apparent transcription factor called *Manx*, is transcribed early in development. It is expressed in presumptive tadpole larval tail cells and may be a key regulator in tail formation. These preliminary results show that it will be possible to define genetic regulatory differences in related species with disparate developmental modes.

Although both embryos and experiments were different than in the case of direct-developing sea urchins, it is clear that the early developmental differences between these closely related ascidian species involve a spectrum of evolutionary alterations in maternal and embryonic gene expression. However, there are notable differences in the kinds of changes that have occurred. Reorganization of cell lineages to produce new founder cells has not occurred in ascidians, although differences in cell fate underlie the failure of some cell lineages to produce functionally differentiated cell types. Cell cleavage patterns and early morphogenetic events appear to be unmodified in ascidians, as does gastrulation. Differences appear as the differentiation and morphogenesis of the tadpole begins. Sea urchins and ascidians have not achieved changes in early development by identical mechanisms.

Vertebrate Life Histories and Embryos: Frogs and Salamanders

Frog anatomy was fully attained by the oldest fossil frogs of the lower Jurassic almost 200 million years ago, and has not changed much since. Yet frogs are highly and spectacularly varied, even sometimes weird, in their reproductive modes. Development is highly modified in some, even to the extent of elimination of the tadpole in favor of direct development. Duellman and Trueb have listed twenty-nine distinct reproductive modes for frogs. These range from the conventional development of a tadpole in standing water to a nonfeeding tadpole with abbreviated development and even direct development without any tadpole stage at all. In many tropical environments the ecological demand seems to be to avoid standing water with its tadpole-hungry predators. Some frogs construct foam nests in water to shelter their eggs. Many others avoid water as much as possible and build a variety of nests in the ground. In some of these species, tadpoles are washed out of ground nests into streams by heavy rains. In others, terrestrial direct developers stay put under shelter until they emerge as small frogs. Some frogs specialize in arboreal reproduction, with tadpoles living in such places as the water that accumulates in the leaf bases of bromeliads. Some of the more bizarre adaptations include marsupial frogs with brood pouches in their backs, gastric-brooding frogs, in which development takes place in the stomach of the mother, and even frogs with viviparous development. As in the case of sea urchins, closely related species can differ dramatically in developmental mode. Individual species, however, produce a single style of embryo, indirect or direct in development.

Two of the extreme modifications of frog development have become subjects of serious investigation. The South American marsupial frogs are a branch of the familiar hylid tree frogs, but have evolved methods of brooding eggs on or within the back of the mother. *Gastrotheca riobambae,* studied by Eugenia del Pino and her collaborators, has provided a store of information about their reproductive physiology and development. In most brooding and marsupial frog species, the eggs are large, and development is either indirect via tadpoles or direct to froglets. In the more primitive mode of brooding on the exposed back of the female, there is no pouch. The related marsupial frogs have elaborate pouches evolved from flaps of back skin. In the most advanced forms, the pouches are closed, with only an aperture at the rear. Scalan and co-workers used immunological cross-reactivities among frog albumins to obtain phylogenetic data on *Gastrotheca* and related brooding frogs. By use of the albumin molecular clock, they estimated that *Gastrotheca* diverged from more primitive brooding hylid frogs 40–60 million years ago. Evolutionary divergence times among the Andean *Gastrotheca* species range from 2 to 9 million years, and direct development has originated more than once among the lineages of brooding frogs. Wassersug and Duellman took a different approach to working out the phylogeny of brooding and marsupial frogs, using morphological and developmental features. They suggest that in the primitive state, the ancestors of the brooding lineages had tadpoles with a complete array of mouth structures adapted for suspension feeding. The various lineages of brooding frogs produce feeding tadpoles, nonfeeding tadpoles, and direct-developing froglets. Nonfeeding tadpoles show a gradation of mouthparts from a complete suspension-feeding apparatus to none. This retention of different elements by different lineages shows that loss of structures was independent among the lineages. Developmental mode does not correlate with brooding structures. Of the marsupial frogs with completely enclosed brood chambers, *Fritziana* and *Flectonotus* produce nonfeeding tadpoles, whereas *Gastrotheca* species include both direct-developing species and species that produce feeding tadpoles. The more primitive brooding frogs without pouches or marsupia, *Hemiphractus, Stefania,* and *Cryptobatrachus,* are all direct developers, and are thus more derived developmentally than are some marsupial frogs.

The marsupia represent more than just handy pockets for eggs. Elaborate behaviors and physiological features have evolved with these structures. *Gastrotheca riobambae* mate on land rather than in water. The male is stationed on the female's back in the usual frog amplexus. He opens her pouch with his hind legs. As she releases eggs from her cloaca, he pushes the eggs into the pouch with his feet, and they are fertilized along the way by sperm deposited on the female's back. At birth, the female uses her feet to release the tadpoles or froglets from the pouch. The operation of the pouch is regulated by the hormone progesterone. The closing of the pouch prior to reproduction can be stimulated by injection of gonadotropin; however, this hormone has this effect only if the ovaries contain large follicles. Progesterone has a direct effect on pouch closing. Gonadotropin evidently acts by stimulating the follicles to secrete progesterone. Progesterone is also required for the retention of developing embryos. Empty follicles remain in the ovary for a significant portion of the incubation time. Del Pino and Sanchez found that the removal of ovaries of marsupial frogs during the first few weeks of incubation caused abortion, although this effect did not occur in later stages of brooding. Other tissues may secrete hormones involved in late maintenance. The analogy to control of mammalian gestation is striking.

Brooding frogs produce eggs ranging in size from 3 to 10 millimeters in diameter and exhibit substantial modifications in oogenesis, even in species that produce tadpoles. In brooding tree frogs with large eggs (in the 10-millimeter range), del Pino and Humphries found multinucleate oocytes to be the rule. In the various species examined, previtellogenic oocytes were found to contain from 4 to 3,000 germinal vesicles (oocyte nuclei). As oocytes mature, excess nuclei are lost, and finally, perhaps through some mechanism related to position in the cytoplasm, only one remains. Macgregor and del Pino suggested that multinucleate oocytes provide a mechanism for achieving high maternal RNA production and ribosomal DNA amplification, albeit differently than in frogs such as *Xenopus* that utilize a system of high amplification of nucleoli within a single oocyte nucleus. Del Pino and coworkers found that *Gastrotheca riobambae* has mononucleate oogenesis, and does not amplify its ribosomal DNA. *Gastrotheca* has a slow rate of development relative to both *Xenopus* and *Flectonotus,* consistent with a lower store of ribosomes. Clearly oogenesis is as evolutionarily labile as other aspects of development in these lineages.

A last interesting point about *Gastrotheca riobambae* is that although it produces a feeding tadpole, it doesn't do so in what we think of as the typical manner. The eggs of most frogs run to about 1 millimeter in diameter. The egg of *Gastrotheca riobambae* is 3 millimeters in diameter. The consequences of this size, and possibly of brooding in a pouch, are striking. Del Pino and Elinson and del Pino and Escobar have outlined early development in this species. Cleavage resembles that of other frogs, but following gastrulation, the pattern of development deviates greatly. Instead of forming a body axis as a cylinder involving the entire mass of the embryo, a disk of cells, called the embryonic disk, forms on the prospective dorsal side. The embryo develops in two dimensions on the surface of the egg, in a manner analogous to that of birds or reptiles. This species also forms an extraordinary set of structures, the bell gills, as extensions from the first and second branchial arches. The bell gills cover most of the embryo and apparently play a role in exchange of materials with the mother.

In direct-developing frogs, the typical tadpole stage is lost altogether. The best-known direct developers are species of *Eleutherodactylus,* originally studied by W. G. Lynn and more recently by James Hanken and his coworkers. *Eleutherodactylus* is a member of an entirely different family of frogs than *Gastrotheca* and its relatives. For collectors of records, the genus

Eleutherodactylus, with its 450 known species, is the largest living genus of land vertebrates. *Eleutherodactylus* not only fails to develop most tadpole structures, but has also undergone a streamlining of development in which adult elements appear along a novel trajectory. Thus, limb buds appear early in relationship to the timing of other features. Some elements of the larval skull never appear, and many regions of the skull proceed directly to the adult condition. A number of other structures, notably the lower jaw and hyobranchial skeleton, initially assume a mid-metamorphic condition and are remodeled. Metamorphic remodeling is a common feature of amphibian development, and has been proposed to reflect important developmental constraints. The aortic arches proceed directly to the adult condition as well, but by a route unlike the ancestral pathway for forming these structures. Thus development of adult features is advanced into earlier development. As in direct-developing sea urchins, early development has been remodeled, but the ultimate product of the complex alternative pathway is an unremarkable frog.

Although direct-developing frogs are better studied, the production of terrestrial eggs that develop directly into terrestrial juveniles is the prevalent developmental mode in salamanders. However, in contrast to the phylogenetic range of developmental modes in frogs, direct development in salamanders is confined to a single family, the plethodontids. This family contains two-thirds of all living salamanders. Direct development has arisen independently in at least five plethodontid lineages. Wake and Hanken have argued that the resultant alterations in life history, repatternings of development, and losses of larval constraints have been major factors in the great evolutionary success of this family. The diversity of patterning processes in plethodontids may make them important subjects for studies of the evolution of regulatory gene expression.

HOW GENERAL ARE EVOLUTIONARY CHANGES IN EARLY DEVELOPMENT?

The largest known eukaryotic cell was the egg of the now sadly extinct *Aepyornis,* the elephant bird of Madagascar. According to the report of one appreciative nineteenth-century French traveler, these eggs (they still turn up intact in subfossil condition) had a volume equal to that of 13 quart wine bottles. Richard Elinson argues that egg size in itself constrains developmental mechanisms. In fertilization, small eggs (a few millimeters or smaller) tolerate entry by only a single sperm, and the resulting early embryonic cells divide completely. Large eggs either direct sperm through a restricted channel or tolerate polyspermy to ensure fertilization in the midst of a giant cytoplasm. Mechanisms have evolved to ensure that only one of the sperm is used. Transitions between these two styles have occurred in terrestrial vertebrate evolution. The primitive state was a monospermic egg. Reptiles and primitive mammals evolved large polyspermic eggs. The secondarily tiny eggs of marsupial and placental mammals are once again monospermic. These changes in fertilization mechanisms paralleled much more drastic alterations in developmental processes.

Terrestrial vertebrates have never evolved free-living terrestrial larvae. Their young are always miniature adults. Elinson has suggested that as vertebrates moved onto the land, they had two options: small eggs that retained the primitive pattern of holoblastic or complete cleavage, or large eggs that exhibited meroblastic or incomplete cleavage. The production of a very large nutrient-filled egg would allow the embryo to become large before having to face the hostile world. Because cell movement has to accommodate a large mass of uncleaved egg, meroblastic cleavage requires different gastrulation mechanisms than does holoblastic cleavage. Instead of an involution of the vegetal cell, large eggs form a cap of cells at the animal pole that gives rise to the embryo proper. The transition from relatively small amphibian eggs to the large meroblastically cleaving eggs of reptiles was a major one. Because their eggs were so large, Elinson proposed that to maximize growth rates, the ancestral amniotes had to evolve vascularized tissues to provide food and oxygen to the embryo, thus the extraembryonic membranes of the amniotic egg. Mammals and birds inherited these large eggs and their modified forms of gastrulation. We still see them in living monotreme mammals such as the platypus and echidna of Australia. In marsupial and placental mammals, small holoblastic eggs re-evolved in response to feeding within the uterus. The return to holoblastic cleavage did not, however, lead to a return to the primitive amphibian mode of gastrulation. Reduction in egg size in vertebrates appears to have been confined to a very few lineages.

EXPERIMENTAL EVOLUTION: MANIPULATING EGG SIZE

Egg size appears to have important evolutionary consequences for development, and changes in size pose a general evolutionary problem. In marine invertebrates, there seems to be a general rule of evolution of large eggs in conjunction with the evolution of direct development. These changes have occurred independently among sea urchin lineages and among other phyla as well. The enlargement of eggs appears to be strongly selected to provide sufficient in-house nutrition for eggs that will not produce feeding larvae. Such parallel evolution of features offers clues as to the genetic changes required for the transition. Facultatively feeding plutei produced by sea urchins with moderately large eggs are poised to make the evolutionary switch to the direct-developing mode. If selective pressure to feed were removed, these embryos could readily dispense with the elaborate feeding structures of the larva, as long as a workable pathway to the juvenile sea urchin were retained. The initial genetic change required appears to be a relatively easy one: the production of a large egg. That step decouples developmental patterns for a feeding larva from those for a juvenile adult. Subsequent evolutionary steps will include the loss of some larval developmental pathways and adjustments to early development that allow direct progression from egg to adult.

Some developmental consequences result from egg size alteration rather than from genetic differences, and these too contribute to the evolutionary result. The longer duration of cell cleavage Parks and co-workers observed in *H. erythrogramma* and the higher cell numbers in its gastrula result from changes in the nucleocytoplasmic ratio in a larger egg.

Two similar species that differ in body size often have other differences in shape and physiological features as well. Differences in size are almost always accompanied by changes in shape that result from the fact that body parts often grow in size with a nonlinear relationship to other body parts. These differential growth properties are referred to as allometry. As Sinervo shows, there is a general way to separate the coincident effects of size from differences that stem from selection on other traits. Sinervo calls this approach "allometric engineering." Selective effects can be directly factored from allometric ones by experimentally altering the size of one species to match the other. Differences due to allometry will disappear, and nonallometric features will remain. This prediction holds for embryos as well. Sinervo and his collaborators have applied the elegant approach of experimentally changing the volumes of eggs. In their first study, they examined embryos of two related sea urchin species that both develop by way of pluteus larvae but differ by sixfold in egg volume. Sea urchin embryos are famous for their ability to develop relatively normally from separated cells of two- or four-cell stages. Quarter-sized embryos of the larger-egged species were obtained by separation of embryonic blastomeres at the four-cell stage. These quarter-sized embryos developed a larval form similar to that of the smaller-egged species. In the second study, lizard eggs were manipulated to have larger or smaller volumes. Hatchlings from reduced eggs of the larger-egged population resembled the normal hatchlings of the smaller-egged population in some respects, but not in others. Both allometric and nonallometric changes were found. Experimental approaches analogous to these may reveal more about the role of egg size changes in the evolution of development.

REGULATION: THE UNDERLYING FLEXIBILITY
OF DEVELOPMENT

We must return here to the apparent paradox of groups that have retained the same larval form unaltered over hundreds of millions of years of evolutionary time readily shifting to a different mode of early development. Is the stability or the change more remarkable? If early development is to be substantially modified, the initial evolutionary tinkering has to occur within the conservative ancestral ontogeny. That is, the first steps toward a new developmental pattern have to be accommodated within the ancestral pattern in order to produce a viable embryo.

There is a concept of internal adjustment, called canalization by Waddington, in which developmental pathways are genetically buffered against disruptive changes. The result is that within limits, environmental or mutational disruption is resisted, and embryonic cells follow their normal programs in development. Developmental catastrophes caused by minor perturbations are thus avoided. Since nonideal conditions are frequent in nature, such a property of embryos should be strongly selected. Canalization probably involves multiple mechanisms, including redundancy of related functions or cells. A seemingly related phenomenon, called developmental regulation, has also been long recognized by experimental embryologists. Developmental regulation is the capacity of embryos or parts of early embryos to resist experimental perturbation and still yield normal embryos when cultured. Typical regulative behavior is displayed when individual cells from early-stage embryos are cultured separately, as in the case of half-embryos produced by separated cells of a two-cell embryo. Not all embryos regulate well. Those that exhibit rigid localization patterns often cannot reprogram portions of the embryo. This kind of development is called mosaic. Spiralians, ascidians, and soil nematodes are largely mosaic. Regulative and mosaic behavior are not mutually exclusive, and both phenomena are observed in most embryos, usually depending upon cell type or stage of development. Nevertheless, regulation is often possible even if maternal localization processes are disrupted or if cells have to follow an altered cell lineage pathway. This phenomenon has been difficult to explain mechanistically and is still poorly understood. Its importance is that it makes evolutionary modifications to early development possible.

The relationship between embryonic regulation and evolutionary flexibility is nicely illustrated by indirect-developing sea urchins. Their embryos develop in an invariant manner, as described earlier. Nevertheless, these embryos are highly regulative when experimentally manipulated, as when the separated cells of a sea urchin embryo develop into small but normal plutei. When experimental surgery is done, even more striking regulatory phenomena appear. Some of these are shown in figure 7.12, which shows the kinds of cells that can produce skeleton-forming mesenchyme under normal and altered circumstances. Figure 7.12A illustrates normal development, with the micromeres differentiating into skeleton-forming cells in the gastrula. If these cells are removed, as in figure 7.12B, cells arising from the macromeres will

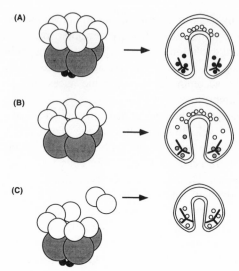

Figure 7.12.   A high degree of developmental regulation is shown by typical indirect-developing sea urchins. *(A)* The contribution of sixteen-cell fates to the gastrula in the normal embryo. Micromeres (black) give rise to skeleton-secreting mesenchyme cells (black). *(B)* When the micromeres are deleted, skeleton-secreting mesenchyme cells arise from a novel cell lineage generated by the daughters of the large macromeres (both macromeres and skeletogenic mesenchyme cells are shaded). *(C)* A complete gastrula also may be formed by two isolated mesomeres. Again, skeleton-secreting mesenchyme cells (not filled) arise, but from yet a third pathway. (After Raff, Wray, and Henry 1991.)

differentiate into skeleton-forming cells. Finally, if a pair of mesomeres is separated from the embryo and cultured, as in figure 7.12C, they can still produce a complete pluteus. These results show that although normal development is highly stereotypic in these evolutionarily conservative embryos, they have an innate capacity to generate novel cell lineages that finally produce larval cell types normally derived from quite different precursors and cell lineages. The same cell types are generated in the direct-developing sea urchin *H. erythrogramma* as in indirect developers, but by different routes. The evolutionary transition would have occurred within the regulatory capacity of the indirect-developing ancestor. If early development were internally constrained, change would be a rare proposition, because any change would be disruptive. We thus have to conclude that mutational alterations occur within a system that can accommodate striking perturbation and still carry out a harmonious ontogeny. In addition, the great evolutionary flexibility of early development supports the concept of the developmental hourglass. The freedom of early stages to evolve without affecting the phylotypic stage has important consequences for the way we perceive developmental constraints and the structure of ontogeny.

8

It's Not All Heterochrony

"Supposing the Elders of a Carpet Snake clan decided it was time to sing their song cycle from beginning to end? Messages would be sent out, up and down the track, summoning song-owners to assemble at the Big Place. One after the other, each 'owner' would then sing his stretch of the Ancestor's footprints. Always in the correct sequence!"

"To sing a verse out of order," Flyn said somberly, "was a crime. Usually meant the death penalty."

"I can see that," I said. "It'd be the musical equivalent of an earthquake."

"Worse," he scowled. "It would be to un-create the Creation."

Bruce Chatwin, *The Songlines*

WHAT PRICE IMMORTALITY?

One of the few scientific concepts to resonate well in popular culture is the idea that the evolutionary history of ancestors is recapitulated in the development of their descendants. There is a wonderful mystery in the idea that the anatomy of our remote ancestors, gill arches, tail buds, and all, unfolds in our own individual development. No matter how we deride the naiveté of Haeckelian recapitulation, and however we redefine the mechanisms that connect evolution and development, there is a shadow of truth in the idea. Nineteenth-century concepts of evolutionary changes in development hinged on recapitulation of ancestral events by the addition of new stages to the adult end of the developmental sequence. However, this is too constraining. Evolution is not limited to terminal addition. Because development unfolds along a time axis, evolutionary modifications of development may also change the relative order of events in what Alberch has called the developmental sequence. An evolutionary change in the timing of developmental events is called a heterochrony.

The ideas of heterochrony developed earlier in this century by Garstang and by de Beer stressed the consequences of relative changes in the rates of maturation of the somatic body and the germ line. Change could proceed by terminal addition to produce classic recapitulation, but it might also occur by reproductive maturity outrunning rates of somatic development. Thus, neoteny occurs if the gonads develop on time but the maturation of the body is slowed. The resulting animal reproduces as a sort of grown-up larva. Of

255

all heterochronies, the concept of neoteny has been the most evocative, and it has been used to explain major evolutionary innovations, including the origin of modern humans and the transitions between body plans. Heterochrony has been advertised as a universal mechanism of developmental evolution. As an explanation of the evolution of developmental processes is necessary to account for changes between and within body plans, heterochrony must be reassessed.

Heterochrony has had a profound effect on macroevolutionary thought, and a certain curious literary appeal as well. The proposition by Louis Bolk in the 1920s that humans are really the fetalized descendants of apes, or more poetically, sexually mature juvenile apes, reached the notice of the reading public in Aldous Huxley's novel *After Many a Summer Dies the Swan,* published in 1939. The novel is a slow read because Huxley used it as a vehicle for a lot of tiresome philosophical lecturing by one of the longer-winded characters in modern literature, the pedantic Mr. Propter. It is a book for a cynical and unhappy time, with a curious plot and a collection of very unsympathetic characters. The story revolves around the attempts of a rich, aging California entrepreneur, Jo Stoyte, to prolong his life. Stoyte exploits farmworkers and owns a cemetery right out of *The Loved One.* He is an archetypical modern philistine: he lives in a castle; he has an awful teenaged mistress, Miss Maunciple, whose passions alternate between Priapus and the Blessed Virgin; and he has a lot of priceless art treasures. Stoyte is also a man of action. He can afford to pay for eternal youth. Thus, he hires a Mephistophelian scientist and physician, Dr. Obispo, to conduct research in his castle on how aging can be arrested. Obispo represents the evil, or at least profoundly misguided, scientist meddling where he shouldn't, which appears to be the stock part for scientists in serious fiction as it is on Saturday morning children's television. Science promises desirable gifts, but at what cost to soul or safety?

Dr. Obispo is convinced that aging is caused by the buildup of toxic steroids in the body. Obispo has noted that two-hundred-year-old carp show no signs of senility. They just swim placidly on. Obispo has found the secret, that the carp don't accumulate excessive fatty alcohols in their guts, and those they do have are not converted to sterols. He devises a way to transplant the carp intestinal flora to the intestines of mice. Senescence is indeed reversed, and the mice live to ripe old mouse ages. Obispo begins experiments on monkeys before he moves on to rejuvenating Jo Stoyte.

There is a second, literary, thread to the puzzle of prolonging life. Stoyte has bought the valuable Hauberk Papers from the last impoverished remnant of a titled British family. He has hired a colorless middle-aged British scholar, Jeremy Pordage, to catalog them. As he reads in his musty workroom, Pordage stumbles on an extraordinary document, the diary of the Fifth Earl of

Gonister, which records in no uncertain terms his long life and prejudices through the late eighteenth and early nineteenth century. He was an intelligent man, if not a nice one. As he aged, he began to meditate on death. As he put it, "Dying is almost the least spiritual of our acts . . ." Like Stoyte, the Fifth Earl intended to do something about it.

In July 1796, the Fifth Earl recorded that the fishponds at Gonister were dug by the monks of the former abbey upon which his estate was built. They were stocked, during the realm of Charles I, with fifty well-grown carp, which his great-grandfather tagged with dated lead discs. Twenty of these carp were still alive in the Fifth Earl's pond 150 years later. He marveled that they were still strong and agile, and noted that the secret of eternal life is to "be found in the Mud and only awaits the skillful angler." Pordage tells Obispo, and they read on together. The Fifth Earl began to reason it out, and concluded that the substance that prevents decay must lie in the digestive organs. He noted that people who eat fish live no longer than others, but precisely because the guts are what gets thrown away, and because people cook their food. By January 1797, the Fifth Earl had entered upon his experiment. He began to eat raw carp guts. At first he suffered from uncontrollable retching. It was only after the tenth try that he could retain a few spoonfuls of this sushi nightmare. He persisted, and soon was eating six ounces of chopped carp guts each day. Over the next few months he grew stronger. By April, the Fifth Earl was riding again. By September he was criticizing Plato. The diary is silent until May 1799, when he expressed his contempt for his recent feminine conquests. By 1820 he was rejoicing on the aging of his despised relatives, and in 1824 he bemoaned the passage of a bill that made slave trading illegal and thus reduced his income. By 1826, he had a new mistress, Kate, his housekeeper. By the early 1830s he was in his 90s. He built a new house, with basement apartments, and he started Kate on the carp guts diet.

A final scandal occurred in 1834. It's not clear just what the Fifth Earl was up to, but his downfall involved the escape of someone named Priscilla from the basement. This apparently was more than could be covered up. There was a riot, and a warrant was issued for his arrest. However, his age, position, and the fear of the scandal of a public trial led to a compromise in which he was put into his family's hands. He had to hand over his properties to his relatives. He refused to enter a private asylum, saying he would rather stand trial. Clearly a trial did not suit the family. In the next entry the Fifth Earl discussed the arrangements for his upcoming funeral. No winged chariot for the Fifth Earl, though. The body of an aged pauper took his place. He and Kate took up residence in their subterranean hiding place. Financial arrangements and arrangements for a caretaker followed. It was the last entry in the diary.

Dr. Obispo decides to head for England. The plot's heavy-handed symmetry between Stoyte and the Fifth Earl includes some more California nastiness, including a murder and its successful cover-up with a faked death certificate, before Stoyte and Miss Maunciple can go with him. The Fifth Earl's old garden is overgrown and seedy. The lawns are untended and the croquet hoops are rusty—genteel gothic. The old Hauberk ladies who live there don't let anyone enter, but Obispo manages to bribe a grandchild into letting him into the place and snatches the key. They find the entrance to the cellar hidden behind a bookshelf. Conveniently, there are working lanterns fueled and hanging on hooks at the top of the stair. They descend. Deep inside are the inhabitants. Their simian forms are covered with reddish hair. They have tatters of clothing. One urinates on the floor. " 'The one with the Order of the Garter,' said Dr. Obispo, raising his voice against the tumult, 'he's the Fifth Earl.' " ". . . two hundred and one last January." The Fifth Earl and his Kate have paid the price of eternal life. It is to grow up. The fetal ape, given sufficient time, has matured. Stoyte considers, and asks Obispo, "How long do you figure it would take before a person went like that?" Obispo laughs.

DEFINING HETEROCHRONIES

Heterochrony has been the single most pervasive idea in evolutionary developmental biology. It was nominated by de Beer as the universal mechanism by which development evolves. Gould devoted his entire 1977 book, *Ontogeny and Phylogeny,* to heterochrony in its historical and evolutionary manifestations, and presented it as the chief mechanism of evolutionary change. This strong bias in favor of heterochrony as an explanatory mechanism has continued. Gould in 1988 says, "Heterochrony is perhaps our best empirical mode for the study of developmental constraint . . ." That is, given that developmental constraints are those attributes of the existing pattern of development that limit or channel further evolutionary change, heterochrony is the chief agent of change arising from within the internal architecture of ontogeny. Other authors have echoed this view of mechanistic universality without much questioning of its validity. The title of McKinney and McNamara's 1991 book, *Heterochrony: The Evolution of Ontogeny,* speaks for itself. In this very thorough presentation of heterochrony and the factors affecting it, these authors set forth their goal to "promote the views that (1) heterochrony is the cause of most developmental alterations and (2) heterochrony can cause major novelties."

In this chapter I will examine the claims made for heterochrony and then try to put them into a more balanced perspective. It appears that heterochrony is more often the result of various processes not in themselves related to developmental timing, and that the uncritical attribution of so many of the phenomena observed in the evolution of development to heterochronic "mechanisms" may be inhibiting a more penetrating investigation of the subject.

Heterochrony was first placed in a modern context in the mid-twentieth century by de Beer in his book *Embryos and Ancestors,* and as noted by Gould, it was de Beer who generalized the term "heterochrony" to mean a change in the developmental timing of a feature relative to the equivalent feature in its ancestor. Gould, in 1988, proposed that observable heterochronies arise from an underlying alteration in developmental timing. He also appropriately noted that no simple chain of causality links events at the phenotypic level to underlying changes in the timing of developmental processes. We will have to examine the question of whether heterochronies, which are claimed to be so highly prevalent, really have as their underlying mechanisms changes in the timing of developmental events. We will have to ask whether changes in developmental timing occur with any prevalence at all. Pretty much any evolutionary change can be explained as a heterochrony. The success of any stock explanation for evolutionary changes in development should make us cautious.

Heterochrony is the dissociation of the relative timing of events in development between ancestral and descendant ontogenies. Heterochrony derived its first theoretical justification as well as its empirical origins from the work of Ernst Haeckel. Recapitulatory phenomena are observed in ontogenies; that is, features of ancestral development are often retained. The concept of recapitulation derived from such observations, and from the need, pointed out by Haeckel, to condense ontogenies as more stages of development were added to the final phases of old ontogenies. Haeckel observed another, and to him a very inconvenient, phenomenon in the tracing of phylogenies, namely, nonrecapitulatory larval adaptations. These represented insertions of developmental features out of the correct recapitulatory time sequence. He termed this phenomenon of change in relative timing heterochrony. Haeckel's theoretical idea that phylogeny drives ontogeny was drained of any value by the advent of Mendelian genetics. Garstang argued in 1922 that modifications could be made at any point in development, and that recapitulation has no universal validity. Morgan pointed out explicitly that a new gene could produce its effect at any stage of development. With the escape from strict recapitulation, de Beer proposed that heterochronies of various sorts are the most important engines of evolutionary change.

Incidentally, heterochrony did not exhaust Haeckel's insights into the kinds of insertional changes that could interfere with strict recapitulation. He noted that spatial displacements also occur in development, and called these heterotopies. Until recently, heterotopies have been neither well understood nor much appreciated.

The most significant theoretical claim that can be made for the concept of heterochrony is that it represents a dissociation between events in ontogeny. The idea of dissociation was put forward by Joseph Needham in 1933. He suggested an engineering analogy, that ontogeny is a machine with metabolism as its main shaft. According to Needham,

> In the development of an animal embryo, proceeding normally under optimum conditions, the fundamental processes are seen as constituting a perfectly integrated whole. They fit in with each other in such a way that the final product comes into being by means of a precise co-operation of reactions and events. But it seems to be a very important, if perhaps insufficiently appreciated, fact, that these fundamental processes are not separable only in thought; that on the contrary they can be dissociated experimentally or thrown out of gear with one another. The conception of out-of-gearishness still lacks a satisfactory name, but in the absence of better words, dissociability or disengagement will be used in what follows. . . . There are many instances where growth and differentiation are separable. It is as if either of these processes can be thrown out of gear at will, so that, although the mechanisms are still intact, one or the other of them is acting as 'layshaft' or, in engineering terms, is 'idling.'

Ontogenies flow along a time line, and give the appearance of a continuous developmental sequence. Imagine what the possibilities for evolution would be if the time sequence actually recorded process sequence as well, and ontogeny were completely linear in terms of process. Interfering with any step would affect all downstream steps. Any change early in development would affect all that follows. Thus, except in the terminal stages of development, every mutational change would require massive changes in the entire program. Evolution could occur only in the last few steps. However, the fact that there has to be a single time line of events does not mean that all processes are linked. Development is more realistically conceived of as containing independent chains of causally linked processes. The whole starts out from a single point, fertilization, with one or a few processes, but expands into a series of parallel pathways. Some pathways will affect others via induction or other signaling mechanisms. However, unless these parallel pathways are completely tied to one another, one may be changed without affecting others. Any such change to a pathway within the stream of pathways results in a developmental dissociation. If such a dissociation still allows the production of a viable embryo, it may serve as a starting point for the evolution of a new ontogeny. It is not clear that this parallel cascade scheme I have outlined really adequately models development, but experimental studies of development show that ontogenies contain bundles of independent and semi-independent processes. Dissociation between them is possible in principle.

If dissociation between processes can take place, then the processes can certainly be slipped relative to one another in developmental time. Such dissociations have been held to be minimally disruptive, and thus heterochronies have been proposed by de Beer and by Gould to provide the primary mechanism of evolutionary change. That focus has been exemplified again and again in studies of the evolution of form in both living and extinct animals. Most studies of heterochrony involve its classic manifestations, what Raff and Wray called "de Beerian heterochrony." These classic heterochronies are global in that they involve the entire body of the animal and are construed as representing a dissociation in rate between two fundamental aspects of animal development, the maturation of the somatic body and the maturation of the gonads. I'll follow the terminology of McNamara, which I'll present here. His definitions provide a good consensus on how the terms should be applied, and take us out of what once was one of the most perplexing terminological morasses in biology. For some reason, the description of heterochronies inspired a colorful and profuse set of terms that included such evocative tags as recapitulation, anti-recapitulation, superlarvation, and tachygenesis among the best of the breed.

The terminology has been made even more confusing by the confounding of heterochronic results with heterochronic processes. The categories of heterochrony defined by de Beer referred to results of poorly defined shifts in timing. Gould, in 1977, provided new definitions of heterochronic results, and related them to two kinds of heterochronic processes, acceleration or retardation of one set of events relative to another. Later definitions by Alberch and co-workers in 1979, by McKinney in 1986, and by McKinney and McNamara in 1991 have presented somewhat longer lists of processes, and have attempted to relate them to general categories of resultant changes.

There are two kinds of global heterochronies, which are the results of shifts of somatic and gonadal development in either of two directions. Shifts in one direction produce paedomorphosis, in which somatic development in the descendant is slowed with respect to gonadal maturation. The descendant will in some respect resemble a sexually mature version of the immature ancestor. It is juvenilized in shape. Shifts in the other direction produce peramorphosis, in which somatic development proceeds beyond what it would have done in the ancestor. These two grand forms of global heterochrony are often what paleontologists score in comparing presumed descendants and ancestors in the fossil record. For example, McNamara, in 1986, evaluated the prevalence of heterochronic changes inferred for a great span of fossil animals from trilobites to vertebrates, and looked to see whether paedo-

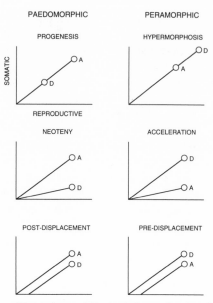

Figure 8.1. Categories of heterochrony. The classic heterochronies of rate of somatic versus reproductive maturation are presented. The relative timing changes between ancestral (A) and descendant (D) ontogenies are diagrammed.

morphic or peramorphic processes prevailed. He found them to be about equally frequent. Heterochronic trends have also been inferred among living animals, such as rodents (Hafner and Hafner), humans and other primates (Gould; Shea), gastropods (Lindberg), and sea urchins (McNamara), and scored as paedomorphic or peramorphic in direction. As shown in figure 8.1, these two grand directions of heterochrony can each be broken down into three specific categories of processes.

I'll start with paedomorphosis, which, as in *After Many a Summer Dies the Swan,* has been the more fashionable category. The most straightforward way to produce something that looks like a sexually mature juvenile is by progenesis, the speeding up of sexual maturity. Examples have been put forward from among many groups, such as aphids and mites, discussed by Gould in *Ontogeny and Phylogeny,* that feed on rich but extremely short-lived food sources such as mushrooms. Fast reproduction is at a premium if such a colonizer is to pass on the most genes, and is achieved by producing sexually mature larval forms when food is plentiful. As food declines and population pressure increases, production of nonprogenic adults that can disperse to new sites is favored. Gould explicitly tied progenesis and other modes of heterochrony to particular ecological circumstances that favor specific life history traits. Evolution of form by heterochrony can thus be driven by selection for life history features as well as for morphological features. McKinney and McNamara review some remarkable examples of sexual progenesis in invertebrates that live in unusual environments. In some parasitic mollusks that live inside the bodies of their hosts, mating is assured by the existence of tiny progenetic males that live attached to the larger female. Selection seems to be for fecundity in a situation in which contact between the sexes would otherwise be difficult. Small progenetic males that live attached to females also exist in some deep-sea fishes.

The most famous heterochronic mode is neoteny. Unlike progenesis, neoteny has been suggested to be favored in stable environments. Again, a juvenilized morphology is achieved, but the process is different. Gonadal maturation proceeds on schedule, but somatic maturation is slowed. This isn't a fast means of reproduction like progenesis. Rather, maturation takes the full length of time, and a large juvenile-like adult is produced. Neoteny, in the classic cases of the Mexican axolotl and other neotenic salamanders, has been posited to be favored in environments in which the juvenile morphology has an advantage. In the case of the axolotl, the lakes in which the juveniles develop present more predictable environments than the surrounding dry high country. Selection favors an aquatic form. McKinney has presented data in support of the idea that members of fossil sea urchin lineages were more neotenic in relatively stable seafloor environments. Neoteny, as I'll discuss below in more detail, has also been proposed as the process by which humans evolved from apelike ancestors.

The final form of paedogenesis, called post-displacement, has quite a different cause. Differences in maturation rates are not involved at all; in fact, the rates can be identical in the descendant and its ancestor. Instead, one or more features of the descendant begin to develop at a later stage with respect to the rest of the organism. If the rates and duration of development are unchanged overall, the late-starting features will not progress as far in development, and those aspects of the final product will be juvenile-like. The disappearance of particular late-appearing bones of the hand or skull in miniaturized vertebrates has been attributed to this cause.

The second broad heterochronic category, peramorphosis, is akin to the recapitulation of Haeckel. That is, peramorphic processes result in the earlier stages of the descendant coming to resemble later developmental stages or even the adult stage of the ancestor. The descendant in such cases recapitulates aspects of the development of its ancestor, but then pushes on to some new final morphology. The most obvious heterochronic process to produce such effects is hypermorphosis, the opposite of progenesis. In this process, the rates of somatic and gonadal development are the same as in the ancestor,

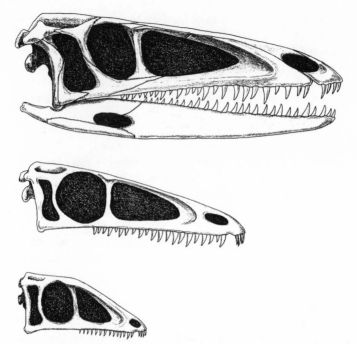

Figure 8.2.   The growth of the skull in the small dinosaur *Coelophysis* exhibits a strong positive allometry. The youngest skull is at the bottom of the drawing. As the head grew, the main change in shape was a pronounced relative elongation of the forepart of the skull. (From *The Little Dinosaurs of Ghost Ranch* by Edwin H. Colbert. Copyright 1995 by Columbia University Press. Reprinted with permission of the publisher.)

but sexual maturation is delayed and growth is extended for a longer total time. The descendant becomes sexually mature later, is larger, and may have novel morphological features that result from allometric growth.

Because growth is very much a matter of temporal parameters such as rate, duration, and onset or offset, growth is a central focus in considerations of heterochrony. Growth can be isometric or allometric. In isometric growth, body parts maintain the same relative size. A large animal thus resembles a small relative in shape. However, growth is often allometric. Allometric growth refers to the unequal growth of body parts in an animal. Figure 8.2 shows the allometric growth of the business end of a carnivorous dinosaur's skull, jaws over brains. Allometric growth is also prevalent among more familiar animals. For example, the claws of lobsters show a strong positive allometry. The claws are small in a small lobster, but come to be huge in a large lobster. In such cases, a change in size, but with retention of the same growth rules, can produce a substantial change in shape. If really big claws are advantageous, they can be produced by hypermorphosis.

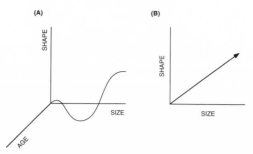

Figure 8.3.   Ontogenetic trajectories. *(A)* The trajectory of an organism or part of an organism in size, shape, and age space. *(B)* A simplified trajectory in which shape is related to size. Size can be used as a proxy for age. (After Alberch and co-workers 1979.)

Allometry plays a role in other kinds of heterochrony as well. Wayne has shown that small breeds of domestic dogs have paedomorphic skull shapes. This has occurred because all breeds obey the same ontogenetic growth curve in head length versus width. Newborn dogs of all breeds have relatively wide heads. Small breeds retain a wide skull shape by ceasing to grow while still small. Large breeds have long skulls because they grow for a longer period of time.

A second form of peramorphosis involving rates of development is acceleration. This process is the opposite of neoteny. In acceleration, the rate of somatic development is increased, but the rate of gonadal development is not. A more developed adult is produced, and given the allometric rules governing growth, a distinct morphology can be obtained. Our breeds of dogs seem to depend in part on variations in prenatal growth rates. An important study by Wayne of the growth of domestic dogs has revealed how different they are from wild canid species. A substantial literature (cited by Wayne) has established that among related mammals, birth weight is proportional to gestation time. This relationship holds true among wild canid species, but not among domestic dog breeds. Both large and small dogs have the same 60–63-day gestation time, but large dogs grow faster during this prenatal period. They exhibit acceleration of prenatal growth relative to smaller breeds. In wild canid species, sizes are determined by postnatal growth rates.

Finally, pre-displacement is the peramorphic opposite of post-displacement. Again, rates of development are not affected, but the time of onset of development of one or more features may shift to an earlier point in development. If duration of growth is the same in the descendant as in the ancestor, the pre-displaced feature will be more developed. It has been pointed out that evolution of the large human brain has involved heterochronic displacements of start or stop points. Of that, more later.

Alberch and co-workers have suggested that the "ontogenetic trajectory" of an organism can be represented on three axes, as diagrammed in figure

8.3A. The three axes are time, size, and shape. Ideally, one would want to plot all three and compare the ontogenetic trajectories of ancestor and descendant. In practice, this is hard to do even if some simple metric for shape is available. Generally, graphs of heterochronic changes are made on two-dimensional axes, as shown in figure 8.3B. Thus, shape or size can be plotted against time (developmental age). The onset, cessation, and rate of development can all be plotted for any feature, and heterochronic inferences drawn from readily collected morphological data.

No ointment provided by nature comes without the requisite flies. Since the practitioners of heterochronic studies are often paleontologists, who cannot determine developmental times for extinct ontogenies, size often serves as a proxy for age, and heterochronic plots are made of shape versus size. The use of size as a proxy for age is common in studies of living organisms as well as extinct ones because it is often very difficult to determine the age of animals in the wild. This fallback is not completely satisfactory. The relationship of size and age might have changed in evolution, but it has been used perforce. There have been attempts to determine ages at death of fossil remains. In some organisms, such as clams, internal growth lines mark daily or annual increments, and their use has been proposed by Jones. McKinney has used the appearance of genital pores to mark maturity in fossil sea urchins. In cases in which such data can be obtained, age, shape, and size can be read from the specimen, and rates estimated.

LOCAL VERSUS GLOBAL HETEROCHRONIES

Because the terminology of de Beerian heterochronies is based on global features of gonadal and somatic maturation and growth, it is easy to come away with a view of heterochrony biased toward whole-body mechanisms and consequences. Explanations of changes in shape come to rest on allometric growth rules. Such a view is one-dimensional; that is, we see differently shaped products arising from timing changes played out on a conserved set of growth rules. There seems to be little doubt that although a good deal of diversity among closely related animals can be generated in this way, the generation of evolutionary novelties will be limited. We will also be hampered in accounting for evolutionary change if our concepts of heterochrony are so impoverished. However, we know that evolution isn't limited to playing out larger and smaller versions of the same theme. Shapes change, and new allometric relationships evolve.

There is no natural constraint on dissociations to produce only somatic versus gonadal effects. Local dissociations in growth should be far more prevalent and may produce novel allometries. The basic concepts of temporal dissociation can be applied to localized regions of the body. Thus, modifications in elements of the limb or parts of the skull can be seen as resulting from heterochronies that occur within some restricted region and exhibiting time dissociations relative to other structures in the localized domain of action. No reference to gonadal maturation is needed at all. Delayed onset of growth or accelerated rate of development can be evaluated within a local reference system. This has been recognized in various ways by students of heterochrony. McKinney and McNamara define "dissociated heterochronies" as those instances in which individual features within the body evolve by different modes. For example, McKinney found that in the same species of irregular sea urchins, test width could exhibit hypermorphosis whereas test height exhibited neoteny. In trilobites, Edgecomb and Chatterton observed that features of the head exhibited evolution by paedomorphosis, whereas the tails of these animals exhibited peramorphosis. Local heterochronic changes can also occur as a result of change in body size. Alberch and co-workers showed that the sequence of appearance of features varies between the salamander genera *Triturus* and *Ambystoma*, with some features that appear at small sizes in *Triturus* development appearing only at large sizes in *Ambystoma*.

HETEROCHRONIES NEED NOT BE LIMITED
TO LATE DEVELOPMENT

There is an enormous predominance of studies of heterochronies in late development. This is not surprising. First, the nomenclature stresses events that occur as maturity is approached. Second, most studies have focused on morphological features, and morphology is richer in well-developed animals than during early development. Third, heterochrony has been the darling of evolutionary biologists, not developmental biologists. Since evolutionary biologists are primarily interested in how adult morphology evolves, they have recorded mainly adult heterochronies. Until very recently, heterochrony was not part of the embryologist's lexicon at all, and studies of heterochrony or any other evolutionary phenomenon were not a part of the discipline of developmental biology. Finally, a substantial part of the literature has been written by paleontologists, who have mostly adult morphology in their material and whose only window on developmental mechanisms has been via the concepts of heterochrony.

Despite the visible effects on discovery that arise from what part of the field the biologists happen to be milling around in, when heterochronies have been sought in early development, they have been shown to be as prevalent as in late development. Some examples include the acceleration of development of the adult rudiment in ascidian larvae described by Berrill, the changes in numbers of cell divisions preceding gastrulation in direct-developing sea urchin embryos described by Parks and co-workers, the early initiation of limb buds in direct-developing frogs described by Lynn, the early onset of and even inversions in the order of morphogenetic events in early development in direct-developing sea urchins that I noted in 1987, and the changes in the relative timing of expression of genes during sea urchin development observed by Wray and McClay. In most cases these heterochronies have no effect on adult morphology, but as noted in chapter 7, changes like these do influence life history in important ways.

If the evolution of early development is constrained, early heterochronies should be rare. In reality, when sought, they are not hard to find. A moment's reflection shows why they should be expected. In early development there are fewer interactions between cells than in later development. In many embryos cell lineages differentiate autonomously, or nearly so, during early development. The relative autonomy of cell lineages varies from group to group. In nematodes, development is highly mosaic, with rigidly determined autonomous cell lineages. There is some cell-cell communication in soil nematode embryos, and induction plays some important roles. However, inductive interactions appear to be far less prevalent in soil nematodes than in vertebrates. Heterochronies should be readily achieved in circumstances in which cell lineage autonomy is prevalent, as long as the dissociations allow production of the correct cell types in a format in which they will be capable of coherent interaction later in development.

Organogenesis begins with gastrulation. New associations of cell sheets occur, and inductive interactions become important in specifying tissue identities and in morphogenesis. A large number of interactive processes occur. Because so many processes are occurring, the period of organogenesis could present numerous opportunities for heterochronies. However, recall that this is also the period of the phylotypic stage, in which we suspect that the interconnectivity of processes constrains evolutionary change. If this is so, heterochronies too should be constrained. Studies of heterochronies in the phylotypic stage appear to be scarce, but Richardson has reexamined a vast amount of descriptive data and has discovered that heterochronies are rampant in development of the phylotypic stage across the range of vertebrate taxa. Heterochronies also occur in larval forms and involve timing changes in the appearance of larval features and those leading to metamorphosis. Thus, acceleration of development of adult structures occurs in direct development, and retention of larval features in neoteny. Neither affects the phylotypic stage per se, and both appear to be possible because early development can be uncoupled from development of the phylotypic stage.

DEVELOPMENT OF THE DEAD

Since so much of evolutionary history is recorded in the fossil record of extinct species, a study of the development of extinct forms would be highly desirable. Despite the impossibility of doing genetics on defunct animals and the difficulties in approaching development when only static objects are preserved, a great deal of information on reproduction and development can in some cases be reclaimed from fossils. Detailed studies of dinosaur nurseries, as discussed in Carpenter, Hirsch, and Horner's book *Dinosaur Eggs and Babies*, show that such data can be recovered from the fossil record. Although preservation of early stages of development is extremely rare, preservation of postlarval growth stages of many animals, including corals, bryozoans, graptolites, trilobites, eurypterids, brachiopods, snails and clams, nautiloids and ammonites, sea urchins, and various vertebrates, has occurred with reasonable frequency.

The importance of studying inferred fossil developmental series is that they may allow us to derive both descriptive and quantitative data that can be used to infer the roles of developmental processes in the evolution of extinct animals. These data should facilitate evaluations of past evolutionary trends. They can also make possible comparative studies of past and living representatives of a clade, such as the attempts that have been made to infer growth and maturation rates of extinct species of hominids. Finally, if we can read the developmental data from the fossil record properly, we might obtain information about mechanisms of morphological change in extinct animals. Because the various modes of heterochrony have been regarded as mechanisms of evolution, the identification of specific modes of heterochrony in fossil lineages could be seen as supplying the mechanisms underlying changes in those lineages.

There is a downside to hunting fossil heterochronies. Because the developmental age of fossil organisms can seldom be determined, size has to be used as the indicator of age. Sophisticated and informative size-based heterochronic inferences are possible in some cases, and the lack of direct data on age doesn't in itself prevent good studies from being done. Such studies may be more effective when living relatives are available to provide information about growth. Nevertheless, dependence on size may pose unanticipated difficulties. A recent study on heterochrony in living water striders by Klingenberg and Spence shows that even among a closely related set of species, size can be a very poor proxy for age. In very remote extinct groups, inferences will, as usual, be more problematic.

A second problem to be considered in studying fossil heterochronies is that in order to determine the direction of change, we need to know the phylogeny of the clade being studied. Inference of heterochronic mode is highly sensitive to perceived phylogeny. Fink has emphasized the evaluation of heterochronies in a phylogenetic context, a position consistent with the general principle that any evolutionary change in development must be placed within a phylogenetic framework to be properly understood. The determination of exact phylogenies is always problematic in fossil forms. Direct demonstration of genealogy is not possible, but cladistic analyses of morphological features and data on location and time of occurrence allow good inferences to be drawn when the record is reasonably complete. When the record is not very complete, phylogenetic resolution may be quite coarse. The relationship of *Archaeopteryx* to later birds is a case in point. Lineages like that of birds may be so important that we want to study evolutionary modes despite the low resolution of the phylogeny. As potential ancestors must exist before their descendants, the stratigraphic range of fossil species can be used to infer temporal ordering of the forms being compared. Outgroups may provide important data as well. In the case of birds, small carnivorous dinosaurs provide outgroup data.

A final problem in studying fossil heterochronies is the very incomplete developmental data available. Such data may be available for some species in a lineage, but rarely for many. Even with incomplete developmental data, however, some inferences are possible. A descendant species might have an adult distinct from its presumed ancestral form, but have an earlier stage in its development that is similar to the adult of the ancestral form. Such a relationship would be a recapitulation, and would be scored as a peramorphic heterochrony.

THE MEANING OF PAEDOMORPHOSIS

Paedomorphic transformations have been well documented in both living and fossil lineages. Paedomorphosis has been of interest as a possible mechanism for the evolution of novel forms because it has been thought to provide an escape route from more specialized morphologies. In some cases (discussed by Gould in 1977), selection has been for small size per se. Hanken and Wake have noted that miniaturization is commonly associated with certain environments, such as those of marine invertebrates specialized for life on unstable substrates or living interstitially, as between sand grains. Hanken and Wake have noted that almost every metazoan clade contains miniaturized members. The consequences of miniaturization have been striking. Because paedomorphic animals are often smaller than their ancestors, they are subject to different selective pressures, and they may undergo simplifications of structure, show increased morphological variability, and even exhibit morphological novelties.

There are some intriguing examples that support this generalization. Hanken has shown that in highly miniaturized salamanders the arrangements of the wrist bones have become highly variable. The extent of variation is wider within a species of these miniaturized animals than among all twelve related genera of larger salamanders. Miniaturized annelids have been shown by

Rieger to have lost segments and even coelomic linings. Seilacher has suggested that miniaturization in oysters has produced novel modes of shell morphogenesis subsequently inherited by large descendants. The morphological changes in the miniaturized forms are apparently not selected by adaptive or ecological pressures, but may become so in larger descendants. Such variation has been noted in a number of taxa, and may have produced novel morphologies ultimately important in the origin of new clades. Losses of complex features associated with body plan definitions have occurred, and body plan reorganization may be possible in these reduced animals.

New groups often appear suddenly in the fossil record, giving the impression that they have arisen through some cryptic mode of evolution. The fossil record is far from perfect, and the hypothesis of punctuated equilibrium put forward by Eldredge and Gould in 1972 suggests that much of evolution is both rapid and occurs at the periphery of the main geographic range of a species (the reader can find critiques of this concept in Levinton and in Futuyma). Both factors would lower the potential for fossilization of the transitional steps between ancestor and descendant.

If paedomorphosis, especially through progenesis, is commonly involved in the evolution of new clades, the small size of the transitional forms would also reduce the chances of their preservation. Various major clades have been proposed to have originated through paedomorphic evolution, including vertebrates from ascidian larvae (Garstang), insects from neotenic millipedes (de Beer), flightless birds by retention of juvenile features of flying birds (de Beer), and a number of marine invertebrate clades by retention of various juvenile features (McNamara). Paedomorphic evolution has been proposed to correlate with particular ecological situations and to have major consequences for life histories.

Because paedomorphosis has been implicated in a range of important evolutionary events, it is worthwhile to ask whether dissociation of rates of development is really an adequate description of what is occurring. Animals generally have complex life histories that include a mode of premetamorphic development distinct from postmetamorphic development. Paedomorphic evolution substitutes attributes of the earlier developmental mode for those of later ones. What paedomorphosis does, in effect, is to select between existing developmental modes in the repertoire of the ancestral species. Clearly, the deployment of existing developmental equipment provides for a more rapid (and functionally integrated) response to selection than inventing a novel ontogeny. Selection in these cases is not likely to operate on the specifics of development, but rather, on the need to meet some functional demand—for example, to shorten generation time, as in progenesis. In these cases, the heterochronic explanation is that relative timing has been affected. Perhaps it would be more accurate to look at such changes as dissociations between developmental pathways—that is, as the selection of one developmental pathway over an alternative pathway. The mechanism of the evolutionary change may lie in mutations affecting expression of the relevant selector gene that determines the switch into a pathway.

The most famous of all heterochronies is that of neoteny in the axolotl and other ambystomatid salamanders. Raff and Wray have pointed out that the mechanism of retention of larval morphology does not involve a change in timing of growth. The hormonal basis for amphibian metamorphosis has been extensively studied and discussed by Etkin, Dodd and Dodd, White and Nicoll, and Duellman and Trueb. Indirect-developing amphibians undergo metamorphosis as a result of body tissue responses to a dramatic rise in thyroxin. This hormone is produced by the thyroid gland in response to maturation of the hypothalamus. In neotenous amphibians, metamorphosis does not occur. Some species are permanent neotenes whose target tissues have lost the ability to respond to thyroxin. Other species, such as the axolotl, are facultative neotenes. Their target tissues are still capable of responding to thyroxin, but they normally don't undergo a rise in thyroxin during development. In these species, as suggested by Norris and Gern, neoteny has resulted from a failure of the hypothalamic release. Tompkins has shown that in the axolotl, this failure results from a single-gene change. Not all tissues require a high level of thyroxin to undergo metamorphosis, however. Ducibella demonstrated that serum proteins and hemoglobin change to their adult forms in the axolotl in a sort of cryptic metamorphosis. Because gonadal maturation does not require a thyroxin rise, it too is dissociated from the events that do require it.

Although a morphological heterochrony has occurred in the axolotl, the mechanism has not been one of differential timing. Some tissues, including the gonads, have achieved an adult state. Other tissues have not received the "selector" signal and thus have retained a premetamorphic set of developmental processes.

THE MEANING OF PERAMORPHOSIS

Peramorphic heterochronies supposedly change timing to produce "more adult" development than seen in the ancestor. Some evidently are what they are advertised to be. In a number of cases, such as in large breeds of dogs versus small ones, or in gorillas versus chimps, larger sizes are attained in larger species by faster growth, an acceleration. Differential growth is generally of more interest than mere size. Differential growth of body proportions appears to involve more growth along one axis than another. A heterochronic outlook would see differential growth rates as crucial, but it's not clear whether amount of growth or rate of growth is what is being regulated.

Hypermorphosis can result from two kinds of causes. In cases in which it results from prolonged growth to a larger size, the rules of allometric growth control the final shape. These situations are what McKinney and McNamara refer to as "size-based heterochronies." Their most important attribute would seem to be the nature of the control over the allometric relationship. The issue then is how to relate allometry to heterochrony, an accomplishment that is a trifle complicated by the fact that we don't know what governs allometric growth. Coordination of growth to produce consistent allometric relationships within a species requires that the growth of various body parts be tied into the hormonal system of the animal. Growth along an axis will be tied to the response of the tissue in the structure to whatever systemic or local factor is controlling its growth. The presumption that links heterochrony to allometry is that it is growth rates in each structure that are being regulated. However, hypermorphosis tied to allometry can result from duration of growth, or amount of growth, rather than relative rate effects. Its being considered a heterochrony is purely a matter of labeling, not necessarily one of cause.

The more interesting examples of hypermorphosis from the point of view of the generation of evolutionary novelty are those that produce new shapes, not merely allometric projections to a larger size. The evolution of the rotulid sand dollars shown in figure 8.4 is a case in point. This clade has a very localized geographic distribution in north and west Africa and a time range from the late Miocene (on the order of 20 million years ago) to the present. There is a strong directional trend to their evolution. The earliest species show slight marginal indentations in the posterior lateral edges. In later species these become much more pronounced, and the scalloped edges appear at a smaller size (presumed to be earlier in development) than in the earlier species. One late subspecies produces an entirely scalloped margin. McNamara has interpreted these results as a peramorphic trend through time (peramorphocline) caused by hypermorphosis. The marginal gaps of rotulids are a special case of sand dollar structures called lunules, which can be formed internally as perforations of the test, or marginally as gaps. Seilacher showed that lunules result from changes in growth relationships along different plate axes. There is no question but that local changes in pattern formation and growth have produced a novel morphology. By definition, this is an example of hypermorphosis, and there is a peramorphic trend through time. An unconsidered question is whether attributing the changes in this lineage to heterochrony has yielded any mechanistic insight.

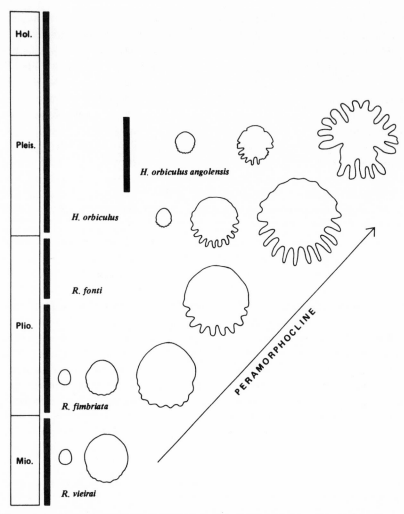

Figure 8.4.   Heterochrony in the evolution of rotulid sand dollars. The evolution of test shape
follows a peramorphic trend. (From McNamara, 1988. Heterochrony and the evolution of echinoids. In *Echinoderm Phylogeny and Evolutionary Biology*. C. R. C. Paul and A. B Smith
(eds.), Clarendon Press, Oxford, pp. 149–64. By permission of Oxford University Press.)

PRE-DISPLACEMENT AND POST-DISPLACEMENT

The kinds of heterochronies discussed so far involve changes in relative rates.
However, when processes start or stop is likely to be just as significant
as rate changes in producing evolutionary changes in development. This

observation is particularly striking when one examines the evolution of direct development. The predominant impression one gets when looking superficially at direct-developing embryos is of loss or truncation of the larval features of indirect development, early onset of adult features, and in some cases, an acceleration in rate of development. These are all heterochronic phenomena. It is only when one looks carefully at the details of early development—for example, as discussed in chapter 7 with regard to the sea urchin *Heliocidaris erythrogramma*—that one sees the underlying complexity of processes and realizes that many of the features that shape the modified pattern of development do not involve timing as a primary mechanism (for example, changes in localization of determinants, in cleavage patterns, in numbers of particular cells produced, in allocation of cellular materials, in gastrulation mechanisms, and in which members of multigene families are expressed). Nevertheless, heterochronies have occurred in the evolution of various direct-developing embryos. These include changes in the timing of gene transcription, of appearance of complex morphological features, of individual cell cleavage events, and of cellular commitment in cell lineages. Examples of pre-displacement of morphological features in the evolution of early development include the adultation of ascidian larvae, in which adult digestive structures appear early in the tadpole larva, the early onset of deposition of the adult skeleton in direct-developing sea urchins, and the early appearance of limb buds in direct-developing frogs. This last example, noted in *Eleutherodactylus nubicola* by Lynn, was contrasted with the indirect-developing frog *Xenopus laevis* by Raff and Kaufman. The limb buds appear after the closure of the neural tube and differentiation of regions of the brain and eye in *Xenopus,* but before brain and eye and at about the time of neural tube closure in *Eleutherodactylus.* In the indirect-developing frogs *Xenopus* and *Rana,* limb buds appear after rapid growth of the tail has been initiated, but before it in *Eleutherodactylus.* The mechanisms for these dissociations in onset are not known, but it is likely that each of these structures is an independent modular entity in the sense I will discuss in chapter 10. Modules such as limb bud, brain, and tail would be expected to have distinct triggers and independent courses of development. That the triggering events can be shifted in time is a significant and important observation.

The closest approach to learning how triggering of onset might be modified in evolution has come from studies of the so-called heterochronic mutations of the nematode *C. elegans,* which I will discuss in the following section. These mutations affect the timing of occurrence of particular cell lineage events. Since these mutations affect decisions about cell fates, they are affecting the onset of specific differentiative pathways within the embryo. Cell lineage decisions show evolutionary changes in timing in both sea urchins and nematodes.

GENETICS OF HETEROCHRONY

Heterochrony is not a developmental process, but rather, an evolutionary process that affects development. As these changes are heritable, they must in some way involve the modification of genes that act in development. One approach to learning what the genetic basis of heterochronies might be has been taken in the nematode *C. elegans*. Ambros and Horvitz showed in 1984 that mutations that affected certain hypodermal cell lineages that produce structures of the epidermis resulted in a modified pattern of cell division and differentiation in those lineages. The phenotypes that resulted from those mutations could be interpreted as heterochronic descendants of the wild-type ancestor. Two kinds of mutations were observed. Ambros and Horvitz noted that recessive mutations in *lin-14* caused precocious appearance of adult cuticular structures in the larval stages. These mutations also caused the affected cell lineages to omit their early pattern of cell division and to produce late division and differentiation patterns too early. Semidominant mutations had the contrary effect of causing a neotenous condition in which the affected cells were retarded in lineage-specific behavior. These mutant cell lineages continued to reiterate an early pattern of cell division and differentiation.

As discussed by Ambros and by Ambros and Moss, the *lin-14* gene acts as part of a series of interacting genes *(lin-4, lin-14, lin-28, lin-29)* that regulate the differentiation of certain postembryonic cell lineages. Mutations in *lin-14* and the other genes in its pathway of action cause timing displacements in cell lineage differentiation in *C. elegans* larvae. The *lin-4* gene regulates *lin-14* activity, which in turn regulates *lin-28* and *lin-29*. The *lin-28* gene controls vulval cell division. The *lin-29* gene controls terminal differentiation, the larva-to-adult switch. In *lin-29* mutant animals, this switch fails to occur, and the larval molting cycle continues. Thus, these genes control the timing of several distinct processes.

Ruvkun and Giusto showed that *lin-14* encodes a nuclear protein found in somatic cells up to the first larval instar. Failure to turn off the gene at that time causes reiteration of early larval cell lineage events. Temporal regulation is under the control of 3' elements of the *lin-14* gene (see papers by Arasu and co-workers, Lee and co-workers, and Wightman and co-workers). The lin-14 mRNA remains more or less constant during development, but *lin-4* activity regulates the normal decline of the lin-14 protein. The *lin-4* gene encodes small RNAs that are complementary to seven short sequence domains in the 3' elements of the lin-14 mRNA. Interaction of the lin-4 RNAs with these domains appears to block translation of the lin-14 mRNA. Although the *lin-4/lin-14* system provides temporal regulation, Ambros and Horvitz showed that *lin-14* levels also can be viewed as alternate states of a selector gene. At high levels, the protein promotes early larval cell lineage behaviors, and at low levels, the later pattern.

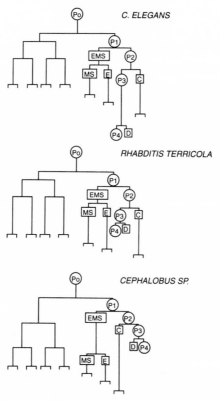

Figure 8.5.   Heterochronies in early cell lineages of soil nematodes. The early cell lineages of *Caenorhabditis elegans* are compared with those of two other species. The focus of the figure is on the daughter lineages of the P_1 cell. In *Rhabditis terricola,* the P_3 cell divides earlier in relation to the MS and E cells than in *C. elegans*. In *Cephalobus* sp., P cell divisions are greatly accelerated relative to those of MS and E cells. In addition, the orientation of cleavages of the P_2 daughters is inverted (anterior is to the left). (From Skiba and Schierenberg 1992, *Developmental Biology* 151:597–610, with permission of Academic Press.)

Soil nematode species are conserved in body plan and in cell lineages in development. In studies of postembryonic cell lineages among nematode species, Ambros and Fixsen reported changes in cell lineage patterns as well as heterochronic changes. In comparisons among nematode species done to date by Skiba and Schierenberg, early cell lineages were found to be heterochronically shifted among soil nematode species. In only one of the species was there an alteration in cell lineage pattern as well. A diagram of heterochronic cell lineage changes among soil nematode species is shown in figure 8.5.

TIME AND GROWTH CONTROL

Much of the literature on heterochrony deals with growth and allometry, but little is known about the genetics of heterochronies manifested in later development. Molecular biology is yielding new information on the proteins that influence growth and body proportions in mammals. Although we still don't have a secure grip on the mechanisms underlying allometry, the shape of the solution is beginning to reveal itself. Relative growth of body parts seems to be a product of the functions of a large number of growth-promoting substances. The relevant effects of these factors can be summarized in a few general statements. First, there are circulating hormones that promote overall growth. Second, circulating hormones that have overall effects on growth can also have specific local effects on particular tissues. Third, circulating hormones can stimulate the localized production of other growth factors in particular tissues. Finally, many of the properties of relative growth arise from the autonomous control of growth in individual structures exercised by local expression of a number of growth factors.

Growing up requires growth hormone (see reviews by Kelly and co-workers and by Shea). In mammals (in which growth has been extensively studied), the growth hormone gene is expressed in a specific cell type, the somatotropes of the anterior pituitary, and the hormone then enters the circulation. Its action is required for postnatal, but not for prenatal, growth. As shown by Rimoin and co-workers, growth hormone–deficient humans are of normal birth weight, but exhibit proportional dwarfism. Growth hormone acts primarily by stimulating the production of insulin-like growth factor-I in the liver and its subsequent secretion as a general growth factor. Mice in which growth hormone production has been genetically ablated are dwarfed. Behringer and co-workers have crossed these mice to a transgenic line that expresses IGF-I under a metallothionine promoter. The progeny, which express IGF-I in the absence of growth hormone, grow to normal size. The consensus view is that IGF-I acts primarily in postnatal growth. The related IGF-II acts primarily in prenatal growth. This is probably an oversimplification because the regulation of synthesis of these factors is complex, and both act through the same receptor. For our purposes, the generalization is sufficient. In their investigation of pygmies, Merimee and co-workers noted that although small in stature, these people produce normal levels of growth hormone. However, in adolescence, pygmies fail to produce the pulse of IGF-I seen in adolescents of larger peoples. Analogously, plasma levels of IGF-I have been found by Eigenmann and co-workers to be proportional to body size in breeds of dogs.

A lack of these general growth hormones results in proportioned dwarfism. Modifications of their expression could also give rise to heterochronic responses. For example, a change in level of growth hormone could cause a change in rate of growth. A change in the onset of the IGF-I pulse could change the timing of the growth pulse relative to other events in maturation. The effects of growth hormone increases were explored by Shea and co-workers in studies of the allometry of skeletal and organ growth in giant transgenic mice. The giant mice were the progeny of a transgenic line transformed with a gene for growth hormone placed under the control of the mouse metallothionine promoter. As a result, growth hormone levels in these animals were very high, and their IGF-I levels were two to three times normal. The mice weighed twice as much as normal mice. The timing of growth was unchanged, but the overall growth rate was accelerated. The giant mice shared the growth trajectory of normal mice, but extended it. Most body organs grew isometrically, but some changes in growth allometries were observed, notably in the spleen and liver, which were larger than expected. The most interesting change was that brain growth showed a striking negative allometry, and the brain ended up half the expected size for the larger body. It appears from these results that prenatal and postnatal brain growth are under the control of a set of factors other than growth hormone. Changes in skeletal shape in the giant mice were allometric extensions of the same growth curve as in normal mice. Thus there was increased growth of the skeleton in response to the elevated levels of growth hormone and insulin-like growth factors, but its shape was regulated by the intrinsic growth regulation system.

Growth hormone has its major effect through its initiation of a rise in circulating IGF-I, but it also stimulates local production of growth factors. Nilsson and co-workers showed that growth hormone causes local production of IGF-I in the chondrocytes of growing bone and thus increasing proliferation of these cells. Differential responses of this type in various parts of the skeleton could obviously contribute to differential growth.

A number of circulating hormones have tissue-specific effects mediated by the expression of hormone receptors and the organization of the corresponding second messenger systems in the target tissue. There are many examples among vertebrates. The sex hormones, testosterone and estrogen, have profound effects on the development of the somatic parts of the gonads and of secondary sexual characteristics. Effects as different as growth and cell death can be elicited by hormones. In frog metamorphosis, thyroxin causes growth of leg tissues, but death of tail tissue cells. The developmental effects of some targeted circulating hormones may have powerful effects on evolution. Androgens and estrogen, for example, are not only involved in sexual differentiation, but also in behaviors that are subject to sexual selection. Such an effect is seen in songbirds. Work by Nottebohm and his collaborators (see Arnold; Nottebohm; Nottebohm and Arnold) showed that the development of male song in songbirds depends on the development of several linked centers within the brain. Lesions in these centers cause a loss of ability to sing. The startling feature of the development of these song centers, noted by Schlinger and Arnold, is that it depends on the local expression of estrogen in the male brain.

Estrogen also has a surprising role in growth regulation in both male and female mammals. Because there is now a pool of hundreds of millions of people who have access to specialized medical clinics, individuals bearing various rare mutations in human development and growth are being reported with increasing frequency. A patient homozygous for an unexpected mutation in the estrogen receptor has revealed that estrogen plays a major role in the cessation of bone growth. Smith and co-workers have reported that this patient, a man in his late twenties, has failed to cease growing. The epiphyses of his long bones are open, and the bones continue to elongate as they would in an adolescent. Examination of his estrogen receptor gene revealed a mutational change in one of the exons, and he failed to respond to treatment with estrogen. Thus, the estrogen receptor of this patient is nonfunctional. The involvement of the estrogen receptor in the cessation of bone growth provides an important clue to how change in size might evolve via control of time of termination of growth.

It may well be that allometry is primarily the result of local controls that govern the extent or rate of growth in individual body structures. Before discussing the evidence for local control of growth and the growth factors involved, we should consider a point made by Bryant and Simpson in their review of growth in developing organs. Differential growth is not necessarily caused by differential growth rates. The standard quantitative way of dealing with growth is through growth curves plotting size as a function of time. Time is thus taken as the key variable, rather than considering the amount of growth to be a function of some internal property of the growing structure. Bryant and Simpson argued that the sigmoidal curves typical of growing structures could be explained by positional information resident in the structure. Under this scenario, mitotic stimulation would be governed by the numbers and identities of cells already present. Bryant and Simpson suggested that in the imaginal discs of *Drosophila*, which are self-contained in their growth and differentiation, new cells intercalate between existing cells. According to this hypothesis, cell division is stimulated by discontinuities in positional information. The values of the new cells eliminate the discontinuities, and growth ceases. Whether or not this hypothesis is correct, it puts us on notice that mechanisms of growth control unrelated to growth rate can exist.

There is a marvelous experimental tradition on intrinsic growth regulation in vertebrate development. Bryant and Simpson have effectively reviewed these classic experiments, and I follow their summary. The basic protocol for such experiments is the transplantation to a host of a body part from a different developmental stage of the same species, or from another related strain or species that has a different growth trajectory or reaches a different size. A number of vertebrate organs have been studied in this fashion. The result of transplanting an organ into a host of a different developmental stage is that the transplanted organ develops at the rate of the donor, not the host, and reaches its normal size. Cross-species transplants behave in the same way. Two examples will illustrate this phenomenon. Baby rat bones transplanted by Felts to a mouse host attained the size and shape characteristic of the rat donor. In a particularly spectacular experiment done by Twitty and Schwind in 1931, embryonic limb buds from a large species of salamander, *Ambystoma tigrinum,* and a much smaller species, *Ambystoma punctatum,* were exchanged. The transplanted limb buds gave rise to normal limbs of the size typical of the donor. These and other studies suggest that both systemic and local growth factors are required for normal growth. The surprise lies in the degree of autonomy of growth of individual organs.

Local growth regulation requires that specific growth factors and their receptors exist in locally distributed patterns. Five prominent growth factor superfamilies are listed by Mercola and Stiles: the epidermal growth factor, insulin-like growth factor, transforming growth factor-β, heparin-binding growth factor (including the fibroblast growth factors), and platelet-derived growth factor superfamilies. Each superfamily includes several families of factors. The TGF-β superfamily, for example, includes TGF-βs, inhibins, activins, Müllerian inhibiting substance, and decapentaplegic and bone morphogenic protein families. Some of these families have several members, each with differing specificities. Among the several BMPs, one has been shown to affect the axial skeleton, whereas another affects development of limb bones (see chapter 10). There are still other growth factors that don't fall into these superfamilies, such as the nerve growth factors.

The striking finding by Shiang and co-workers that a mutation in the transmembrane domain of fibroblast growth factor receptor-3 (FGFR3) causes human achondroplastic dwarfism exemplifies the role of growth factors and their receptors in local regulation of growth. This condition results in dwarfism, relative macrocephaly, and skeletal deformities. Not only are there several FGFs, but, as reported by Patstone and co-workers, by Johnson and Williams, and by Partanen and co-workers, genes encoding four related fibroblast growth factor receptors have been characterized in mammals and chicks. They have distinct sites of expression in development, and they have splicing variants. The existence of a diversity of receptors indicates that they have distinct functions. The distribution of FGFR3 has been reported by Avivi and co-workers, Chellaiah and co-workers, and Peters and co-workers as more restricted in mouse development than that of FGFR1 and FGFR2. These investigators noted that FGFR3 mRNA has a high level of expression in several organs, in the central nervous system, and in the cartilage rudiments of growing bone. Isoforms of FGFR3 have distinct binding specificities, with one specific to FGF1 and another to both FGF1 and FGF4. The implication is that multiple locally expressed growth factors and receptors have potent effects on growth and provide highly specific regulatory patterns. Although it is not known in detail how the growth factors regulate the growth and shaping of the skeleton and other organs, observations on the remodeling of bone (reviewed by Mundy) support their playing a complex role. Bone remodeling has a specific sequence of resorptive and depositional steps, and requires the action of at least five locally acting growth factors.

The local control of growth of body features suggests that dissociations should be frequent in evolution. Dissociations are visible in two ways: in mosaic evolution, in which body parts evolve along distinct trajectories, and through changes in allometric relationships. Evolution through dissociation can include changes in location of the expression of a growth factor, in its timing, in its level, and in combinatorial expression of factors. Equivalent sorts of changes can occur with local receptors. Nondissociative evolution can occur as well, including co-option of a growth factor system to a new function, homeotic transformation of a feature, changes in the number of cells involved, changes in induction of local primordia, and changes in patterns of cell migration to produce new local geometries.

HETEROCHRONY AS A RESULT, NOT A PROCESS

Heterochrony has been touted for its explanatory power as an evolutionary mechanism that modifies development. That explanatory power depends on whether the cited examples really do arise from temporal dissociations of developmental events. Dissociations of developmental events do occur. The question is whether they arise primarily from the unlinking of temporal controls in development. There has been surprisingly little (essentially none!) in the way of critical tests of the basic assumption that observed heterochronies are dissociations of processes in time. I think this uncritical attitude has come from both the commitment we have made to heterochronic explanations and the difficulties that we are presented with in coming to grips with the underlying mechanisms. The finding of so-called heterochronic mutations in the nematode *C. elegans* has been hailed as the first well-understood example of how timing mechanisms operate in evolution. However, with the more detailed information now available on the genetics of *C. elegans,* a heterochronic interpretation is not by any means inevitable.

I presented the *lin-14* system and its mutations from the heterochronic perspective. Let's consider it from a different point of view. This control system functions by regulating the amount of the *lin-14* product. The amount of this nuclear protein determines which pattern of cell lineage behavior is expressed. If we view the mutations as affecting selector gene function, they are causing a kind of homeotic transformation. The effects are most dramatically visualized by their temporal behavior, but the primary control may be over cellular states. Soil nematodes have rigidly defined cell lineages, and cell lineage behavior is critical to development in these organisms. The evolutionary import of changes introduced by mutations like those in *lin-14* will depend on the aspect of the mutant phenotype acted upon by selection. It might be timing, but it could be some other result in the hierarchy of changes, such as number of cells of a certain kind produced.

Regier and Vlahos have pointed out that evolutionary changes in the synthesis of RNAs, proteins, or other substances in development may extend beyond changes in location or rate of production. Changes in amount per se may also be important. This observation leads to a simple but relevant equation: rate = amount/time. If we observe a rate alteration, it may have arisen from a change in timing or from a change in amount. Amount is not limited to molecules. Amount can also refer to growth or extent of shape change. What we have to realize is that just because we observe a heterochronic result—a pattern that can be scored as a paedomorphic or peramorphic change—that does not mean that a timing mechanism had to have been involved at all, or that selection was necessarily operating on any aspect of timing. Perceived heterochronies can thus be manifestations of the way our conception of things can prevent us from grasping the quite different evolutionary processes that are actually at work. An example is worth considering.

McNamara has presented a lucid review of heterochronies in the evolution of test shape in sea urchins. Regular echinoids have globular tests composed of about 100–3,000 rectangular plates. These plates originate in rows on the top or apical side of the test. As new plates are added, older plates are displaced toward the oral side; therefore, the oldest plates surround the mouth. The largest plates are at the equator of the test. Raup showed that the fastest growth occurs at the top end of the rows, and that growth is very slow at the oral end. The large size of the equatorial plates stems from a combination of their being older than the apical plates and having a faster growth rate than the oral plates. The shape of the test is affected by the relative width and height of all the individual test plates. Heterochronies in plate growth should thus underlie the evolution of test shape. That is just how McNamara interprets the evolution of shape in the highly modified irregular sea urchins, the heart urchins and the sand dollars. Regular echinoids are globular because their plates all have about the same ratio of width to height. Sand dollars are very flattened. They have extremely narrow plates at their equators. Heart urchins similarly have highly modified plates, with some being very large relative to others. These plate differences produce highly modified test shapes. Differences in size among plates are concluded to result from differential rates of plate growth.

Explanations based on heterochronies in growth rates are consistent with the data, but are they informative? If we instead think in terms of amount, we see that it may not be relative growth rates, but relative sizes and relative amounts of plate growth, that determine shape. If the regulation of growth of sea urchin test plates matters, the problem translates into one of accounting for the processes that control plate growth. We know that sea urchin plates are composed of crystallographically continuous magnesian calcite. Each plate (see Smith's excellent book *Echinoid Paleobiology* for a full discussion of sea urchin skeletons) is built as a complex three-dimensional meshwork. The plates are built by mesenchyme cells that exhibit a specialized pattern of protein synthesis and secrete calcium. Although this system provides a relatively simple model (see Harkey and co-workers for a review of the genes involved in skeleton secretion) and has been well investigated by embryologists, we know little about calcium secretion, and nothing about pattern formation or regulation of growth. If we are satisfied that we have understood what is going on by placing heterochronic labels on these phenomena, we will miss out on learning about what really regulates shape.

The answers we accept as providing "mechanisms" depend upon the hierarchical level that we are examining and upon our disciplinary bents. There are good reasons for the emphasis on heterochronies. First, paleontologists, and many biologists, simply don't have data on the genetic regulation of growth in their species, living or fossil. Second, many studies, particularly those of paleontologists, focus on long-term trends in whole-organism morphology. Often these studies are motivated by an interest in phylogeny, functional morphology, or the relationship of evolutionary trends in shape to environmental changes over geologic time. In cases such as these, global heterochronic appellations may provide the only approach to associating evolutionary trends with developmental processes. However, it is easy to fall into an overemphasis on heterochrony. One more example will be informative. The bizarre extinct Triassic reptile *Tanystropheus longobardicus* was about four and a half meters long. It had a lizardlike body, and its neck made up half of its total length. Tschanz concluded that the neck exhibited strong positive allometry and had evolved by heterochrony via hypermorphosis. However, *Tanystropheus* had twelve neck vertebrae instead of the ancestral eight, and each vertebra was highly elongate. Thus, we could also give a nonheterochronic interpretation of *Tanystropheus*, in which a meristic change in vertebral number and a change in the amount of growth along the long axis of the cervical vertebrae could account for the evolution of its long neck.

HUMAN EVOLUTION

How puissant is timing change as a universal mechanism for the evolution of form? A brief examination of the application of heterochrony to human evolution will be useful in illustrating the dangers of attempting to equate it with the entirety of the evolution of development.

Gould in 1977 and McKinney and McNamara in 1991 provided extensive and important reviews of the perceived role of heterochrony in human evolution. Both placed heterochrony at the head of the designer's table in generating the body morphology and brain of modern humans. The curious thing about the two discussions is how disparate are the heterochronic processes they advanced as important. Gould presented the pattern as paedomorphic, with neoteny as the significant process, whereas McKinney and McNamara saw the pattern as peramorphic, with hypermorphosis as the major process. Neither discussion is simplistic, and both recognize that other complexities and processes in addition to global heterochronies have been significant. Yet they differ in what they purport to explain. We need to consider these excellent presentations of the evidence for heterochrony as a universal mechanism in human evolution to see whether such explanations really explain, or if they are merely formalistic descriptions.

The first comprehensive heterochronic theory of human evolution was elaborated by Bolk in the 1920s. Bolk provided a long list of presumably paedomorphic features, ranging from the flat shape of the human face through hairlessness, brain size, position of the foramen magnum, structure of the hand and foot, and the prolonged period of infant growth and dependency. This listing established a tradition followed by other writers. In his 1981 book *Growing Young,* Montagu, for instance, not only listed presumptive neotenous physical traits, but also presented a whole suite of complex behavioral traits that I think could only be very dubiously attributed to neoteny. These included friendship, curiosity, creativity, sense of humor, dance, and song! The poetry of the morphologically youthful human ape seems to have overwhelmed critical thought in some instances.

Gould in *Ontogeny and Phylogeny* disparaged what he called the "enumerative tradition," and he rejected Bolk's obsolete and anti-Darwinian ideas on evolution. However, Gould built his heterochronic hypothesis of human evolution on Bolk's basic insight into human heterochrony. Gould says the following (p. 365): "I believe that human beings are 'essentially' neotenous, not because I can enumerate a list of important paedomorphic features, but because a *general, temporal retardation of development has clearly characterized human evolution. This retardation established a matrix within which all trends in the evolution of human morphology must be assessed*" (emphasis in the original). In setting up this paedomorphic "matrix," Gould did not deny that other heterochronic processes also occurred, perhaps most notably hypermorphosis in the length of our legs. However, the overall context of human evolution was to be taken as globally paedomorphic.

The case for neoteny was strongly criticized by Shea, who accepted heterochronies as the means of evolution of human form, but rejected global neoteny. He argued that prolonged growth does not require paedomorphosis and, as I will discuss in a moment, that many of the supposed paedomorphic features of humans actually represent novel growth trajectories. McKinney and McNamara rejected most of the supposed neotenic features put forward in earlier considerations of human evolution. They reinterpreted many of those features as being due to heterochronies, but to hypermorphosis rather than neoteny. They plotted several features to illustrate the role of hypermorphosis. Overall growth is one. Gorillas grow faster and much larger than chimps or humans. Chimps and humans exhibit similar rates of growth, but humans continue to grow for several years beyond the cessation of chimp growth. Sexual maturity comes earlier in chimps and gorillas than in humans. Hypermorphosis or neoteny?

The work of Count and of Holt and co-workers showed that prenatal growth curves for all primates, including humans, are the same. In nonhuman primates brain growth ceases at about the time of birth. In humans it continues for nearly another two years. As with many cases of heterochrony, labels come into play as one decides what the temporal reference points are. Gould has argued that the prolonged fetal growth rate represents a retardation of development in humans, resulting in paedomorphic development of the brain relative to other primates. McKinney and McNamara have also supported heterochrony as the basis of brain development, but have argued that its continued growth represents hypermorphosis by prolongation of brain growth relative to birth time. Or, perhaps with equal justification, we could conclude, as did Portman (cited by Gould), that birth time has been pre-displaced relative to other features of growth. Passingham has argued that we can look at brain growth in another way. He has shown that the rates of brain growth are similar for all mammals, including humans. What differs is the rate of body growth. It is thus possible that absolute rates of brain growth are constrained by the properties of neural growth and integration. Without some compensation, the enlarged human brain would not be possible. Selection has to be for growth patterns that allow the development of a large brain, but also for production of an infant skull that can pass through its mother's birth canal.

One of the strongest pieces of evidence put forward by Gould for human neoteny is the paedomorphic shape of the human skull. Its superficial resem-

blance to the skull of a juvenile ape is striking. Young apes have proportion-
ally rounder skulls and smaller faces than the grownups. Their skulls are
much more similar to those of human adults and babies than to those of
adults of their own species. The reason for this deviation in the skulls of
adult apes is that ape brains grow little after birth, but their faces grow a
good deal, resulting in larger jaws and teeth than in humans and pronounced
brow ridges over the eyes. The shape of the adult human cranium reflects
the presence of a large brain. The human brain is not merely a holdover of
the juvenile ape brain. It's much bigger, and its size affects the overall form
of the skull. That is particularly so in modern humans, because modern
human teeth and jaws are greatly reduced.

The shaping of the face and the large brain appear to be dissociated in
evolution. Neanderthals had much stronger jaws, and thus larger faces and
brow ridges, than modern humans. Nonetheless, their brains were as large
as ours. This is really not so astonishing. Human evolution has been highly
mosaic. The greatest surprise in physical anthropology was the finding that
hominids evolved a body adapted for bipedal locomotion before they evolved
an enlarged brain. The evolution of facial features in humans has involved
more than simple growth trajectories. The growth of specific features also
involves remodeling of bone (see Enlow; Moore and Lavelle). According to
Bromage, remodeling fields have four attributes: size, shape, placement, and
rate of activity. In a study of facial remodeling in the Taung skull, that of a
three-year-old *Australopithecus africanus* child, Bromage concluded that the
pattern of remodeling was similar to that in chimps, but that the rates varied.
Other elements of patterning have changed in the evolution of the modern
human face.

Humans also differ greatly from apes in important features of the upper
respiratory tract and the associated features of the base of the skull. Laitman
has noted that the upper respiratory tract of the human newborn resembles
that of other primates. At about two years of age the human larynx begins
to descend in the neck. It ultimately occupies a much lower position in
humans than in any other mammal, and produces the voice box required
for human speech. This clearly nonpaedomorphic feature has influenced the
evolution of the base of the skull. Shea has reviewed the particulars of skull
growth in apes and humans. Development of the apparently paedomorphic
features of the base of the human skull follows a different developmental
pathway than in ape species. The results of Laitman and Heimbuch on development of the basicranial angles of the floor of the skull show that these
angles change less in chimp development than in humans. They change more
in gorillas than in humans. The superficial paedomorphic appearance of the
human basicranium arises from a growth trajectory unlike those of apes and

the concomitant development of the human larynx. On the basis of their studies of basicranial growth, Dean and Wood concluded that no global heterochrony could be involved.

The foramen magnum is the opening at the base of the skull through which the spinal cord passes from the skull to the spine. In humans it lies directly beneath the skull, reflecting the vertical position in which we hold our heads. In adult apes it lies farther back on the skull, and opens rearward at an angle off vertical. Apes don't hold their heads upright. The position of the foramen magnum has been advanced as an important paedomorphic feature. In both apes and humans, it lies under the skull in neonates. In apes, the remodeling of the skull during growth moves the foramen magnum into its rearward position, but in humans this does not occur. The result is a neotenic retention of the neonate pattern in humans. Has a heterochronic process been involved in this evolutionary change? The foramen magnum has to be somewhere. If it were in some location on the human skull different from the neonate or ape position, we could still attribute the change to a different heterochronic process, hypermorphosis. It all depends on what we are after. As a description, paedomorphosis makes clear the difference between the human and the ape foramen magnum. We could also describe what has happened as the loss of a morphogenic process involved in remodeling the ape skull. We don't know what genetic or morphogenetic mechanisms were changed. Dissociations occurred, but dissociation in control of timing was not necessarily the cause.

Finally, there is the brain itself. The human brain is much larger than the chimp brain, and some of its features, such as the amount of cortical surface and the degree of folding (noted by Jerison), and the percentage of neocortex (noted by Passingham), are consistent with allometric scaling to the larger brain size. However, the human brain is not merely a scaled-up version of the chimp brain. There appears to have been a major reorganization of the human brain relative to that of chimps. The extent of that reorganization has not been fully defined, but it appears to involve the acquisition of the pronounced left/right asymmetry associated with the evolution of handedness and the origin of the speech center in Broca's area of the left brain (see Falk, Passingham, and Passingham and co-workers for discussions of brain lateralization), the rewiring of connections in the brain (as exemplified in the study of Galaburda and Pandya), and a mosaic pattern of changes in cell numbers and types in specific portions of the brain (see Armstrong). It is impossible to attribute all of these changes to any global heterochrony. It is all too evident from the review of Goodman and Shatz that the establishment of neural connections is the result of a number of complex processes involving pathway selection, target selection, and address selection. Repatterning can hardly be merely a function of a global heterochrony.

To attribute brain evolution to numerous local heterochronies is both to lose the elegance of a global heterochronic explanation and potentially to obfuscate the causes of the complex changes in pattern. Local heterochronies might well have produced some of the modifications of the human brain. However, given the range of phenomena observed, nonheterochronic dissociations, co-option, and the forging of novel neural connections have obviously been at work as well.

There are clear qualitative processes that can account for the evolutionary remodeling of brain development without recourse to timing changes. A brief consideration of three makes the point. First, large numbers of neurons die during normal development. As discussed by Patterson, as many as 80% of neurons die in the cat retinal ganglion, and 40% in the chick, but none in fishes or amphibians. Which neurons die or survive is determined by their success at finding target tissues. An experimental increase in mass of target tissue promotes a higher neuron survival rate in response to the higher levels of neurotrophic factors produced by the target tissue. Evolutionary changes in size of parts of the brain might thus depend on cell mass or level of production of neurotrophic factors, independent of time dissociations. A second process is axon guidance. Nerve cells establish proper connections by following molecular guidance cues and making contact with other appropriate nerve cells. The molecules involved in these guidance behaviors and their roles in establishing connections are discussed by Baier and Bonhoffer, Dodd and Jessell, Harrelson and Goodman, and Goodman and Shatz. Evolutionary changes in the patterning of the nervous system will be very sensitive to which molecules are expressed and to the sites of expression of the guidance molecules independently of time. The third example shows that brain reorganization can arise from the localization of hormone receptors. Prairie voles are monogamous; montane voles are not. The bonding behavior of prairie voles is affected by the peptide hormone vasopressin. A rise in hormone level is stimulated by sexual behavior. Insel and co-workers showed that by itself, injection of vasopressin stimulates long-term bonding in prairie voles, but not in montane voles. When they mapped the sites of vasopressin receptors in the brains of these animals, they found that vasopressin receptors are present in different locations, and stimulate different brain circuits, in the two species. This finding represents a substantial remodeling of the brain in two closely related species, and one surely little related to rates of growth.

TO RECAPITULATE

An accounting for human evolution that began as ineluctable has ended up as ineffable. The evolution of nearly all features of the human body has been attributed to heterochrony. It really doesn't matter that the adherents to this view disagree as to which heterochronic processes predominate. The real question is whether this outlook gives us a mechanistic insight or merely describes the results of various underlying processes expressed in developmental time. I'm not arguing that dissociation is not an effective process of evolutionary change. Human evolution has required significant dissociations between events. But not all dissociations are primarily heterochronies, and the heterochronic patterns that have emerged may not have resulted from dissociations in timing of processes. Some no doubt have. Accelerations of growth rates and displacements of initiation or cessation of events seem like good candidates for temporal dissociations. However, heterochronic results can arise from other kinds of dissociations: dissociations of cell numbers, of amounts of growth, of direction of growth, new cell-cell contacts or connections in neural pathways, or functions expressed in a new location. All occur in the ordered stream of development, and so will inevitably produce changes in timing as well as in size, shape, and function.

Dissociations between processes in development occur frequently in evolution. Because ontogeny occurs in a temporal stream, any change in developmental processes is likely to result in a heterochrony. Nonheterochronous processes will thus at some level in the hierarchy of events produce heterochronous results. Heterochronies at basic mechanistic levels also occur, but these do not account for the preponderance of heterochronic outcomes. If heterochrony is to be useful as a concept in evolutionary developmental biology, heterochrony as a process should be kept distinct from heterochrony as a result. In some cases, such as cell lineage heterochronies, the underlying dissociation may be one of time. In other cases it may not be. However, whether heterochrony is considered as a process or as a result will often depend upon the kind of analysis being performed. Whether the ultimate cause of a heterochronic pattern matters depends on what level of mechanistic explanation is sought. For an ecologist, knowing that an organism is progenetic or neotenic may be sufficient to analyze the role of life history in selection. In that case, progenesis or neoteny, by whatever ultimate cause, would be a process at the level of life history. For a paleontologist, the pattern may be all that is available, and so a heterochronic explanation may provide the best ordering available for what amounts to incomplete data on development combined with a direct historical view of morphological consequences. For a developmental biologist, the nature of the dissociative mechanism will ultimately lie in gene expression, cell lineages, pattern formation, and other morphogenetic processes. The heterochronies observed in nematode and sea urchin cell lineages suggest that timing mechanisms do occur at these levels. So do other processes.

It should be kept in mind by anyone interested in the role of heterochrony that anything that grows will have a rate of growth. The current understanding of the mechanisms of growth control that I've sketched for mammals shows that both global and local controls are involved. As the molecular mechanisms become better known, we will have a more certain view of allometric growth and the evolution of shape change in adult development. It is already becoming apparent that these controls and evolutionary changes in them may not be primarily matters of rate. What emerges is that although heterochrony is a common phenomenon, it is not the universal evolutionary mechanism. The limits of its utility in experimental investigations of the evolution of development remain to be fully investigated.

9

Developmental Constraints

When you're young, all evolution lies before you, every road is open to you, and at the same time you can enjoy the fact of being there on the rock, flat mollusk-pulp, damp and happy. If you compare yourself with the limitations that come afterwards, if you think of how having one form excludes other forms, of the monotonous routine where you finally feel trapped, well, I don't mind saying life was beautiful in those days.

Italo Calvino, *Cosmicomics*

THE DIVINE WATCHMAKER

One of the most influential books in early-nineteenth-century biology was William Paley's *Natural Theology,* first published in 1802. Paley used the intricacy of anatomical structures ranging from eyes to earwigs to argue that "these provisions compose altogether an apparatus, a system of parts, a preparation of means, so manifest in their design, so exquisite in their contrivance, so successful in their issue . . ." as to admit no question but that there is design in nature. Paley saw perfection such that "the hinges in the wings of an earwig, and the joints of its antennae, are as highly wrought as if the Creator had nothing else to finish." Paley asked his reader to imagine finding a watch on a path. He suggested that if the finder would for a moment consider, he would realize that such a complex and purposeful device could only have arisen by intelligent design. Paley considered living things as natural complex objects that had an obvious purposeful design, and that thus required an intelligent designer, God. Although now hackneyed, Paley's example of the watch was extremely powerful in its time. Paley's famous analogy of a watch requiring a watchmaker is still fondly recalled by modern creationist debaters, who, in what Russell Doolittle has referred to as the "timeworn watch caper," love to hold up their watches before the audience and ask if they can imagine a watch assembling itself.

Another problem was posed for pre-Darwinian biologists by the vast diversity of living forms, whose discovery was one of the major rewards of eighteenth- and nineteenth-century voyages of exploration. That the effort to classify all the newly discovered organisms drove European biology during that period is clear from Mayr's *The Growth of Biological Thought.* Shortly before the period dominated by *Natural Theology,* the concept of a great chain or ladder of being provided an organizing principle. This concept posited a linear ordering of living beings from lowest, such as plants, through increasing degrees of animal complexity to humans, and finally to supernatural entities. Although it's hard for biologists to understand such a point of view now, the great chain of being implied no evolutionary connection between the organisms in the chain. This was a concept of hierarchy of structural perfection, but not of descent. The chain of being also suggested a principle of divine plenitude, in which all possible life forms were created by a beneficent Creator. This idea had interesting consequences in suggesting that plenitude could not allow any creature to become extinct. Thus, Thomas Jefferson, who included mammoths in his list of the North American fauna, reflected on the probability that mammoths still roamed unobserved in the unexplored far northern territories. "It may be asked, why I insert the Mammoth, as if it still existed? I ask in return, why I should omit it, as if it did not exist? Such is the economy of nature, that no instance can be produced of her having permitted any one race of her animals to become extinct; of her having formed any link in her great work so weak as to be broken."

Early in the nineteenth century, it became apparent that many once-living species of animals had become extinct. Extinction broke the confidence of scientists in the principle of plenitude. A more sophisticated comparative anatomy did the rest. As the anatomies of invertebrate animals were described, it became apparent that the great chain of being was inadequate to explain them. Their disparate anatomies simply could not allow a single linear chain of organization or classification. Biologists thus had to go beyond descriptive taxonomy and seek patterns of relationships that reflected an expected underlying order to life. Before Darwin's *Origin of Species* was published in 1859, idealistic conceptions of the ordering of diversity analogous to those of *Natural Theology* held sway. Darwin recognized the role of variation among individuals in providing the basis for the operation of natural selection. In so doing he provided the insight that organisms do not possess the perfection of design ascribed to them by Paley. He also drew the critical conclusion that living things are the products of long evolutionary histories: that there is a genealogical link between extinct ancestors and their living descendants.

At about the same time Paley was crafting a view of nature that would pervade British biology for several decades, Baron Cuvier was establishing the science of comparative anatomy in France, and was astonishing his contemporaries with his reconstructions of fossil mammals and his revelation of great extinct faunas succeeding each other in the dim prehistory of the planet. To Cuvier, anatomy was predictive because it possessed an underlying principle of correlation of parts. A jaw with large canines and molars adapted for shearing meat must have belonged to a carnivore with all its other body parts bearing commensurate adaptations for catching prey: hooves and antlers would hardly be sensible in a reconstruction of such an animal. Cuvier realized, in a far more sophisticated way than Paley, that all organisms, living or fossil, exhibit a profound integration of body structures. Even though Cuvier well appreciated the historical succession of animal faunas, he rejected the evolutionary theory of his contemporary, J. B. Lamarck. To start with, he saw no evidence for evolutionary transformations between faunas in the fossil record, but second, and more fundamentally to Cuvier, the functional and morphological unity so ubiquitous in anatomy demanded stability. Any evolutionary change in major body features would disrupt the unity of the whole.

With the Darwinian revolution, the cozy Paleyian view of nature was forever exploded, and "nature red in tooth and claw" took its place among the disturbers of Western society's inner peace of mind. The deeper problems of integration and stability posed by Cuvier also have continued to haunt evolutionary theory to the present day. Representatives of different guilds among evolutionary biologists have articulated this issue in different ways. To an adaptationist, the problem is one of changing one highly functional complex to another without producing a poorly adapted transitional morphology—in T. H. Frazzetta's words, how to modify a machine while it's running. Sometimes we can see how it's done. The jaws of mammals are very different from those of reptiles, and are hinged to the skull at a different point. The transformation from reptilian to mammalian jaws is revealed in a 100-million-year fossil history of the transition between these classes preserved in the rocks of South Africa. In an amazing way, the intermediate mammal-like reptiles possessed both functioning reptilian and mammalian jaw articulations. Later, the remnants of the old reptilian jaw articulation were banished to become the small bones of the mammalian middle ear.

CONSTRAINTS

To paleontologists, the problem is how to explain the large-scale patterns of life's history in terms of changing global environments, catastrophic extinctions, and long-term rules inferred from our short sampling of existing life. To those of us attempting to unravel the mechanisms by which animals change form in evolution, the issues emerge in the guise of the tension between the demands of natural selection and the internal rules that govern the expression of genes and the development of embryos. The nature of the existing developmental system somehow constrains or channels acceptable change, so that selection is limited in what it can achieve given some starting anatomy. In this chapter I will discuss the idea that there are internally imposed constraints on evolution, their proposed origins, and their proposed effects.

WHY ARE THERE NO CENTAURS OR SIX-LEGGED GREYHOUNDS?

In many cases, we can recognize constraining factors in evolution. For instance, why there are no animals with wheels is easily explained. It is not possible to connect nerves and circulatory systems to the tissue of a free-turning wheel. Questions such as why there are no 200-foot gorillas are readily answered on the basis of strength of materials and scaling. Because body mass increases roughly as a cube of length, but strength of bone and muscle only as a function of the square of its diameter, a 200-foot ape could not withstand the stresses on its limbs and so would be a rather inert spectacle (a revealing discussion of the constraints of size for big animals can be found in Alexander's *Dynamics of Dinosaurs*). Questions about missing body plans or unexplored adaptations are more difficult to answer. The question of why there apparently never have been any vertebrates with more than four limbs shows why.

The tetrapods have a 400-million-year history that encompasses a great deal of anatomical disparity. One could imagine great advantages in having, say, six legs. Centaurs, after all, could run like horses but had their hands free to shoot arrows or carry away hapless maidens. Six-legged tetrapods might be able to burrow more efficiently or run faster, or, as in centaurs, they might more easily evolve limbs differentiated for running, grasping, and archery. More legs might have facilitated the evolution of larger and heavier bodies: imagine the possibilities for a six-legged sauropod dinosaur. Is this oversight on the part of tetrapods due to some constraint in the way their genomes are expressed in development? Some developmental constraint might make it impossible for a vertebrate embryo to generate more than two pairs of limb buds. This could be the case if the regulation of limb number in the vertebrate embryo results not from local interactions within the limbs but from a global field involving the components of the body axis that contribute to the development of limbs: the lateral plate mesoderm, somites, and body wall ectoderm. I'll discuss the development of limbs and the evolution of limb number and identity further in chapter 10. Much of the history of limb origins can now be understood, but not the apparent limitation in limb number. Perhaps no constraints exist, but the vagaries of history are such that once a four-limb pattern was established, it worked sufficiently well that no other pattern could pass from the crude initial stages into satisfactory operation. Perhaps there was functional selection. One can imagine that six-legged tetrapods would have had intrinsically inefficient gaits, or perhaps there is a neurological inability of the vertebrate nervous system to coordinate six legs. Problems of gait or neurological wiring seem to be eliminated by comparison with arthropods. Centipedes do quite well with numerous legs attached to a sinuous body. Thus the side-to-side sinuous gaits of primitive tetrapods would seem possible in a multilegged version as well.

IS SELECTION ADEQUATE?

Current evolutionary biology holds that natural selection acting on randomly generated variation produces the biological order we see around us. Natural selection operates externally to the organism, and in principle, it should not matter what particular internal rules of gene organization or developmental machinery are in place. Population genetics, which has dominated evolutionary theory for the past few decades, has operated with this scenario as its worldview. Classic population genetics regarded genes as separate beans in the genomic beanbag, and devised an effective mathematics to deal with selection on individual genes. The real world is far messier than this, and evolution often involves continuously varying traits such as size. The approach of quantitative genetics (see Stearns and also Lande for introductions) recognizes that phenotypes depend on the individually small effects of large numbers of undefined genes as well as on nonheritable environmental factors. The course of phenotypic evolution under selection can be constrained by the covariance of many genes. In selection studies on small populations, selection limits, in which a trait reaches a plateau and can't be pushed further, are quickly reached. In large populations in which more variation is present, selection limits are not reached as rapidly. The important conclusion is that genomic constitution can affect selection.

There is no doubt that existing order constrains the variation available, but a purely selectionist approach (probably not supported by many) would nevertheless consider that such apparent constraint is merely a matter of historical accident and probability. Some variations are more likely to occur than others in any particular genome. However, nothing would be really forbidden, and given enough time, sufficient variation could arise for selection to transform a horsefly into a horse.

This hypothetical conception of the freedom of selection to shape the evolution of development has been opposed by the hypothesis that internal organization constrains evolutionary possibilities in a systematic way. Gould and Lewontin raised the issue of the limits of selection by considering a feature of cathedral architecture. Spandrels are the tapering triangular spaces formed at the intersections of two rounded arches meeting at right angles. Gould and Lewontin pointed out that such panels may be painted with saints and other narrative elements incorporated as vital parts of the overall decorative design of the building, and thus present a part of a harmonious whole. However, the spandrels were not put into the architecture to bear their part of the painted design. Rather, they are the geometric consequence of the intersection of the arches. Gould and Lewontin argued that biological analogies abound, and that not all features need be explained by selection. Seilacher also suggested that constructional or architectural constraints exist in biology and produce features that lie outside of the domain of direct selection. The structure of a paper cup serves nicely as an analogy. The seam down the side has no bearing on function at all as long as it holds up when liquid is poured into the cup. The seam offers no selective advantage in the world of paper cups. It is there simply because a cup is made by sealing the edges of a sheet of paper to form a cylinder or cone, and is thus the drinking vessel analogue of the belly button, a feature that is not important in the adult but marks a critical developmental event. Allometric growth in which body parts exhibit a particular ratio of growth with respect to body size is often conserved among related organisms. The advantage of conserving particular growth relationships in evolution is again not obvious, and the conservation of allometric relationships may relate more to some internal constructional rule than to selection.

The limitations of selection have become apparent from other lines of evidence about evolutionary phenomena that take place at the genomic level. Among these is neutral evolution, something that Kimura showed is likely to account for much of the evolutionary change observed at the DNA level. In neutral evolution, changes in base sequences and other features of the DNA occur that are invisible to selection at the phenotypic level. Neutral changes can spread through the genome of a species by stochastic means. Dover has argued that a process called concerted evolution has also produced much of the evolution observed at the genomic level. In concerted evolution, processes such as recombination, gene conversion, and unequal crossing over homogenize members of multigene families within genomes. These processes don't necessarily produce results corresponding to anything selection would have favored. The effects of concerted evolution may also include coevolution with other genes whose products interact with the genes undergoing concerted evolution. Finally, in selfish DNA evolution (first recognized in 1980 by Doolittle and Sapienza as well as by Orgel and Crick), certain genetic elements replicate themselves. No functions are known for such sequences, and no function beyond the ability to replicate and transpose in the host genome is required for their proliferation and maintenance. No doubt selection favors "parasitic" sequences that can increase in number over those that cannot. However, an increase in the number of such elements in the genome can occur whether or not it benefits the host organism.

Stuart Kauffman has suggested that complex interactions among genes provide an order to cell differentiation quite independent of selection. I'll discuss that suggestion in the next chapter. In his book *The Origins of Order*, Kauffman takes a hard look at the limits of selection from a different perspective, and has generated his own list of untidy loose ends about selection. One particularly intriguing item is the phenomenon he calls "missing phenotypes," the idea that there are a lot of potentially highly adaptive features and body plans that do not exist. Raup and Michelson pioneered this concept by using the basic rules of growth of mollusk shells to generate the full range of possible shell morphologies by computer. Mollusk shells are all cones that expand and coil within varying parameters. Most of the possible shell shapes produced by computer evolution are not found in living or fossil mollusks. Either such forms perform poorly in the selective marketplace, or something in the internal constitution of the organisms prevents their generation. Natural selection should allow the exploration of most of phenotypic space. However, if there are constraints that prevent selection from operating freely, we would not expect phenotypic space to be filled. Kauffman dislikes this argument because even if no constraints at all exist, evolution might be little more than a random walk through a mostly empty phenotypic space, with much of it simply unexplored. The observation of missing phenotypes would then be the consequence of chance, not necessity.

The missing phenotypes argument can be presented in a somewhat different and more interesting way. Alberch has proposed a thought experiment. Consider a simple situation in which the morphology of a set of organisms can be described by two variables, x and y. On a graph of y versus x, species appear as bounded clusters of phenotypes separated from other bounded clusters by empty two-dimensional phenotypic space. This hypothetical situation resembles what we observe for a set of related species in the real world. In the hypothetical world, we can do an experiment in which we allow one such bounded cluster to vary freely with all selection removed. We wait long enough and apply enough of our favorite mutagen so that a large amount of variation is produced and accidents of prior evolutionary history are minimized. Alberch has suggested that two very different outcomes are possible. The first hypothetical outcome is predicated on the assumption that there are no developmental constraints in the system. In that case there will be an essentially continuous distribution of genotypes. If no discontinuities apply between genotype and phenotype, a continuous distribution of morphologies will appear when the resultant phenotypes are plotted. We simply do not yet know how phenotype maps to genotype, so we can't be sure about the ultimate distribution in phenotypic space, but discontinuities will be reduced from those expected under a selective regime. Phenotypes generated under this regime need not be well bounded.

The alternative assumption is that the system contains inherent developmental constraints. These would affect the ensuing processes, and we would not observe a continuous outcome. We might get more phenotypes than we began with, but the phenotypes would be as ordered as in the sample we started with. The observed phenotypes would be bounded and discontinuous, similar to the pattern we observe in nature. The reason Alberch advanced for this hypothetical result is that internal order is so strong that even in the absence of selection it limits the extent of transformation and imposes an extensive amount of patterning on possible outcomes. That ordering must derive from the hierarchical structure of developmental processes and from the discontinuities between stable nonlinear dynamic systems. The phenotypes we observe in nature should reflect some balance between internal constraints and external selection. The empty phenotypic space houses Platonic shadow phenotypes that might be attainable under other rules than those that have operated in metazoans since the Cambrian radiation.

This concept of constraint doesn't deny selection, but rather, indicates that what is selected for and how the system responds to selection do not necessarily map on each other very closely. In the case of the paper cup, selection is for a particular function. There are many possible topological and material solutions, with the paper cup being only one of them. Once the paper cup has originated, selection will change features of its shape, size, decoration, and price (its ability to compete with other kinds of cups), but not the internal constraints of material and manufacture.

WHENCE CONSTRAINTS?

There have been some slippery phenomena and hypotheses in science, and their continued investigation has led to rather diverse and even strange outcomes. The notorious N-rays (hyperimagination in dark rooms), polywater (human sweat in small glass capillaries), and cold fusion (critical thinking eclipsed by hopes of striking it rich) proved ignominious for their "discoverers." On the other hand, other improbable phenomena, such as reports of stones and even chunks of iron falling from the sky, and outlandish concepts, such as continents drifting apart, bore up under ridicule and were ultimately validated as major discoveries about nature. What do we make of the hypothesis of developmental constraints?

If constraints exist, they can come from only three sources. Identifying these formal sources doesn't explain how constraints operate, but should allow a more precise way of defining them. The first kind of constraint that we will consider arises from the rules of physics operating on organisms. Such physical limits form the subject matter of biomechanics. Effects of gravity on large organisms, effects of hydrodynamic properties on small ones, strengths of materials, geometrical constraints on surface-to-volume ratio, and efficiency of energy conversion all affect how organisms are constructed. In development, the physical constraints that are likely to be interesting are those that affect morphogenetic processes. There may be only a few ways of stretching a cell sheet or forming a tube. Quite aside from the particulars of gene regulation and cell biology, such physical limits, if they exist, could bound the universe of morphogenetic possibilities, and thus the diversity of phylotypic stages. We'll examine the school of thought that has considered these constraints to be the primary limits on available patterns of developmental mechanics and evolution of form.

A second source of constraints may lie in the organization of animal genomes. Genetic constraints may be consequences of genome size. Numbers of genes are not the issue, as even the smallest animal genomes contain enough DNA to encode as many expressed genes as larger genomes. However, as I'll discuss below, genome size affects properties such as cell size and division rate. Other genetic constraints might arise from the neutral or selfish DNA features of a genome. The existence of transposable elements can have a substantial effect on mutation rates (for example, as observed in *Drosophila* by Kidwell). Numerous dramatic homeotic mutations in flower parts are caused by transposable elements. Other constraints will arise from lack of certain genes. Genes encoding pathways for particular metabolites are the most obvious example, but the kinds of regulatory genes available could potentially constrain additions of novel developmental pathways. Regulatory gene families are widely distributed phylogenetically, and novel members have arisen numerous times. An inability to evolve a new gene is rather hard to document. If we look at the converse, it seems that the phenomenon of gene duplication and divergence is so frequent that an inability to generate an appropriate member of a regulatory gene family is unlikely. What is more conceivable is that limits might arise from difficulties in undoing established interactions among regulatory genes. It is hard to disentangle this idea from the final category of constraints, those arising from organizational features within ontogeny.

Finally, we face the category of constraints that most of us think have exercised the primary effect on metazoan evolution. These are constraints imposed not by genes, or materials, or mechanical possibilities, but by the resistance of existing integrated organizations to modification or reorganization. In this rendition of constraint, the same genes, cell types, and structural materials are integrated in distinct ways in each particular ontogeny. It would have been possible in evolution to get there from a simpler ontogeny, but difficult to disintegrate a pattern once it had become firmly established. Constraints of this kind have been visualized as arising from two interrelated and perhaps difficult to separate causes. One posited cause has been called "phyletic constraint." Phyletic history would have entailed the establishment of a particular ontogeny to the exclusion of other options. Subsequent evolutionary changes might be possible given the existing developmental rules, but as a result of evolutionary history, the starting variation simply would not be there to allow some kinds of evolution.

Constraints may also arise as a consequence of ontogenetic rules themselves. In that case, particular changes cannot occur because the requisite variation cannot be generated. This idea is appealing because if one can work out the rules governing some particular developmental process, they should allow concrete predictions about what should be possible in evolution. Holder has done that for tetrapod limbs. Using the "known" patterning rules, he worked out "permitted" and "forbidden" kinds of limbs. For example, a limb with a reduced digit 3 but with the others retained is forbidden. Indeed, no such limb pattern is known for any living or fossil animal. Once again, however, this is an example of unfilled regions of phenotypic space: are the missing patterns the result of constraint, random walk, or selection? We can't know.

The Cambrian radiation was possible despite the intense constraints hypothesized for living organisms because it represented the initial establishment of integrated body plans. The first metazoans may have had relatively few organizational constraints. As they became more elaborate in their ontogenies, constraints would have appeared as a direct consequence of both growing complexity of organization and the specific histories of each evolving lineage. Transitions would have become effectively impossible. Despite the freezing of body plans, a huge range of morphological variation and developmental innovations, as seen in vertebrates, remained possible. As further evolution occurred, additional constraints would have continued to appear. The initial constraints on the basic vertebrate body plan would not have included any affecting limb development. However, once limbs had originated and acquired an elaborate program of development, constraints on limb development would have inevitably appeared. These would have reflected the constructional rules of limb development, as suggested by Holder. In addition, limits would have been posed by the phyletic constraints of particular lineages. The bird wing offers an example of the latter. The loss of digits and the fusion of some hand elements in bird wings concretely limit the variation available to selection regardless of any rules of vertebrate limb construction.

Maynard Smith and his co-authors attempted to give the concept of developmental constraint some substance by defining kinds of constraints and how they might be recognized. Aside from limits imposed by the materials from which living organisms are built, they suggested that developmental constraints might arise if a particular developmental mechanism produced a phenotype that was more advantageous than those produced by other mechanisms. The mechanism would not have originated because of its selectively advantageous phenotype, but selection would see to it that such a mechanism came to represent a local optimum to the exclusion of other mechanisms. These authors also suggested that constraints might arise for adaptive reasons; that is, a change in what might ultimately be an adaptive direction could be blocked because the intermediate steps were nonadaptive. Finally, they considered genetic constraints to arise from the limited capacity of an existing genotype to produce new genes necessary for a transition.

Alas, actually demonstrating developmental constraints is a tough proposition, because we are looking for something that has kept evolution from following one course rather than another. In looking at the long evolutionary conservation of developmental patterns, we have an uncomfortable mental sensation that there must be something operating within the developing system that we could, at least in principle, describe as preventing modification. Maynard Smith and his co-authors proposed three ways to look for evidence for the operation of constraints. First, they suggested that if actual organisms occupy only a limited portion of the possible phenotypic space, constraints may be operating. Of course, so might selection, thus this is a rather weak criterion. Second, they suggested that the repeated occurrence of a given kind of variant independently acquired among related species reflects a common constraint. For example, there is an ordering of evolutionary losses of digits in salamanders. Experiments carried out by Alberch and Gale showed that reductions in digit number in salamander limbs induced by treatment with the mitotic inhibitor colchicine resembled digit reductions occurring in the evolution of progenetic species with reduced limb buds. Thus, digit 5 was the first lost in both experimental animals and species with four digits. More extreme reduction to two digits mimicked the pattern of digit loss in *Proteus anguinus,* which retains only digits 1 and 2. The result is quite consistent with the existence of a constraint on loss, and should reflect the modifications permitted by the underlying rules of limb patterning. Finally, Maynard Smith and his co-authors suggested that constraints could be revealed by exposing animals to powerful selection and observing what kinds of responses are obtained, or perhaps more significantly, not obtained. A study done by Maynard Smith and Sondhi was cited to make the point. They selected for left/ right bias by selecting mutant flies that possessed the left but not the right ocellus. No heritable variation for left/right asymmetry could be found, although loss of anterior or posterior ocelli could be readily selected for. The message is clear: As left/right asymmetry cannot be selected for, a constraint on the evolution of left/right symmetry exists. This is a potentially important kind of experimental test, but it may be limited in its application, because lack of response might mean only that no heritable variation happens to exist in the test case, not that it couldn't exist.

ON THE OTHER HAND

As developmental genetics reveals more about the developmental mechanisms that underlie the patterns we observe, some constraints become clear, but complexities often remain. The case of left/right asymmetry illustrates the dilemma. At the start of *Drosophila* development, the patterning of both anterior-posterior and dorso-ventral axes is established by molecular mechanisms that do not distinguish left from right, and there is no early left/right symmetry system. Other arthropods, notably crabs, are asymmetrical, with one claw larger than the other. However, these animals grow gradually through a series of molts. *Drosophila* grows quite differently. Left/right symmetry might be constrained in flies and other holometabolous insects because the adult body arises from bilaterally symmetrical imaginal discs. The fully grown adult emerges from the pupa in a single developmental step. The problem with being sure that a developmental constraint has been identified is that there might be quite unexpected selective pressures that maintain a particular pattern. Thus, in the case of left/right symmetry, sexual selection for symmetry appears to be quite strong in animals with complex social behaviors, such as insects, birds, and mammals. The work of Perrett and co-workers suggests that humans prefer symmetrical faces. Thornhill has shown that some insects, such as female scorpion flies, strongly select for mates with bodies that are very left/right symmetrical. The female scorpion flies don't do this by looking at their suitors. They prefer the sex-attractant pheromones presented by more symmetrical males. Good symmetry might advertise superior health or reproductive capacity. In such cases, selection is on behavior and not on developmental rules per se.

Left/right symmetry is complex even in animals with strong external bilateral symmetry. The external bilateral symmetry of mammals is primarily a manifestation of the axial body, limbs, and sense organs. It is reflected in some internal organs, such as the lungs and kidneys, but not in others, including the brain, stomach, spleen, and heart. The prevalent internal symmetry of mammals is that in which the heart lies on the left side of the body. The opposite symmetry, situs inversus, also occurs. One genetic cause of situs inversus lies in mutations of the *inv* gene described by Yokoyama and co-workers. Animals homozygous for this mutation do not survive, suggesting important internal constraints on the rules of symmetry. However, situs inversus has other causes, and individuals exhibiting situs inversus do survive. How well situs inversus animals might do in nature is not clear, but the preferred outcome may be favored by selection.

GENOMIC CONSTRAINTS

The lungless plethodontid salamanders that I used to capture under logs and rocks in the woods during school holidays are humble in appearance, but in terms of numbers and biomass are the vertebrate lords of the forest in the northeastern United States. Halliday and Verrell cite estimates that in some woods the total mass of salamanders exceeds that of birds and mammals together. As I had not yet heard of DNA in my salamander-catching days, I had no idea of the disproportionately huge genomes they carried around on their little stubby legs. Although plethodontids are small in size and relatively primitive in overall body structure, according to Sessions and Larson, they have genome sizes on average ten times that of humans. This feature of plethodontids is an expression of the so-called C-value paradox, in which similar or even simpler animals can have larger genomes than more complex related forms. The largest animal genome going is that of a lungfish, almost fifty times the size of the human genome. Primitive grasshoppers have larger genomes than the more sophisticated drosophilid flies. Most of the metazoan genome is composed of extraneous DNA that lies in sequences other than genes. Some genomes have more "junk" DNA than others. Cavalier-Smith has noted that the unicellular yeast uses 70% of its genome to encode its proteins, humans use about 9% of theirs for that purpose, and salamanders only 1%–5%.

Gains in genome size occur over evolutionary time via selfish DNA processes. There is direct evidence (cited by Cavalier-Smith) from fossil lungfish bone cells that lungfish cell volume, and thus probably genome size, has more than doubled over the past 300 million years. These gains may be selectively advantageous or at least neutral in many instances. In species that cannot replicate selfish DNA rapidly enough, or reach other barriers, selection should work against its maintenance. Having a large genome has consequences outside of the properties of the genome per se. Larger genomes result in larger cells. Because cells containing large genomes replicate their DNA more slowly than cells with a lower DNA content, large genomes might constrain organismal growth rates. Cell size will also determine the cell surface-to-volume ratio, which can affect metabolic rates.

Constraints introduced by large genome size have been predicted and sought in salamanders. Most of plethodontid DNA are, as discussed by Larson and Chippindale, of the middle-repetitive class. The amplification of sequences may be driven by selfish processes, and may have little effect on animals with slow growth or metabolic rates. Sessions and Larson have tested for possible constraints of genome size on growth and differentiation rates by amputating the limbs of various plethodontid salamander species ranging over a fivefold span of genome sizes. Rates of differentiation in the regenerating limbs were strongly inversely correlated with genome size, and growth rates less so. The relatively weak association that they observed between genome size and growth rate suggests that large genome size might set a maximum limit on developmental rates. Below such a hypothesized boundary, various relationships between genome size and growth rates may be possible.

It's not clear which connections are causal in the C-value constraints game. Big genomes might just reflect tolerance to large amounts of junk DNA in the nucleus, or there might actually be a function for the apparently excess DNA. Cavalier-Smith suggested that larger cells require a greater nuclear membrane surface to accommodate RNA transport. Noncoding DNA might thus function as a nuclear skeleton controlling nuclear size. Pagel and Johnstone noted that genome size, nuclear size, cytoplasmic volume, and developmental rate all correlate. They examined these correlations statistically and found that genome size was still correlated with developmental rate after the effects of nuclear and cytoplasmic volume were deleted. However, genome size was not correlated with cytoplasmic volume after controlling for developmental rate. These results are consistent with the junk DNA explanation, but not with a DNA nucleoskeleton.

Amphibians, with their large range of genome sizes and their small bodies, provide a good case for seeking constraints generated by genome size. Roth and co-workers have observed that in both frogs and salamanders, larger genome size results in larger cells. In turn, larger cells result in a simplification of brain morphology. Thus, quite independently of the demands of function, internal features such as genome size can affect the morphology and organization of complex animals. Plethodontid salamanders share the basic vertebrate nervous system and brain, but they have very little space in their small skulls and spinal cords. Nevertheless, they carry out some remarkable neural functions. There are paradoxical compromises between plethodontid nervous systems and genome sizes as well, as illustrated by Roth and Schmidt. These salamanders occupy a variety of caverniculous, aquatic, terrestrial, and arboreal habitats. They possess a full range of sense organs, and most remarkably, a spectacular insect-catching mechanism consisting of a projectile tongue that can reach out in ten milliseconds to half the animal's trunk length (snout to vent is the way herpetologists express it). Plethodontid tongues have been extensively studied by Lombard and Wake, and provide an important resource for evolutionary studies. The studies by Alberch described in chapter 6 on the developmental transformations of the branchial cartilage associated with the projectile tongue illustrate how other developmental constraints operate in this complex system.

The projectile tongue mechanism is accompanied by binocular vision and a neural system with projections from the retina onto both the same (ipsolateral) and the opposite (contralateral) side of the visual cortex of the brain. This arrangement resembles that of mammals with good depth perception. What is surprising is that the visual system, the optic tectum in the brain, is reduced in complexity. The reduction is particularly noticeable in species with large genomes and, consequently, cells with large volumes. The larger cells require a reduction in cell numbers just to fit everything in. Although genome size constrains cell numbers, it does not seem to adversely affect function. Frogs are also highly visually oriented creatures, and they have about a million central nervous system visual neurons. Plethodontids can have as few as 60,000. They shoot bugs well all the same.

A large genome exacerbates the effects of small body size, and it's clear that cell size introduces a constraint on cell numbers in salamanders. The reasons for maintaining large genomes in the face of these effects are not obvious. Genome size varies over a sixfold range in plethodontids. It would seem that genome size reduction is possible because some of the most highly miniaturized species have relatively small genomes. Perhaps they have passed the limits imposed by larger genome and cell sizes. There are consequences and compensations. Plethodontids are sluggish, and the low metabolic rates introduced by large cell volume may be advantageous to sit-patiently-and-wait hunters that can afford long fasts. Vision at a distance is reduced to two handbreadths, but since these animals are ambush hunters that strike at short range, that probably doesn't affect their efficiency much. A considerable portion of the brain is devoted to the narrow binocular vision field, and the lateral parts of the field are reduced. There are other changes as well, such as an increase in the relative volume of the vision centers in the brain and an increase in cell packing density.

Genome size appears to generate constraints at extremes of developmental or metabolic rates or with small body sizes, but such constraints don't seem terribly hard to escape. Escape can be through structural or functional compensation or through reduction in genome size. Evolutionary reductions in genome size have occurred, for example, as shown by Sessions and Larson, in over 30% of plethodontid salamander lineages.

CONSTRAINT BY NUMBER OF CELL TYPES

Metazoans are by definition multicellular, but both the number of cells and the number of cell types vary among animal clades. Valentine and co-workers have touched on a very interesting potential constraint on body plan evolution. They suggested that the number of cell types possessed by the members of an animal clade could be used as a measure of organizational complexity. They presented an intriguing plot of numbers of cell types versus estimated times of origins for major animal clades. The plot showed a rapid initial rise in number of cell types from sponges and cnidarians through various bilaterian phyla and into the vertebrate classes. Three interesting features are apparent. First, the simplest bilaterian is estimated as having 30 cell types. Second, that number is attained very rapidly and early in metazoan history. Third, the number of cell types has continued to rise throughout vertebrate history, from about 70 in Ordovician jawless fishes through about 150 in amphibians to over 200 in modern humans. Number of cell types may have been a constraint on the evolution of phyla, but it was overcome rapidly, just before the Cambrian radiation. Number of cell types should not have been a constraint on the evolution of new phyla following the Cambrian.

The question is whether the increase in cell types has been selected for. Valentine and co-workers modeled the curve of increasing cell types using a Markovian process (a chain of events that are constrained by prior states of the system). Their model yielded the same kinetics of rapidly increasing numbers of cell types, indicating a time for metazoan origins not earlier than 600 million years ago. Since the increase can be attributed to a nonselective model, it's impossible to say whether it underlies the evolution of increasing vertebrate morphological and behavioral complexity. The extra cell types could be merely carried along as neutral freight, or a nonselective origination process could produce cell types that might be subsequently co-opted for a selectively advantageous function.

FROZEN CONTROLS?

Constraint in gene organization is a clouded topic at best, but disturbing observations loom up like logging trucks on a foggy mountain road. The Hox genes have presented the most puzzling instance of deeply conserved gene order. In all phyla so far examined (arthropods, nematodes, and vertebrates), the *Antennapedia* and *Bithorax* homeotic gene homologues are clustered, they have the same transcriptional orientation and order of activation, and their transcription is collinear with the body axis. The conservation of a set of clustered genes over half a billion years is difficult enough to accept, but collinearity with body axis defies credibility. Yet it's true. Not surprisingly, the hypotheses as to why these features are conserved have revolved around regulation of gene action or around the collinearity of the genes in the complex with their order of expression along the body axis. Tight clustering could be required for transcriptional control operating via a global chromatin organization in the cluster. The linear transcriptional hypotheses discussed by Dollé and co-workers, by Peifer and co-workers, and by Gaunt and Singh suggest that sequential activation of the genes is regulated by their order through a mechanism involving the organization of chromatin in the complex. Consistent with hypotheses of a role for chromosome organization, Pattatucci and Kaufman have observed that regulation of at least some genes of the *Drosophila* Antennapedia complex is sensitive to proper chromosome pairing. Collinearity between genes and body axis is harder to explain. The collinearity hypothesis discussed by Krumlauf suggests that the order of the Hox genes itself defines the anterior-posterior axis. Lewis, in 1978, was the first to suggest that the collinearity of the complex reflected its function as a combinatorial code to produce segmental diversity. It is extremely difficult to account for the collinearity of genes and body axis on the basis of any known developmental mechanism.

Another approach to the apparent constraint in the Hox cluster is to consider it as having intermingled transcriptional regulatory sites. As discussed by Rubock and co-workers and by Renucci and co-workers, subsequent inversions or other rearrangements of the complex would disrupt the established regulation of one or more genes. Randazzo and co-workers suggested that such an effect would constitute a selective "prisoner effect." There have been some revealing recent tests of this hypothesis. The three homeotic genes of the Bithorax complex were genetically separated by Tiong and co-workers, but still functioned normally. In addition, Chouinard and Kaufman, as well as Randazzo and co-workers, showed that mutations in individual genes of the Antennapedia cluster could be rescued. This was done by injecting mutant fly eggs with DNA containing individual genes lacking adjacent sequences. The progeny of the genetically transformed flies correctly expressed the injected genes. If intermingled transcription control sites were present, the introduced genes lacking these regions would not have been expected to function correctly. The experimental results thus fail to bear out the hypothesis that the genes are linked because they form a single transcriptional regulatory unit.

The most informative analysis of the evolutionary latitude of the Hox cluster to date was made by Randazzo and co-workers in 1993 in a comparative study of the Antennapedia complexes of two species of *Drosophila*. The structure of the Antennapedia complex of *Drosophila pseudoobscura* was determined and compared with that of *Drosophila melanogaster*. Both molecular clock estimates of divergence times and skillful phylogenetic reasoning were used to infer the pattern and direction of evolutionary changes in the cluster. Molecular clock estimates made by Beverly and Wilson yielded an estimated divergence time of 46 million years ago for these two species. This relatively short-range phylogenetic comparison revealed an unexpected evolutionary flexibility.

As expected, the clusters of the two *Drosophila* species are highly similar in gene composition, arrangement, and conservation of the very long transcription units characteristic of the *D. melanogaster* Antennapedia complex. However, some revealing differences were found. The *Deformed* locus of *D. pseudoobscura* is inverted with respect to the *Deformed* gene in the *D. melanogaster* cluster. The *D. pseudoobscura* orientation of this gene is the primitive one shared with mammals (all Hox genes transcribed in the same order). It was inverted sometime during the 46 million years that separates the *D. melanogaster* lineage from its congeners. In addition, the *z2* gene is absent in *D. pseudoobscura*. Both species share a number of nonhomeotic genes in their Antennapedia complexes. These genes include a member of the immunoglobulin superfamily and several genes related to those encoding insect cuticular proteins. The homologues of these genes are not present in mammalian Hox clusters, so they appear to represent interpolations introduced during evolution of the insects or even the drosophilids. The differences noted between the two *Drosophila* species indicate that the Hox complex can tolerate substantial change over moderate periods of evolutionary time and within a common body plan. Comparisons with mammalian Hox clusters show that there has been a considerable fluidity to the Hox cluster. The dynamic aspects of the cluster include the kinds of genes present, the number of genes, the direction of transcription, the spacing between genes and the sizes of individual genes, the presence of a single, apparently primitive, collinear Antennapedia-plus-Bithorax complex in mammals versus a divided one in *Drosophila,* and even multiple replications of a number of clusters in vertebrates. As the Hox clusters of other phyla become known in detail, we may observe that surprising changes have occurred in other lineages as well.

The absolute evolutionary conservation of Hox gene order might in part be an illusion resulting from a limited phylogenetic sample. The Hox gene cluster of the nematode *C. elegans* contains recognizable homologues to the clusters of arthropods and vertebrates, but there is one inversion of order. The *C. elegans* gene *ceh13* is a homologue of the anterior group gene *labial* of *Drosophila,* but is embedded among more posterior Hox genes (see Schaller and co-workers, and Salser and Kenyon).

In contemplating the rapid radiation of animal body plans during the Cambrian and their subsequent stability, both Gould and Jacobs have suggested that a form of internal genetic constraint may act as a brake on body plan evolution. They argue that an originally flexible developmental regulatory system allowed experimentation with very basic patterns of development. Once patterns were established, genetic regulatory systems became rigidly fixed. After that, a powerful genetic constraint limited evolution to changes within existing body plans.

We should like to see whether there is any evidence that evolution at higher levels of biological organization has slowed since the Cambrian radiation. Jacobs has collected and graphed the numbers of new orders of segmented (mostly arthropods) and nonsegmented phyla that made their appearance in each geologic period from the Cambrian onward through the course of Phanerozoic time. Such quantitative treatment of originations or extinctions is common practice among paleontologists as they seek periods of particularly high rates of extinction or evolutionary innovation. Jacobs's tabulation indicated that the rate of innovation in segmented phyla was highest in the Cambrian, and drastically lower since. On the other hand, his data for nonsegmented groups indicated that higher taxa have continued to arise in these groups since the Cambrian. As new ecological opportunities have not differed in an obviously taxon-specific way, these data suggested to Jacobs that regulatory controls assumed different patterns in segmented and nonsegmented groups. Jacobs suggested that serially organized body plans depended on a simple serially organized set of homeotic selector genes. Initially these genes were all closely linked, allowing the formation of hybrid genes and homeotic transformations based on expression pattern changes in adjacent genes. As these genes became more dispersed (I'm not sure what he meant by this) in the genome, the existing patterns became set and new patterns became difficult to generate. He further suggested that nonsegmented taxa evolved new kinds of interactions between Hox genes that did not restrict them to the rigid serial expression of selector genes generating segmental identities along the body axis, as in arthropods. This hypothesis seems to be weak in its particulars, but the general principle is worth considering. Some regulatory gene architectures may be more rigid than others, even if the same genes are involved.

CONSTRAINTS IMPOSED BY THE LIMITS OF STRUCTURE

Each generation views biology in its own brightest light. In recent times that light has come from genetics and molecular studies of gene function. Thus, the current discipline of developmental biology revolves around the role of genes in establishing patterns in the developing embryo. To some extent this approach must be an oversimplification of nature. Genes don't act directly to make morphology. They act within an existing epigenetic environment (the egg or cells of the embryo), and they encode proteins that have precise functions within the cell. These proteins are then available to serve in pattern formation and morphogenesis. In some cases we know that protein products of regulatory genes play direct roles in pattern formation. The clearest such case is the role of the bicoid protein in establishing the pattern of expression of gap genes, notably *hunchback,* to set up the anterior-posterior patterning

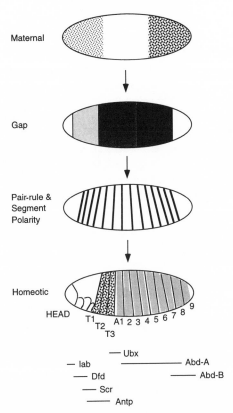

Figure 9.1.   Domains of regulatory gene expression in the establishment of segmentation in the *Drosophila* embryo. Maternal mRNAs establish the anterior and posterior poles. Once zygotic gene expression begins, the gap genes establish broad domains and activate precise patterns of expression of pair-rule genes. Through the action of the pair-rule genes, the broad domains of the gap genes become resolved into fourteen parasegments. Final segmental identities are established by the Hox genes.

of the *Drosophila* embryo (fig. 9.1). There is a cascade of gene action in both time and space along the body axis. The bicoid protein is a transcription factor, and it directly activates transcription of the *hunchback* gene in nuclei suspended in the common cytoplasm of the early *Drosophila* embryo. The interactions among gap genes further refine a pattern consisting of a few zones of specific gene expression. The gap genes in turn influence the pattern of expression of the pair-rule genes, which establish the boundaries of the parasegments of the larval fly. Cellularization of the embryo occurs late in this process, during expression of the pair-rule genes. The study of this process has focused on the effects of the proteins encoded by regulatory

genes in activating or repressing other genes so as to refine the pattern of
gene expression in the epithelium to sharply bounded zones a few cells wide.
The translation of the molecular domains into morphological domains is not
fully understood, but the generally accepted hypothesis is that the pattern-
forming genes establish domains (the body segments) within which the homeotic selector genes act. The selector genes activate downstream genes that
execute the morphogenesis and differentiation of the segments.

One school of developmental biologists derides this gene-focused program
as epiphenomenological. These "structuralists" envisage morphology as arising from the properties of physical rules that generate structure in developing
systems. This is a long-standing perspective on development that arises from
pre-Darwinian idealistic morphology and continued to be held by anti-
Darwinists. J. Bell Pettigrew published a wonderfully illustrated three-
volume treatise in 1908 entitled *Design in Nature*. Unlike modern structuralists, Pettigrew was arguing on behalf of "First Cause and Design." Pettigrew
based his argument for common rules of form in nature on the commonalties
of particular forms in nature, such as the spirals present in nonliving and
living structures. Thus, "the dendrites of minerals and metals, the frost pic-
tures on window-panes and pavements in winter, the lightning flash, &c.,
are amazingly like the arborescent forms seen in the branches and leaves of
plants and trees, the division and subdivision of blood-vessels and lymphat-
ics, the branching of nerve cells . . ." and so on. Common principles im-
pressed in matter underlie spiral galaxies and spiral shells and horns alike.

The structuralist hypothesis comes in both strong and weak versions. The
strong version has been forcefully propounded during the past decade by
Goodwin, who has rejected both the concept of a genetic program regulating
morphogenesis and the neo-Darwinian concept of selection driving the evolution of form. Goodwin bases his conception of development on the pioneering
work of Turing on pattern formation resulting from the properties of diffusing
and reacting chemicals in a field. This concept of development regards an
embryo as a field in which a series of periodic domains is generated by the
reaction and diffusion kinetics of morphogens that specify the identities of
cells in those periodic domains. In this perspective, genes do not make binary
decisions that determine subsequent states. Instead, they provide reactants,
set boundary conditions, and affect details of the periodicities predicted by
the generic field theory. Goodwin argues that no amount of information about
genes and their products can predict the emergent properties of morphogenetic
fields. As an example, he has suggested that the properties of a cleaving
embryo illustrate the operation of a set of permissible transformations arising
from surface energy functions resident in the field. Gene sequences only
influence the specifics of cleavage patterns in individual species. Goodwin
has rejected any compromise, such as that proposed by Wolpert in which the

positional information generated by a Turing-type field would be interpreted as positional information by receiving cells and would produce morphogenesis through appropriate patterns of gene expression. In Goodwin's model the progression from a simple to a complex pattern results from progressive bifurcation of domains in the morphogenetic field. Gene products play a role only in stabilizing particular pathways of morphogenesis.

The consequences that Goodwin's hypothesis poses for evolution are somewhat startling. Goodwin has stressed that there are well-defined constraints on the transformations that fields can undergo. The range of patterns produced may be rich, but it is finite. In other words, in evolution, morphogenetic processes are "never displaced significantly from these generic states." Genetic changes available to selection would be limited to modifications of initial states or boundary conditions within an established field system. Goodwin would thus redefine homology in nonhistorical terms. We are used to a concept of homology in which similarities are the consequence of an inherited genetic commonality. Goodwin's structuralist definition returns to a pre-Darwinian concept of similarity arising as a result of shared morphogenetic principles. To show my reader that I'm not exaggerating, let me quote Goodwin's most extraordinary statement of this view, made in 1982: "That no such historical continuity is necessary is made clear by the field analysis of the generative principles operating to produce the mouth of ciliate protozoa such as *Tetrahymena,* and the dorsal lip of the blastopore in amphibia, both of which are ascribed to the same type of singularity (a saddle point) in a cortical (surface) field, and so are homologous under the original definition of the term."

Goodwin has suggested that the pattern of the Cambrian radiation is precisely consistent with structuralism. Body plans would have arisen from the hierarchical structure of morphogenetic rules and their limited potential. That some morphologies are historically older and others newer would have no real meaning in the exploration of morphogenetic space. That is, the order of appearance of morphologies observed would not be causally derived from selection operating on a phylogenetic sequence of genetically related forms.

These structuralist constraints are posited to arise not from the genetic organization of animals, but from the rules that govern field behavior. Are we bound to accept such a source of constraint in developmental processes? That depends on how the findings of experimental developmental biologists are interpreted. As we learn more about systems such as the *Drosophila* egg and the limb bud, which have been prime candidates for determination by the operation of morphogenetic fields, it is becoming apparent that these systems operate by more local nuts-and-bolts systems of interactions between molecules or between cells. I see no support in the data for a strong version of structuralism.

There is also a weaker version of structuralism that might well account for a significant source of constraint. This weaker structuralist concept arises from two considerations. The first, argued by Newman, is that there are generic processes that give rise to generic morphologies such as segments. These generic processes include "standard" physical processes, including adhesion, surface tension, and phase separation. A gene-driven ontogeny could have arisen to make such processes more stable and precise. The other consideration, argued by Mittenthal, is that there are only a limited number of possible morphologies consistent with particular functions. Thus, a tubular gut is probably the most efficient form for a digestive organ, and may have arisen several times independently. Furthermore, Mittenthal has pointed out that there are a limited number of morphogenetic processes. Deformations of a sheet, for example, must involve processes that change the shape of the sheet, but maintain its connectivity. Such deformations include pits, bulges, grooves, and hollow extensions that maintain a single contiguous surface. Other processes can occur that destroy connectivity. These include processes that pinch off vesicles or tubes from a sheet. There are only so many ways in which a tube can be generated: by invagination and pinching out of a sheet, by ingrowth from an end, like a finger being pushed into a rubber sheet, or by cavitation from a solid matrix. All of these are used in animal development.

This weaker form of structuralism suggests that there is a limited universe of animal body plans and ontogenetic methods of achieving them. This universe was evidently explored early in generating the spectrum of body plans appearing in the Cambrian radiation. The same body plans have continued to be exploited ever since because there are no or few other possibilities. It is not clear just how big a universe of body plans might be predicted as possible from Mittenthal's set of standard processes. Given the very few things that cells can do, and the great diversity of structures that cells produce, I doubt that thirty-five phyla have exhausted the combinatorial possibilities inherent in the standard processes. Perhaps the supposed few extinct phyla of the Cambrian represent explorations of other possibilities that simply didn't measure up to the fatal eye of the grim editor.

A caution to structuralism is necessary because we can't be sure that we know the limits of morphology, nor that the forms of living animals extend to the limits of the possible. Bacteria have high internal pressures, and so generally assume forms of minimal surface area: spheres or rods with rounded ends. Square bacteria would appear to be impossible. Nonetheless, while investigating bacteria that contain gas vacuoles and float on Sinai desert brine ponds in 1980, Walsby discovered that the most common forms were astonishingly square. This organism, called *Haloarcula,* or the salt box, looks like a miniature postage stamp. It is flat and has precise corners and flat edges. A few years later, Takashima and co-workers discovered a related triangular bacterium living in hot salt ponds in Japan. These organisms resemble two-dimensional creatures out of Abbott's *Flatland,* but life teems in some odd and inhospitable environments, and it can assume unpredictable forms.

CONSTRAINT BY COMPLEXITY

At least one creationist tract, in extolling the second law of thermodynamics as proof of the impossibility of evolution, has noted that no matter how long the sun shines on a pile of scrap metal, a jet airliner will never assemble itself. This seems to be a fairly safe contention, but by and large an irrelevant one to evolutionary biology. No one seriously entertains the possibility of complex objects arising spontaneously. Evolutionary increases in complexity occur because the genome can respond to selection for new functions. For example, bacteria have shown themselves capable of acquiring plasmids that bear genes encoding resistance to various antibiotics. Because genes can be duplicated and the copies can diverge to yield new functions, increasing genetic complexity is readily generated. The fact that a large proportion of the genes in animals belong to multigene families bears testimony to this process. There appears to be little constraint on the addition of DNA to most genomes.

The question arises as to whether the complexity of genetic and developmental processes in itself becomes a constraint to further evolution. Unfortunately, in dealing with the consequences of organization that preclude change, we have to avoid becoming entangled in the wait-a-bit thornbushes and lawyer vines of untestable hypotheses. It's easy to propose that things that have been conserved in evolution have not changed because of internal constraints, but it's difficult to prove such a contention. The reasons are twofold. First, selection may be operating to retain a developmental feature, or the feature may simply be conserved by inertia. Stasis and change in the marine larval forms discussed in chapter 7 show that these factors are far from negligible. Second, experimental tests with living embryos may only poorly reveal evolutionary causes. One might, for example, attempt to measure the entrenchment of particular developmental processes by interfering with them by mutation or other means. Experiments of this type have been the mainstay of developmental genetics, and they have produced both expected results and surprises. Some knockouts of genes are lethal. Severely abnormal phenotypes are produced, and subsequent development is blocked. Most of the known important genes operating in early development have been found this way. Other knockouts have had disappointingly small effects, even when the gene is expressed relatively early in development.

We can put the small-effect experiments aside. There may be redundancy because a related member of a gene family can substitute for the lost gene, or we may have misunderstood what the gene's role is. It might be required in some selective regimes and not in others. The experiments that produce lethal effects demand notice. Cascades of genes such as the early patterning genes of *Drosophila* are certainly consistent with the idea of constraint from complexity. The problem with such a conclusion is this: there is no doubt that knocking out a step within a well-integrated developing system is unlikely to improve the workings of that particular ontogenetic machine. However, such an experiment on a single ontogeny doesn't tell us whether other processes could be successfully substituted for those already in place. A replacement of one process by another is what occurs in the evolution of early development. We are only beginning to acquire the tools to do the needed evolutionary experiments. As biotechnological experiments in which genes for growth hormones or other regulatory processes are emplaced within organisms become more sophisticated, we will see how such replacements might be effected.

Hypotheses of developmental constraint arising from the complexity of developmental processes have been presented by a number of authors, notably Alberch, Arthur, Rasmussen, Riedl, and Schank and Wimsatt. Most of these hypotheses envision constraints as measurable consequences of what Riedl called burden. He defined burden as "the responsibility carried by a feature or decision." Riedl envisioned burden as involving both the hierarchical position of a process or event and the number of functions or parts provided to the organism by that event or process. The hierarchical aspects of burden are most visible in concepts of constraint. After all, development progresses from earlier to later events, and von Baer's laws still retain a strong hold on the evolutionist's views of development. A number of analogies have been used to express the idea of constraint arising from burden, but this concept can be summarized readily. Later events depend upon earlier ones. Early events or processes are more general in scope and have wider ramifications than do later ones. As a consequence, these early processes are more entrenched than later ones. Some interesting predictions follow. First, no less entrenched process can precede a more entrenched one in development. Second, early processes will be more resistant to evolutionary change than later ones because their degree of entrenchment limits their evolutionary freedom. As a result, mutational disruptions of these more entrenched early processes are more likely to be lethal. Third, evolutionary changes will be much more frequent in late development than in early development in all animals. And finally, the entrenched processes of early development will be evolutionarily older than those of late development.

These vertical constraints sound like reasonable predictions, but they have fared about as well as Pacific island cargo cults in delivering the goods to expectant believers. The counterexamples are legion and phylogenetically wide ranging. The diversity of vertebrate early developmental modes shows that the first prediction, that less entrenched processes can't precede more entrenched ones, is not borne out. The second prediction, of the resistance of early development to change because it is more burdened, is belied by the rapid and frequent evolution of direct development. That the idea that early development cannot evolve readily has been so strongly held is a tribute to the isolation of developmental biology from natural history and life history studies, which provide so many examples of highly modified patterns of early development. The third prediction sounds straightforward, but it's not. As we have seen, what one means by an evolutionary change will be qualitatively different in early and late development. If the developmental hourglass is correct, both early and late changes will be observed frequently.

Finally, the suggestion that early developmental processes are evolutionarily older than late ones is a direct marriage of von Baer's laws with phylogenetic theory. It has been proposed by Nelson that the order of ontogenetic appearance of features could be used to decide which character states in a phylogeny are more primitive and which are more derived. As pointed out by Kluge and Strauss in their critique, this approach assumes that ontogeny is hierarchical and that the order of ontogeny corresponds to the order of evolutionary relationships. The general validity of this application of vertical developmental constraint is dubious. Given the prevalence of heterochronic changes and nonrecapitulatory patterns in development, it simply can't be so. The best single counterexample I know to illustrate the absurdity of any early-equals-ancient proposal is the existence of the amniotic egg. It occurs about as early as can be in development, and it represents a major feature in the organization of later development. Yet, the amniotic egg arose only halfway through vertebrate evolution.

It is especially important to recall that the models of burden that we have been considering are largely those of vertical constraint. This is a very aristocratic idea: ancestry is all. However, late development does not map closely on early development. The substitution of one early set of processes for another has occurred in the sea urchin *H. erythrogramma* and in numerous other cases in other groups as well. The idea of burden is, however, anything but ridiculous. It may simply not arise primarily from sequence through developmental time. The developmental hourglass suggests that the most telling constraints arise from a different cause. These are horizontal constraints, and they arise from the degree of interactiveness among developmental modules at any particular slice of time during an organism's ontogeny.

Vertical and horizontal constraints should not be conflated. The distinction is important. Vertical constraints arise from the causal linkage of processes along a developmental time course. That is, one can postulate that there are chains of events in which an earlier step causes another downstream from it, and so forth. Such chains can be linear or branching. These are the links postulated in vertical hierarchy models, notably that of Arthur. Constraint is claimed to arise from the dependency of later steps upon those preceding. Horizontal constraints arise from interconnections among features or processes at any stage of development, that is, within a slice of time. These constraints arise because inductive interactions between functional domains (modules) within an embryo have powerful effects on subsequent events. The interacting modules need have no vertical connection. Constraint will be proportional to the number of interactions.

Some predictions are possible. If vertical constraints predominate, the earliest stages of development should be the most highly constrained. The consequence of horizontal constraint should be that early development is little constrained in evolution, but that some later stage is highly constrained as horizontal interactions reach some critical limit. That point in development is defined as the phylotypic stage, and it is inevitably the most highly conserved part of ontogeny. Constraint in this case occurs not because of vertical burden, but because of the maximal interlinking of modules at what becomes apparent as the phylotypic stage. The developmental hourglass is a simple graphic representation of this hypothesis.

Observations drawn from the distinct investigative traditions of evolutionary biology and developmental biology are not obviously consistent with each other in predicting whether vertical or horizontal constraints will predominate. Evolutionary biologists' observations of the variation of early stages of development among related animals support the notion that horizontal constraints prevail in determining patterns of evolution in ontogeny. However, the observations of developmental geneticists are very largely consistent with the importance of vertical constraints. By these lights, early development has two important properties. First, it is under gene regulation. Precise gene cascades exist to pattern the early embryo. In many cases (for example, over the 100 million years separating houseflies from *Drosophila*) these gene cascades are highly conserved. Transcriptional controls over the expression of particular structural genes are intricate and specific to certain cell types. As discussed by Davidson for genes expressed in sea urchin embryos, these controls can involve multiple transcription factor–promoter interactions. Thus, each of a set of related genes may have distinct promoter elements and correspondingly precise patterns of expression in particular embryonic cells. Such transcriptional precision implies functional constraints on which genes are expressed. Oddly enough, as Henry, Klueg, and I found for orientations of embryonic axes, and as Wray and McClay and Kissinger and I have shown for expression of various genes, highly regulated patterns in one species may be quite different in a related species. It's not that the regulated patterns of cell commitment or gene expression are unimportant for development within a species; rather, it is that other patterns can be substituted without loss of coherent and even conserved modes of early development.

The second significant property of early development with respect to vertical constraints is that mutations in genes expressed early in development often have profound effects. Mutations in maternal pattern-specifying genes cause global deficiencies in *Drosophila* embryos. Mutations in gap, pair-rule, polarity, and homeotic genes all cause major defects in the domains of their actions. Most of these mutations are lethal. The view of ontogeny that we get from these observations says that there is a linear series of connected steps, and that the deletion of any step blocks subsequent development. And yet we know that evolutionary changes in early development are not precluded. Escape from vertical constraints has occurred frequently in evolution. It inevitably requires that one set of early events is substituted for another, but that vital outputs to organize the phylotypic stage are nevertheless generated. To make an adequate synthesis from the disparate observations presented by evolutionary and developmental biology will require considering the arguments raised in the next chapter.

ALLOMETRY: DONE IN BY BIG ANTLERS?

It seems fitting to end a listing of possible causes of developmental constraints on a note of finalism. One way to look for such constraints, suggested by Maynard Smith and co-workers, is to seek examples in which a given kind of developmental variant has been independently acquired among related species. Such a repetition would reflect a common constraint shared by the ontogenies of the species. Occurrences of this kind are known to result from allometric properties deriving from growth relationships, which are often shared among related species. Because allometries are expressions of a particular ratio of differential growth properties among parts of an animal, independent size increases can cause parallel changes in shape. In many cases, allometries are strongly positive for particular features; when overall body size increases, such structures get very much larger.

The intriguing thing about allometric growth laws shared by distinct species is that they might arise from developmental rules that do not necessarily bend to selection. If that is the case, then it should be possible to cite examples of animals at size extremes in which allometric rules have resulted in nonadaptive proportions in body structures. A candidate for such an animal is the now lamentably extinct giant deer known as the ''Irish elk.'' The remains of this deer are known from roughly 12,000-year-old deposits in European peat bogs. The deer were big, and their antlers were indecently big. Various morality tales have been told about how the 12-foot, 90-pound rack of antlers that the males had to grow each year was nonadaptive and was the cause of their extinction. Was the Irish elk a casualty of runaway allometry? Gould did an allometric study of Irish elk specimens in various museums and baronial halls (Irish elk antlers have long been regarded as ultimate trophies), and concluded that there was indeed a strong correlation between body size and antler size. Selection could have operated on either trait, and "entailed the other as a correlated consequence." Plots of antler size versus other body dimensions showed that although the Irish elk was larger than any living deer, its antlers obeyed the same allometric rules as the antlers of various species of living deer. Thus, allometry might have been in the driver's seat in this fatal developmental vehicle. However, deer antlers are used for sexual display in living species, and Gould argued that it seems likely that they were so used by Irish elk bucks as well. If that is the case, the obedience of allometric rules may have been merely coincidental with strong sexual selection. Geist has proposed that the open environments of Pleistocene Europe and selection for the metabolic ability to produce large offspring, able to run at birth, led to sexual selection for males with the best ability to convert forage to biomass. Big antlers were the sign of such superior males. The big antlers might have caused the downfall of these huge deer, but not necessarily because of inexorable developmental constraints. Consider that the moose is also very large, but its antler span falls below the size expected for such a massive deer. Moose live in dense woods where a large antler spread would be a real inconvenience. Smaller antlers would be advantageous to animals that had to move among the closely spaced trunks of the northern pine forest, and no developmental constraint appears to have blocked their evolutionary reduction in size in the moose. As the climate warmed, the open country of northern Europe became heavily forested.

The lesson of the Irish elk, as in the case of the long duration of various styles of marine embryos, is that we have to be careful in our use for documentation of constraint of things that don't change over long evolutionary times. Morphologies are arguably maintained over vast periods of evolutionary time by forces other than internal constraints on their ability to evolve. If the moose didn't exist, we would be tempted to cite Irish elk antler allometry as an example of a powerful developmental constraint that overrode selection with ultimately lethal consequences to the lineage. Incidentally, the fact that deer antlers are not so tightly constrained as to be unable to respond to selection does not mean that powerful internal constraints that limit selection, even disastrously, do not exist. It's just that we shouldn't look for simplistic answers.

10
Modularity, Dissociation, and Co-option

Medusa . . . was once a beautiful maiden whose hair was her chief glory, but as she dared to vie in beauty with Minerva, the goddess deprived her of her charms and changed her ringlets into hissing serpents. She became a cruel monster of so frightful an aspect that no living thing could behold her without being turned to stone.

Bulfinch, *The Golden Age of Myth and Legend*

MOSAIC BODIES

Our hominid lineage has an almost 5-million-year history in the fossil record. However, humans have created representational art for only the past 25,000 years. Representational art appears with striking brilliance and suddenness as the product of Cro-Magnon painters working deep in the caves of southern Europe at the height of the last Ice Age. As art was absent among Neanderthals and more archaic humans, these paintings may be the visible manifestation of an evolutionary breakthrough in the organization of the human mind. The purposes of the cave paintings are lost to us, but these paintings may well have played a role in complex magical or religious ceremonies. Their positions deep in inaccessible cave chambers suggest that they were not merely decorative. Another indication that the paintings may have served ceremonial functions comes from the occasional hybrid creature depicted in them. Perhaps the most famous of these is the ''sorcerer,'' a Cro-Magnon cave painting from the Trois Frères Cave in France. The sorcerer is part human and part animal, with a bird's face, a fox's tail, and a reindeer's body and antlers. His arms, hands, legs, and feet are human. He stares disconcertingly toward the viewer. Reasonably enough, the sorcerer has been interpreted as the image of a shaman dressed in animal skins. But there may be a deeper meaning underlying this hybrid creature. As suggested by the work of Lewis-Williams on African rock art, the sorcerer may not be simply a picture of a shaman dressed in animal skins, but an image of something altogether different. Evidence from African rock art suggests that what appear to be naturalistic illustrations of people and animals may really be representations of shamanistic hallucinatory visions in which the shaman is literally transformed. Lewis-Williams and Dowson suggest that similar altered mental states may have given rise to Upper Paleolithic art as well.

The creation of mosaics between humans and animals has a two-hundred-century-long tradition, and the animal-human hybrid is a type of "monster" envisioned by many cultures. The Minotaur, sphinxes, centaurs, mermaids, and all the rest combine portions of a human body, generally including the head, with the body parts of one or more powerful animals. The ancient Egyptians and Indians went in for the odd exceptions in which human bodies were combined with the heads of birds, lionesses, or elephants. As pointed out by Grant, mythological human-animal hybrids contained a powerful religious symbolism. Moderns may find other symbolisms in these creatures. For instance, Sigmund Freud interpreted the severed Medusa's head to represent the female genitals and to relate to fear of castration (which strikes me as a decidedly lurid and untestable hypothesis).

It is important to note that the creators of these wonderful human-animal constructs designed mythical organisms of clear functional unity despite the disparate origins of their body parts. One of my favorites is an elegant triton, or merman, who appears on a 2,500-year-old classical Greek vase in the Nicholson Museum at the University of Sydney. This vibrant creature has a human head and arms, and he seems to be enjoying a good joke at the expense of a fish he is holding. However, he is not exclusively real at the human end. He also has a well-formed fish body, complete with scales, fins, and a lateral line organ. He thus possesses not only the full panoply of human senses, but also the lateral line sensory system so critical to fishes but absent in terrestrial vertebrates. Although an impossible product of the human imagination, he is nevertheless a whole, integrated being in whom the impossible mix of human and piscine features actually looks plausible.

The modularity of human-generated hybrid monsters reveals two important features of the way we conceive of the living world. The first is that the human imagination is really very limited. We create hybrid monsters built of familiar body parts that contain symbolic attributes, but we almost never create novel life forms. Even the most imaginative designers of movie aliens rarely get beyond thinly made-over takeoffs on arthropods, cephalopods, hominids, and other familiar designs. The second and more interesting aspect is that our hybrid beings have a functional modularity. Although centaurs and tritons are impossible creatures, we find them harmonious and even beautiful. Their body parts are all highly efficient modules for the expression of the attributes projected upon them by their creators. Human designers of novel creatures can select components of known symbolic or functional value. Evolution cannot do it in just that way, but there is nonetheless a remarkable parallel between our imagined hybrid monsters and the products of evolution. Evolution also produces modular creatures, and natural selection, that ferocious and demanding machine of evolution, can operate independently on individual body parts. This feature of evolution is sufficiently obvious to have been termed "mosaic evolution." Mosaic evolution is what we observe in creatures like *Archaeopteryx* with its wings and feathers tied to a dinosaurian skull with teeth, or in early whales with their admixture of streamlined body and sizable legs. Moreover, the products of evolution, like the products of our imagination, are constrained by what exists.

EVOLUTIONARY MECHANISMS

I have discussed the tension between externalist and internalist views of evolution in chapter 9, which dealt with the concept of developmental constraints. Evolutionary biologists have in their hands a demonstrably effective evolutionary process in Darwinian selection. For instance, Gingerich has summarized the rates of evolution observed in the fossil record and under experimental selection. Some fossil rates of change are very fast, but rates of change under experimental selection exceed the fastest rates observed in the fossil record by four orders of magnitude, and testify to the efficacy of selection. (Fossil rates are expectedly slower than experimental ones because they are time-averaged.) Studies by Endler have demonstrated that a number of traits of guppies are under strong selection and are altered rapidly. In experimental situations, color patterns in the males, which are under positive sexual selection and negative selection by predation, respond to changes in predation in as few as six generations. Boag and Grant, as well as Gibbs and Grant, have observed that strong directional selection rapidly shifts body and beak size in one of the finches, *Geospiza fortis,* of Daphne Major, one of the Galápagos Islands. During a pronounced drought during the late 1970s, selection favored large birds with deep beaks. This episode produced a brutal selective regime that killed off over 80% of the finches and left the island carpeted with mummified birds. As plants set no seeds, the most edible seeds were eaten first by hungry finches, and finally only birds capable of dealing with the tough seed cases and large, hard seeds that remained could survive. Change in the direction of larger birds and beaks was as fast as any evolutionary rates known. Once the rains returned to the island, selection again shifted direction, and smaller birds and beaks were favored. Again, rates of morphological evolution were very rapid. Selection is clearly efficacious and can produce very rapid modifications. (Compelling summaries of work on Darwin's finches and related studies that demonstrate the power of natural selection can be found in Grant's *Ecology and Evolution of Darwin's Finches* and Weiner's *The Beak of the Finch.*)

Genetic variation arises, and selection sorts the variant phenotypes. Those phenotypes that best match the selective pressure leave more offspring, and thus a greater proportion of their genes, to the population. Selection has been shown to be effective both in models of gene distribution in populations and in experimental trials with real organisms. It provides explanations that conform to the well-understood mechanisms of transmission genetics and to observations of gene sequence evolution. The question for us is whether selection working on random variation is a sufficient explanation for the evolution of animal body plans and their macroevolutionary modification. Are microevolutionary processes extended over long spans of time sufficient to cause the evolutionary patterns seen in macroevolution? Can the events of the Cambrian radiation be played out in a vial of fruit flies?

Selection is often seen as providing all of the order observed in biological systems. Evolutionary developmental biologists nonetheless observe that variation appears to be nonrandom in developing systems. It is fair to suspect that the internal organization of the existing system must significantly affect the evolutionary outcome because it can constrain the range of possible phenotypes available for sorting by selection. The idea of an internal order biasing the course of evolution has been effectively expressed metaphorically by Jacob, who styled the evolution of complex organisms as a process of tinkering that uses existing structures to produce solutions that are biologically adequate, if not the best in engineering terms. The evolution of wings in vertebrates has always involved tinkering with the existing forelimb, not the invention of a novel kind of appendage. Evolutionary developmental biologists have no specific hypothesis on how the internal aspects of evolution work. The problem lies in determining whether features of internal organization create internal processes of evolutionary change or merely supply accidents of history.

A long-standing dichotomy has been imposed on discussions of internal versus external selection in evolution. Internal factors were often suggested by opponents of Darwinian evolution as alternatives to natural selection. Because of the demonstrable power of the selectionist paradigm, internalists have never done well in pressing their views, and they have been in large part discredited for having backed such dubious horses as racial senility and Lamarckian inheritance. The value of continuing this artificial conflict between hypotheses of internal and external factors in evolution seems questionable to me. Darwinian selection is firmly established, and the internal architecture of animal genomes and ontogenies probably also plays a role. Our goal is to define a simple set of principles by which internal processes might function and by which they might interact with the external process of selection. Internal rules should not be expected to supersede Darwinian selection, but rather, to complement it in predicting the behavior of evolving ontogenies. Internal and external rules can be posed as analogous. Darwinian selection includes both pattern and process. Pattern lies in the features of the

ancestral population and the variation available. Variation results from mei-
otic recombination, random mutation, and other events. However produced,
that variation is then available for the operation of selection. Some variants
are selected over others, and the genetic and phenotypic properties of the
population are affected. If sufficient variation is available, and selection favors a shift in phenotype, evolution will occur, and a new descendant pattern
will result. For evolutionary changes to occur in development, there must be
an external process of selection on the resultant phenotype. However, if the
internal order of the system constrains both variation and the processes that
sort variation, the resulting pattern may not map very closely on the external
selective regime. The internal sorting processes will govern the behavior of
the system under selection and strongly influence the kinds of changes that
appear in the evolutionary descendant.

PRINCIPLES OF EVOLVABILITY

Given the complexity of metazoan animals and their ontogenies, it might
seem that very little evolutionary alteration should be possible. Developmental pathways should be so finely tuned and so complex that although they
might accept some modification at the terminal end of development, they
would be fatally derailed if changes were introduced too early into the ordered
stream. We have seen that this is not the case: evolutionary changes have
occurred in early as well as late development. We thus need to understand
how complex and apparently interdependent systems can evolve at all. Four
elements are proposed that make it possible for ontogeny to evolve. The first,
modularity, is an ineluctable feature of biological order. It is arguably the
most crucial aspect of order in living organisms and their ontogenies, and is
the attribute that most strongly facilitates evolution. We can also identify
three processes that produce nonrandom variation within the existing modules
that can lead to new internal patterns of order. These processes are dissociation, duplication and divergence, and co-option. They provide internal, nonrandom evolutionary variation. Such variation is sorted both by the internal
requirement to maintain integrated function and by external selection on resultant phenotypes. Because of the internal requirement that modified ontogenies
be functional, only a subset of all theoretically possible phenotypes will be
generated.

I will start by defining my terms. Modularity refers to the basic pattern of
order characteristic of organisms. Organisms are constructed of units, or
modules, that are distinct in genetic specification, autonomous features, hierarchical organization, interactions with other modules, location, time of occurrence, and dynamic properties. The role of modularity as a key to biological order has been pointed out by Riedl in his book *Order in Living*

Organisms. Riedl posed the problem of biological order in terms of information theory. The information content, or determinacy, of a complex anatomy is orders of magnitude higher than that of the genome. Riedl proposed that a very large part of that determinacy arises from the nature of order in the anatomy of a living organism. He proposed that order arises from the employment of standard parts that are arranged hierarchically and that exhibit connectivity with each other. Thus, cells are composed of organelles, tissues of cells, and organs of tissues. This scheme, although somewhat static, applies to ontogeny as well. Developing systems have a modular organization, although we should consider developmental modules as dynamic entities representing localized processes (as in morphogenetic fields) rather than simply as incipient structures in anatomy (morphogenetic modules, such as organ rudiments). I'll discuss the properties of developmental modules and their roles as sources of evolutionary variation in a moment.

The existence of a modular organization allows the operation of the three internal evolutionary processes. The first, dissociation, is the process of unlinking features within a system. Heterochrony is the most commonly cited form of dissociation. In heterochrony, the dissociation is one of timing between individual developmental events. There are, as we will see below, other forms of dissociation as well. The second process is duplication and divergence. In essence, this is the means by which both molecular and morphological serial homologues are produced. An existing structure can be duplicated in a variety of ways, depending on the hierarchical level of organization. Duplication yields a copy that can be lost or undergo mutational changes. Once altered, the duplicate can play a role related to but distinct from that of the original. The duplication event is most likely neutral, but the novel functions of the copy may not be, so diverged copies may enter the selective marketplace. The third process is co-option, the taking over of an existing feature to serve in a new function or structure. These three processes are not developmental mechanisms. They are instead evolutionary processes that operate on development. Whatever the external selective pressures on development, evolutionary modifications of developmental features will be constrained by the modular organization of the system and will occur by means of these internal evolutionary processes. Internal responses to external selection will thus be ordered.

MODULARITY

Modules are discrete subunits of the whole. They may have a readily apparent morphological structure, such as that of neural plates or limb buds, but we should regard them as dynamic entities and not as stable anatomical structures. Developmental modules have several discernible properties. First, they have an autonomous, genetically discrete organization. Second, like Riedl's standard parts, they contain hierarchical units and may in turn be parts of larger hierarchical entities. Third, they have physical locations within the developing system. Fourth, they exhibit varying degrees of connectivity to other modules. Last, developmental modules undergo temporal transformations. Thus, all of the first four features can change dynamically during the course of development. Developmental modules also undergo transformation in evolution, and thus can be identified as modified homologous elements in comparable ontogenies.

The Genetic Organization of Modules

My statement about the nature of modules comes from the enormous body of molecular biological and developmental genetic data derived from a large range of developmental modules. Those that have been investigated come from early as well as late development. Early modules include such entities as the domains of the *Drosophila* body axis destined to become the segments. These modules are delineated early in the embryo by highly specific patterns of gene expression that specify the position and polarity of the segments. (Detailed summaries are presented in recent books by Gilbert and by Lawrence.) The Hox genes are then expressed in a segment-specific manner and activate specific groups of downstream genes that execute the development of each segment (see fig. 9.1). The modularity of these segments is graphically demonstrated by the homeotic transformations caused by mis-expression of the Hox genes. For instance, mis-expression of the *proboscipedia* gene in *Drosophila* produces a leg in place of mouthparts. That is, it elicits the pattern of gene expression characteristic of a leg module, but in a quite inappropriate place. The resulting fly suffers from the double embarrassment of looking very strange and starving to death. Other modules include specific early cell lineages, such as the micromeres of sea urchin embryos that differentiate into the larval skeleton-forming cells. These cells express a characteristic set of genes involved in the secretion of the skeleton (reviewed by Harkey and co-workers). Similarly, Jeffery and Swalla have shown that the myoplasm-containing cells of the ascidian embryo that are destined to give rise to the tail muscles of the tadpole larva require the localization of a specific cytoskeletal protein and later express muscle-specific genes. The inner cell mass and the trophoblast cells of the mammalian embryo constitute other early developmental modules. These will give rise respectively to the fetus and the extraembryonic membranes. Distinct patterns of gene expression characterize each of these early modules. Modules can be readily identified in later development as well, and these too have discrete gene expression features. Late modules include, as reviewed by Bryant and Simpson, the

imaginal discs of insects that give rise to definitive adult structures such as
a leg or wing. Such modules also include vertebrate organ rudiments (e.g.,
salivary gland, pancreas, or lung), which are built of characteristic submodules of adjoining mesenchymal-epithelial sheets combined in an interacting
organization. As I'll discuss below in detail, organ rudiments show highly
patterned and characteristic gene expression.

The definition of modular units of gene expression still has its imponder-
ables. The most readily definable such unit is probably the differentiated cell
type. Cell types exhibit a stable and characteristic pattern of gene expression.
Several thousand genes are typically expressed in each cell type, and most
of those genes are shared with other cell types. However, a number of histospecific genes are expressed that encode proteins specifically involved in the
function of one cell type. Many cell types are extraordinarily stable in evolution: sperm, muscle cells, neurons, intestinal cells, and others have had a
half-billion-year duration. Others, such as some specialized cell types of
vertebrates, have more recent origins, but have been conserved for a few
hundred million years. The cell types of embryos also have long evolutionary
durations. Vertebrate embryos share notochord cells, neural tube cells, neural
crest cells, and others. The zebrafish represents a lineage that diverged over
400 million years ago from the tetrapod lineage. However, it shares these
embryonic cell types with tetrapods, and, as is now emerging, the two lineages have common patterns of gene expression as well.

Stuart Kauffman has suggested that cell types can be regarded as having
an internally imposed order arising from interactions among the genes expressed in them. In Kauffman's model, cell types are defined by networks
in which thousands of genes turn one another on and off. Each gene will
have only a few inputs, perhaps two, that serve as on/off switches to control
its activity. At any time, this Boolean network exhibits a particular state of
on and off genes, but the system is not static. It exhibits a cyclic pattern of
states of gene activity, with particular states eventually recurring over time.
This cycle is called an attractor, and it has a dynamic stability. Any one
attractor is discrete from any other, and is said to occupy its own "basin of
attraction." In this model, each particular cell type is an attractor, and obeys
the rules of dynamic state systems. Each cell type is thus self-ordered and,
given the properties of interactions, can be quite stable. Kauffman has noted
several interesting properties of such systems as applied to cells. First, the
number of attractors, and thus cell types, is not infinite, and is proportional
to the number of genes in the genome. Kauffman estimates that a genome
with about 100,000 genes would produce a few hundred attractors. Second,
not all attractors will be produced; that is, not all potential cell types will
be realized. Third, one attractor can be flipped into another. However, the

attainability of new states is limited. Thus, in development, a branching pathway of generation of more attractors is to be expected. Finally, not all genes are expressed as part of any attractor.

There are a number of genes that are expressed during development but are not expressed in any stably differentiated cell type. Kauffman has suggested three interesting ideas based on that observation that have considerable implications for the evolution of development. First, the cell types of embryos can be considered attractors, even though they are not permanent cell types. They exhibit transitions into other cell types as development proceeds. Kauffman's second suggestion is that genes that act as developmental regulators and are not expressed in differentiated cell types function in making transitions between basins of attraction, that is, between cell types. Finally, he proposed that mutations can affect developmental pathways without affecting cell types. This suggestion is an interesting one because it predicts that alternative developmental pathways can occur in evolution without any change in the component cell types of the organisms. Indeed, that is what we commonly see in the evolution of new phenotypes. Numerical simulations have suggested that such mutations can alter only a few pathways of differentiation, which would allow the evolution of a new transition without altering all pathways. I find this suggestion particularly intriguing in that we have observed what appear to be analogous phenomena in the evolution of direct-developing sea urchins, as discussed in chapter 7. Embryos of *Heliocidaris erythrogramma,* the direct developer, have most of the same cell types as the closely related indirect developer *H. tuberculata.* However, their pathways of differentiation differ, with the major difference lying in the pathways to the founder cells for skeletogenic cells and coelomic lineages in the two species. The implication is that the difference results from a change in the regulatory gene action involved in decisions affecting cell differentiation pathways, but not cell types.

At present, Kauffman's proposal for cell types as dynamic systems of gene interaction remains untested. However, there has been a good deal of study of gene pathways involved in developmental decisions. I have discussed some examples: the genetic and inductive events that specify photoreceptors in the insect eye (in chapter 4), the control of timing in nematode cell lineages (in chapter 8), the determination of pattern in limb buds (this chapter), and the pathways of sex determination (which I will discuss in chapter 11). In these and other cases, there is a strong commonality. Pathways of expression of regulatory genes that determine which of two possible pathways of differentiation a cell will follow involve a short train of genes whose activities force the cell into one of two alternative pathways of differentiation. Execution of the differentiated state then involves a large number of genes lying downstream of the determinative pathway.

Standard Parts

Riedl has provided a thorough analysis of the role of standard parts in biological order. In his scheme, most of the anatomical complexity of organisms is generated through the repetitive use of standard parts. Standard parts occur at all hierarchical levels. For example, individual α- and β-tubulin protein molecules combine into a heterodimer, which is the standard part for microtubule assembly. Microtubules and other elements combine as standard parts to form the axoneme, the shaft of a cilium, which is composed of a characteristic pattern of nine fused microtubules surrounding a central pair of singlet microtubules. Cilia, in turn, are standard parts in ciliated cells. Assemblies of ciliated cells are the standard parts of features such as the ciliary bands in embryos. This kind of hierarchical organization is so commonplace that it seems unremarkable. Nevertheless, it lies at the heart of metazoan organization, in which cells and groups of cells are the standard parts of multicellular life.

Standard parts are crucial in both development and its evolution. The deployment of standard parts can generate a more complex anatomy than could be encoded in a one-to-one fashion by the genome, and it reduces enormously the complexity of the command system in ontogeny. Standard parts at lower hierarchical levels are generally capable of self-assembly. That is, under standard conditions within the cellular or organismal environment, particular proteins will assemble into supermolecular structures based on the intrinsic properties of the protein chains. No further genetic instructions are necessary in these cases. If Kauffman is correct, the dynamic states of interacting genes also generate domains of attraction, or cell types. However, the results of developmental genetics show that self-assembly in itself is an incomplete model for ontogeny. Self-assembly phenomena in development operate within bounded conditions, beginning with the initial required conditions provided by the egg. The contexts change drastically as development proceeds. Although lower hierarchical levels within the pathways that produce standard parts involve self-assembly, higher-level events appear to require external instructions as well. These higher-level constructs of standard parts and regulatory processes in development are modules. Overall, development involves both self-assembly and the actions of regulatory genes, such as those involved in determining switches between cell fates. The locations, numbers, and interactions of standard parts also enjoin the actions of genes that lie outside the construction of the parts themselves.

The fact that development employs both standard parts and multiple simple controls on their deployment makes the evolution of development possible. Since the same standard parts are maintained across evolutionary modifications of development, genes do not have to be invented each time an ontogeny is changed. This does not mean that the genes involved in building standard parts are of no interest, but it suggests that the real evolutionary action lies in other genes. This idea goes back to the dichotomy between structural and regulatory genes proposed some years ago by Allan Wilson and his co-workers. The distinction was made most dramatically by King and Wilson, who noted that humans and chimps differ by only 1%–2% in their DNA, but enormously in their morphology. The DNA differences are consistent with their being congeneric species, but on the basis of morphology, chimps and humans have been placed in different families. If structural genes by and large build standard parts, and regulatory genes determine how these parts are used in development, then regulatory genes will be the objects of some passion. And indeed they are. In terms of evolvability, if conservation of standard parts is important, so should be conservation of regulatory genes. We'll get to that point below. However, the vital point in thinking about standard parts in evolution is that they are used over and over in varying ontogenies. Their patterns of use depend at least in part on the expression patterns of a relatively small number of regulatory genes.

Physical Location within the Developing System

One of the most characteristic features of any embryo is its increasing complexity of structure as development progresses. This complexity is highly spatially ordered. Modules are not static anatomical features, but they have specific locations within the overall pattern. Spatial patterning of modules in ontogeny begins with the initial pattern of information in the egg cytoplasm. The proper function of maternally loaded determinants that initiate pattern formation in the embryo depends upon correct physical localization. Ectopic expression of such substances produces developmental abnormalities. As modules form in development, they do so in specified positions within the overall pattern. Developmentally correct interactions between modules require proper spatial relationships between them, although spatial relationships can be fleeting as tissues move past one another. In some cases, two separate cell sheets may come together to form a stable interacting system that forms a new module. Organ rudiments and limb buds form in this way.

Connectivity to Other Modules

The common genetic strategy of metazoan development appears not to involve large numbers of genes acting in a long, rigid series to control a predetermined "developmental program." Instead, there are short, tightly linked cascades of genes that force downstream genes into particular pathways of expression. These are what we commonly define as regulatory genes. Some act as binary switches. Many are transcription factors with particularly powerful actions. For instance, the transcription factor myoD, as shown by Davis and co-workers, causes cells in which it is expressed to express muscle-specific structural genes and differentiate into muscle cells. These regulatory cascades may act as proposed by Kauffman to mediate choices between dynamic attractors; that is, a module can be forced into a course of differentiation by the expression of powerful switch genes. I'll return to this property later in the chapter.

Regulatory cascades are often activated by epigenetic links between domains. These epigenetic links take a common format although the players in each drama differ. A reasonable general statement is that cells can produce extracellular signaling molecules. Responding cells have receptors for these molecules. The receptors pass on a signal via one of many possible intracellular signaling systems. That signal activates expression of a chain of genes that results in production of transcription factors that determine the path of differentiation of the receptive cell. Modules can thus have profound influences on the genetic state of neighboring modules. Embryologists have long recognized that these inductive interactions are a primary part of development.

Temporal Transformations

Developmental modules are dynamic entities. They have transitory identities and play changing roles in development. Modules form, and can then divide, transform into different modules, enter a terminally differentiated state, or undergo programmed or stochastic death. Cell lineages illustrate all of these dynamic properties, and a cell lineage can be considered a temporally connected series of cellular modules. Lineages arise at precise moments in development and have recognizable genetic properties. For example, the cells that give rise to the central nervous system of *Drosophila* arise from totipotent cells of the ventral ectoderm. About a quarter of these cells become neuroblasts and give rise to neural cell lineages. As discussed by Doe and Goodman, a characteristic set of cell divisions to produce neurons ensues. The remainder of the neighboring cells become hypodermal skin cells. Whether a cell enters the neural pathway depends on the action of the *Notch* gene. The Notch protein has been proposed by Heitzler and Simpson as a receptor of intracellular signals, probably from the product of another neurogenic gene, *Delta*. Lineages divide to produce precisely determined daughter cell types. Thus, lineages not only transform dynamically, but can also bifurcate to generate more cellular modules. The downstream identities of cells in cell lineages depend on the expression of regulatory genes. Thus, in the lateral hypodermal cell lineages of the nematode *C. elegans,* Ruvkun and Giusto have shown that the timing of expression of the nuclear protein product of the *lin-14* gene determines the correct timing of lineage differentiation. This gene is part of a short cascade of genes. The entry of cells into a terminally differentiated cell type depends on the action of switch genes. In the case of the gonad and germ cells of *C. elegans,* described by Kimble and White, four blast cells are involved, Z1, Z2, Z3, and Z4. The Z2 and Z3 cells arise from a different embryonic founder cell than Z1 and Z4, and produce germ cells, whereas Z1 and Z4 produce gonadal cells. The ability of Z2 and Z3 to form germ cells requires that Z1 and Z4 give rise, through two more cleavages, to the distal tip cells. This is what occurs in *C. elegans,* and the adult has two functioning gonads. In the related soil nematode *Panagrellus redivivus,* the Z4 granddaughter cell, Z4pp, undergoes a programmed cell death instead of giving rise to a distal tip cell. *Panagrellus* consequently has only a single-armed gonad. The Z1 and Z4 cell types are rigidly programmed, but as suggested by Sternberg and Horvitz, the program has been altered in evolution. The result is a drastically different final anatomy produced by modifications to cell fates within an overall conserved dynamic module.

We need to pause a moment here, because in their incipient stages of development, what I've been calling modules intergrade into an older embryological concept, the morphogenetic field. Fields were prominent subjects of investigation in the first few decades of this century. They are regions of an embryo that show no overt differentiation, but are destined to give rise to particular structures in development. Various morphogenetic fields were recognized, including limb, nose, lens, heart, and gill fields. Experiments showed that fields have striking properties. If divided, they give rise to duplicated structures, and if damaged, they can regenerate lost parts. If transplanted into another site or animal, fields produce supernumerary structures in the host. De Robertis and co-workers have noted that many classic properties of fields are consistent with modern observations of the domains of expression and modes of action of Hox genes and other major regulatory genes. For example, limb fields on the flanks of the *Drosophila* embryo are first marked by expression of the gene *Distal-less.* Gilbert and co-workers have elaborated the properties of these genetically refurbished fields, and envision them as "units of developmental and evolutionary change." These ideas are not far from the properties incorporated into the definition of modules that I have used. However, I have not used the term "field" extensively here. Fields are indeed modules, but the term "module" is broader and more flexible, and not all modules are fields. Over the hierarchical range of domains in developing systems, "module" allows the designation of more kinds of units than "field."

Modules become visibly predominant in later development. The rudiments that give rise to the appendages and body organs are each modular in organization and dynamic in time. Dynamic changes take place in interactions

between submodules, in their locations, and in the initiation of new constituent submodules. By the time the drama has been acted out, a suite of new
morphological entities has been generated, each exhibiting the features of
modularity. The development of the vertebrate limb bud, which I'll get to in
a moment, illustrates how a complex modular structure undergoes temporal
changes in gene expression patterns to generate the ultimate adult structure.

DISSOCIATION

The idea that developmental processes are dissociable from one another in
evolution rests upon the fact that a great deal of parallel processing goes on
within ontogeny. What appears to the casual observer to be a unified flowing
stream of events, an orderly unfolding of ever more complex features, can be
readily dissociated by an experimenter as well as by evolutionary processes.
Dissociation was implicit in the very founding of experimental embryology
by Roux. His program was to perturb individual components or particular
processes in the embryo and observe the effects of that perturbation. His
underlying expectation was that other parts of development would proceed.
If that were not the case, all perturbations would produce uninterpretable
monsters or death. However, if unitary components or processes produced
definable and discrete effects, then their roles in development could be dissected out. Thousands of experiments done by developmental biologists along
the lines of this very simple paradigm have yielded mainly discrete and more
or less interpretable results of this second kind.

   The extent to which the unmodified portion of an experimentally manipulated embryo develops normally reflects the degree to which regulation is
possible. Embryos have been regarded classically as either mosaic or regulative. This distinction was generally observed in cell deletion studies of embryos. In some embryos, typically those of spiralians, nematodes, and ascidians, removal of a cell or a group of equivalent cells early in development
results in an embryo in which the structure produced by the deleted cells
doesn't form, but the remainder of the embryo develops essentially normally.
In these mosaic embryos, individual cellular modules become autonomously
determined early in development. In other embryos, such as those of sea
urchins or frogs, removal of cells early in development has little effect. In
these regulative embryos, other cells can compensate and produce the cell
fates that would have derived from the missing cells. Autonomous modules
form somewhat later in these embryos. Thus, modules appear during early
development, and can have noticeably different levels of functional integration. The difference between mosaic and regulative development does not lie
in a dichotomy of developmental patterns; rather, it is a question of timing.
Most cells of regulative embryos also become committed and unable to as-

sume alternative fates as development proceeds. Modules occur in all embryos, and their functions are generally dissociable.

Dissociations occur between temporal events and processes (heterochronies), between spatial relationships, and finally, in interactive links between modules. I have already explored the basics of dissociation as they apply to heterochrony in chapter 8. Early events in a time sequence do not necessarily cause later events. Such time sequences can be rearranged in an evolutionary descendant. These dissociations have occurred frequently in evolution, and their consequences have been documented. As we saw in chapter 8, heterochronies most often arise from causes that do not involve the regulation of timing per se, but nonetheless produce results in which relative timing is altered. Heterochronies can result from other kinds of dissociation. Thus, how we score a dissociation will depend on the hierarchical level of interest.

Spatial dissociations, called heterotopies by Haeckel, also occur. They have received relatively little theoretical attention until recent times because of the overemphasis lavished on heterochrony. Spatial dissociations include such diverse events as shifts of gene expression from one cell type or group of cells to another, homeotic changes, production of serial homologues, production of repeated structures, changes in location of structures relative to the body axis or some other frame of reference, and changes in relative proportions of structures. Clearly the causal mechanisms by which such dissociations occur differ, but their results are clearly definable changes in spatial relationships.

The third kind of dissociation involves interactions between modules. Dissociations of processes have been demonstrated experimentally in both early and late development and in both large embryonic modules and differentiating cellular modules. Simple dissociations within what appear to be unitary structures indicate that more than one underlying process is involved. For example, homeotic transformations of the appendages of insects represent dissociations between location and identity. One set of genes establishes the placement and polarity of the segments, a second set of genes their identity. Extensive parallel processing events in differentiation can also be demonstrated, and may provide rich possibilities for evolutionary dissociations. A particularly well defined example of molecular parallel processing in early development, discussed by Nüsslein-Volhard, is the establishment of the anterior-posterior and dorso-ventral axes in the *Drosophila* embryo. These axes define the ensuing major partitioning of the body of the fly. The axes are specified by two distinct localized molecular systems established in the egg. Such axes are a crucial feature of early development. In many kinds of embryos, including sea urchins and frogs, one axis, the animal-vegetal axis, is established maternally, and a second axis, the dorso-ventral axis, is established by one of a variety of mechanisms. These can include a separate maternal system or some zygotic event, such as processes triggered at the site of sperm entry.

The highly modular nature of late development also lends itself to numerous parallel processing events. One such case is spermatogenesis, the process of differentiation of the highly specialized sperm cell. Individual morphogenetic processes in sperm assembly can be readily separated. For example, Henry Hoyle and Elizabeth Raff showed in *Drosophila* that they could block growth of the flagellar axoneme, but even so, a second microtubule-mediated process, elongation of the mitochondrial derivative, proceeded. As Elizabeth Raff has summarized the overall process, the first steps of spermatogenesis, from the gonial cell through mitosis and through the primary spermatocyte stage, follow linear processes. Mutations in these steps block all downstream events. However, once the cells are committed to meiosis, a very different effect of mutations is observed. Several processes occur: meiosis occurs first, then nuclear shaping, elongation of the mitochondrial derivative, and axoneme assembly all occur more or less simultaneously. Mutations that block any one of these processes do not block the others. It is significant that many parallel processes have been revealed by mutations that result in dissociations of developmental events, because these give us an indication that evolutionary dissociations too should have genetic causes.

Dissociations between relatively unconnected modules are easily comprehended. However, dissociations within interactive systems may be possible as well, although they are likely to be more difficult to accommodate within the requirement of maintaining a functioning ontogeny. Although modules are relatively isolated from one another in late development, there is a great deal of interaction between them earlier. As separate modules form in substantial numbers in mid-development, inductive interactions between groups of cells become prevalent. Modules that will be quite separate from one another late in development can do a fair bit of interacting during mid-development because modules may be in contact transiently and later separate. While they touch, inductive interactions occur, some of which are unexpected. For example, in a pioneering study, Antone Jacobson showed that although the retina is the major inducer of the eye lens in late development, there is a good deal going on before. In the late gastrula and neurula stages, it is first endoderm and then heart mesoderm that induce the lens. That happens because of the geometry of amphibian gastrulation. The endoderm initially underlies what will eventually become the brain and its derivatives. As gastrulation proceeds, heart mesoderm advances until its forward edge moves beneath the presumptive lens cells. During subsequent neurulation, the neural plate closes and becomes the neural tube. The optic vesicle evaginates. As the neural tube closes, the future lens cells come into contact with the presumptive retina, which continues the lens induction process. The early inductions by endoderm and heart mesoderm are surprisingly effective. Jacobson found that when he removed the retina experimentally, 42% of the embryos still formed lenses. Other organs such as the ear and the nose also have multiple transient inducers in the embryo.

Dissociation appears paradoxical as a creator of developmental novelty because nothing new is added. In the case of some heterochronic dissociations, such as neoteny in the axolotl, a novel developmental pathway and life history have resulted from the loss of a feature of the ancestral system; in the case of the axolotl, it was thyroxin production. Without dissociation, there would be no way for a novel element to be introduced into a developmental pathway, regardless of its source. No feature can be subtracted from or added to an ontogeny with too much linkage to allow some dissociation to occur. Those features of development that cannot be dissociated may be the features that define the phylotypic stage, and thus the conserved elements of the body plan.

DUPLICATION AND DIVERGENCE

Serial homologues are often posed as one of the great puzzles in evolution. Yet, when properly viewed, serial homologues can be seen as a major way in which novel features and additional complexity are introduced. Meristic changes, that is, alterations in numbers of a standard part, are one of the most common occurrences in evolution. We can perhaps visualize them best at the gene level, and also see best there how new information is added in evolution. In 1970 Susumu Ohno proposed a major process in evolution, spelled out in the title of his book, *Evolution by Gene Duplication*. Existing genes can be duplicated by unequal crossing over between sister chromatids at meiosis. Once a second copy is produced, it is redundant and possibly even deleterious, and it is likely to suffer mutations and eventual inactivation and conversion to a pseudogene. This may be the fate of most individual duplicated genes or of genes borne on duplicated chromosomes. Eukaryotic genomes are veritable junkyards of unwanted former gene copies. However, that is not their inevitable fate. A duplicated gene can undergo mutations, but still produce an active protein. This somewhat different protein may serve a new function. If it does, it becomes subject to selection in its new role, and a gene family is born.

Consider two patterns. In one, A, A', A'', A''', a set of related elements are arranged in a row. In a second linear organization, A, B, C, D, the elements are unrelated. The first pattern is the familiar one of serial homology. This pattern is more economical of information than the second, more complex pattern. Serially repeated structures are common among metazoans. Arthropods are very obviously built this way, and indeed, tagmosis is characteristic of their evolution. These similar modules have similar developmental genetic underpinnings. There is no reason why the second pattern couldn't be used, but reinvention of whole modular assemblages is far less likely than duplication and divergence of existing modules and the co-option of duplicated modules for new uses. The examples given above show that modules can be specified by relatively few switch genes. It is only that event that needs to be displaced spatially to produce a serial homology.

These days it would be easier to write a chapter section on unique regulatory genes than one on regulatory genes that occur in families. There is by now a huge collection of known regulatory genes, and most of them sooner or later turn out to have serial homologues. The profound lesson of developmental genetics and molecular biology has been that quite distinct organisms utilize the same regulatory gene families to achieve related but divergent ends in development. The pervasiveness of molecular serial homologies is little short of astounding. However, it's what we should have expected as the most parsimonious means of assembling novel regulatory systems in relatively short evolutionary time spans. The metazoan radiation was arguably the consequence of the most encompassing example of duplication and divergence of regulatory genes.

I'm not going to attempt to present the diverse and wondrous zoo of regulatory gene serial homologues here. Instead, I would like to note the range of such genes and briefly mention a few families and their diverse functions. It is important to remember that, in general, developmental regulatory genes are genes that regulate basic cellular functions. They include the whole gamut of cellular control genes. Starting from the outside in, there are families of genes that produce extracellular signals to other cells. The *wnt* and transforming growth factor-β gene families discussed below are important examples. As has been shown (by McMahon and his students) for the *wnt* genes in mouse development, the signals represented by the proteins produced by different members of the family are expressed in different sites in the central nervous system and other tissues. Deletion of individual signals causes some structures not to form. There are analogously many families of receptors for exogenous signals, as well as for the proteins involved in intracellular signaling systems required for signal transduction. Signal transduction results in transcriptional activation or repression in target cells. Finally, there are numerous transcription factors.

Families of extracellular signaling molecules are large and diverse. For example, bone growth is dependent on the production of bone morphogenetic proteins. These proteins are members of the TGF-β growth factor superfamily. The BMP family has seven canonical members, BMP-1 through BMP-7 (see Rosen and Thies, and Wozney and co-workers), which have distinct sites of expression in developing bone. There are also related proteins, the

Gdf proteins reported by Storm and co-workers, which indicate that the family is even larger. As Tickle has noted, genetic lesions in different members of the BMP family in mice produce quite distinct skeletal phenotypes. Thus, BMP-5 has been shown by Kingsley and co-workers to affect the axial skeleton, but not the limbs. Loss of GDF-5 function, on the other hand, has been shown by Storm and co-workers to cause limb defects.

Many transcription factors also occur in large families. For example, the myoD family (reviewed by Weintraub) contains myoD, myf-5, and myogenin, which are powerful determinants of muscle cell differentiation. They belong to a larger family of helix-loop-helix transcription factors, which includes several *Drosophila* genes, such as genes of the *achaete scute* complex that establish the pattern of the peripheral nervous system, *daughterless* and other sex-determining genes, and *extramachrochaetae,* a negative regulator. In mammals, the members of the family include E12 and E47—which, when combined as heterodimers with myoD, promote muscle cell differentiation—and a negative regulator, Id. myoD and myf-5 determine the identity of myoblasts, and myogenin acts downstream of them to activate the muscle structural genes of the differentiated muscle cells. The helix-loop-helix proteins form both homodimers and heterodimers. The Id protein was shown by Benezra and co-workers to act as a negative regulator by forming inactive complexes with other proteins of the family.

One last family example: The homeodomain protein superfamily (see Treisman and co-workers) also encodes a number of related transcription factors. The superfamily contains 9 classes of homeodomain protein genes. They comprise 23 genes or families of closely related genes. Some of those families include several member genes. For example, the Antennapedia and Bithorax families include, respectively, 5 and 3 members in *Drosophila,* and altogether 38 members in the four Hox clusters of mammals. Many, perhaps all, have specific and distinct functions.

The duplication and divergence hypothesis effectively describes the creation of novel features within genomes. At about the time that Ohno wrote, the sequencing of proteins was beginning to yield a good look at the universe of structural genes, and some real surprises turned up. One of the most interesting of these was the case of α-lactalbumin. Lactose biosynthesis involves the linkage of galactose to glucose by the enzyme galactosyltransferase. Lactalbumin itself has no enzymatic activity; however, it has the ability to bind to galactosyltransferase. In this state, the complex can bind glucose with high affinity, an ability that galactosyltransferase alone lacks. This complex results in the unique ability of the mammary gland to synthesize lactose. The mammary gland is a relatively new tissue, with its origin in the Mesozoic radiation of mammals. It seems likely that some cutaneous gland existing in mammal-like reptiles, along with preexisting hormonal systems, was co-opted in the evolution of this novel organ. Novel structural proteins also evolved, of which lactalbumin is one. Brew and Hill have inferred its origin from lysozyme. Lysozyme hydrolyzes the peptidoglycan of bacterial cell walls, and its ability to bind to the polysaccharides of this material presages its duplication and divergence to yield a protein with a different sugar-binding function.

Duplication and divergence have always been evident in morphological evolution. That is, after all, what one is seeing in such simple evolutionary changes as in the numbers of ribs in a scallop shell, or numbers of body segments in centipedes, or numbers of vertebrae in snakes, or numbers of teeth in the jaw. It is what we observe in meristic traits. Muscle cells, gill arches, bristles, feathers, and toes are all repeated standard parts with divergent patterns of expression based on locality in the overall system.

CO-OPTION

It is indisputable that co-option of preexisting features into new features has occurred during the course of evolution. The issue of homology turns on it. Co-option is really the only adequate answer to the painful fact that it is difficult to account for the origins of incipient structures in evolution. This difficulty played a role in the first major scientific attack on Darwin, which was launched by St. George Mivart in his book *Genesis of Species,* published only a few years after *The Origin of Species.* Mivart was not reticent in pointing out that partially formed structures that could not function effectively could not have been selectively advantageous. However, co-option offers a way around the problem. Preexisting structures that could perform a particular function might be co-opted for a new use and be selectively advantageous. Traditionally, the term "preadaptation" has been used to refer to such structures. Unfortunately, this term bears an obviously dubious semantic load in suggesting that somehow features can be preselected for some future utility. Gould and Vrba suggested a way out by presenting some new definitions. They suggested that the term "adaptation" be defined as a feature currently used and currently under selection, and that the term "exaptation" be used to replace the disgraced "preadaptation." Features co-opted for a new use could have had various origins. One is that they might have been adaptations in the ancestor and then co-opted for a new use in the descendant. The example of feathers is apt. Feathers probably originally evolved as insulation in some small, warm-blooded dinosaur. They were adaptive in that role, but were co-opted for a second function because of their aerodynamic properties. They have also retained their original role. Other preexisting features might have been nonadaptive in the ancestor, but as exaptations co-opted for some use in the descendant, have become adaptations.

Co-option has been demonstrated for a number of morphological features. Wings evolved three times within vertebrates, and each time the forearm and its components were co-opted for an entirely novel function. Ribs have assumed the function of legs in snakes, the old reptilian jaw articulation elements have become the bones of the mammalian middle ear, and the sesamoid bone of the wrist has become the panda's "thumb." In some examples of co-option, as in the case of forelimbs transforming to wings, the old function has been lost. However, in many other examples, the old function is retained. The dual use of feathers in birds retains the old insulating function and adds the new aerodynamic function. Co-option has also occurred frequently at the genetic level. Genetic co-option has generally occurred in such a way as to maintain old gene functions and simultaneously generate novel ones. In the next chapter, I'll discuss how that has occurred in one of its most transparent manifestations, the recruitment of enzymes for use as crystallins in eye lenses.

The exciting news from the era of developmental genetics has been that important developmental regulatory genes have undergone patterns of evolution involving both duplication and divergence as well as co-option. A whole suite of genes first discovered as being important in pattern formation in *Drosophila* development have turned up again and again as regulators of development in vertebrates, nematodes, echinoderms, and even cnidarians. This result was startling because body plans and development are so distinct among these animals. No generalization of developmental biology (there is no theory of development as of yet) suggested that regulatory genes would be so conserved and so widely used. Yet they are, and, as in the cases of structural genes, they have been co-opted for novel uses. There doesn't appear to be any unique class of genes utilized in this manner: transcription factors, signaling molecules, transmembrane receptors, and second messenger components have all been co-opted again and again throughout metazoan phylogeny.

A CO-OPTION EVENT AT MORPHOLOGICAL AND GENE LEVELS

In rereading Ohno's book, I noted that he had applied his principle of duplication and divergence to a particularly interesting example, the origin of the vertebrate jaw. In 1970, too little was known for him to make much headway in explaining the connection between morphological and molecular evolution, but it is evident now that a novel structure that had a profound effect on the evolution of feeding, aggression, and conversation in vertebrates had its origin in evolutionary co-options of both preexisting genes and structures. The trail of evidence has survived remarkably unobscured for over 400 million years. The earliest vertebrates were jawless, and had vacuum-cleaner mouths useful for a sort of low-level mud-sucking existence. Presumably they sucked

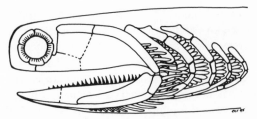

Figure 10.1. Homology of gill arches and jaws in the primitive Devonian fish *Acanthodes*. The elements of the upper and lower jaws are homologous to the gill supports. (After Gregory 1951.)

up small invertebrates in a leisurely way. However, some ability to grasp prey must have been helpful. The precipitous and almost entire replacement of the agnathans by the enormous Paleozoic radiation of jawed vertebrates illustrates the point. Once jaws and teeth evolved, they opened niches for large carnivorous fishes and other specialists.

Where the lower jaw came from is obvious from a look at the jaw structure of some of the most primitive fossil jawed vertebrates. The jaw in these forms is very similar in structure to the hyomandibular and branchial arches that support their gills. In the Devonian fish *Acanthodes,* shown in figure 10.1, the lower jaw is plainly a homologue of the gill arches. The jaw contains the same set of segments as the gill arches, and the mandibular rays show that the jaw supported an opercular flap like those of the gill arches. The musculature of the lower jaw in primitive living fishes is part of a homologous series with the muscles of the gill arches. Here the most anterior of a set of serial homologues has been co-opted for a novel function. The co-option has been accompanied by divergence in structure. Analogous events occurred later in the evolutionary history of jaws as well. Lauder and Liem, for example, have noted that whereas primitive ray-finned fishes have only a single mechanism for depressing the lower jaw, more advanced ray-finned fishes have evolved a second mechanism linking the hyoid to the mandible. This change provided an independent biomechanical pathway for opening the mouth, permitted the second mechanism to diverge, and allowed the evolution of novel chewing functions.

The origins of the jaw and the serial homology of the pharyngeal arches are also visible in the development of these structures in mammals and in the expression of the serially homologous Hox genes in this region, as elucidated by Hunt and co-workers. The structures of the pharyngeal arches— including the lower jaw, which arises from the first arch in the series of six—derive from neural crest cells that migrate along defined tracks from the crest of the closing neural tube to a variety of destinations. Much of the skeleton, as well as arteries, muscles, and nerves, of our faces arises from

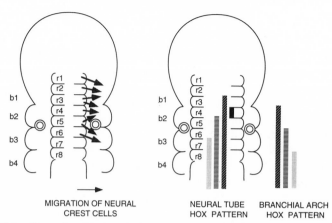

MIGRATION OF NEURAL   NEURAL TUBE   BRANCHIAL ARCH
     CREST CELLS   HOX PATTERN   HOX PATTERN

Figure 10.2.   Hox codes of the neural tube and of the neural crest–derived cells of the branchial arches. Concentric rings on each side of the embryo are otic placodes that give rise to ears. r = rhombomeres; b = branchial arches. The diagram on the left shows pathways of neural crest migration from the rhombomere regions. The diagram on the right plots the expression patterns of genes of the HoxB cluster. The top of each bar indicates the anterior boundary of expression of a gene in each tissue. Note that the branchial arch patterns are offset by one rhombomere width from the neural tube patterns. The gene expression patterns of neural crest cells are influenced by their branchial arch destinations. The genes, in order of anterior boundaries, are *HoxB 2, HoxB 1* (confined to rhombomere 4), *HoxB 3,* and *HoxB 4.* (After Hunt and co-workers 1991, Sechrist et al. 1993, and Bronner-Fraser 1995.)

the neural crest. The lower jaw arises from Meckel's cartilage, which is a serial homologue of the cartilages arising in pharyngeal arches two through six. These ultimately produce the bones of the middle ear (once parts of the mammal-like reptile jaw) and the styloid process, as well as the hyoid and the laryngeal cartilages of the throat. (Reviews of the development and anatomy of this region can be found in Gilbert's and in Larsen's recent developmental biology texts.)

A Hox code has been shown by Hunt and Krumlauf and their co-workers, as well as by Lufkin and co-workers, to underlie the patterning of the neural tube and the neural crest cells that derive from it. As figure 10.2 shows, each arch, including the first arch that produces the lower jaw, has a unique expression pattern of Hox cluster genes. This molecular geography has some interesting features. First, each pharyngeal arch corresponds in position to two rhombomeres in the nervous system. Second, the anterior boundaries of the gene expression patterns in the nervous system are displaced from those of the pharyngeal arches because some rhombomeres do not produce significant numbers of neural crest cells. Third, mutations in Hox genes experimentally expressed in the pharyngeal region were observed by Chisaka and Capecchi and by Lufkin and co-workers to affect structures formed in the pharyngeal region. Finally, although the arches are serial homologues, the Hox gene expression patterns that specify each of them are distinct. Each arch has its own "address." However, the gene expression programs that establish these modules are likely to be similar. Serial homologues should play out similar gene expression programs with a similar machinery. The locations and numbers of serial homologues must be specified by genes whose expression establishes pattern in a larger domain that includes the repeated modules.

The genes that establish the "segments" in vertebrates are not yet well defined, although they apparently include some of the genes (such as the *wnt* homologues of the *Drosophila* gene *wingless* that have been studied by McMahon) that establish segment polarity in *Drosophila*. This scheme has been well defined in early *Drosophila* development in the action of a cascade of genes that sets up the positions and polarities of body segments. Only then are the identities of individual segments specified by combinatorial codes resulting from the action of the Hox genes.

DUPLICATION, CO-OPTION, AND REDUNDANCY

One of the interesting expectations we might have of duplicated genes is that they will retain similar functions, at least early in their histories. They might even have redundant functions. A study of redundancy in regulatory genes by Li and Noll presents us with a revealing instance of what redundancy means, and shows how evolutionary changes in gene regulation can rewire gene expression patterns. The genes that Li and Noll examined were three paired-box- and homeobox-containing genes (*paired, gooseberry*, and *gooseberry neuro*) involved in pattern formation during early development in *Drosophila*. Despite differences in coding sequences and their normal functional differences, in appropriate experimental circumstances these genes could replace one another. These genes share similar DNA-binding regions and may have originated by duplication events. Although functionally equivalent, they have evidently acquired new functions in the normal course of development through changes in their cis-regulatory domains and consequent distinct expression patterns. As noted by Li and Noll, redundancy may result from partial retention of an old function by duplicated genes. We'll see this effect again in chapter 11 in the evolutionary co-option of enzymes used as crystallins in eye lenses.

LIMB BUDS: EVOLUTION OF A MODULE

A mathematically inclined evolutionary biologist of my acquaintance once complained that developmental biologists like to explain everything by use of examples. Perhaps that is because there is so much diversity in ontogenies, but it is also because the examples contain the real observations and so make developmental generalizations plausible. A detailed example illustrating the operation of the principles I've just outlined is more than desirable here. The tetrapod limb bud is one of the best-studied morphogenetic entities. Morphogenesis and pattern formation of the limb bud have long been studied, and more recently, links have been made between patterns of expression of regulatory genes and the resulting morphology.

Limb buds are modular in structure, and they fit the definition of developmental modules very well. They have a clear identity, including a demonstrably discrete genetic organization. They contain a rich suite of hierarchical units. They have distinct physical locations along the body axis. They have a limited degree of connectivity to other modules. Finally, they undergo temporal transformations.

A huge experimental literature has been inspired by limb ontogeny. Fortunately, both developmental and evolutionary aspects have been well reviewed by Hinchliffe and Johnson, by Shubin and Alberch, and by Tabin. The lateral plate mesoderm of the limb bud region of the body wall initiates formation of the limb field, and will do so even if moved to a new site on the body. The induction of the limb field, and the limb bud to which it gives rise, is only beginning to be understood. This understanding has been stimulated by a dramatic experiment by Cohn and co-workers. The results of an experiment done in the 1920s by Balinski, in which he found that transplantation of the otic vesicle from the head of a newt embryo to the flank of another induced formation of a limb, had long been puzzling. More recently, it has been found that the otic vesicle expresses a fibroblast growth factor (FGF). Cohn and co-workers showed that the same forelimb duplication was produced in chick embryos by implantation of an acrylic bead soaked in FGF-1, FGF-2, or FGF-3. The experiment doesn't show which of the several FGFs is the natural inducer. Further, FGF cannot be solely responsible for normal limb bud induction, because most of the FGF-induced limbs were reversed. Tabin, in reviewing these experiments, has suggested that retinoic acid may induce expression of a member of the Hox family, possibly *HoxB 8*. The expression pattern of this gene is consistent with a role in establishing a posterior signaling domain called the ZPA (see below) in the prospective bud, and thus a correct polarity.

Two kinds of mesenchyme cells contribute to the limb bud. Cells from the somites produce limb muscle cells, and lateral plate mesoderm produces cartilage. This dual origin of mesoderm provides an extraordinary example of dissociability in development within a module. Chevallier and co-workers and Kieny and Chevallier have shown that if the somites are deleted, the limb mesoderm provided by lateral plate mesoderm will produce a limb with normally formed skeleton and tendons, but lacking muscles.

The limb bud mesoderm induces the overlying ectoderm to form a thick-

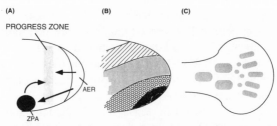

Figure 10.3.   Dynamic submodules in the limb bud. The diagrams from left to right show three states of submodular organization in the developing limb bud. *(A)* The signals that integrate the activities of the apical ectodermal ridge, zone of polarizing activity, and progress zone. *(B)* The five submodules of activity of genes of the HoxD cluster. The smallest region (in black) expresses *HoxD 13, HoxD 12, HoxD 11, HoxD 10,* and *HoxD 9.* The next larger zone (in dark stipple) expresses all except *HoxD 13,* and so forth. *(C)* As development proceeds, domains of ossification appear, and the pattern is reified to form the concrete elements of the limb.

ened ridge, the apical ectodermal ridge, which assumes a vital role in the limb bud. If it is removed, further development of the limb ceases. If the ridge is experimentally broadened by grafting an extra piece of ridge anterior to the existing ridge, the limb bud broadens and forms extra digits. The apical ectodermal ridge can be transplanted to another location on the body, where it leads to formation of an ectopic limb at that site. A veritable transplant industry developed among limb bud aficionados, and this protocol provided the major premolecular route to finding out how limbs develop. It was an amazingly informative approach. Various transplant and deletion experiments have shown that the apical ectodermal ridge and the limb bud mesoderm form submodules that signal each other. A failure of signal production by either of the interacting submodules causes cessation of limb development.

Overall, the early limb bud contains major interacting submodules. Figure 10.3A shows the submodules of a limb bud and the signals that pass between them. The apical ectodermal ridge (AER) interacts with the other submodules that reside in the mesoderm: the zone of polarizing activity (ZPA) and the progress zone. The mid-anterior part of the limb bud contains most of the mass that executes the formation of limb elements, but is apparently a passive player. The ZPA determines the anterior-posterior patterning across the limb bud. That ordering will ultimately be visible morphologically in the identity of the elements of the hand or foot. The progress zone was first hypothesized by Summerbell and co-workers, and separates more proximal regions of the limb bud from distal regions. Cells in the progress zone are proposed to remain in an undifferentiated state until they are stimulated to proliferate and are assigned fates. As the bud matures and the limb forms, cells move out of the progress zone and come to lie behind it, where they then differentiate into limb structures. The first cells to leave the progress zone will thus be involved in making the femur or the humerus. Cells that persist longer in the progress zone acquire more distal identities, with the last cells producing finger or toe elements.

The AER demonstrates how powerfully a submodule can influence the actions of other submodules. It maintains an active progress zone in several ways. As Reiter and Solursh noted, it stimulates cell proliferation in the progress zone, and as Globus and Vethamany-Globus found, it keeps those cells in an undifferentiated state. Summerbell and co-workers showed that the AER further acts to assign progressively more distal positional values to the cells of the progress zone with time. Finally, Vogel and Tickle showed that the AER is critical in maintaining the ZPA.

The ZPA submodule resides in the mesoderm at the posterior edge of the limb bud and specifies anterior versus posterior positional values in limb structures. Its existence was inferred from tissue transplantation experiments such as those done by Saunders and Gasseling in the 1960s, and others done later by Tickle and co-workers and by Summerbell. If a small block of tissue is transplanted from the posterior margin of a limb bud to the anterior edge of a host limb bud, a duplication of digits occurs. The properties of the duplicated digits indicate that the host limb bud has acquired two zones of posterior identity. Recall that a chick wing has been reduced in evolution from the ancestral pattern of a hand with all five digits to a structure consisting of digits 2, 3, and 4. In the wings resulting from transplantation of posterior tissue, the digits formed are 4, 3, 2, 2, 3, 4. This result suggested to Wolpert that a diffusible morphogen was involved. This notion has had such a powerful impact on interpretations of how the ZPA works that we need to consider it a bit. A morphogen is a substance that diffuses through a tissue from a source. Cells along the concentration gradient interpret the positional information. In the case of the limb bud, the higher the concentration, the more posterior the identity of the limb element formed. Under this hypothesis, the chick wing result has a simple interpretation. The experimental limb, instead of having a single posterior-to-anterior gradient of the ZPA morphogen, has two gradients, one posterior and one anterior. Mirror-image gradients yield mirror-image digits.

IDENTITIES OF MOLECULES SIGNALING BETWEEN LIMB BUD MODULES

The striking properties of the ZPA led to a search for "the morphogen," and by the early 1980s it seemed to have been discovered. The application of retinoic acid to the anterior part of a limb bud was found by Tickle and co-workers in 1982 to mimic the effects of a transplanted ZPA. That, of course, could have been merely an attractive artifact. The crucial observation supporting the in situ role of retinoic acid came from the observation of Thaller and Eichele that there is a graded distribution of retinoic acid from posterior to anterior in the limb bud. Unfortunately for a good story, it's been downhill from there. Retinoic acid very likely functions in an important way in the limb bud, but newer observations are not consistent with its being the morphogen. The body of data reviewed by Tabin suggests that the endogenous levels of retinoic acid in the limb bud are inadequate to produce the patterns forced by the exogenous application of retinoic acid by experimenters. Two experiments show why. Wanek and co-workers implanted a bead containing retinoic acid at the anterior edge of a limb bud. They subsequently removed a wedge of tissue adjacent to the bead and implanted it in the anterior edge of a host limb bud. The implant acted as a ZPA even after the exogenously supplied retinoic acid had disappeared. They argued that rather than acting as a morphogen, retinoic acid acted as an inducer of the ZPA. Once induced, the ZPA produced the true morphogen. This conclusion was supported by the experiments of Noji and co-workers, who found that retinoic acid, but not a ZPA, placed at the anterior edge of a limb bud induced the expression of the β-retinoic acid receptor.

The displacement of retinoic acid from its cherished role as the morphogen has not removed the need for a ZPA-generated signal molecule, and indeed, a striking new candidate has appeared. The gene *hedgehog* was originally discovered in *Drosophila,* where it functions to determine the polarity of body segments. The *hedgehog* gene encodes a secreted protein that influences neighboring cells through the actions of other downstream genes that produce signaling molecules. Riddle and co-workers showed that a homologue, which they dubbed *sonic hedgehog* after a cartoon character (following the creative tradition of mutation naming fostered in the *Drosophila* genetics community, but not in the nomenclaturally more stodgy mouse and *C. elegans* communities), exists in the chick and is expressed precisely in the region of the ZPA. If Riddle and co-workers expressed *sonic hedgehog* ectopically in anterior cells of the limb bud, they were able to produce the same mirror-image duplications as produced by a transplanted ZPA. The implication is clear that the sonic hedgehog protein is the morphogen, although that remains to be shown. It may well be that the sonic hedgehog protein acts to induce signal molecules in neighboring cells, and does not itself have a graded distribution, as expected of a morphogen. The distribution of the protein in the limb bud appears to be limited to the region of the ZPA. It thus must act through other signaling proteins to influence the rest of the limb bud. Incidentally, *sonic hedgehog* has been co-opted for several inductive roles. It is expressed in the notochord and in the floor plate of the neural tube. It also apparently functions in chick gastrulation, as it is expressed in Henson's node.

Some of the other molecules that mediate signaling between submodules are known. The AER, the ruler of so much else in the limb bud, produces several signaling molecules. A very potent one with multiple actions is fibroblast growth factor-4. Niswander and co-workers showed that it could, when applied to a limb bud from which the AER had been removed, restore almost all of the functions of the AER in maintaining the progress zone and the ZPA. The resulting wing was almost normal. The remaining minor abnormalities might well have resulted from the absence of other factors. Fallon and co-workers, in a set of experiments similar to those done by Niswander and co-workers, also showed that a related molecule, fibroblast growth factor-2, alone could restore almost all of the functions of a deleted AER. Again, an almost normal limb was formed. There is some dispute between research groups about whether both fibroblast growth factors are present in limb buds. The extent to which either promotes development indicates that the two growth factors play such closely related roles that one can assume most of the roles of the other in experimental situations. The defects in digits that result when the AER is replaced by either growth factor suggest that both could play a role. These details remain to be sorted out. What is remarkable and important is that a single or small number of signaling molecules can have such a profound effect on a neighboring module. This phenomenon has been observed in signaling between other modules in development as well.

The AER also has been found (by Lyons and co-workers and by Niswander and Martin) to produce members of the bone morphogenetic protein family. Proteins of this family were discovered through their effects in promoting condensation of cartilage and bone. The early presence and localization of BMP-2 and BMP-4 in the limb bud implicate them in signaling between submodules, although their roles are not yet understood.

Signaling between the AER and the progress zone may be mediated by two other genes related to the *Drosophila msh* class of homeotic genes involved in mesoderm differentiation. These genes, *msx-1* and *msx-2,* have been studied by Coelho and co-workers and by Robert and co-workers. The *msx-1* gene is induced in the mesenchyme of the progress zone by a signal from the AER. It appears to be required for cell proliferation in the mesenchyme. In the *limbless* mutation, an AER is lacking and proper limb outgrowth fails to occur. Expression of *msx-1* is severely impaired in the absence of the signaling submodule, but is restored along with limb outgrowth if a genetically normal AER is grafted onto a *limbless* mutant limb bud. In the *talpid* mutant, a very wide AER is formed, and a correspondingly broad domain of expression of *msx-1* is elicited. The AER expresses another member of the *msh* family, *msx-2*. It appears to be a component of AER signaling to the progress zone. As a nuclear transcription factor, it would not itself be a signaling molecule, but could act upstream to regulate production of signaling molecules such as the fibroblast growth factor secreted by the AER.

GENES AND THE EXECUTION OF PATTERN

As development progresses, the limb bud module behind the progress zone assembles another set of submodules, the zones of cartilage condensation that will lead to the bones of the arm or leg. As discussed by Oster and co-workers and by Shubin and Alberch, the developing limb is the product of global organizers and local interactions. These interactions produce local patterns that can be recognized as homologous and conserved over several hundreds of millions of years through the evolution of the tetrapods. Cartilage condensations progress from the most proximal element in a distal direction. Condensations obey very definite rules (see fig. 10.3C). The first condensation is a singular one. It is followed by a binary bifurcation to produce the next most distal elements—in the arm, the radius and ulna. A series of segmental bifurcations follow, which produce the separate elements of the hand and digits. These condensations are actual physical entities. They are the result of cellular behaviors in response to local molecular environments, including the presence of fibronectin, and they manifest the localized expression of extracellular matrix proteins. The patterns of these later submodules include both the structures of the individual elements and the spatial relations between them. As physical objects, they reify the molecular patterns generated in the limb bud. These patterns must be specified by domains of localized gene action earlier in limb bud development. Thus, submodules are connected during development of the limb bud by signal molecules, and as a result, exhibit localized patterns of gene expression that translate local positional information into a pattern of transcription that directly determines local submodule identity.

There are three axes along which positional information is passed in limb buds: anterior-posterior, proximal-distal, and dorso-ventral. The molecular architecture of the dorso-ventral axis is poorly understood, but Parr and McMahon recently discovered that expression of the *Drosophila wingless* gene homologue *wnt-7a* regulates dorso-ventral development of the mouse limb. Gene knockout of *wnt-7a* causes very peculiar-looking transformations that include foot pads on the dorsal side of the foot. (Loss of expression of *wnt-7a* also affects digit differentiation along the proximal-distal axis, suggesting an interaction between axial control systems.) The nature of the positional information that specifies the pattern of the limb elements along the anterior-posterior and the proximal-distal axes is suggested by recent studies of Hox gene expression in the limb bud. Two Hox gene clusters have been shown to have different and complementary patterns of expression. Yokouchi and co-workers showed that genes of the HoxA cluster are expressed in a proximal-distal order, effectively partitioning the limb bud into zones from tip to base. Dollé and co-workers observed a second pattern in the expression of genes of the HoxD cluster. The most 5′ gene of that cluster is the one expressed most posteriorly in the animal. That is in keeping with the general order of expression of genes in the Hox clusters reflecting their physical order in each cluster (see fig. 6.2). The most 3′ member of any cluster is most anterior in its expression boundary along the developing animal axis. Dollé and co-workers found that the HoxD expression pattern in the limb bud included the five genes located most 5′-ward in the complex. All five were expressed at the posterior edge of the limb bud in the vicinity of the ZPA, and the most 5′ gene, *HoxD 13*, was expressed there only (see fig. 10.3B). The rest of the genes were expressed more anteriorly in a nested fashion, with *HoxD 12* occupying a somewhat larger zone than *HoxD 13*, and so forth, until only *HoxD 9* occupied the entire limb bud. Thus, five distinct anterior-posterior zones of HoxD gene expression occurred across the limb bud. The combination of HoxA and HoxD domains of expression results in a Cartesian grid of positional information. The question is whether the localization of Hox gene expression informs cells in each region of the identities to be assumed by cartilage condensations. Three lines of evidence strongly support that idea. First, experiments conducted by Izpisua-Belmonte and co-workers and by Nohno and co-workers have shown that duplication of digits caused by application of retinoic acid to the anterior edge of a limb bud is accompanied by ectopic expression of the *HoxD 13* gene on the anterior side of the limb bud. Second, in a different study, Izpisua-Belmonte and co-workers have shown that in the *talpid* mutant of mice, the absence of anterior-posterior polarity in the limb bud correlates with mis-expression of the terminal *HoxD 13* gene, which is expressed all across the mutant limb bud. Finally, in their study of *sonic hedgehog* as the putative ZPA morphogen, Riddle and co-workers have shown that mis-expression of *sonic hedgehog* at the anterior edge of the limb bud causes both a duplication of digits and the ectopic expression of the *HoxD 13* gene. It would appear that the sonic hedgehog morphogen generates a pattern of five zones of distinct HoxD gene expression, which in turn define five modules in which distinct morphogenetic patterns of chondrogenesis occur. It is, of course, still a substantial step from Hox gene patterns to concrete patterns of cell behavior and ultimately particular bones.

The execution of the chondrogenic pattern requires the expression of further genes that will refine the pattern and actually determine the shape of the bones to be formed, as well as generate the requisite biochemistry. Interesting candidates are beginning to appear. Storm and co-workers have identified three genes related to the bone morphogenetic protein family that are expressed in the limb bud. Mutations in one of these genes, *GDF-5*, correspond to the *brachypodism* mutation in mice. This mutation causes a reduction in foot length by fusing some skeletal elements. Among its effects on skeletal development in the foot is a reduction in the ability of isolated mesenchyme to form cartilage condensations. These genes are expressed in wildtype animals in regions of distal precartilage condensations. Kingsley has suggested that the normal function of bone morphogenetic proteins is the patterning and formation of bones and other structures during development. The patterns of expression of the eleven or so known members of the family are complex. The roles they play in patterning appear to be varied and specific. It is especially striking that whereas the *brachypodism* mutation of *GDF-5* affects limb development, but does not affect the development of the axial skeleton, the *short-ear* mutation of *BMP-5* has contrary effects. Null mutations in *short-ear* alter the size and morphology of the ears, sternum, ribs, and vertebral processes. Null mutations in *brachypodism* reduce bone lengths in forelimbs and hindlimbs.

COMMONALITIES IN LIMB DEVELOPMENT ACROSS PHYLA

It would be a long stretch to suggest that *Drosophila* legs or wings are homologous to the limbs of vertebrates. Arthropods appear in the Cambrian fossil record complete with elaborate jointed legs. Evidence described in the next section of this chapter supports a later and independent origin for vertebrate appendages from lateral fin folds originally evolved as stabilizers in swimming. Insect wings, as I'll discuss in chapter 12, apparently arose in Paleozoic insects by evolution from legs. In *Drosophila* development the processes that shape legs and wings from imaginal discs are similar, and quite different from the morphogenetic processes that build vertebrate limbs. It is thus rather surprising that both genes and principles of appendage development are shared between the two phyla.

In *Drosophila,* both legs and wings arise morphologically by telescoping from imaginal discs that are divided into anterior and posterior compartments. These compartments correspond to submodules as defined above. These submodules interact with each other during appendage development. A large body of work has established that, as in insect body segments, the posterior compartment of a developing appendage expresses the homeobox-containing gene *engrailed.* This gene is involved in posterior compartment specification and signaling. The boundaries between the anterior and posterior compartments may thus have organizing properties. The operation of the genetic interaction has been established by Basler and Struhl, who showed that cells of the posterior compartments of wings or legs express and secrete the hedgehog protein. This protein is "read" by cells of the boundary region between anterior and posterior, and causes them to synthesize and secrete two other signaling proteins.

The story is complicated by the fact that wing and leg discs have dorsal and ventral compartments as well as anterior and posterior ones. In both the dorsal and ventral compartments of the wing disc, the *hedgehog* signal causes the *decapentaplegic* gene to be expressed in the anterior-posterior boundary regions. Its product is a member of the transforming growth factor-β family. The same events occur in the dorsal compartment of the leg disc. In the ventral compartment, however, a different signal molecule, the product of the *wingless* gene, is expressed. Although the pattern of gene expression in the leg is more complex, the decapentaplegic and wingless proteins appear to play the same role. Both of these boundary molecules interact with the anterior and posterior compartments and are required for organizing both compartments. Ectopic expression in the anterior compartments causes duplication of posterior compartment structures. Failure to express these genes results in failure of the cells of either compartment to differentiate.

The parallels between genetic controls in the vertebrate limb and in *Drosophila* appendage development are notable. Leg fields in both *Drosophila* and vertebrates express the *Distal-less* gene. The role of the hedgehog protein as a primary signal from posterior to anterior submodules in both phyla suggests three interesting things. The first is that appendage development depends upon the interactions of submodules within both of these quite different appendage primordia. The second is that the genes for the initial signaling system may have been independently co-opted for appendage development in both phyla. However, it may be more parsimonious to infer the contrary, that is, that appendages have a deep underlying and (prior to the findings of developmental genetics) completely unexpected homology. Appendages may have arisen from some primitive body projections in the common ancestor of vertebrates and arthropods, and these may have primitively required expression of *Distal-less, hedgehog,* and other shared regulatory genes. A look at the appendages (parapoda) of polychaete annelids for *hedgehog* expression might be a very enlightening way of testing this idea. Finally, *hedgehog* may not itself be a morphogen in the vertebrate limb bud, but may control the actions of downstream signaling genes. If this is the case, the sonic hedgehog protein will not diffuse out of the posterior compartment. Instead, it will signal more anterior cells to produce a member of the *wnt* family (the vertebrate homologues of *wingless*) and/or a member of the transforming growth factor-β family *(BMP-2 versus its insect homologue decapentaplegic).*

LIMBS AS SERIAL HOMOLOGUES

Vertebrate forelimbs and hindlimbs are clearly serial homologues, but just what that means in terms of vertebrate evolutionary history has been a matter of real confusion to evolutionary biologists over the past century. The solution may be simple, if unexpected. As reviewed by Tabin, the paired appendages of vertebrates appear to have evolved from a pair of continuous fin folds along the body that acted as a stabilizer in swimming in primitive jawless "fishes." (This hypothesis is not established by any means. As Coates has noted, other ideas, such as the derivation of fins from gill arches, have been put forward, and there is no strong evidence for an ancestral fin fold.) The multiple Hox gene clusters of vertebrates had their origins prior to the evolution of fins or limbs. Studies by Holland and co-workers and by Pendleton and co-workers have shown that there are two Hox clusters in cephalochordates and three in agnathans. "Spare" Hox genes would thus have been available for co-option in fin patterning in primitive vertebrates. The Hox gene codes for the regions of the body axis from which forelimb buds and hindlimb buds arise in development are quite different. However, the forelimb bud has the same Hox code as the hindlimb bud, suggesting to Tabin that the forelimb bud might literally have evolved from the hindlimb bud. Tabin and Laufer have proposed an explicit model of how this might have occurred as a transformation via a homeotic mutation in which the hind appendage pattern replaced the forelimb pattern: one module, complete with its characteristic Hox code, replaced another. Ahlberg has suggested that something like this may have occurred in the coelacanth. The coelacanth's single posterior dorsal fin and its single anal fin are not built, as expected, like its anterior dorsal fin, but instead like the paired forefins and hind fins. This observation is consistent with a homeotic transformation of some non-paired fin buds to a paired fin developmental program.

It's not known when in evolutionary history the hypothetical homeotic event that gave rise to the vertebrate forelimb occurred. There is some fossil evidence suggesting that it may have happened in association with the origin of tetrapods. Recent discoveries of very early tetrapods from the late Devonian by Ahlberg and by Coates and Clack have been surprising in showing that the earliest tetrapods were still aquatic animals. They had functional internal gills and less developed terrestrial legs than one would have expected. New fossils from Scotland indicate that these old-timers may have had a limblike hind appendage combined with a finlike fore-appendage. The next step of evolving a true forelimb may then have occurred by homeotic transformation of the forefin bud.

If the homeotic transformation event postulated by Tabin and Laufer took place in the tetrapod lineage, we might expect to find a different and more anterior Hox gene expression pattern in forefin buds than in hind. Unfortunately, that is not the case. A study of Hox expression in the zebrafish fin bud by Sordino and co-workers shows that if a homeotic transformation occurred, it did not involve Hox patterning. The pectoral fin bud expresses the same HoxD gene set as the pelvic fin and as tetrapod limb buds. The pattern of expression of HoxD genes in the fin bud casts an interesting light on the origins of tetrapod limb elements. The fin of lobe-fins, such as lungfishes, has a linear axis of condensation of skeletal elements, the metaptery-gial axis. Shubin and Alberch have shown that the metapterygial axis of the tetrapod limb is linear through the leg, but makes a sharp turn from the posterior to the anterior side of the limb in condensation of the elements of fingers or toes. In 1991, Coates pointed out that in mouse limb development the boundary of HoxD expression follows the metapterygial axis. Sordino and co-workers have found that the HoxD boundary does not make that turn in zebrafish fin development. This observation suggests that digits and their underlying genic patterning are a novel feature in tetrapod evolution. Since teleosts like the zebrafish are highly modified, it would be desirable to observe the HoxD expression pattern in the fin development of a more primitive fish.

WHY NOT EIGHT TOES?

Tetrapods have varied the numbers of digits borne on their feet. In many cases, as in horses with their single-digit hooves, reduction has been accom-panied by extensive specialization. In other cases, a primitive hand or foot with five digits has been retained, a trait we share with many salamanders. Five digits had long been taken as the primitive condition for tetrapods, because all known fossil and living tetrapods had either five digits or feet in which such a five-digit pattern had been modified. However, in 1990, Coates and Clack published the surprising discovery that the oldest amphibians didn't have the canonical five fingers at all: *Acanthostega* had eight (fig. 10.4). This finding had two important consequences. It showed once again that inference of primitive traits based solely on taxa at the ends of branches can be mis-leading. We need the fossil record. Second, it showed that all models of limb origins based on the sign of five were in need of no small revision.

How should we interpret the eight digits of *Acanthostega*? Tabin suggested that this ancient polydactylous foot did not have eight different digital identi-ties. Rather, he proposed that *Acanthostega*, like living tetrapods, had only five posterior-to-anterior positions specified by the HoxD code. By this inter-pretation, the "extra" digits represented duplications of some of the five digits present in the modern tetrapod foot. This is the situation observed in polydactylous mutants. For example, humans with a sixth toe don't have a new digit, but a duplicate big toe. Tabin suggested that the extra digits were selected to provide *Acanthostega* with a wide foot useful in swimming. As tetrapods became more terrestrial, the extra digits became superfluous and were lost, but the same Hox code was retained. I'm not convinced that the digit pattern of the *Acanthostega* foot represents digit duplications within a five-zone Hox code. It may well be that the primitive limb bud had an eight-zone code involving other Hox genes no longer used in specifying digit patterning.

One way that *Acanthostega* might have generated those extra digits is

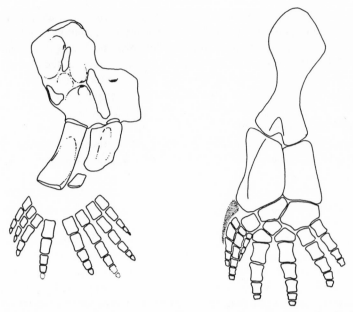

Figure 10.4.   The earliest known tetrapod limbs had more than five digits. The forelimb of *Acanthostega* (left) has eight digits. The hindlimb of *Ichthyostega* (right) has seven digits. Both are ancestral tetrapods from the upper Devonian. (Reprinted with permission of *Nature*. M. L. Coates and J. A. Clack. 1990. *Nature* 347:66–69. Copyright 1990, Macmillan Magazines Limited.)

through an anterior duplication of the ZPA in the primitive limb bud. Such a duplicated ZPA would have yielded a polydactylous foot like that observed in the old-timers. However, in the zebrafish, which represents a bony fish outgroup, Krauss and co-workers have shown that the fin bud has only a single posterior region in which *hedgehog* is expressed. The postulated anterior ZPA would have meant that lobe-fins had a differently organized fin bud than teleosts. The fossils can't tell us much about that, but living representatives of primitive outgroups of the rhipidistian radiation, such as the sturgeon, might, and would be well worth examining.

MODULES SET UP BY GENE SWITCHES

One of the puzzles we confronted in chapter 7 was the ease with which a set of mechanisms underlying early development could be replaced by another in evolution. These alterations in early development are striking because quite distinct early developmental pathways can result in the production of the same cell types, structures, or even the final developmental result. Direct- and indirect-developing sea urchins illustrate all of these features. Why should it be possible for one mode of early development to be substituted for another in evolution, as occurred in *Heliocidaris erythrogramma?* The answer would seem to lie in the simplicity of the signals and switch gene commands that initiate later modules. We should not think of the early stages of development as merely direct linear precursors of subsequent stages. Those early stages may only need to generate a system capable of producing, at the correct time and place, the simple signals necessary to initiate subsequent differentiative events that give rise to coherent later modules. Somehow, the attainability of later modules in development must be separable from the particular mechanisms utilized earlier. This capability appears to lie in a crucial attribute of the way modules are generated. Modules are genetically defined, and they exhibit complex states of gene action. However, as I'll show in a moment, the genetic states that define particular modules are often reachable through the action of single switch genes. That fact means that alternative earlier pathways need only result in the correct spatial and temporal expression of the requisite switch gene(s) to produce correct later modules.

Single switch gene determination of developmental modules occurs widely and at various points in development. I don't propose to try to detail all of developmental genetics to show that genetic transitions into modules involve a few, or even a single, critical genes. Segmental identification and the differentiation of segment-specific modular structures in the insect body axis, in the vertebrate brain, and in the specification of individual organ primordia all illustrate the point that simple signals and genes that make binary choices are crucial. A few examples will make the point. In *Drosophila* early development, specification of the posterior end of the embryo depends on a marvelous cascade of genes that produce single determinative signals. As reviewed by Lehmann and Rongo, there is a set of genes, *cappuccino, spire, mago nashi,* and *staufen,* that are involved in the transport of the maternal mRNA produced by a crucial control gene, *oskar,* to the posterior pole of the egg. The *oskar* gene has been shown by Ephrussi and Lehmann to be responsible for germ cell formation. In a definitive experiment, they showed that by replacing the 3′ posterior localization signal of the oskar mRNA with the 3′ anterior localization signal of *bicoid,* they could obtain anterior expression of *oskar.* When they did, germ cell precursors were formed ectopically at the anterior end of the resulting embryo. *Oskar* controls the action of a small number of genes—*vasa, tudor,* and finally *germ cell-less,* which operate immediately downstream—to determine the germ plasm. The gene *germ cell-less* has been shown by Jongens and co-workers to produce a nuclear protein that is required to execute germ cell development.

Another crucial maternal mRNA is also localized to the posterior end of the *Drosophila* egg. The product of the *nanos* gene has been shown by Wang and Lehmann and by Gavis and Lehmann to be specifically localized by a 3' localization tag on the mRNA. Ectopic expression of the gene product in the anterior end of a *Drosophila* embryo causes formation of a second abdomen. Thus, two complex developmental modules in the *Drosophila* embryo are initiated by single-gene inputs.

The second example of a simple input that determines differentiation of a complex module comes from the demonstration by Roberts and co-workers that the knockout of the *Hox-11* gene in mice causes the spleen to fail to differentiate. (This gene is not a part of the Hox cluster.) The action of a single selector gene appears here to cause an organ rudiment to form and to differentiate into a specific modular identity.

The third example is analogous to the preceding one. The gene *eyeless* (a homologue of the vertebrate gene *Pax-6*) has been shown by Halder and co-workers to be a master regulatory gene for eye development in *Drosophila*. Mutations in this gene cause eye loss, and it encodes a transcription factor. These observations are consistent with *eyeless* having a regulatory role. However, the observation of Halder and co-workers that ectopic expression of the gene causes the formation of complete extra eyes is a dramatic demonstration of its role in establishing a module. *Eyeless* apparently lies at the top of a gene cascade that may involve as many as 2,200 downstream genes that execute the pattern of differentiation of the eye.

A similar case of an apparent master regulator that sets up a cascade of events leading to a large and complex module has been described in mice by Shawlot and Behringer. The dorsal lip of the amphibian gastrula was discovered to be the organizer for head development through transplantation experiments done by Mangold and Spemann early in this century. Transplanted dorsal lips induced the production of a second head in hosts. The homologue of this organizer in chicks is Henson's node, and its equivalent is the node in mammal embryos. Shawlot and Behringer investigated the role of a homeobox-containing gene, *Lim1*, by examining the development of mouse embryos homozygous for an incapacitated copy of the gene. In normal embryos *Lim1* is expressed in mesoderm at the anterior end of the primitive streak, analogously to its expression in the dorsal lip and anterior-dorsal migrating mesoderm. In *Lim1*⁻ mice the node fails to develop, and so does the entire head anterior to the otic (ear) vesicle. The trunk of the body is almost normal, as expected since its development is induced by a trunk organizer that appears later in development than the head organizer. Some potentially downstream regulatory genes normally expressed in the forebrain and midbrain (*Otx2* and *engrailed*) fail to be expressed, but *Krox20*, whose transcription marks rhombomere 3 of the hindbrain in normal mice, is expressed. Thus the domain of *Lim1* regulation is precisely bounded.

A final example is derived from the extremely mosaic development of

C. elegans. The tails of male *C. elegans* include a set of nine pairs of sense organs called rays. Each ray is built from three cells, the ends of two neurons and a support cell, and has a specific morphological and functional identity. Chow and Emmons have shown that the identity of rays is determined by levels of expression of two Hox gene homologues, *mab-5* and *egl-5*. Specification of rays results from cell-autonomous expression of these genes. Since the cells arise from stereotypic cell lineages, the precise identities of the cellular modules are generated by the gene dosages of just two transcription factors that act as selector genes.

Observations like these suggest that although modules may have complex internal gene expression states, equivalent to those suggested by Kauffman for cell types, expression of only a single gene may be required to direct them to particular attractors. The product of such a selector gene (the originally recognized selector genes were the homeotic genes of *Drosophila*) commits cells expressing the selector gene to a state defining the formation of some specific module. The consequence for the evolution of development is important. If only simple signals are required to trigger complex modules, small changes in the expression of selector genes can have major consequences for development. And, perhaps paradoxically, because defined modules result from selector gene action, changes in selector genes will likely yield a modified ontogeny that features the same modules as its ancestor, with changes in timing, size, number, or location of such modules being the most frequent outcomes. Substantial evolutionary modification can take place without reinventing the basic components, cell types, tissues, and organs of the animal body. In most cases, changes remain within the parameters of the ancestral body plan.

SUMMING UP

Earlier in this chapter I suggested some features of modular organization and internal evolutionary processes that operate to constrain the effects of external selection, and I've exemplified the organization and evolutionary behavior of modules with a deeper focus on the limb bud. I'd like to summarize the skeleton of the argument. The scenario for evolution involving both external and internal processes goes something like this: External selection may favor any number of phenotypic traits. Larger size, more effective propulsion, or a protective covering might all be desirable features in the external world. Selection can act only on the phenotypic variation presented to it as a result of ontogeny. The organization of the ontogenetic process is thus of pivotal importance. Developing organisms are modular in organization. Modules span a hierarchy from molecules to organ primordia and body segments. External selection acting upon development will be presented with the results of internal evolutionary processes acting on existing modular features. All the crucial features of modules—state, number, location, and interactions with other modules—are potentially changeable. Changes in these properties result in a system that is still modular, but presents a modified set of features to external selection. The changes in modular features are not achieved directly by external selection, nor are they random.

Alterations in phenotypes presented to external selection will result from internal sorting of genetic variation. Thus, suppose that greater body flexibility were favored by external selection. Greater flexibility might be attained by making the body longer. Such a change might well depend on the kind of modularity available in the organization of the body axis. One of the most common kinds of evolutionary change seen is in numbers of similar modules. Greater length in annelids or primitive arthropods was attained by addition of segmental units. Vertebrates have changed numbers of somites, vertebrae, and ribs. Starfishes have varying numbers of arms. In all of these, an existing module has undergone duplication. As the new module can't occupy the same position as the original module, dissociation from the prior ordering in a system of other modules occurs to some extent. Dissociation in timing of module appearance is equally likely, and can itself modify relationships of size or interaction. New serial modules can diverge in organization and can be co-opted for new functions.

We should, given the view of internal order and internal evolutionary processes proposed here, expect that regulatory genes will be evolutionarily conserved. Their duplication, divergence, and co-option for new roles in the patterning of development is the most parsimonious means available for introducing changes in development. Most of what goes on in the development of a new descendant species will utilize the same standard parts as the parent species. Novel form will arise mostly from the modification of existing modules in development. The differences will arise from the effects of modified co-opted regulatory genes on other hierarchical levels of modular organization. Because, as discussed by Müller, substantial threshold effects can result from relatively minor alterations in developmental interactions, relatively subtle genetic changes may lead to striking developmental modifications. The paradoxical story that is emerging from the discoveries of phylogenetically widespread regulatory gene families is that regulatory genes that are more or less similar control quite distinct ontogenies. The wide range of possibilities in gene utilization allow this to occur. Conserved receptor-ligand systems can be linked to quite different second messenger systems. Conserved transcription factors can produce new patterns of gene action if transcription factor binding sites of target genes have changed.

At present, studies of the evolution of developmental regulatory systems are being conducted by means of comparisons across large phylogenetic gulfs. It is important to find the big picture by surveying for similarities between flies, mice, and nematodes. However, to understand the nuts and bolts of how developmental changes come about, we will have to make comparisons of regulatory genes and their actions between closely related species so that small step changes can be dissected out. Given appropriate means for the experimental expression of modified regulatory genes, as is now done in a rudimentary way in *Drosophila,* it will be possible to rewire elements of the regulatory circuitry of development to see whether the consequences resemble those seen in evolution in the related species. An experimental evolutionary biology in which we test hypotheses about changes in developmental regulatory systems by means of experimental manipulation of transgenic animals is not far off.

11

Opportunistic Genomes

Evolution is, as an engineer, an opportunist, not a perfectionist.

Stanislaw Lem, *His Master's Voice*

A TERMITE HISTORY OF THE WORLD

Our perceptions of the world are filtered through our cultural experiences. There are polygamous cultures, cultures that eat raw fish on rice for breakfast, and cultures able to watch entire games of cricket. Fifty years ago the linguist Benjamin Lee Whorf proposed, based on his study of Hopi, that the language we speak provides more than merely a grammar; that as much as culture, it determines the way we perceive the world and interpret external events. Well over five thousand different human languages have generated a richness of human creativity and expression and have maintained the subtle and not so subtle patterns of thought that distinguish human cultures. All this is achieved with a common genome. Languages, cultures, and individual behavioral specialization are means by which our genome allows immensely wide human developmental variation. Whatever the biases introduced by culture and language, how much more must our perceptions and most fundamental ability to comprehend reality be determined by the features of our biology and evolutionary history. We are vertebrates, mammals, primates, descendants of a long line of hunter-gatherers. It is difficult to imagine how intelligent beings arising from a different evolutionary history and possessing different sensory systems and a different central nervous system might comprehend the world. It has been attempted in literature with varying degrees of success. I like best Stefan Themerson's *Professor Mmaa's Lecture*.

Professor Mmaa is a termite—highly erudite, but a termite all the same. His chief intellectual concern and the subject of his lecture is the nature of the "homo." Termites cannot bear the intense, painful light of day or the dryness of the open air. Thus, the homo is most intensively and successfully studied by termite investigators in the part of his life cycle in which he is in a horizontal, subterranean, and "microbiologically socialized" form. When termite scientists finally succeed in capturing a living homo, a crazy train of events beyond their imagining is set in motion, which ultimately results in

362

catastrophe to termite civilization. After all, even with application of the best foresmell, not all consequences of meddling with an irrational and wrathful nature can be predicted. Meanwhile, termite life, culture, and politics swirl violently around Professor Mmaa. The magnificent Smelofactory Orchestra plays its unparalleled olfactory symphonies amid debates as to whether modern smells should be performed on public occasions. In the University, "Growing old evenly from one end and regenerating from the other, the great biochemical clock—the pride of the University, for during all its centuries of existence it had never been slow or fast—had already chimed its old-world tunes with the discreet scents it had accumulated during the past hour." This is clearly a superior culture, perhaps superior to ours, and certainly different in some interesting ways.

Unlike the clumsy homo, the termites need not build specialized tools. They produce specialized individuals through polymorphic technology: armies of thick-skulled, deaf and blind halberdiers with huge jaws, factories of coarse food processor termites, libraries of refined book termites who recite their remembered texts, schools of termite professors and philosophers, domestic wet nurses to feed them, and nuptial flights of reproductives, the future kings and queens. Crises in polymorphic technology do arise to trouble domestic tranquillity. "It was by no means an infrequent phenomenon for a military termite to possess uneliminated vestiges of the instinct to run away; for a statesman to be equipped with a gynecologist's syringe; not an hour passed without the Queendom being provided with a teacher furnished with the specially restricted intellectual horizon of an ordinary constable, not a quarter of an hour without the arrival of an abdomener or a wet nurse with not completely undeveloped optics."

In this chapter, I want to explore some manifestations of the flexibility, opportunism, and fluidity of animal genomes that facilitate evolution. I'll consider four examples, alluded to in the tale of Professor Mmaa: caste determination, sex determination, evolution of eyes and eye lens proteins, and finally, the extent to which evolution is reversible. These rather disparate topics are all indicators of the evolutionary flexibility allowed by metazoan genomes. These important phenomena confront us with evidence of the profound alterations in genetic mechanisms underlying common developmental outcomes, and with a view of the stability and loss of developmental information. Genomic fluidity and opportunism have about as many faces as an old politician wooing the voters. Some instances are manifested by obvious phenotypic effects, but remarkable changes in gene expression can occur silently. As an example, consider the expression of different actins among the sea urchin pluteii shown in figure 7.8. Extensive alterations in gene expression have had no perceptible effect on development. These and many other silent changes in gene regulation provide latent raw material for evolution.

SEX AND "POLYMORPHIC TECHNOLOGY"

In humans, as in termites and other animals that reproduce sexually, a certain polymorphic technology lies at the heart of things. Sexual differentiation of the somatic body is a crucial part of development, and indeed, produces quite different male and female beings within a species. Sexual differentiation can be under chromosomal, environmental, or social control.

Although not endowed with the intellect and conscious action of Professor Mmaa and his cohorts, real termite societies contain just about as bizarre and complex a set of characters. The castes of termites are summarized by E. O. Wilson in his book *The Insect Societies*. One caste is the larva, an immature form without wing buds. A second caste, the nymph, is derived from the larva and develops wing buds. A nymph can go on through successive molts to develop wings, eyes, and functional gonads. Termites are traditionally divided into more generalized lower termites and more specialized higher termites. Higher termite species possess a distinct worker caste, whereas in more primitive species worker functions are carried out by nymphs or pseudergates, what Wilson refers to as "child labor." Pseudergates are a sort of worker caste derived either from larvae or from regression from nymphal stages through molts that reduce the wing pads. There are two other castes with highly specialized morphologies, soldiers and reproductives. Soldiers have a purely defensive role in the colony, and do not work, reproduce, or even feed themselves. They have specialized defense structures, such as large jaws, heads modified as plugs to close off openings, or nozzles on the head from which they can discharge a nasty defensive liquid.

Primary reproductives are the colony-founding males and females derived from winged adults. Secondary reproductives differentiate from nymphs upon removal of the primary reproductives. Tertiary reproductives are more larvalike supplementary reproductives. The relationship among the castes in lower termites is diagrammed in figure 11.1. Noirot has suggested that the ancestral developmental pathway is from larva to nymph to winged primary reproductive. Development of the other castes has been superimposed on that primitive pathway. Caste differentiation in higher termites is similar to that in lower termites, but is more restricted as to which larval molts can differentiate into specialized castes, and as to which sexes can become soldiers. In lower termites both sexes can become soldiers; in various higher termite species only one sex can do so.

Sex determination in termites is chromosomal, although, as discovered by Syren and Luykx, it has some decidedly odd features. The males of many primitive termites have sex-linked translocation heterozygosities that result in long chains or rings of chromosomes in meiosis. For example, the Florida termite, *Incisitermes schwarzi,* has a diploid number of 32 chromosomes. Of

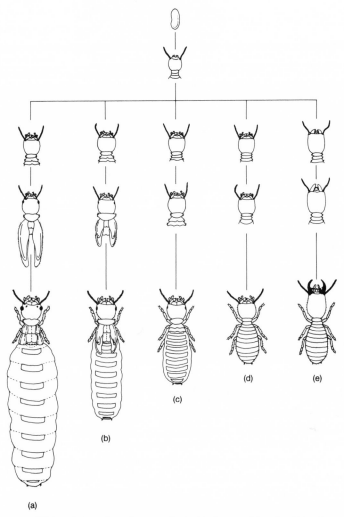

Figure 11.1.   Differentiation of termite castes in the black-mound termite. *(A)*, primary queen; *(B)*, secondary queen; *(C)*, tertiary queen; *(D)*, worker; *(E)*, soldier. (From Raff and Kaufman 1983; drawn by E. C. Raff. Reprinted with permission of Indiana University Press.)

these, fully half act as sex chromosomes, 8 as X and 8 as Y in males, and 16 as X in females. Not all of the sex-linked chromosomes necessarily function in sex determination per se, but the sex linkage of this many chromosomes has unusual consequences for the population structure of a colony. Luykx noted that members of a colony are more closely related to their siblings of the same sex than they would be to their own offspring, if they had any. Most colony members are sexually undifferentiated and free of the cares of passing on their own genes. Whether such relationships increase the altruism of nonreproductive colony members and thus provide the selective basis for the bizarre chromosomal machinery is not clear. One has to be skeptical of that idea, because not all termite species manage their chromosomes in this way. Some carry multiple sex chromosomes, but other species do not, and there is even variation among populations within a species.

In termites, sex determination is separate from differentiation into a functional reproductive. Caste determination governs sexual differentiation regardless of chromosomally determined sex. There were once suggestions that caste itself might be genetically determined, but early experiments, particularly the work of Light in the 1940s, showed that apparently any larva or pseudergate can transform into a secondary reproductive or a soldier. The discovery of intergrades between castes indicated that some epigenetic mechanism was at work. The nature of that control emerged from a series of ingenious and deceptively simple experiments by Marten Lüscher in the 1950s. Of particular significance was his elegant experiment that showed how reproductives inhibit nymphs or pseudergates from becoming reproductives themselves. That such inhibition occurs was clear from the simple observation of "workers" maintained in a little artificial colony. In the absence of reproductives, some workers began to differentiate into secondary reproductives. If those were removed, still others followed suit. Lüscher showed that workers get their caste orders directly from the reproductives, and he also demonstrated how the information is transmitted. Two little colonies of workers were separated by a mesh. Reproductives were tied into the mesh so that their abdomens faced one colony and their heads the other colony. The colony in contact with the head end exhibited no repressive effect. The colony facing the abdominal end of the reproductives, however, showed inhibition of differentiation of secondary reproductives. The inhibition was most effective when male and female reproductives were both present in the mesh.

Like many significant phenomena in biology, this is not an intellectual affair, but a far more earthy one: Termites engage in anal feeding. Some termites, such as the soldiers, are unable to feed themselves. This behavior is also the way termites pass on from adult to larva the elaborate symbiotic flagellate protists that actually digest the cellulose the termites eat. Termites lack any cellulase of their own, and die of starvation if cured of their flagellate "infections." The evolutionary origin of anal feeding behavior is probably causally related to the origin of pheromones in termite social evolution (the original work on this was done by Cleveland in the 1920s and 1930s, and is summarized by Wilson). Primitive sociality may have arisen from the need to transfer protists, which are very sensitive to exposure to oxygen, from the gut of an adult directly to the larvae.

Lüscher's experiment indicated that a rich set of social interactions exists, in that contact between workers and the anal end of reproductives is critical for the transmission of some chemical substance that controls differentiation. Chemical signaling agents by which animals influence the behavior or differentiation of other members of their species (or even of other species) are called pheromones, and are very widespread in nature. Lüscher's work showed that both male and female reproductives produce a pheromone that represses reproductive differentiation, and that soldiers likewise produce a pheromone that prevents soldier differentiation. The presence of soldiers prevents workers from differentiating into soldiers; once soldiers are removed, workers start to become pre-soldiers. In higher termites, Bordereau and Han showed that the control of soldier differentiation may also be stimulated by the presence of reproductives.

Not all termite pheromones are passed on by anal feeding. Soldiers of the higher termite *Nasutitermes* have foreheads developed into elaborate tubular snouts from which they can squirt noxious chemicals at enemies. Lefeuve and Bordereau showed that a pheromone that inhibits soldier differentiation is produced in the frontal gland of the head in *Nasutitermes*. The frontal gland produces defensive and alarm substances as well as the differentiation pheromone; this is no little sophistication in a policeman.

Pheromones allow the social interactions among individuals to impinge on the machinery of development. Although termite pheromones have powerful and specific effects, they do not directly determine the course of differentiation. Instead, they control caste differentiation by interacting with the juvenile hormone system, which is involved in the maturation and development of all insects. Juvenile hormone is involved in decisions made throughout insect development; its functions have been reviewed by Nijhout and Wheeler and by Riddiford. It is best known as the hormone that, through its presence in immature stages, maintains the larval state during successive larval molts. Adult development and metamorphosis ensue upon its disappearance. One prospective mode of action for such a hormone might be in binding to receptors that can act as transcription factors expressed in particular stages or tissues. The juvenile hormone receptor has not yet been identified, and it is not known just how juvenile hormone works. Juvenile hormone is a long aromatic molecule, a sesquiterpenoid, and may have a receptor related to the steroid and thyroid hormone receptor superfamily described by Evans.

Lüscher first showed that implants of the corpora allata, the glands that produce juvenile hormone, into workers of lower termites caused some of the recipients to develop into soldiers. Natural insect hormones were difficult to obtain, but as chemically synthesized juvenile hormone analogues became available, they opened up a whole new world of termite pharmacology, with its own phantasmagorical effects. Entomologists without skills as microsurgeons could apply juvenile hormone analogues directly to termite larvae or nymphs. Wanyonyi noted two effects of direct application of juvenile hormone. If juvenile hormone levels were low, nymphs differentiated into reproductives. If levels were high, nymphs differentiated into soldiers. Lüscher also noted that hormonal effects depended on the timing of the application relative to the time of the last molt. Nijhout and Wheeler have suggested that the effects of juvenile hormone depend on critical temporal "windows" that govern the effects of presence or absence of the hormone. The interaction between timing and level of juvenile hormone is crucial.

Distinct male and female patterns of somatic development are the norm among animals. These alternative somatic states result from alternative regulatory gene cascades early in development, and are reinforced by hormonal control mechanisms later in life. Social insects have elaborated upon this elementary pair of male and female states, and there is a mechanistic interrelationship between larval development, sex determination, and caste determination. Termites represent the extreme and outré in the expression of alternative developmental states. What termites show us is that developmental control mechanisms involving juvenile hormone and ecdysone, the primary control molecules in insect larval development, have been co-opted to perform novel roles in caste differentiation. Elements of regulatory systems are conserved in evolution, but what they do may be greatly changed.

According to Jarzembowski, termites belonging to modern families are present in the early Cretaceous fossil record, about 120 million years ago, and many more termite families are present in the past 50 million years of the Cenozoic, including a number preserved in amber. Termites had evolved from presocial ancestors into fully social insects with complex caste determination perhaps by the middle of the Mesozoic era.

Until recently, it was assumed that although one could not put anything past insects in the way of strange behaviors, mammalian sex determination would be conservative and predictable, and that socially regulated castes resembling those of insects would be out of the question. It's not so. The downright most desperately ugly mammal on Earth is the naked mole rat; it has been called a sausage with fangs. It is also the only eusocial mammal. Naked mole rats live in East Africa in extensive underground colonies. As in insect societies, the colony has only one female reproductive, and brood care and other jobs around the colony are carried out by nonreproductive workers. There are several related species of mole rats, all bearing fur coats and consequently better looks. Although these other species exhibit complex social behaviors, they are not eusocial. The first discoverers of naked mole rats thought that they were the young of a larger, furry species. Although these animals have been known since 1842, it was not until 1981 that Jarvis recognized that naked mole rat colonies contained only one breeding female.

Her discovery, and the ease with which naked mole rat colonies can be kept under observation in captivity, has made it possible to examine the remarkable evolutionary convergence between termite and mammalian sex and caste regulation (see also Sherman and co-workers).

Naked mole rat colonies are huge, with tunnel systems extending up to 3 kilometers. The tunnels connect nest sites and tuber beds upon which the colony feeds. A colony can have over 250 inmates. The animals are completely adapted for underground life. They are cylindrical, and a full 25% of their muscle mass is devoted to their earth-moving jaws. In even the most talkative humans, less than 1% of muscle mass is in the jaw. Naked mole rats have poor physiological body temperature regulation. They consequently live in a thermostable environment and huddle a lot for warmth. Like termites, naked mole rats digest cellulose with the help of intestinal symbiotic microorganisms, and pass these microorganisms to their offspring through the feeding of fecal pellets.

Naked mole rats are diploid and develop as males and females. However, there is only one reproductive queen at any time. By a combination of bullying and pheromones, she prevents other females from entering estrus. When a queen dies, other females become reproductively competent. They fight to the death until a new queen is established. Thus, as in termites, there is, at least for females, a reproductive caste. Many males mate with the queen. Both males and females perform colony tasks. Nonbreeding animals help in child care and in digging tunnels; they are analogous to the termite worker caste. Large nonreproductives perform other digging tasks such as ejection of dirt, and defend the colony against snakes and intrusions by other colonies. These large animals may constitute a sort of worker-soldier caste. The roles of individual animals in the colony change as they grow.

Allard and Honeycutt have used the mitochondrial 12S rRNA genes to infer a phylogeny of mole rats. The results suggest that the mole rat family, the Bathyergidae, had a Miocene radiation, 20–40 million years ago, and that there has been parallel evolution of social systems among the species. The naked mole rat evolved its highly specialized mode of life and eusocial behavior much more recently than termites. The convergence is striking considering the differences in the starting material. Although the pheromones and proximal hormonal controls are mammalian, they are clearly producing social and differentiative effects analogous to those in termites.

In general, we view development as a way to build a single organism from an egg. This is certainly true for most organisms studied by developmental biologists, but it is a limited outlook from an evolutionary perspective. There are many colonial animals, including bryozoans, and some cnidarians and ascidians. These are made up of genetically identical, multiple, repeated individuals. The individual colony members, such as polyps of colonial cnidarians, include morphologically distinct and specialized individuals. Eusocial insects and mammals achieve even more spectacular differentiation of individuals. In eusocial species, alternative modes of development produce castes that allow novel life histories to evolve, and profoundly affect animals' adaptations, distributions, speciation patterns, and susceptibility to extinction. The reproductive female in social species creates specialized extensions of herself through caste development. These animals go beyond the tissue or organ grades of organization to produce extracorporeal alternative selves. The reproductive individual controls the biological state and behavior of these developmental extensions of itself by pheromones, analogous to hormonal regulation within the body.

FLUIDITY OF SEX DETERMINATION MECHANISMS

Sexual reproduction is costly, and it's not clear why it is advantageous. However, it exists, and so do a wide range of genetically complex sex determination mechanisms (see review by Hodgkin). Since sexual reproduction is fundamental to most metazoan animals, we might expect an underlying homology of sex-determining mechanisms among metazoan lineages. That would be in accord with the evolutionary commonality observed in other old and ancestrally shared molecular processes, such as the ribosomal machinery for protein synthesis or the widespread utilization of homeobox-containing genes in axial determination during development. Yet, paradoxically, that is not the case.

Sex determination involves both determination of the sexual identity of the germ line and determination of somatic sexual identity. These processes proceed largely independently of each other. Several different sex determination systems exist. Some animals use chromosomal sex determination systems, and others use environmental sex determination. Genetic sex determination occurs at fertilization and is dependent on the individual's karyotype. Environmental sex determination follows fertilization and depends on temperature, population density, or some other external cue.

A third process, dosage compensation, is required in genetic sex determination. In individuals of a species like ours, in which the female has two identical sex chromosomes (XX) and the male has two different ones (XY), the level of transcription of the X chromosomes is different in cells of the two sexes. Experiments in which imbalances in gene dosages have been created demonstrate that in many animals the imbalance in transcription levels is deleterious and needs to be reconciled between the two sexes. The solutions and mechanisms vary. Butterflies and birds (as reported by Baverstock and co-workers and by Johnson and Turner) seem to lack dosage compensation. *Drosophila* up-regulates expression of sex-linked genes in the male (Lucchesi and Manning), whereas mammals and *C. elegans* down-regulate expression of sex-linked genes in the female. In *C. elegans,* Plenefisch and co-workers found a set of genes that lowers the amount of transcription of sex-linked genes, whereas in mammals, one of the two X chromosomes of each female cell is inactivated (reviewed by Grant and Chapman). Dosage compensation has to be coordinated with sex determination to avoid cell lethality. As pointed out by Hodgkin, in *Drosophila* and *C. elegans,* a master regulatory gene controls both. In mammals there is a curious separation of controls: dosage compensation is determined by the X chromosome, but sex is determined by the Y chromosome.

Chromosomal sex determination mechanisms have been investigated in molecular detail in *Drosophila, C. elegans,* and mice. There is a kind of cybernetic convergence of process among these mechanisms. In essence, sex determination processes operate via the actions of genes that act as binary switches. Thus, a certain input causes gene A to occupy either of two states, active or inactive. The state of gene A determines the state of gene B, active or inactive. A short cascade of such switches leads to downstream gene states that produce sexual phenotypes. The sex determination mechanisms of flies, nematodes, and mammals share this kind of binary choice architecture, but achieve it with different, nonhomologous underlying genetic machinery. In *Drosophila,* Cline, as well as Bell and co-workers, showed that the master regulator for sex determination and dosage compensation is a gene called *Sex-lethal.* The X-to-autosomal-chromosome ratio controls the transcription of the *Sxl* gene. That control involves an intricate balance (see Cline; Erickson and Cline; Keyes and co-workers; Younger-Shepherd and co-workers). A set of helix-loop-helix transcription factor–encoding genes, including *sisterless-a* and *-b* (members of the *achaete-scute* complex of genes that also function in establishing the nervous system), are found on the X chromosome and activate *Sxl* transcription. Their positive regulation is counterbalanced by the negative effect of another helix-loop-helix gene, *deadpan* (also involved in neurulation).

When the *Drosophila Sxl* gene is active, it encodes a protein that splices its own RNA transcripts to produce active Sxl protein. This protein is also involved in sex-specific RNA splicing of the downstream regulatory *transformer* genes (reviewed by Baker). "Default" splicing of *Sxl* (from a late promoter active in both males and females) and of *tra* transcripts is carried out in the male to produce inactive RNAs. The *tra* genes produce proteins that are involved in sex-specific processing of the next downstream regulatory gene, *doublesex.* Males and females both express *doublesex,* but the transcript is alternatively spliced under the control of the *tra* genes. Production of sex-specific *dsx* transcripts leads to the activation of sex-specific genes that execute the sexual phenotype.

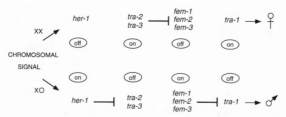

Figure 11.2. Cascades of gene action in somatic sex determination in *C. elegans*. The activity of the terminal gene in the sequence, *tra-1*, determines sexual identity. A high level of activity results in female (hermaphrodite) somatic tissue, a low level in male differentiation. (After Hodgkin 1992.)

Although *C. elegans* has a master gene, *xol-1*, that responds to the X/autosomal signal, Miller and co-workers, as well as Nonet and Meyer, found that it is not homologous to the *Sxl* gene of *Drosophila*. These genes and the entire downstream cascades of sex determination differ so substantially between the two species as to indicate independent evolutionary origins. In *C. elegans*, there is a cascade of gene activities, but it operates by alternate inhibitions of transcription of a series of genes (fig. 11.2). In hermaphrodite (XX) animals, *xol-1* is shut off at the start of the cascade, allowing the transcription of the next genes in the series, the *sdc* genes. The products of these genes block transcription of the *her-1* gene. In males (XO), *xol-1* is on, and its product blocks *sdc* gene activity. Since no *sdc* product is present, *her-1* activity is high. The genes downstream of *her-1* are activated or repressed in alternate ways in hermaphrodites and males. A number of genes are involved, but transcriptional control, rather than RNA processing, lies at the heart of the cascade of gene activities.

Germ line determination involves some different controls than does somatic sex determination. In *C. elegans*, the gene cascade of somatic development operates in the germ line, but in addition, the *fog* genes (discovered by Barton and Kimble and by Schedl and Kimble) decide between spermatogenesis and oogenesis. Much of germ line determination in *C. elegans* seems to be independent of the somatic cells. In *Drosophila*, the situation is more complex, as both germ cell autonomous gene expression and signals from somatic cells are required (see reviews by McLaren and by Steinmann-Zwicky). *Sxl* is involved, but apparently in a manner distinct from its action in somatic sex determination. Other genes involved in somatic sex determination, such as *sisterless* and genes operating downstream of *Sxl*, are not involved. The lack of evolutionary conservation is as striking as in somatic sex determination.

Sex determination in mammals is different from that of *Drosophila* and *C. elegans* and involves more cell interaction and hormonal controls (see the excellent reviews by Gilbert, by Hodgkin, and by McLaren). Sex determination in mammals is driven by a gene on the Y chromosome, called *SRY,* that encodes the testis-determining factor. This gene was discovered because there are human XX males and XY females. Page and co-workers and Vergnaud and co-workers made the crucial discovery that XX males had part of their Y chromosome translocated to a different chromosome, and that XY females lacked a portion of the Y chromosome. Chromosomal localization allowed Sinclair and co-workers to isolate *SRY* and show that it encodes an apparent transcription factor. The mouse homolog, *Sry,* has also been isolated. Koopman and co-workers showed that its expression in transgenic XX mice causes them to develop into males (albeit sterile, because other Y-linked genes are required for spermatogenesis). Female sex determination is not simply a default state. McElreavey and co-workers have postulated that there is a separate gene that stimulates ovary production. In normal males, the *SRY* product would inhibit expression of the ovary-stimulating gene. In females, the ovary-stimulating gene would act to determine the female developmental pathway. Such a gene has been identified by both Arn and co-workers and Bardoni and co-workers on the basis of its ability to feminize chromosomally male individuals when present in an extra dose.

Mammals begin development sexually indifferent. The indifferent gonads are associated with two sets of ducts, the Müllerian and Wolffian ducts, that terminate in the cloaca (future urethra in the male or uterus in the female). The presence or absence of testis-determining factor determines the course of sexual development. If testis-determining factor is present, male development proceeds, and the mesoderm of the indifferent gonad differentiates into a testis. Two kinds of secretory cells differentiate in the testis. The Leydig cells release testosterone, which causes development of the penis and development of the Wolffian duct into the vas deferens and seminal vesicles. The Sertoli cells produce Müllerian-inhibiting substance, causing the Müllerian duct to degenerate. If testis-determining factor is absent, female development is initiated, in which case the mesoderm of the indifferent gonad gives rise to the ovary. Two secretory cell types, thecal cells and granulosa cells, differentiate and secrete estrogen, which stimulates the Müllerian duct to develop into the uterus, oviduct, cervix, and vagina. In the absence of testosterone, the Wolffian duct degenerates. Binding sites for the SRY transcription factor have been found by Haqq and co-workers in the upstream DNA of the genes for aromatase (an enzyme that converts testosterone to estrogen) and Müllerian-inhibiting substance. The implication is that expression of these genes is directly regulated by the SRY protein.

By probing with a conserved region of the *SRY* gene, Tiersch and co-workers showed that the *SRY* gene is present in twenty-three species of agnathans, bony fishes, reptiles, birds, and mammals. The gene has been conserved for over 400 million years in vertebrates. However, it is specifically associated with the Y chromosome only in mammals. Its function in nonmammalian vertebrates is not clear, and it may have been co-opted from some other function for sex determination in mammals.

Chromosomal sex determination is universal in mammals and birds and is ingrained in our thought as general, but it's far from universal among vertebrates. A large proportion of reptilian species exhibit temperature-dependent sex determination (see Janzen and Paukstis for a listing). In these species, the sex of embryos is determined environmentally by the incubation temperature of the eggs. Lineages have shifted between modes in evolution. Although environmental sex determination appears radically different from genetic sex determination, it seems to utilize much of the same machinery (see Crews, Crews and co-workers, and Wibbels and Crews). The crucial evidence is that treatment with estrogen can cause an embryo developing at a male-specifying temperature to develop as a female, whereas treatment of an embryo developing at a female-specifying temperature with an inhibitor of aromatase results in development of a male.

In genetic sex determination, fertilization affects gonad-determining genes. These initiate a cascade of gonad formation, hormone release, and production of sexual phenotype, as sketched out above for mammals. Crews has suggested that in temperature-dependent sex determination, temperature affects both the expression of enzymes that control steroid sex hormone biochemistry and the expression of the receptors for those hormones. The two critical enzymes are aromatase in females and testosterone reductase in males (which converts testosterone to dihydrotestosterone). The hormone/receptor system then activates the gonad-determining genes and the downstream cascade leading to sexual differentiation. This hypothesis is attractive in that temperature-dependent sex determination uses the same genetic machinery as does genetic sex determination. In temperature-dependent sex determination, temperature acts in place of signals from chromosomal elements to activate sex hormone/receptor systems. A mapping of the two modes of sex determination on a cladogram of tetrapod vertebrate groups shows that shifts between modes have occurred in evolution.

Hodgkin has shown that shifts between sex determination regulatory systems can be readily generated experimentally in *C. elegans* by mutating particular genes in the sex determination pathway. These manipulations of existing sex determination genes can produce analogues of several kinds of sex determination systems, including temperature-dependent sex determination, but the changes all result from tweaking parts of a common genetic pathway. That's pretty much what one would expect evolution to do. Although alternative modes of sex determination in reptiles cause alterations with large potential ecological and life-history consequences, they represent an "easy" kind of genetic change. Differences in sex determination systems are not limited to the sex determination genes, but also involve the role of sex hormones. In vertebrates, hormones determine secondary sexual features. In insects, such as *Drosophila,* decisions are made on a cell-by-cell basis. *Drosophila* individuals in which somatic loss of an X chromosome has occurred during development have patches of XO cells in a body whose cells are mostly XX. The XO patches develop male body structures, showing that no circulating sex hormones exist to regulate somatic sexual phenotype throughout the animal.

The differences between the mechanisms of sex determination of flies, nematodes, and vertebrates present a gulf almost as profound as that separating their body plans. As these model systems represent randomly chosen branch tips on the phylogenetic tree, it seems likely that equally distinct sex determination mechanisms will be found in other phyla. Distinct genes and gene action pathways have evolved independently to achieve the same ends. They show that genetic systems one would expect to be highly constrained may be evolutionarily fluid.

EVOLUTION OF EYES

As a child, one of the profundities that I was offered by my elders was that "the eyes are the windows to the soul." I wondered a lot about some of the corollaries to that intriguing principle. Was it true for blind people? Is the soul right there behind your eyeballs? No wonder they were always warning me about sharp scissors. Should I wear sunglasses when I tell a lie? On reflection, my grandmother's aphorisms were not totally off the mark. In highly visual creatures such as ourselves, eyes release social signals as well as view the world. A flat stare is a mark of aggressiveness in humans and dogs. Wide-open pupils are more sexually arousing than closed pupils. If eyes are not the windows to the soul, they are tools of sociobiology. Perhaps they are even interesting windows to the genome.

Eyes have evolved to exploit the information provided about the world by the broad spectrum of photon wavelengths emitted by the sun. In vision, photons interact with a photopigment to produce a chemical signal sensed by the optic nerve. The atmosphere does not interact with visible light, so there is a window through which photons of certain wavelengths can reach the Earth and interact with matter on the surface. The shorter-wavelength photons of gamma rays and X-rays are absorbed by the atmosphere. That is just as well for living beings, considering that such photons are so energetic as to destroy the molecules, including those of the eye, that absorb them. Some insects, such as honeybees, have eyes that can detect near-ultraviolet, and flowers in turn have ultraviolet colors on their petals that serve as nectar guides. Such flowers, when photographed under ultraviolet light, look quite different than they do to us by visible light.

One may ask, as has Neil Comins in his imaginative book of might-have-beens, *What If the Moon Didn't Exist,* why eyes that see infrared or radio waves don't exist. Radio-wave eyes are theoretically possible, but because radio waves have wavelengths on the order of a millimeter, eyes the size of football fields would be required to give good resolution. They would go far beyond the Irish elk's antlers in unwieldiness. Infrared eyes pose a more interesting problem. They wouldn't have to be much different in size from those that respond to visible light, and they also might offer some real advantages at night. Objects that are cooling radiate infrared photons. Furthermore, warm-blooded animals do likewise, and would glow in the dark. Infrared eyes would be difficult for warm-blooded creatures to manage because their internal body heat would interfere. However, most animals are cold-blooded, and should be able to keep their infrared eyes cool enough at night to give them an advantage over warmer enemies or prey. Rattlesnakes and other pit vipers have primitive versions of infrared eyes in the form of facial pits. These organs detect infrared radiation from warm-blooded prey and enable the snakes to hunt small mammals effectively at night. The pits have a low resolution, but high-resolution infrared eyes would be possible if a sufficient number of photoreceptors were arrayed behind a lens.

Photoreception in animals has several roles. One is simple detection of light levels. To a larva that is about to settle and undergo metamorphosis, detection of a lit or dark area of substrate may be critical. To a sessile filter feeder, a sudden drop in light intensity provides an unmistakable early warning of the unwelcome attentions of a predator. Longer-term light level changes also have physiological roles in regulating biological rhythms in many phyla. Finally, some eyes can form an image, and these are found in the most behaviorally sophisticated groups: arthropods, cephalopods, and vertebrates. These animals use vision to evaluate their environments, to locate prey or evade enemies, and for a gaudy range of social purposes. A survey of the detailed descriptions of visual systems of the metazoan phyla presented by Brusca and Brusca, Coomans, Eakin, and Salvini-Plawen and Mayr shows that 21 out of 34 phyla have eyes during some stage of the life cycle. Some of these eyes are quite complex and have surprising evolutionary histories.

The most primitive animals with eyes are the cnidarians. Some have simple eyes lacking lenses, but other medusae have well-developed eyes on the edges of their bells. The Cubomedusae (box jellies), whose highly toxic stings are such a notorious threat to swimmers on Australia's north coast, have up to 24 eyes that are linked to the nerve net and enable them to orient accurately in light. These eyes are complex, with an epidermal cornea, a spherical lens, a multilayered retina, and a region of nerve fibers. There are about 11,000 sensory cells in each eye. As diploblastic cnidarians are the deepest branch of the metazoan tree, the complexity of their eyes is surprising. If cnidarians were indeed part of the Ediacaran fauna, it suggests that eyes long predate the Cambrian radiation of bilaterian animals. At the base of the bilaterian branch, the flatworms have a complex nervous system that includes a well-developed cerebral ganglion. Eyes are present, but consist of simple pigment-cup ocelli lacking lenses. The retinal cells project into the pigment cup with their ends facing away from the light; this arrangement represents the inverse type of retina. The eyes of the flatworms might well represent the primitive photoreceptors basic to bilaterian animals at the verge of the radiation of the coelomates.

Several phyla of small burrowing or parasitic organisms, including the pseudocoelomate phyla, have no or poorly developed eyes. Rotifers, gastrotrichs, kinorhynchs, nematodes, and nematomorphs have at best simple ocelli. These animals may have undergone simplification in structure, including that of their photoreceptors. This may be the case as well for two small phyla that are closely related to the arthropods. Tardigrades, the half-millimeter "water bears," have five-cell sensory eyespots. Pentastomatids, which are parasites in vertebrates, are most convincingly highly modified crustaceans. They have undergone a loss of organs, including eyes. Other larger and more complex animals with sedentary adults do most of their photoreception as larvae. That is the case for echiurids and sipunculids among the protostomes, brachiopods and bryozoans among the lophophorates, and ascidians and hemichordates among the deuterostomes. Echinoderms, despite their large size, predatory habits, and activity on the seafloor, have at best rudimentary eyespots. Most other phyla have invested in eyes in a big way.

Among annelids, the highly cephalized, highly active, predatory marine polychaetes have well-developed eyes. These run the gamut from eyecups that have photoreceptors but no lenses to large eyes that have lenses, a cuticular cornea, a pigment and photoreceptor cell layer beneath the lens, and optic nerves. Image perception seems likely in the latter case. Mollusks too have a range of eyes. Again, these run from eyecups to eyes with lenses in gastropods (snails) and in bivalves (clams). A third type of photoreceptor, called an aesthete, is found in polyplacophorans (chitons). Aesthetes lie at the termini of nerve strands that penetrate the shell, and form a little eye resting in a pigment cup in the fabric of the shell. The primitive cephalopod *Nautilus* has a pinhole eye, whereas octopuses and squids have highly developed visual systems with image-forming lenses, and see color as well. Octopus and squid eyes are built much like those of vertebrates, with a cornea, iris, and lens. Unlike that of the vertebrate eye, the lens is of fixed focal length. It is clear that other molluscan eyes also have image-forming abilities, although not always achieved in the way we might expect. Scallops have brilliant blue eyes lining their mantle edges. Land has shown that they form images, not with a lens, but by focusing light from a reflective layer that underlies the retina.

Arthropods have three kinds of eyes: simple ocelli, complex lensed ocelli, and compound eyes. The compound eye has the capacity to form an image, although in a different way than a simple eye. Each facet represents a unit of the compound eye, called an ommatidium, with its own lens and retina. Each ommatidium retains an image as a discrete unit. Each lens has two parts, a corneal lens and, underlying it, a crystalline cone. The narrow image from the crystalline cone is projected onto the rhabdomere, or light-transducing photoreceptor cell, of the ommatidium. As discussed by Horridge, these eyes can be appositional or superpositional. In appositional compound eyes, a pigment layer isolates each ommatidium from the others, which improves bright light resolution. In superpositional eyes, the ommatidia are not so isolated, so they can function at reduced light levels, albeit with some loss of resolution. In some species, the style of the eye is fixed; others can switch styles with lighting conditions. Onychophorans, featured in so many discussions of phylogenetic relationships between annelids and arthropods, have annelid-like eyes with a cornea, lens, retina, pigment layer, optic nerve, and optic ganglion. Despite the mitochondrial 12S rRNA sequence phylogeny of Ballard and co-workers, which implies that onychophorans are arthropods, there is no trace of compound eyes in these animals. Compound eyes are old in arthropods, appearing with the first Cambrian trilobites. The annelid-like eyes of onychophorans raise doubts about the 12S rRNA result.

Typical trilobites had compound eyes similar to those of living insects or crustaceans. Some trilobites, however, as noted by Clarkson, had a rather different eye with an aggregate of distinct lenses, called a schizochroal eye. Sections of these lenses showed a doublet structure separated by a peculiar wavy surface. In examining the structures Clarkson had discovered, Riccardo Levi-Setti, a physicist with an interest in trilobites, found that trilobites had evolved a novel solution to spherical aberration in lenses. A calcite lens was combined with an underlying chitin lens to create a compound lens matching the classic seventeenth-century solution devised by Christian Huygens to the spherical aberration problem. The preservation of such lenses in the fossil record has been documented by Clarkson and by Towe. These elegant schizochroal eyes persisted to the final extinction of the trilobites. Levi-Setti has regretfully noted that "it is unfortunate that the genetic information of such a perfected visual apparatus became lost to further evolution in the animal kingdom." Extinction does not necessarily remove only inferior solutions. In this case a superior functional design was lost because it was confined to an unlucky lineage.

We are highly visual creatures, and vertebrate eyes in all their expressiveness draw more empathy from us than any invertebrate eye possibly can. Vertebrate eyes include the most versatile and highly resolving visual systems in nature. The most primitive chordates, the cephalochordates, represented by the living *Branchiostoma (Amphioxus),* have, as noted by Lacalli and co-workers, an anterior photoreceptor possibly homologous to the vertebrate paired eyes. (They also have simple pigment-cup ocelli located farther to the rear of the "head.") The eyes of the most primitive living vertebrates, the lampreys, are similar in construction to those of more advanced vertebrates. Vertebrate eyes have as a light port a transparent cornea, which refracts incident light. The amount of light that enters is controlled by muscles that open or close the pupil by dilating or contracting the iris. Light is focused on the retina by a lens that can be adjusted by muscles that change its curvature or its position with respect to the cornea. The vertebrate retina contains two kinds of photoreceptor cells, rods and cones. Color vision depends on the cones, with rhodopsins of three different absorption maxima, blue, green, and red, forming the basis of color discrimination. Cones have a high threshold to light stimulation and provide the basis for high-resolution vision. Rods have a lower threshold to light and are effective in dim light. Vertebrate retinas range from all cones, as in birds, to primarily rod-containing, as in nocturnal animals.

Ancient primitive vertebrates actually had two kinds of eyes, a pair of main lateral eyes on the face and a pair of medial ones, the pineal and parietal eyes, on the midline or the roof of the skull. There is a certain mystique about the "third eye," and some living vertebrates still have such a medial photoreceptor on the top of the skull. In lizards, the parietal eye is well developed with a cornea, lens, and retina. The photoreceptor arrangement of the medial eyes is everse (the ends of the photoreceptor cells face the direction of the light), rather than inverse as in the main lateral eyes. The medial eyes may have had a visual function in primitive vertebrates, whose fossil skulls show large medial openings in the skull roof. Living vertebrates do not form visual images from their third eyes. The complex parietal photoreceptors of lizards are covered by a large translucent scale. When light shines on a lizard's head, neural activity can be detected in the parietal retina. Instead of relating to vision, however, the functions of the pineal and parietal organs are endocrine, and are related to circadian or longer time rhythms in the life cycle and to sexual maturation. The pineal gland in mammals, no longer a light-collecting organ, produces such hormones as melatonin, serotonin, and norepinephrine. Eakin and his collaborators showed that behavioral changes occurred in lizards from which the parietal eyes had been surgically removed.

One of the striking features of the mid-Cambrian Burgess Shale fauna is how many of these animals have eyes. The eyes are obvious on the fossils themselves; their presence in restorations doesn't reflect guesses or our preference for having eyes on our creatures. Vision accompanied the Cambrian radiation. A number of the animals of the lower Cambrian Chengjiang fauna illustrated by Chen and Erdtmann and by Hou and Bergström have prominent eyes. Eyes are present in the "normal" members of the Cambrian fauna, such as trilobites and other arthropods. Probable compound eyes are present in some of the oddities as well. *Opabina* has five large eyes on its head, mounted over the frontal nozzle with its terminal claw. *Nectocaris,* with its arthropod-like front half and its chordatelike rear half, has a pair of eyes. The giant predator *Anomalocaris* also has a pair of large eyes. The earliest fossil vertebrates, Ordovician in age, have both lateral and medial eyes.

There are a number of other, different senses, enabling animals to detect heat, gravity, touch, chemicals, magnetic fields, and sound. At least some of these were present in Ediacaran trace fossil makers as well as in the Cambrian metazoans. What is significant here is that photoreception too is an ancient metazoan sensory system. The eye was held up by Darwin's critics as an insuperable difficulty for natural selection—too complex to have arisen by "chance." Darwin's answer was a wonderful chain of logic: "Reason tells me, that if numerous gradations from a simple and imperfect eye to one complex and perfect can be shown to exist, each grade being useful to its possessor, as is certainly the case; if further, the eye varies and the variations be inherited, as is likewise certainly the case; and if such variations should be useful to any animal under changing conditions of life, then the difficulty of believing that a perfect and complex eye could be formed by natural selection, though insuperable by our imagination, should not be considered as subversive of the theory." Obviously, Darwin had no trouble in finding and holding up for view a range of simple to complex eyes in living animals. How many times did eyes evolve?

The widespread distribution of eyes among long-diverged living phyla confirms the antiquity of vision as a major animal sense. About thirty years ago, Eakin, on the basis of anatomical studies of the photoreceptor cells of many phyla, proposed that a grand evolutionary pattern of descent with modification was present. Eakin distinguished two main types of receptor cells, ciliary and rhabdomeric. Photoreceptor cells have evolved extensive membrane surfaces to house the rhodopsin light reception/transduction system. This has been achieved in two ways. The first is by increasing ciliary membrane surface area by elaboration of projections from the membrane surrounding the cilium, or by increases in photoreceptor ciliary number. The second is by elaboration of extensive microvillar membranes. The pattern initially distinguished by Eakin was that deuterostomes and lophophorates had ciliary photoreceptors, whereas protostomes had rhabdomeric photoreceptors. Since many rhabdomeric receptor cells have a cilium, the distinction may be arbitrary. The basic biochemistry of photoreception is universal in animals.

Martin and co-workers, for example, have found through DNA hybridization that rhodopsin genes homologous to those of mammals are found in fishes and frogs as well as in insects, crustaceans, octopods, and even protists. Opsins (as discussed by Applebury and Hargrave and by Nathans and co-workers) are members of a family of membrane-bound receptors in which there are seven membrane-spanning domains. Rhodopsins bind a retinaldehyde chromophore for absorption of light. Once a rhodopsin molecule has absorbed light, it interacts with a G-protein to activate a second messenger cascade in the photoreceptor cell. The cascade ultimately acts on a sodium channel, and a nerve signal is generated. Animals that are widely separated in phylogeny, and have very different eyes, share a common light activation system, rhodopsin, and a G-protein second messenger system.

To return to Eakin's proposal for two great evolutionary lineages of eyes, a ciliary one in the deuterostomes and a rhabdomeric one in the protostomes: is it so? If eyes originated once, then the diversity of eyes must present a distribution of features that allows a diverging evolutionary pattern to be traced. Unfortunately, Eakin's elegant and parsimonious idea founders on the much stranger reality of the rampant convergent evolution documented by Salvini-Plawen and Mayr. The simple division of ciliary and rhabdomeric photoreceptors does not hold. Some deuterostomes have rhabdomeric photoreceptors, and some protostomes ciliary ones. There are just so many ways for cells to generate a large membrane surface to house rhodopsin molecules. Physics dictates the set of mechanisms that animals can use to generate an image, but the details can vary and reveal independent, convergent solutions. For example, there are four ways in which lenses are constructed. In vertebrates and some mollusks, the lens is composed of an aggregation of many whole cells. In cephalopod mollusks and some annelids, lenses arise from cellular processes. In cephalopods, these processes derive from the cornea. Note that in appearance the finished eyes of cephalopods and vertebrates are very similar. This is part of the striking convergence between cephalopods and fishes described by Packard. In development, the cephalopod eye arises very differently from the vertebrate eye, and it is distinct at the cellular level as well. Lenses in arthropod compound eyes are extracellular secretions, as they are in some mollusks, annelids, and onychophorans. Finally, lenses can be formed from refractile bodies within cells, as in some annelids, mollusks, pogonophorans, and larval crustaceans.

Bilaterians need not have inherited more in the way of eyes (as morphological entities) than patches of light-sensitive cells like those of living flatworms. Again and again, eyes have arisen within phyla or classes. The pinhole eyes of primitive cephalopods, the Paleozoic relatives of the living *Nautilus,* evolved into the sophisticated eyes of octopuses and squids. Analogously, complex eyes have evolved among clams and among snails. The solutions have often been similar, but not always, as shown by the eyes of one group of clams, the scallops, with their unique reflector focusing system. Arthropod compound eyes are different from the eyes of any other group, and distinct novelties, such as the schizochroal eyes of trilobites, have arisen as well. Salvini-Plawen and Mayr tallied all the independent events and estimated that the lineages leading to living animals have invented eyes somewhere between 40 and 60 times.

These numbers, of course, fail to impress upon us just how much genetic change was entailed. On one side, there has been a strong conservation of photoreceptor biochemistry, as well as of deep-seated regulation by *Pax-6*. However, given the basic underlying biochemical and regulatory machinery, it is clear that in each of the independent streams of eye evolution, numerous independent gene co-options took place, as did assembly of new patterns of gene expression and, in some cases, evolution of new structural genes.

CO-OPTION OF EYE STRUCTURES AND GENES

The inference of multiple origins for eyes requires not only that this complex structure be engineered by evolution, but also that this be done again and again. It would seem likely that eyes originated early in metazoan history, and in many lineages. An innovative but straightforward computer simulation by Nilsson and Pelger shows that such intuition is not always a very good guide. Nilsson and Pelger assumed the existence of light-sensitive cells and began with a primitive light-sensing organ. This structure was flat and consisted of a clear surface layer, a layer of light-sensitive cells beneath that, and beneath both, a dark pigment layer. The model allowed the transparent layer to deform. Each step was limited to a 1% change. For a change to be accepted, it had to improve function. Because optics are well understood, function could be related to structure.

The result was a rapid change that proceeded through a shallow cup, to a deeper cup, and finally to a sphere. In all of these, the clear layer thickened to form a vitreous body that filled the cavity. Further steps formed a localized region of higher refractive index—a lens. Using very conservative rules for change, the number of 1% steps required to progress from a flat field of cells to a complete eye capable of forming an image was estimated at about 400,000. If a one-year generation time is assumed, eyes could have appeared in a fraction of the time required for the Cambrian radiation. If the basic biochemical components for photosensitivity are present, they can be effectively and rapidly built into an eye. In the computer simulation, a lens simply condensed from the clear layer. The actual origin of lens proteins presents an amazing window on gene co-option and the fluidity of gene regulation.

Eye lenses are transparent and are filled with a high concentration of a small number of proteins called crystallins. The name suggests that lens proteins somehow form crystals. They do not. Delaye and Tardieu showed that the lens is essentially a fluid of nonattracting hard spheres. The crystallins need only be stable globular proteins. That very limited functional constraint has had unexpected evolutionary consequences (see Piatigorsky and Wistow, Wistow, Wistow and co-workers, Hodin and Wistow, and de Jong and co-workers). One of the major vertebrate crystallins, α-crystallin, is a small heat-shock protein. A study by Horwitz shows that α-crystallin functions as a molecular chaperon and prevents heat denaturation of proteins. It may well have been selected for this function as well as for its role as a crystallin. The β- and γ-crystallins are related to microbial sporulation proteins.

These relationships of crystallins to other protein families show that lens proteins had their origins via gene duplication and divergence. In itself, this is hardly astonishing. The surprises came when the study of lens crystallin biochemistry was extended to a phylogenetically broader sample, and it transpired that there were a number of taxon-specific crystallins. These taxon-specific crystallins were identified: they were familiar enzymes. A massive co-option of genes as lens crystallins was discovered, and it sheds light on how genomic fluidity is exploited by selection.

Crocodiles and a few birds use ε-crystallin, which turns out to be lactate dehydrogenase. Most birds and reptiles use δ-crystallin, argininosuccinate lyase. Other taxon-specific crystallins (ζ, η) are identical to NADPH:quinone oxidoreductase and aldehyde dehydrogenase I, respectively. Still others are related to, but not identical to, several other common enzymes, including enolase. No selective advantage can be ascribed to using any particular one of these enzymes. If one maps the distribution of ε-crystallin on a phylogenetic tree of bird orders, it suggests that losses and possibly gains have occurred within lineages of birds. Idiosyncratic distributions of crystallin enzymes, such as the use of aldehyde dehydrogenase I by elephant shrews and the use of NADPH:quinone oxidoreductase in a few groups of mammals, show that crystallins have been recruited within the past 65 million years. The repeated co-option of metabolic enzymes as crystallins represents a remarkable transience in gene usage.

The co-option of metabolic enzymes for use as crystallins requires that transcriptional regulation be altered. As suggested by Wistow (fig. 11.3), a number of independent solutions have evolved. These include modification of existing promoters, insertion of new promoter elements, insertion of enhancers, gene duplication, and modification of lens-expressed duplicated genes. There may be no single lens-specific promoter, but there may be some common motifs. In at least three cases documented by Wistow, enzyme genes expressed in non-lens tissues have GC-rich upstream motifs associated with general expression, but upstream TATA boxes in the homologue expressed

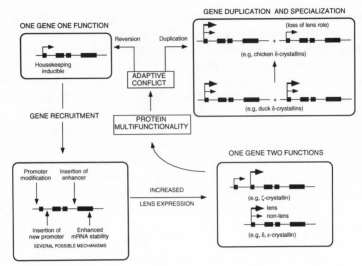

Figure 11.3.   Gene regulation and the recruitment of enzymes as lens crystallins. Introns are represented by thin lines, exons by black boxes. Transcription start sites are indicated by right-angled arrows. Small arrowheads indicate low-level, non-lens expression; large arrowheads indicate lens expression as crystallins. (From G. Wistow. 1993. *Trends in Biochemical Sciences* 18: 301–6. With permission of Elsevier Trends Journals.)

as a crystallin. In one of these cases, the same gene is used, but has alternative splice sites, so that the CG-rich promoter is used for liver expression and a downstream TATA-associated promoter is used for lens expression. Wistow and his colleagues (see Lee and co-workers, Richardson and co-workers) have discovered that the ζ-crystallin of guinea pigs, which is the recruited enzyme NADPH:quinone oxidoreductase, has acquired a lens-specific promoter. The nature of the new promoter was astonishing.

There is a small family of transcription factors that bears a common DNA-binding motif called a paired box. Some, in addition, contain a homeodomain. These *Pax* genes are involved in brain and eye development. As discussed in chapter 10, one of these genes *(Pax-6* in vertebrates and *eyeless* in *Drosophila)* was found by Quiring and co-workers and by Halder and co-workers to be a master regulator in early eye development in insects and vertebrates. The gene is present in flatworms, nemertines, and cephalopods as well. Eyes great and small, primitive and specialized, use *Pax-6* in a great demonstration of the long conservation of the function of this gene in development of animal neural systems and photoreceptors. This commonality doesn't mean that the compound eyes of insects and the camera eyes of vertebrates are homologous, but that photoreceptors share a long-conserved master regulatory gene present in the ancestral neural precursors of eyes.

This finding has a major bearing on our inferences as to how often photoreceptors arose in evolution. The detailed forms of eyes may have many origins, but the underlying photoreceptors may have arisen only once in the bilaterians.

The ζ-crystallin promoter has been found by Richardson and co-workers to contain a binding site for the Pax-6 protein, and the Pax-6 protein is essential for lens-specific expression of the ζ-crystallin gene. The αA-crystallin gene has also been found by Cvekl and co-workers to exhibit binding of the Pax-6 protein to its promoter. The αA-crystallin promoter sites do not match the consensus for Pax-6 protein binding sites, indicating an independent recruitment of this transcription factor in lens protein expression.

The number of co-option events in the evolution of lens proteins, as well as the variety of changes in cis-regulatory regions of the involved genes, is important because it indicates that the genome is fluid with respect to the migration of transcriptional regulatory regions. There is a constant movement of transcription factor binding sites among gene promoters. This is seen most easily in the lens because the low selective constraints on crystallin function allow changes to persist and to be readily observed. These changes are only the tip of the iceberg.

There is yet one more evolutionary twist to eyes. The eyes of octopuses and squids are remarkably convergent in form and optical function with vertebrate eyes, but they evolved independently from the primitive eye of the ancestral flatworm-like bilaterian and are fabricated differently in development. As in vertebrate eyes, light passes through the cephalopod lens and is focused on the retina. In cephalopods, the retina is derived from an invagination of the epithelium and contains two layers of photoreceptors. These layers are surrounded by a reflective lining, which also forms the iris. Like those of many vertebrates, the lens proteins of octopuses and squids are recruited enzymes. The major cephalopod lens protein is S-crystallin, which is related to glutathione S-transferase; a less prevalent Ω-crystallin, related to aldehyde dehydrogenase, is also present (see Chiou; Tomarev and co-workers).

The bodies of some deep-sea squids are speckled with galaxies of tiny lights. These light organs exhibit an extraordinary convergence with eyes, although they arise differently in development. The center of the light organ is the photogenic tissue, which contains light-emitting bacterial symbionts. This tissue arises from an epithelial invagination analogous to that which produces the retina. The photogenic tissue is surrounded by a reflective lining and a pigment layer that is derived from the ink sac. These layers regulate the intensity and direction of the light emitted by the photogenic tissue. The light is directed toward the ventral side of the animal and is refracted into the surrounding water by a lens. Unlike the eye lens, which is epidermal in origin (as described by McFall-Ngai and Montgomery), the light organ lens derives from muscle, a mesodermal tissue. Montgomery and McFall-Ngai have found that the muscle-derived light organ lens does not have a muscle-like protein composition, but rather consists primarily of a single aldehyde dehydrogenase–like protein, Ω-crystallin. The eye lens and light organ lens derive from different tissues in development, but they have recruited one lens protein in common. Are these independent events, or is there a common regulatory basis for these two organs despite their different tissue origins? S-crystallin is expressed at a very high level in the eye lens, and Ω-crystallin at a low level. In contrast, in the light organ lens, no S-crystallin is made, and Ω-crystallin is expressed at a high level. Proteins used as crystallins have been recruited by both organs, but a different spectrum of transcription factors may function in each lens type, resulting in different levels of expression of genes with lens promoter specificity. It will be interesting to learn whether the convergent evolution of eyes and light organs involves other molecular convergences. If *Pax-6* is involved in making a light organ, the convergence would bear the hallmark of a homeotic event, and the light organ would be recognized as a serial homologue of the eye.

DOLLO'S LAW

One of the classic horror devices in both written tales and movies is the transformation of the protagonist or some other major character into a primitive and beastly animal form: Mr. Hyde, The Fly, Count Dracula, were-wolves, sundry unfortunate princes turned into frogs. The most sophisticated of these characters is Gollum, the sinister counterpoise to the Hobbit hero of Tolkien's *The Lord of the Rings*. Gollum himself came from Hobbit stock, but once the evil ring had taken possession of him, he devolved from Hobbit form into an amphibious predator, a dark-adapted creature with large eyes, paddlelike feet, and a pathological aversion to sunlight. Gollum's devolution was both physical and moral. He became cunning, sneaky, murderous, disgusting. These literary reversions are emotive steps of various lengths down the Ladder of Creation to the beastliness that lies below the civilized human veneer. For dramatic effect, the changes transform an existing person over a short time. In real life, evolutionary reversals must be produced in each new generation as the modified outcomes of an ontogeny that reverts to an ancestral one in some aspects.

Primitive features do occasionally reappear in the progeny of more advanced living animals. Such atavisms occur naturally and sometimes attract considerable attention: people born with heavy body hair, or with short tails, or with a row of extra nipples running down each side of the chest and abdomen; horses with three hooves on each foot instead of one. Such cases nowadays generally suffer fates little worse than a brief notoriety as subjects of lurid stories in the tabloids that provide the heady reading matter of supermarket checkout lanes. Their fates were not always so benign. In the 1870s the Italian physician Cesare Lombroso propounded his theory of the criminal man, in which criminals were considered to be evolutionary atavisms driven by their primitive nature to commit acts that Lombroso supposed would be normal for savages or apes, but are crimes in a civilized setting. He claimed that he could recognize physical stigmata that would reveal these criminal atavists. As Gould nicely detailed in his discussion of Lombroso's fallacies in *The Mismeasure of Man,* most of these stigmata were extremes on the tail of the normal curve of human variation. Long arms, greater skull thickness, darker skin, and the supposed prehensile feet of prostitutes are not the kinds of discontinuous variations that characterize differences between species or represent true atavisms. The most ironic outcome of Lombroso's approach was Thorstein Veblen's well-crafted contention that the railroad and banking tycoons of the Gilded Age of the 1890s were not the pinnacle of evolution they so fondly believed themselves to be. In that era of social Darwinism, attainment of wealth was thought to be a sign of biological and evolutionary superiority. In *The Theory of the Leisure Class,* Veblen suggested that the aggressive and ostentatious financier ruling class instead represented an archaic throwback to a barbaric predatory race, and was thus akin to lower-class criminals. It goes without saying that his views had little noticeable effect on the subsequent organization of society.

Atavisms are generally defined as evolutionary reversals. By that definition, they should represent recapitulatory reversals of development. Determining whether that is the case is difficult because we lack sufficient developmental or genetic data on most atavisms to discern their mechanisms (see Hall). Furthermore, some atavisms can be shown not to be recapitulatory. The question is whether evolution can be reversed and lost features regained. If that is the case, we will have identified a likely source of homoplasy among evolving lineages, as well as a potential source of variability for evolution.

The past revealed by nineteenth-century paleontology was orderly. There was a distinct temporal flow to the fossil record, and it was clear that once animals became extinct, they never returned. Eventually a progressionist view of the fossil record was fused with an evolutionary perspective of life. It also became evident that once an anatomical feature was lost in the course of evolution, it was not regained. This principle of evolutionary irreversibility was propounded by Louis Dollo (of *Iguanodon* fame) and is known as Dollo's law (see Gould for a history). Evolutionary loss is not surprising: examples, such as degeneration of pigmentation and eyes in cave fishes, are common and presumably result from loss of selection for the function or structure.

Unless the genes involved play other roles, they will be released from selection, and will undergo mutational degradation and permanent loss.

A consequence of Dollo's law is that when lost functions are regained, the most probable mechanism is through evolution of genes that produce convergent or parallel structures. This is clearly the case in the return of terrestrial vertebrate groups—reptiles, birds, and mammals—to aquatic habitats. These returns to the sea have been accompanied by the return of various fishlike anatomical features. It is unlikely that the genes governing development of a hydrodynamically streamlined body shape and fins were saved for a rainy day through more than 300 million years of terrestrial evolution. Rather, fishlike morphologies arose by the evolution of novel genetic pathways. Indeed, a detailed comparison reveals that the old structures were not regained. In the case of whales, it is especially interesting that the muscular machinery for producing an undulating motion is not fishlike. Fishes swim using side-to-side body motions generated by the primitive body wall musculature. Whales produce an up-and-down undulation. A block of dorsal muscles provides the power stroke. Their flukes are oriented horizontally, rather than vertically like the tail fins of fishes. This fact was noted by Ishmael, the narrator of *Moby Dick*, who refused to accept that whales are not fish despite their lungs and hot blood, and concluded: "A whale is *a spouting fish with a horizontal tail*." Ichthyosaurs, although built very much like dolphins, evolved vertical tail fins like those of sharks rather than the horizontal flukes of whales.

Fins or paddles have appeared independently in whales, marine turtles, and penguins as well as in the extinct marine reptiles, ichthyosaurs and plesiosaurs. The structures of all of these fins are different, and none reproduces the ancestral rhipidistian lobe fin (see Carroll). Similarly, the pseudodontorns, the huge extinct Cenozoic seabirds discussed by Feduccia, had impressive bony toothlike structures arming the lengths of their bills. These projections, although reminiscent of the reptilian teeth of Mesozoic birds, were not true teeth, and are thus not a real evolutionary reversal. No retention of genes for tooth development was required to produce them. I'll return below to the issue of birds' teeth, in conjunction with a reconsideration of experiments that have attempted to retrieve the ancient program of tooth development in avian oral tissue.

In 1939 H. J. Muller pondered the genetic consequences of evolutionary reversal. G. G. Simpson in 1953 extended Muller's ideas and united them with paleontology. Both argued that there is no theoretical block to evolutionary reversion because back mutations, recombination, and suppression of mutations all occur regularly. DNA sequence data show that reversions frequently occur on the microscopic scale of point mutation. However, reversals are not expected if significant genetic change has accumulated. It is statistically improbable for an evolving lineage subject to mutation and selection to return to an earlier complex genetic state via the accumulation of random change. As they diverge, genetic systems become differently integrated. Even if a reverse mutation occurs, it may not be reintegrated into the new genetic background as it was in the ancestral genome. Finally, the environment will have changed in complex ways, and a return to the same conditions that acted upon the ancestor is unlikely. Nevertheless, the direction of selection can be reversed, and the above considerations must be taken relative to the degree of genetic relatedness. Simpson concluded that evolution is "more or less reversible to a condition of the immediate ancestry in which the genetic system was essentially the same." Thus, reversals over relatively short evolutionary distances probably occur, and we should be able to find examples of such reversals.

TURNING BACK THE CLOCK: EVOLUTIONARY REVERSALS

We (Marshall, Raff, and Raff) have reexamined Dollo's law from the perspective of genomic behavior. We used data on the inactivation of silenced genes and on the tolerance of genes to mutation to quantify the probability of evolutionary reversal. Our results show that reversals are possible over significant periods (0.5–6 million years), but not in time frames over 10 million years unless the gene in question is maintained by active selection. These results are important in evaluating claims of reactivation of ancient developmental pathways.

The simple model that we used to examine the genetic sequelae of Dollo's law was that features can be lost in evolution through the silencing of a gene critical to their development. Evolutionary silencing of a gene means that the gene continues to exist in the genome, but no longer produces a transcript. Silencing does not require that the gene be deleted, nor that a deleterious mutation occur in its coding sequences. Any mutation in its control circuitry that results in loss of expression or expression below a threshold level will silence an otherwise intact gene. Once that has happened, the gene is no longer maintained by selection, and degradation of genetic information is likely to ensue. In this simple case, the pathway to production of the lost feature decays with the degradation of the silenced gene. This model, however, does not describe all cases of gene silencing. Genes no longer expressed in one pathway need not suffer loss of information if they also function in some other pathway in development. Thus, there is a mechanism for cryptic retention of apparently lost genes over long intervals of time.

Frequent and dramatic modifications of gene expression patterns in various tissues, including loss or gain of expression, have been documented by Dickinson and by Rabinow and Dickinson. These, along with the recruitment of novel eye lens proteins, show that changes in cis-acting regulatory elements are common and that the control circuitry of the genome is fluid. Thus, in principle, silenced genes could be reactivated by any process that restores or replaces the damaged control site. The morphological result would be a violation of Dollo's law. If a process or structure is to be regained by reactivation of a silenced gene, the encoded protein must still be functional. Rates of inactivation of coding sequences can be estimated. We considered as given that loss and later recovery of expression of genes is possible in evolution. We then estimated how long a time would elapse before a protein encoded by a silenced gene would no longer be functional because of accumulation of deleterious substitutions. The critical data for calculating the probability of survival of a silenced gene were observed rates of degradation of pseudogenes (stand-ins for silenced genes) and the tolerance of proteins for damage by point mutations and frameshifts. The reader is spared the calculations here.

Our calculations showed that many genes may remain silenced for up to 6 million years without mutational inactivation of the encoded protein—a surprisingly long time. This result indicates that reactivation of genes can occur over time spans exceeding the time required for speciation. With the passage of time, the number of individuals in a population that still retain a potentially functional coding sequence in a silenced gene may fall to only a small percentage of the population. Nonetheless, resurrection of the gene could occur following a change in expression pattern if sufficient selection favored its return. Population genetics studies indicate that even with only a 1% selective advantage, a favored allele goes from a frequency in the population of 0.01 to 0.25 in 350 generations; for a 10% selective advantage, only 35 generations are required (Strickberger). Thus the potential time for resurrection may exceed the half-life for inactivation.

The reactivation of a complex morphogenetic pathway seems intuitively unlikely because such pathways should involve a number of genes, and the probability of reactivating multiple genes should be small. Nevertheless, evolutionary reactivation of morphogenetic processes has been observed, although the genetic mechanisms involved are not well understood. There are examples of experimental reversals among lizards and fishes, as well as natural reversals among axolotls and related salamanders. There are parthenogenetic fish species in which males are absent. Turner and Steeves, as well as Schartl and co-workers, have shown that males can be generated experimentally in such species by treatment with testosterone. Male body proportions, pigmentation, complex insemination apparatus, sexual behavior, and even spermatogenesis can be restored. Analogously, Wibbels and Crews showed that treating the eggs of a parthenogenetic lizard with a potent inhibitor of aromatase results in development of fertile males. In these cases, silenced genes controlling complex differentiative pathways were reactivated.

The time frame of silencing was evidently very short, as these unisexual species probably originated within the past 10,000 years.

Axolotls are permanently neotenic because a single-gene change has resulted in the loss of production of thyroxin, and thus failure to activate other genes involved in metamorphosis. Nonetheless, axolotls undergo metamorphosis if exogenous thyroxin is provided. Thus, the downstream genes for metamorphosis are still functional. Permanently neotenous species occur among other groups of salamanders in which metamorphosis has apparently been silenced for a much longer time, and as noted by Dent, these species fail to respond to exogenous thyroxin. Shaffer has studied a complex and species-rich radiation of ambystomatid salamanders from the Mexican highlands that has given rise to metamorphosing, facultative, and permanently neotenic species. What is significant about this radiation is that the phylogeny indicates that lost traits have reappeared; that is, neotenic species have occasionally given rise to metamorphosing species. The branches between cladistic events in many cases span less than 0.5–1 million years. The surprising conclusion is that loss of a complex trait is readily reversible over these evolutionary time spans and "flickers" with respect to speciation events.

I earlier discussed the evolution of direct-developing larvae from more complex indirect-developing forms. Structures lost by direct-developing larval forms include prey capture structures, mouthparts, a functioning gut, complex ciliated bands, and elements of the larval skeleton. The presumption would be that once these structures are lost, they obey Dollo's law and don't reappear. However, the transition times are not necessarily long. DNA distances (determined by McMillan and co-workers and by Smith and co-workers) indicate that the shift to direct development in *Heliocidaris erythrogramma* occurred in 10 million years or less. Some species of sea urchins develop partial plutei. These are essentially direct-developing embryos that still have larval skeletons and associated arms. Because of the complexity of the pluteus larva, it has been assumed that species that exhibit such nonfeeding partial plutei represent intermediate stages in the loss of the larval form. However, given the considerations outlined above, they may represent a partial recovery of some larval developmental processes. Recovery of the larval skeleton is possible because both direct- and indirect-developing sea urchins secrete adult skeletal elements late in development. Many of the structural genes expressed in secretion of the pluteus skeleton are also expressed by cells that secrete the adult skeleton (Drager and co-workers). The genes encoding skeletal proteins are preserved because they are used. However, the genes regulating the larval skeletal pattern must be reactivated. These possibilities leave us with a dilemma, because there is no easy way to distinguish whether partial plutei represent intermediates in the loss of larval features or direct developers that have reexpressed features of the feeding larval form. Moreover, the possibility of regaining features poses a general challenge for interpreting the polarity of evolutionary change in development over relatively short time spans.

Some studies have concluded that lost larval features or life history modes have been recovered. Plankton-feeding larvae are thought to be primitive in marine polychaete annelid worms. Rouse and Fitzhugh have concluded that in the polychaete annelid family Sabellidae, brooding is primitive. A cladistic phylogenetic analysis suggests that swimming larvae have arisen from these in some clades. As both styles of larvae in this family are nonfeeding, this apparent reversal of life history need not involve substantial recovery of feeding larval structures. However, Reid has suggested from a cladistic analysis of the snail family Littorinidae that planktotrophic larvae have arisen three times in a family that primitively has nonfeeding larvae.

Apparent reversals in early developmental mode may result from four causes. First, the phylogeny may be incorrect, and a correct tree would show that no reversal has occurred. Second, the apparent reversed state may actually be primitive, but the derived (direct-developing) condition has arisen in a majority of lineages, and these many convergent events lead us astray. Third, the apparent reversals may represent evolution of new feeding structures, not necessarily homologous to those of the primitive state. Finally, true evolutionary reversals may have occurred. Therefore, apparent evolutionary reversals in developmental mode must be evaluated critically.

The recovery of long-lost features is potentially an important source of information about the evolution of developmental regulatory mechanisms, and that potential has motivated some interesting, albeit misinterpreted, experiments. Experiments done on the development of the chick leg by Hampé in the late 1950s had a powerful influence on ideas about retention and reactivation of developmental genetic programs in evolution. Modern birds have a greatly reduced fibula that does not extend the full length of the tibia. The Jurassic dinosaur-bird *Archaeopteryx* had a much more reptilian leg, with a fibula that articulated at both knee and ankle. Hampé embedded a mica flake in a chick embryo between the adjoining regions of the limb bud destined to produce the tibia and the fibula. The experimental limbs developed a fibula as long as the tibia and bearing apparent articulation surfaces at both ends. Hampé concluded that normally in birds, the fibula is reduced because of competition between the tibia and fibula territories. He thought that he had blocked this competition in his experimental chicks, thereby restoring the ancestral condition. The inference was that ancient developmental programs could be retained in silent form for well over 100 million years. Müller repeated and reinterpreted the Hampé experiment in 1989. He discovered that the apparent lengthening of the fibula was due to experimentally induced shortening of the tibia. No fibular articulation formed, and there was no replay of a cryptic *Archaeopteryx* limb development program.

There is another famous case of purported experimental recovery of an ancient developmental program. The last toothed birds date from 80-million-year-old late Cretaceous rocks. Kollar and Fisher performed an experiment to determine whether mouse oral mesenchyme grafted to chick pharyngeal epithelium could induce the avian epithelial cells to produce teeth. The result was spectacular. Enamel organs, and even in a few cases molariform "teeth," appeared. Other investigators (Cummings; Lemus and co-workers; Fuenzalida and co-workers) have reported conflicting results in obtaining enamel organs or enamel deposition from avian oral epidermis cells co-cultured with mouse oral cells. The interpretation of these experiments is clouded by the possibility that the mouse dermis may have been contaminated by mouse epithelial cells. Such hidden contamination has been shown by Henry and Grainger to be a confounding problem in other studies of induction using tissue chimeras. The possibility of some response by chick tissue to inductive signals produced by mammalian tissue is consistent with the results of Dhouailly's experimental epidermal grafts between different classes of vertebrates. However, teeth contain a specialized protein, enamelin. For Kollar and Fisher's interpretation of their experiment to be correct, the enamelin gene would have had to survive in silenced form in birds for 80 million years. Although the homologue of the mammalian enamelin gene has been detected in reptiles by Lyngstadaas and co-workers, they did not detect it in birds. The probability of restoring a silent gene after more than 50 million years is near zero, even in lineages with slowly evolving genomes.

There are intimations of immortality here. Genes no longer expressed in one pathway can remain intact if they also function in some other developmental pathway. Thus, there is a plausible mechanism for the cryptic retention of genes over longer intervals than those suggested by the rate at which accumulating mutations cause degradation of coding sequences. Limb loss has occurred in whales, snakes, lizards, and amphibians. Digit loss has occurred in numerous vertebrates, including mammals. In his discussion of mechanisms of evolutionary limb loss in vertebrates, Lande showed that relatively weak selection could lead to limb reduction and virtual loss in as little as 1 million years. He noted that atavistic recoveries, such as of hindlimbs in whales, occur even in lineages in which the structure was lost up to 40 million years ago. He suggested that either the decay of requisite genes in these cases was slow enough to allow reappearance, or the required genes were preserved because they were required in other developmental pathways.

Recoveries of long-lost structures in several lineages have been interpreted as resulting from the survival of genes in other roles. Among living cats, *Lynx* possesses primitive felid features in its dentition. The carnassial teeth of the lower jaw (M1) in *Lynx* exhibit a third cusp at the hind edge of the tooth. Fossil evidence indicates that this structure was lost and then regained in this lineage. *Lynx* also exhibits the reappearance of the second molar, M2. This tooth has been absent in felids since the Miocene, about 20 million years. Kurtén has suggested that the recovery of a cusp on the rear edge of the carnassial and the reappearance of M2 are correlated, and result from a "reactivation of the molarization field," which brings M2 above the threshold of realization. Suppression of M2 probably did not involve the loss of any structural genes, since the same genes are presumably required for the morphogenesis of all teeth. In this case, recovery of a lost structure after a long period of eclipse may not represent activation of any silenced gene, but rather, alterations in the level of gene activity controlling the size or strength of the molarization field.

Other reversals have been noted by Raikow and co-workers in complex structures, including the reappearance of "lost" muscles in the limbs of some birds. Wright studied the experimental reversal from the normal three toes to a more primitive four toes in guinea pigs by selective breeding. This reversal is likely to be a consequence of the continued maintenance of an ancestral developmental pathway that can produce more toes in guinea pigs than normally appear, and can be elicited in the appropriate genetic background. Like the *Lynx* molar, toe number is a meristic trait: once the anlage is provided, the "toe program" is played out automatically.

Gene maintenance through multiple functions is not limited to morphological features. There is a growing set of examples of proteins that serve multiple functions, and thus provide a basis for their possible evolutionary retention after loss of one or a subset of their functions. The recruitment of metabolic enzymes to serve as structural proteins of the eye lens in both vertebrate and invertebrate groups is a striking example. This phenomenon illustrates the relative ease with which a change in the site and timing of expression of a protein may occur, since similar events involving several different proteins have occurred several different times.

The lens protein story is not unique. Other examples of proteins with known multiple functions include the glycosylation signal sequence-binding protein, which has at least three other diverse functions in the lumen of the endoplasmic reticulum (Geetha-Habib and co-workers). The cation-independent mannose-6-phosphate receptor involved in targeting proteins to lysosomes has been reported by Tong and co-workers to also function as a receptor on the cell surface for insulin-like growth factor-II. Deletion of the human *KALIG-1* gene has been reported by Franco and co-workers to cause both sterility and loss of olfaction. This gene functions in the establishment of two distinct nerve pathways: one in olfactory neurons, and the second in neurons that control the production of hypothalamic gonadotropin-releasing hormone, which causes the pituitary to release hormones that stimulate maturation of the ovarian follicle. Finally, the helix-loop-helix proteins discussed above in connection with sex determination play two roles in *Drosophila:* sex determination and control of the early steps of development of the nervous system.

CONSEQUENCES OF FLUID GENOMES

The significance of the disparate tales I've told in this chapter is that the expression of metazoan genomes is enormously flexible and shows an opportunistic fluidity in evolution. The term "opportunism" is not meant anthropomorphically. It refers to the process by which natural selection achieves the possible with the starting material on hand, not to any planning on the part of the evolving organism. This opportunism has several possible consequences for evolution and development. The first is that one genome can express more than a single morphology in a life history. Alternative phenotypes are present in complex life histories, in sexual dimorphism, and in castes. The features found in alternative phenotypes are important in themselves, and if dissociable, can allow novel morphologies to arise through the use of existing features, perhaps in novel juxtaposition. Heterochronic, heterotopic, and homeotic changes are the best-known evolutionary results of such dissociations. Complex features that one might expect to be constrained may be quite readily dissociated. Thus, some processes, such as sex determination, that might appear on the face of it to be homologous across wide phylogenetic spans may have different genetic bases. Because related genes are available for co-option, convergent evolution has often used similar molecular processes. Eye lenses provide one of the clearest demonstrations of this phenomenon.

In addition to its effects on development and evolution, the behavior of the genome has consequences for tracing phylogenetic histories. The fluidity of the genome generates evolutionary innovations, but may confound phylogenetic inferences by creating new patterns of gene integration. The potential for reactivation of silenced genes may seriously affect phylogenetic inferences. Cladistics assumes that evolution produces a nested pattern of synapomorphies. Yet we know that even the most parsimonious cladogram will have a substantial number of false homologies, homoplasies. Short-term evolutionary reversals may generate much of this apparently inevitable noise. Because speciation events can occur within much less than a million years, it is possible that morphological features will disappear and reappear among diversifying lineages. In groups undergoing rapid speciation, morphological innovations will not necessarily produce a simple nested pattern, but instead may produce a mix of features not all neatly tied to line of descent.

Mechanisms that allow DNA sequences to move within the genome also allow gene regulatory regions to do so. Control regions as well as genes can be co-opted for new functions. Most changes will be neutral (as suggested by Dickinson) or deleterious, but some may be selectively advantageous, as in the case of recruitment of new lens proteins. Cavener has noted that the majority of gene expression differences between species result from changes in the cis-acting regulatory elements to which transcription factors bind to control gene expression. Evidently, generation of variation often occurs through the appearance of new transcriptional regulatory sites adjacent to genes.

12

Evolving New Body Plans

Lung-fish . . . Four hundred million years ago they were the cream of
life, lords of creation; pioneers in a new way of living, escaping the
threat of death that lurks in droughts, stagnant pools, poisoned waters,
through breathing air by means of their newly invented lungs. But
they have remained almost unchanged through the ages, carrying on
in the old way of living . . . Life has gone around them, leaving them
behind.

Homer W. Smith, *Kamongo*

PATAGONIAN THINKERS

One of the most successful of all travel books is Darwin's *Voyage of the
Beagle*. It is compelling because it's a good yarn and because we detect in
it the seeds of Darwin's later theory and fame. But Darwin's account is also
compelling because of what it tells us about novelty and perception. Darwin
and Captain Robert FitzRoy shared the captain's cabin throughout the voyage.
Despite enduring crowded conditions over the years of their long journey,
they managed to remain close friends. Both were intensely curious men of
ideas, and Patagonia left its stamp on both of their minds. But how profoundly
differently it did so. To FitzRoy, the superficial geology of Patagonia confirmed the literal truth of the Great Flood. In later years, FitzRoy was unable
to reconcile himself to evolution, and was highly distressed by the public
prominence of Darwin's theory of natural selection. Darwin was to leave his
early pieties behind him and revolutionize the world. To him, Patagonia
whispered new ideas.

Perhaps the most notable of Darwin's observations in Patagonia were that
South America has many unique animals, and that the extinct Pleistocene
giant mammals of the pampas were closely related to the living animals of
the same region. The two-ton tortoiselike *Glyptodon* was essentially a gigantic and very heavily armored armadillo, and *Megatherium* merely an elephant-
sized ground-dwelling sloth, related to the much smaller living tree sloths of
South America. To Darwin, the similarity between living and fossil forms in
the same geographic region suggested a genealogical connection. One of

the links in the chain of his evolutionary theory was forming: descent with modification.

Patagonia offers rich fossil beds documenting the course of mammalian evolution in South America, which have attracted numerous investigators and yielded an amazing view of convergence in mammalian evolution. George Gaylord Simpson was to make the greatest contributions to the discovery of that history. Also of particular interest to Simpson was the intellectual history of Florentino Ameghino, the most prominent of the late-nineteenth- and early-twentieth-century Argentinean paleontologists. Early in his career, Ameghino supported himself by running a bookstore called "El Glyptodonte." In 1902 he became the director of the National Museum in Buenos Aires. Ameghino vigorously described the fossil mammals of Patagonia and named them eponymously for famous Victorian scientists. The results included such sonorous names as *Carolodarwinia, Ricardowenia,* and *Thomshuxleya.* He used the animals he described to construct his own unbridled conception of mammalian evolution. Because he failed to recognize that the similarities between South American mammals and the more conventional mammals of other continents were due to convergence, he concluded that all major groups of mammals, including humans, originated in Argentina and migrated out into the rest of the world. Not surprisingly, he became an Argentine national hero.

South America remained isolated from North America from the end of the Cretaceous (about 65 million years ago) until about 3 million years ago when the Isthmus of Panama rose and allowed the migration of northern mammals into the southern continent. The native mammals of South America were thus the products of a 60-million-year isolation of that continent. Most of the now familiar South American animals, such as big cats, tapirs, peccaries, llamas, and deer, are not natives at all, but invaders from the north. Although the native mammals that occupied similar niches are now extinct, their radiation produced half a dozen orders. The native carnivores were not cats and dogs, but bearlike and saber-toothed tiger–like marsupials. The herbivores converged on the fauna of other continents. Among the surprises, the litopterns produced a horselike set of animals that exhibited an evolutionary toe reduction uncannily like that observed among true horses. Some species had a primary hoofed middle toe with two reduced side toes, just like the three-toed horses. Another species had a single-toed foot without side toes, like that of modern one-toed horses. Whether the genetic and developmental basis for toe reduction was the same as in horses we are unlikely to know. The parallels were not identical, and litoptern dentition was very different from that of horses. Other South American orders converged on rabbits, rhinos, and elephants. The self-contained worlds of isolated continents, like the world of the early Cambrian or the world following mass extinctions, provided the empty niches necessary for evolutionary radiations and for attendant novel features to evolve.

Defining novelty is actually difficult. Certainly a novel feature must differ in some important qualitative way from the ancestral feature. Recent discussions of innovation by Mayr, Cracraft, Liem, Müller, and Raff and coworkers visualize novelty in two distinct ways. In the first point of view, novel features are considered significant if they permit a new function or provide a key element allowing new evolutionary directions to be taken by an evolving lineage; key innovations are thus recognized in evolutionary retrospect. This concept is important because a key innovation might have a relatively low cost genetically, but open possibilities for further genetic change. What makes this concept slippery is that multiple features emerge during an evolutionary history, and several of them may be important in achieving the final adaptation. How is the key innovation to be identified and assigned its proper phylogenetic role? Because the identification is retrospective, one can be left with a speculative "just-so story."

Lauder and Liem have proposed that the causal impact of a supposed key innovation on the subsequent diversification of a clade can be tested by a comparative and phylogenetic procedure that avoids the discomfort of a posteriori explanations. This method requires that the presumed innovation first be defined. A hypothesis about its historical role is then proposed based on the predicted advantageous function. Possession of the innovation is mapped onto a phylogenetic tree derived from a dataset not including the feature in question. If the innovation has caused an evolutionary diversification, a novelty having a single origin in the tree should be present in some substantial portion of the branches, and those branches should exhibit increased structural diversity related to the predicted functional advantages of the novel feature. The function of the innovation can be determined by direct experimentation, and functional biology can be used to test the presumed relationship between form and function in taxa lacking and possessing the innovation. Those data allow statistical comparisons between ingroup and outgroup taxa. If the novelty has had the predicted effect, a statistical difference in form or function should exist, with the ingroup possessing functions not present in the outgroup. Not many presumed novelties have been fully evaluated in this way, but the protocol is important in that it combines the generation of specific functional hypotheses with mapping to a phylogeny.

The second concept of novelty is that it represents a departure from an existing pattern of development. We would attempt to identify the basis of such a novelty by using comparative and experimental means to sort out possible mechanisms of evolutionary change in development. An appropriate phylogenetic framework lets us define the timing and direction of change. From this developmental perspective, a new feature can be novel even if no major clade arises as a result of its appearance. This approach focuses much less on identifying key innovations and more on degree of developmental and structural innovation.

Novel features pose the most interesting problems for understanding how evolutionary alterations in form arise. In this final chapter I present four examples of the evolution of novel features. Each example provides a perspective on the origins of the features that compose a new body plan. My discussion of body plans to this point has emphasized phylum-level disparity, but it's important to note that the term "body plan" is not equivalent in meaning to "phylum" or any other particular level of the taxonomic hierarchy. "Body plan" refers to an underlying anatomical organization that defines the members of a clade and is distinct from the anatomical organization of other clades. Because the evolution of distinct body plans includes the evolution of novel features or novel arrangements of features, distinct body plans are generally recognized as high-level taxonomic categories—as orders, classes, or phyla. The hierarchical arrangement of clades and their body plans comes as no surprise to anyone familiar with the nested arrangement of shared derived features used to construct cladograms. What must be remembered is that the complex of features that we recognize as constituting a body plan derives from genetic and developmental rules that are deeply ingrained in the ontogeny of members of the clade. The four examples in this chapter are drawn from diverse disciplinary approaches. They illustrate the tempo of evolution and anatomical changes involved in the evolution of an aquatic mammalian body plan from a terrestrial one; the origin of insect wings; the co-option of regulatory genes in the radiation of insects; and finally, the transformation of a bilaterian body plan to a radial one in the origin of the echinoderms.

WHALES AND THE RETURN TO THE SEA

Upon the extinction of the great marine reptiles at the end of the Cretaceous, a vacant niche appeared in marine ecosystems. That vacancy was filled within 15 million years by the ancestors of modern whales. Cetaceans are the descendants of rather ordinary terrestrial mammals, and their history is a story of radical morphological change. As Novacek has summarized it, "This return to the sea was so decisive that adaptations to swimming, diving and feeding match or surpass those in fishes and sharks." Among living mammals, whales appear to be most closely related to the artiodactyls (which include pigs, hippos, antelopes, deer, and camels). Both orders are descendants of a group of primitive ungulate mammals known as mesonychids, which were diverse and prevalent about 55 million years ago during the early Eocene epoch. Until recently, the transition in body plan from a four-legged terrestrial ungulate to a fish-shaped whale was very poorly understood. Recent fossil discoveries in Eocene rocks of Pakistan have transformed that puzzle into one of the best-documented examples of a rapid transition between major body plans.

The specialized anatomical features of modern whales include jaw and ear bone structures adapted for directional hearing underwater and strongly divergent from those of land mammals; forelimbs modified as powerful flippers; hindlimbs and pelvic bones reduced to a few rudimentary bones embedded in the body musculature; and unfused sacral vertebrae that allow the free dorso-ventral undulation of the body required for swimming. All of these features of the skeleton have left a fossil record.

Modern whales appeared in the Oligocene, about 30 million years ago. The phylogenetic relationships among extinct and living whales are still ambiguous (as discussed by Berta). A sketch of grades of whale evolution is shown in figure 12.1. The oldest well-adapted, albeit primitive, whales are called archaeocetes. They had evolved into morphologically primitive, but fully functional, whales by 49–45 million years ago. These animals were proportionately longer and smaller-skulled than modern whales. Gingerich and co-workers investigated well-preserved skeletons of the late middle Eocene archaeocete *Basilosaurus* (or king lizard, first described as a giant reptile in the early nineteenth century). Surprisingly, these skeletons still retained sizable and structurally complete hindlimbs. These legs were far too small for any walking, but may have served as copulatory guides, and they certainly confirm the terrestrial origin of whales.

Despite phylogenetic uncertainties, the earliest stages in the transition from terrestrial mammal to whale are now evident. The earliest well-preserved intermediate between mesonychids and whales is *Ambulocetus natans,* described by Thewissen and co-workers from 52–49-million-year-old shallow marine rocks from Pakistan. The cranium of the fossil resembles that of the archaeocetes, but the postcranial skeleton comprises a very odd mix of features. Sturdy forelimbs and hindlimbs were present, and the feet were exceptionally large. The toes bore small hooves, as in mesonychids. *Ambulocetus* was evidently amphibious in habit. It could walk on land, albeit with a sprawling gait analogous to that of sea lions. It swam by dorso-ventral undulation of the hind part of its body. The large hind flippers would have provided much of the power stroke. Whales swim by dorso-ventral undulation of their tail flukes (see Pabst). The epaxial muscles that insert along the spinal column provide the powered upstroke of the tail. *Ambulocetus* shows that spinal undulation arose before the switch from hind flippers to flukes.

The origin of specialized underwater hearing in whales has been investigated by Thewissen and Hussain in the remains of another ancestral archaeocete, *Pakicetus*. This fossil is slightly older than *Ambulocetus*. The ear bone (incus) of *Pakicetus* is intermediate between that of land mammals and that

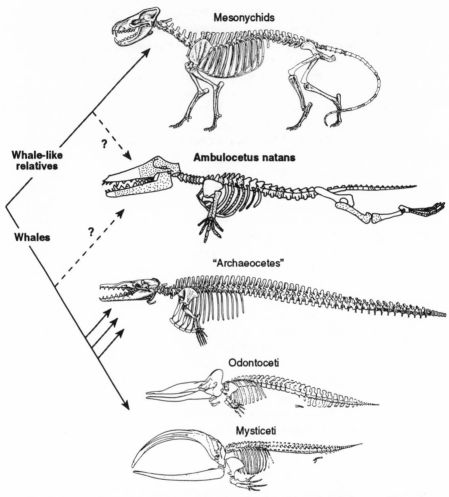

Figure 12.1. A whale phylogeny, showing the pivotal position of *Ambulocetus*. (From A. Berta. 1994. What is a whale? *Science* 263: 180–81. Copyright 1994 by the AAAS.)

of whales. In whales, the large mandibular foramen contains a fat pad that serves a wave guide for transmitting vibrations to the middle ear. The lower jaw of *Pakicetus* has a small mandibular foramen like those of terrestrial mammals. Thus, this whale ancestor was intermediate between land and water in its hearing mechanisms.

The 49–46-million-year-old *Rodhocetus kasrani* from Pakistan represents the next known grade of archaeocete ancestor. This animal, described by Gingerich and co-workers, retains some features of terrestrial mammals, such as external nostrils that open above the canine teeth. Other features of the skull, such as its proportions and the dense auditory bullae, are typical of archaeocete whales. The presence of four sacral vertebrae with an attached pelvis is a retained primitive feature. The femur was relatively short, but *Rodhocetus* could probably move awkwardly on land. Some features of the axial skeleton, including nonfused sacral vertebrae, a shortened neck, and strong dorsal spines, are whale-like. These features indicate that dorso-ventral undulation of the body axis was well developed. Unfortunately, the tail is not preserved, so it is not possible to tell whether flukes had yet evolved, but Gingerich and co-workers have suggested that the rest of the skeleton is consistent with their presence. The transition from mesonychid to primitive whale may have occurred in as few as 3 million years.

The archaeocetes evolved during the Eocene radiation of aquatic mammals. This radiation evidently occurred in and along the edges of shallow seas in what is now Pakistan. The sediments in which the fossils were buried derived from several on- and offshore environments, suggesting ecological and evolutionary diversity among these early forms. *Pakicetus* was discovered by Gingerich and co-workers in river sediments that bear fossils of numerous terrestrial mammals. *Ambulocetus* was found in shallow marine deposits, and *Rodhocetus* in deeper-water shelf sediments. Other primitive archaeocetes contemporary with *Rodhocetus* lived in shallow waters, and were noted by Gingerich and co-workers to retain long hindlimbs and fused sacral vertebrae. The mosaic nature of evolution is clearly visible within particular animals. The combination in *Pakicetus* of ear bones evolutionarily intermediate between those of whales and land mammals with a terrestrial mammal–like lower jaw is a prime example. Although the existing fossil record can tell us nothing about the developmental basis of the origin of the whale body plan, the processes discussed in chapter 10 are evident. The extent of mosaic evolution is consistent with selection acting on a highly modular body organization to dissociate developmental processes among the modules. It is interesting that the directions of dissociation shifted quite drastically in this history. In *Ambulocetus*, the toes of the hindlimbs were much longer than in mesonychids because selection favored a foot with a large paddle surface. Later in whale evolution flukes appeared, and hindlimbs were reduced in size and eventually lost. Body parts were co-opted for radically new functions as both the swimming machinery and the underwater hearing apparatus evolved.

Mosaic evolution is also evident between lineages, with some archaeocetes possessing unfused sacral vertebrae and other contemporary species possessing fused sacral vertebrae. As major anatomical features of the radiating lineages evolved, the cetacean body plan was rapidly assembled. Existing features were modified to produce a novel form of locomotion, and the steps in this process can be seen. Although animals like *Ambulocetus* and *Rodho-*

cetus look ungainly, the intermediate features that they possessed were functional for mammals that were not yet fully marine. As the machinery for dorso-ventral undulation improved, the propulsive surface was transferred from hindlimbs to flukes. Hindlimbs passed from being used for terrestrial locomotion, to use as swimming paddles, to still later possible use as copulatory guides, and ultimately, in modern whales, into oblivion. All this suggests that the evolution of new body plans can occur rapidly and in limited geographic regions, as suggested by Eldredge and Gould. Transitions like that of the cetacean body plan are interesting in their own right, and they also inform us about the nature and rates of body plan origins.

This example and the following ones emphasize the importance of the fossil record in understanding evolutionary transformations. Some cladists have dismissed the fossil record as unimportant (see Patterson). This dismissal is narrowly drawn from the concern of systematists with inferring relationships among living groups. Donoghue and co-workers have effectively shown that those who hold that the fossil record is superfluous discard a powerful source of data. Because cladists focus on shared derived features that link clades, they have little interest in the primitive features revealed by fossils. Like other methodologies, cladistics fails if features are not correctly evaluated. Fossil data are also vital to evolutionary developmental biologists who must be aware of the phylogeny of their organisms, and who must be able to map the evolutionary histories of the features they study. The fossils directly show us primitive features and transitions lost in living representatives of a clade. How would we have ever been able to infer the existence of the eight-toed foot of primitive tetrapods without the fossil record?

WINGS AND THE INSECT BODY PLAN

There are a few primitive insects that lack wings. However, the efficient wings and complex flight behaviors fundamental to insect success have been characteristic features of most insects for over 300 million years. The wing is a structural novelty that has been added to the primitive insect body plan, and is arguably a key innovation that made possible the most bountiful radiation in the animal kingdom. The question of the origin of insect wings is thus of substantial interest. This question was regarded as pretty much settled from the late nineteenth century until recently. The historically dominant view was that insect wings had arisen from paranota—solid, flat projections of the dorsal exoskeleton. According to this scenario, these paranota were at first useful as planes for gliding. Later they became hinged and articulated structures useful as powered flapping wings. The first serious modern questioning of this hypothesis came in 1973 in a paper by Wigglesworth, who argued for a completely different origin for insect wings. Wigglesworth was struck by a 1906 suggestion made by Woodworth that the mobile gill plates on the abdomen of mayfly larvae might be homologous to wings. These gill plates, which are moved by leg muscles, drive water across the gills, and also help in swimming. In an interesting chain of reasoning, Wigglesworth argued that gill plates arose from styli, the remnants of coxal exites of now-vanished biramous limbs (see chapter 4). Wigglesworth thus suggested that the thoracic styli still found on modern bristletails represent the homologues of gill plates or wings. Wings would have evolved from coxal gill plates present on the thorax of an aquatic insect. Wigglesworth cited as supporting evidence the existence of movable gill plates on the abdomens of some living primitive insects (the mayflies). He also noted that the primary flight muscles of primitive insects such as grasshoppers have their origin in the coxal segment of the leg, and that the wing lobes of roaches arise laterally in the leg region and then migrate to the dorsal plate of the thorax. Movable thoracic gill plates might have been first used in passive flight by ancestral insects. This hypothesis reopened the debate on the origins of insect wings and flight.

These ideas have been thoroughly investigated by Kukalová-Peck, who has argued that instead of gill plates having originated from styli arising from the coxa (as proposed by Wigglesworth), wings arose from an exite articulated between the uppermost segment of the ancestral insect leg, the epicoxa, and the flattened subcoxa reinforcing the body wall. The epicoxa in later insects was incorporated into the body wall as sclerites that articulate the wings. The germ of the idea of wings originating from an exite of the primitive multiramous arthropod leg remains. Kukalová-Peck's derivation of insect wings is based on her studies of beautifully preserved fossils of Paleozoic insects and aquatic insect larvae. Late Paleozoic (Carboniferous and Permian) winged insects include primitive representatives of both the Paleoptera and the Neoptera. The Paleoptera generally hold their wings outstretched like airplanes (dragonflies are familiar living members of this group). The Neoptera have mechanisms that allow them to flex their wings back over their bodies and lock them in place (as do most modern insects, such as grasshoppers, beetles, and flies). As there were no fast-moving vertebrate predators at that time, many Paleozoic insects were giants by modern standards. The largest aerial predator was a hawk-sized dragonfly. Ever since I saw my first Coal Age museum diorama, I've had an image of roaches as large and appealing as lapdogs rummaging in the Coal Age underbrush. This doesn't appear to be strictly true, but there were paleopteran insects the size of crows that sucked sap from the cones of giant club moss trees. These primitive insects retained numerous ancestral features lost in most living insects. The Palaeodictyoptera (a prominent Carboniferous and Permian group of paleopteran insects) provide spectacular examples (fig. 12.2). They are preserved as fos-

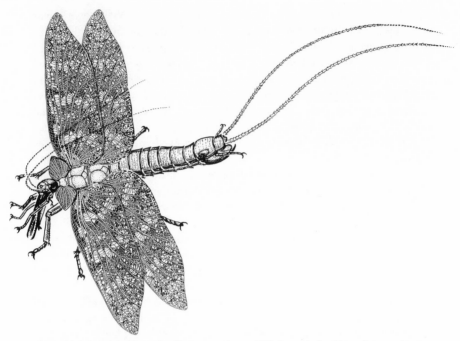

Figure 12.2.   A primitive paleodictyopteran insect. This insect had a 40-centimeter wingspan. Note the small prothoracic wings. Also note that behind the complex sucking mouthparts, the palps retain the form of complete legs. The genetic basis for the evolutionary co-option of legs into mouthparts is indicated by homeotic mutations of anterior gene members of the Hox cluster. For example, in the beetle *Tribolium,* the *labiopedia* mutation converts the palps into complete legs. The palp/legs of the paleodictyopterans may have served as a tripod to support the large head while the insect fed from the sap of Coal Age forest plants. (Drawing courtesy of J. Kukalová-Peck.)

sils in some formations in such clear detail that compound eyes, claws, hairs, exites, and even the color patterns of their large, horizontally held wings are retained. These insects had large mesothoracic and metathoracic wings, as do many modern insects. In addition, they had a third pair of smaller but functional prothoracic wings, giving each segment of the thorax a pair of wings. Palaeodictyopteran insects also retained exites on their thoracic legs and even small abdominal legs. Of their mouthparts, the palps retained a leglike form and terminated in double claws. Surprisingly, this ancient feature has been retained in living bristletails and some sawflies. The trend in insect evolution has been to reduce the leg appendages on head and abdomen and to fuse the first pair of thoracic wings to the dorsal skeleton.

The most studied living insect, *Drosophila,* is also one of the most highly

modified and streamlined. Although studies of *Drosophila* genetics and development have provided a wealth of information and insight, the nature of the
beast has also obscured our understanding of insect body plan evolution. The
extent of modification is most obvious in the head of the larva. In most
insects, the hatchling bears an insectlike head. In *Drosophila,* the embryonic
head structures invert, placing all the mouthparts inside and giving the maggot
a rather blank expression (fig. 12.3). Similarly, *Drosophila* has a long germ
band, retains no traces of legs on its abdomen, and produces adult body parts
from imaginal discs rather than by direct growth. There are living insects
that retain more primitive features. However, even they are modified and
simplified compared with Paleozoic forms. The fossils allow a look at the
primitive insect body plan, and reveal that phylogenetic inferences about
insect origins drawn solely from living species may be misleading or incomplete.

Two particularly significant observations have emerged from our understanding of the anatomies of Paleozoic insects. First, the legs of primitive
"unirames" are multiramous. The insect in figure 12.4A is a Frankensteinian
composite of ancient and modern that compares the reduced states of appendages in modern insects (right) with the primitive state (left). The modern
insect leg has a reduced number of segments (seven) and usually lacks exites.
The primitive leg has eleven segments and is multiramous, with several
small exites serially repeated on the upper segments. Abdominal limbs were
prominent in Paleozoic insects. The primitive fossil hemipterans had reduced
abdominal legs as well as the standard thoracic legs. These abdominal legs
were present at the intersegmental boundaries of all abdominal body segments, and possessed double terminal claws. Each thoracic leg bore repeated
exites (fig. 12.5). Most modern insects lack abdominal legs, although the
remnants are retained in some aquatic larvae. Exites are still evident in the
gill plates of primitive mayflies and in abdominal leglets serving as gills in
primitive damselflies, dobsenflies, and alderflies. The multiramous limbs of
primitive insects were largely lost or streamlined as modern insects evolved,
but they offered structures that could be co-opted for other uses in evolution.

The second major insight from the Paleozoic insect fossil record is
that larvae had articulated and movable winglets on their thoraxes, and
aquatic larvae had them on their abdomens as well. Figure 12.4B shows a
winged larva of a Paleozoic mayfly. Modern mayfly larvae are similar in
general appearance, but their thoracic wing pads are fused with the dorsal
skeleton. The paranotal theory of wing origin arose from the suggestion that
the fixed projections seen in living larvae might have given rise to planes
useful in gliding flight. Robertson and co-workers have found homologies of
neural connections that show that wings arose from a serially repeated set

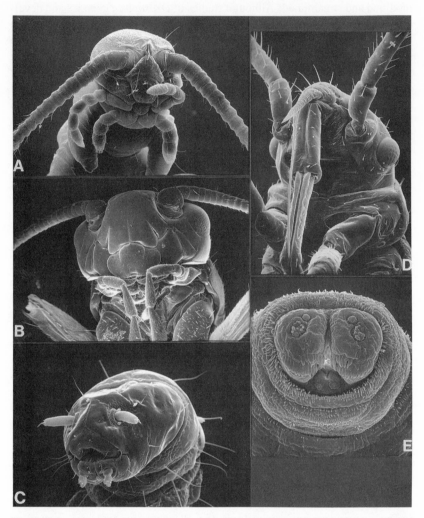

Figure 12.3. The heads of newly hatched insects representing five insect orders. *(A)* The silverfish *Thermobia domestica*, a primitive wingless insect (thysanuran). *(B)* The house cricket *Acheta domestica* (orthopteran). *(C)* The cat flea *Ctenocephalides felis* (siphonapteran). *(D)* The milkweed bug *Oncopeltus fasciatus* (hemipteran). *(E)* The fruit fly *Drosophila melanogaster* (dipteran). (Photographs courtesy of B. T. Rogers, F. R. Turner, and T. C. Kaufman.)

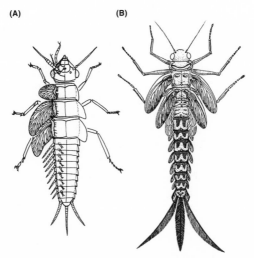

Figure 12.4. Modern and ancient insect appendages. *(A)* A composite that shows the primitive condition in the left half and the modern condition in the right. In the primitive condition, the mouthparts still retain leg morphology, and there are abdominal legs that retain exites. Wings are present on all three thoracic segments. These features are lost in the modern insect. *(B)* An aquatic Paleozoic mayfly larva with abdominal wings. (Drawings courtesy of J. Kukalová-Peck.)

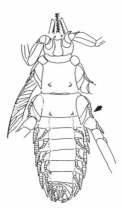

Figure 12.5. The multiramous legs of a Paleozoic hemipteran insect. The thoracic legs are multiramous, and relict abdominal multiramous legs are also present. (From J. Kukalová-Peck. 1987. *Can. J. Zool.* With permission of the National Research Council of Canada.)

of coordinated movable appendages present on both thorax and abdominal segments. Kukalová-Peck has shown that early winged insects had articulated, movable wings as larvae.

The function of protowings before flapping wings capable of powered flight evolved has been problematic until recently. Protowings must have attained a considerable size and supporting venation, articulation, and musculature before powered flight was achieved. Gliding with outstretched protowings has been proposed as one possibility. Wootton and Ellington built model insects equipped with nine rotatable winglets that were 5% of body length. These models glided quite well, with glide angles of 20–30 degrees. A Paleozoic insect equipped with such short winglets could have leapt from a low tree and glided a substantial distance. However, Kukalová-Peck has pointed out that there is no fossil evidence for outstretched projections used for gliding in ancient insects. The outstretched wings of modern insects capable of gliding (dragonflies, mayflies, and migratory butterflies) have locking mechanisms. More primitive insects do not hold their wings in an outstretched position for gliding.

A different scenario for the origin of insect flight is as follows: Primitive insects started out with multiramous limbs inherited from aquatic arthropod ancestors. The exites already present on the ancestral insect leg were articulated, movable appendages under neural control. Over time, leg segments were fused to one another or incorporated into the body wall. These early insects had aquatic larvae similar to those of living mayflies. Development was direct, with a gradual transition from the aquatic larva to the terrestrial adult. There was no metamorphosis. Larval wings evolved by modification of the preexisting leg exites. Wing veins probably evolved from the cuticular ridges that strengthen flat, movable arthropod appendages. As extensions of the blood cavity of the body, they also serve to distribute blood to these appendages. Primitive wings are similar to flattened swimming structures in crustaceans, which are also filled with veins. In primitive living and fossil insects, wings are broadly articulated to the body by several articular sclerites (chitinous plates) evidently derived from the first leg element. In primitive aquatic larvae, exites on the first leg segment were flattened and served in respiration. However, as these larval flaps could move water, they were preadapted for use in locomotion. In essence, the larvae could use these structures in underwater flight. The larval protowings would have served newly emerged adults in weak flight to escape predators. Movable protowings are clearly a preadaptation to the evolution of larger wing surfaces, musculature, and more efficient powered flight.

The environment in which insect flight arose is an important consideration in the protowing scenario. Kukalová-Peck has argued that although Paleozoic winged insects are known from their spectacular flowering during late Carboniferous and Permian times, they arose much earlier. She has suggested that they arose as early as Ordovician to Silurian times, and that stem-group pterygotes (winged insects) existed by Silurian times. These ancestral ptery-gotes were probably equipped with movable protowings, which may have allowed effective escape from predators by powered flapping and skimming away. Kukalová-Peck's scenario puts the origin of flight much earlier than previously thought. Instead of having their origins as gliders among the tall brush and trees of the Earth's first forests in late Devonian times, insects gained their wings in an environment with little elevation. Early Paleozoic land plants were stunted things projecting from pools of standing water. There was little soil or plant litter. Pressure from carnivorous arthropods was probably high. Gliding would have been far less effective than a powered escape from predators. The nature of that powered escape is suggested by a recent study by Marden and Kramer, who observed stoneflies, living primi-tive insects that can skim on the water surface by flapping short wings that cannot support effective free flight. Skimming is a partially airborne form of powered flight that allows stoneflies to support part of their weight on the water. Marden and Kramer found that a short-winged species of stonefly could skim effectively even with wings experimentally shortened to only 20% of their normal length. This experimental evidence supports the inference from the fossil record that ancestral insects with short protowings might have found a selectively advantageous use for these structures intermediate be-tween rowing through the water and flying through the air.

The origin of insect wings has left clues to a genetically complex history in the fossil record and in the developmental genetics of living insects. Cohen and co-workers have used genetic markers to trace the origins of wing and leg primordia in *Drosophila*. The localization of leg fields is patterned by the expression of the wingless and decapentaplegic genes in the embryonic epithelium, as diagrammed in figure 12.6A. The *wingless* gene is expressed at parasegmental boundaries as bands around the embryo (parasegments are the first manifestation of segmentation in *Drosophila* development, and are offset by half a segment from the final segment boundaries). The *decapen-taplegic* gene is expressed in two stripes that run down the length of the embryo on its ventral surface. Limb primordia arise at the intersections of these two sets of gene expression stripes in the three thorax parasegments. The sites of origin of the limb primordia are causally related to the molecular pre-pattern: Simcox and co-workers observed that no imaginal discs formed in *wingless* mutant fly embryos.

The limb primordia can be visualized by their expression of the *Distal-less* gene, which is probably activated by the *wingless* and *decapentaplegic* pro-teins. Anderson showed that the primordia for the wings of the Queensland fruit fly, *Dacus tyroni,* originate anatomically in leg discs. The wing primor-

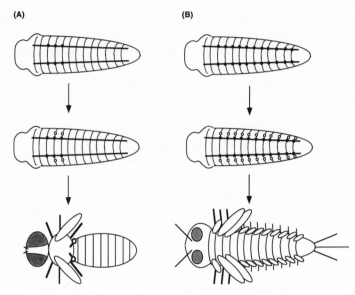

Figure 12.6.   Gene expression patterns in the specification of the leg primordia of insects.
(A) Ventral view of *Drosophila* embryo with *wingless* expressed in bands and *decapentaplegic*
in two longitudinal stripes. Leg and wing primordia appear at intersections of gene action in the
thorax. *(B)* Hypothetical gene expression patterns in establishment of legs and wing primordia
in a Paleozoic insect. Legs and wings arise in both thorax and abdomen.

dia then migrate toward the dorsal surface. They ultimately give rise to a
pair of wings on the second thoracic segment and to a pair of lollipop-shaped
balancing organs, the halteres, on the third thoracic segment. The halteres
are evolutionarily reduced wings. In dragonflies, living paleopterous insects,
Bocharova-Messner showed that the wing bud first appears near regions that
represent portions of the primitive legs incorporated into the thorax wall,
then migrate dorsally to (ultimately) produce wings. The relationship of leg
and wing primordia is thus an ancient one.

The ancestors of beetles and flies must have diverged by the early Carboniferous, about 350 million years ago. (Protocoleoptera are known from the
early Permian. The mecopteroid ancestors of flies are known from the late
Carboniferous; the oldest true fly is from the late Permian.) Figure 12.6B
diagrams a sequence of hypothetical events in an ancestral Paleozoic insect,
illustrating a possible evolutionary pathway from primitive multiwinged insect to modern insect. We don't know whether the *wingless* and *decapentaplegic* coordinate system was present in the ancestral winged insect, but it
very likely was. The beetle *Tribolium,* as a short germ band embryo, retains

the primitive mode of laying down the body segments. Nagy and Carroll found that *Tribolium* embryos also express *wingless* in a segmentally iterated manner. The *wingless*-expressing cells lie in the same relationship to *engrailed*-expressing cells as in *Drosophila,* implying a long-conserved mechanism of gene expression as well as a conserved pattern. In the primitive Paleozoic insect, leg and wing primordia would have arisen in all thorax and abdominal segments. The wing primordia would have migrated dorsally, as they do in dragonflies. The adult would have had legs and wings on all body segments. As Paleozoic insects evolved further, thoracic legs and wings came to predominate. The suppression of abdominal legs and wings is controlled by the homeotic genes of the Antennapedia and Bithorax complexes.

The Antennapedia and Bithorax complex genes act in regionalized domains in *Drosophila* (see Kaufman and co-workers and Sánchez-Herrero and co-workers). The Antennapedia complex genes *Sex combs reduced* and *Antennapedia* control segmental identities in the head and first and second thorax segments. The Bithorax complex gene *Ultrabithorax* regulates the identity of the third thoracic segment. *Abdominal A* regulates segmental identities in abdominal segments one through seven, and *Abdominal B* in abdominal segments six through nine (the tail end of the fly). Expression of the *Distal-less* gene, which identifies limb primordia, has been shown by Simcox and co-workers to occur in the abdomen only in embryos lacking the Bithorax complex. The *Distal-less* gene has been shown by Vachon and co-workers to possess a homeodomain-binding site. Modification of that site causes expression in the abdomen.

Loss of function mutations in *Ultrabithorax* cause the now famous phenotype of four-winged flies, in which the third thoracic segment assumes a second thoracic segment identity, and the halteres return to their ancestral identity as wings. There are important limits, however: the transformed halteres become contemporary *Drosophila* wings, not Paleozoic insect wings. Two kinds of gene evolution have occurred: changes in patterning and changes in downstream genes that execute the pattern. As we are concerned with body plans here, we will focus only on patterning and not on the specific kind of wing that gets made.

Deletion of the Bithorax complex produces a larva in which the segments following the second thoracic segment are morphologically transformed to assume the same identity as the second thoracic segment. As *Drosophila* larvae have no legs on any segment, this transformation yields no abdominal appendages. However, an analogous deletion of Hox cluster genes in the beetle *Tribolium,* which has thoracic legs as a larva, was found by Stuart and co-workers to restore appendages to all abdominal segments. Such deletions are lethal, so these manipulations do not result in adult throwbacks. What they do show is that the evolution of novel regulatory patterns in the

Hox genes led to the modern patterns of limb and wing expression. It appears that these patterns resulted from alterations in the control of expression, not from the evolution of new Hox genes. There are two reasons for this conclusion. First, the *Drosophila* Hox genes have complex promoters and transcriptional regulation. The Bithorax complex contains only three genes, *Ultrabithorax, Abdominal A,* and *Abdominal B,* yet a complex pattern of segmental identities is generated by the transcripts of these three genes. Second, it is now clear that the Hox gene cluster predates the insects. These genes have been co-opted into regulation of complex segmental identities.

GENES FOR LEGS AND WINGS

The appendages of insects are extensions of the embryonic epidermis. As Williams and Carroll have pointed out, there has been extensive evolutionary co-option of regulatory genes in their patterning. That patterning lies in three axes: anterior-posterior (front and back), dorso-ventral (top and bottom), and proximal-distal (from body wall to appendage tip). Williams and Carroll have suggested that genes involved in patterning the parental field were co-opted during evolution to function in derivative fields as well. This evolutionary logic is evident in the determination of limb axes. The segment polarity genes that are involved in the generation of body segments are also required for the anterior-posterior patterning of the imaginal discs that give rise to the appendages.

The dorso-ventral patterning of appendages has also been co-opted from the epidermal pattern. The *wingless* gene expression stripes that define the parasegmental boundaries of the body are interrupted at the level of the stripes of *decapentaplegic* gene expression, and thus at the level of limb primordium formation (see fig. 12.6). The expression of *wingless* is restricted to the ventral cells of the limb primordium. This association is a functional one. Struhl and Basler showed that ectopic expression of *wingless* in dorsal cells of the developing leg disc caused the appearance of ventral cell identities on the dorsal side of the resulting adult leg.

The patterning of the proximal-distal axis of appendages is less well understood, but requires the action of genes that establish the dorso-ventral axis. It is significant that, downstream of the genes involved in setting up the dorso-ventral axis, the establishment of the proximal-distal axis uses genes that are unique to proximal-distal patterning of appendages. The wing has acquired a genetic machinery for its patterning distinct from that of the limb. Proximal-distal patterning of the leg requires *Distal-less,* which is not expressed in the wing. Instead, proximal-distal patterning of the wing requires the action of the genes *apterous, scalloped,* and *vestigial.* All (as summarized by Williams and Carroll) are transcription factors, and thus establish cellular identities and regulate the expression of genes involved in the execution of pattern.

Although the wing is evolutionarily derived from part of the ancestral limb, it is topologically distinct in being planar rather than a tapering cylinder. Williams and co-workers have suggested that an evolutionarily novel aspect of wing development was the formation of a dorso-ventral boundary. Williams, Paddock, and Carroll have shown that expression of the *wingless* and *apterous* genes is required to establish dorsal (*apterous* expression) and ventral (*wingless* expression) wing surfaces, and that establishment of a dorso-ventral boundary is required for wing development. Establishment of that boundary requires the expression of yet another gene, *vestigial*, which Williams and co-workers have shown to be expressed at the boundary. The *vestigial* gene is controlled by a highly conserved enhancer located in its second intron. Both *vestigial* and the similarly expressed *scalloped* gene are transcription factors that may establish the identities and subsequent behaviors of cells at the boundary. Thus, a novel gene expression system has evolved as part of the conversion of a limb element into a novel planar appendage, the wing.

DIFFERENTIATING THE FLIES FROM THE BUTTERFLIES

The butterflies are the birds of paradise of insects. They are also fine subjects for testing hypotheses about the roles of regulatory genes and developmental programs in the evolution of the insect body plan. Sean Carroll and his collaborators (see Carroll; Carroll and co-workers; Panganiban and co-workers; Warren and co-workers) have investigated the regulation of patterning in these insects to explore how differences in patterning have arisen in two related but distinct insect orders. Their results have revealed three major evolutionary shifts in genetic controls.

The classic hypothesis for the evolution of the insect body plan and its intricate pattern of specialized segments and appendages was put forward in papers published in 1963 and 1978 by E. B. Lewis. He suggested that in the evolution of six-legged insects from a myriapod-like ancestor with legs on all segments, a set of genes arose that suppressed the development of abdominal legs. Further, flies have only a single pair of wings on the second thoracic segment, the second pair of wings having evolved into the halteres. Lewis suggested that genes had evolved to direct this conversion in the third thoracic segment. Because mutations in the genes of the Bithorax complex change segmental identities, Lewis proposed that the complex tagmosis pattern of the fly arose in conjunction with a set of gene duplications that produced the complete Bithorax complex. This idea is not tenable now that the Hox gene cluster is known to predate the arthropods and to be shared with other phyla.

That being the case, Carroll and his colleagues addressed the question of how, if they arose first, the Hox genes have influenced the subsequent evolution of the insect body plan.

The function of the Hox genes in the regulation of butterfly development offers a particularly interesting case in comparative biology because, unlike flies, these insects possess abdominal as well as thoracic legs as larvae, and as adults they have four wings. The abdominal legs are the five pairs of stumpy prolegs characteristic of caterpillars; no abdominal legs are present in adult butterflies. Dipterans and lepidopterans are fairly closely related insect orders despite these differences. Carroll and his colleagues cloned from *Precis coenia,* an easily reared butterfly, a number of butterfly homologues of major regulatory genes discovered in *Drosophila.* This work is a particularly nice example of an informed extension of information and methodology from a model system to a nonmodel organism.

The Hox genes were found to be expressed similarly in flies and in butterflies, with the Hox genes *Scr, Antennapedia, Ubx,* and *abdominal A* all expressed in the same anterior-to-posterior domains and in the same parasegmental registers. *Ubx* and *abdominal A* regulate segmental identities in the posterior part of *Drosophila,* and evidently do so in butterflies as well. A fundamental similarity in body plan thus underlies the differences in development. The development of larval thoracic legs in the butterfly is marked by expression of *Distal-less,* as is development of the leg imaginal discs of *Drosophila.* As butterfly development proceeds, however, the pattern of abdominal gene expression deviates sharply from that of *Drosophila. Ubx* and *abdominal A* expression disappear from patches of cells destined to give rise to abdominal proleg primordia. *Distal-less* is then expressed in these cells, marking commitment to proleg development. It is interesting, as pointed out by Warren and co-workers, that the apparently simple route of modifying cis-regulatory regions of *Distal-less* to allow its expression in the presence of *Ubx* and *abdominal A* expression has not been taken; instead, localized repression of the Hox genes has been required. They suggest that *Distal-less* action alone does not lead to limb development, but rather that a whole program must be locally activated. These observations pose interesting questions about how abdominal limbs are regulated by Hox genes in myriapods, other insects, and crustaceans.

Since *Ubx* mutations result in conversion of the halteres to wings in *Drosophila* and other dipterans, one might suppose that the *Ubx* gene acts directly, and that its product is sufficient to cause the wing-to-haltere conversion. If this is so, butterflies, which have a full second pair of wings, should not express *Ubx* in the third thoracic segment. Warren and co-workers tested this hypothesis in *Precis.* It is not the case. *Ubx* is expressed both in the butterfly third thoracic segment and in the wing rudiments. The evolutionary difference between wing and haltere has arisen via the evolution of new responses by downstream genes in these two structures. The ground plan of homeotic gene expression allows divergence in subsequent developmental steps, and thus differently winged insects.

Butterflies illustrate a final level of gene co-option: pattern formation within the wing. The color patterns of butterfly wings are wonderfully diverse, yet Nijhout has shown that they repeat variations on an underlying ground plan. Color patterns develop within regions of the wing bounded by veins and the wing margin. These domains are called wing cells (an unfortunate terminological ambiguity). Molecular pre-patterning takes place in the wing imaginal discs before the differentiation of the scale cells that produce the pigments of the wing pattern. Carroll and his co-workers hypothesized that the genes involved in patterning the entire appendage might also govern the development of the color pattern of the wing. Amazingly enough, that's exactly what they found. *Wingless* is expressed in rays in cells that flank the midline at the distal end of each wing cell. *Scalloped* appears in stripes along the wing veins. *Apterous* appears in chevrons at the distal ends of dorsal wing cells. None of these patterns corresponds to the ultimate wing color pattern of *Precis,* but Carroll suggests that these transcription patterns reflect the underlying general patterning system for butterfly wings. The interpretation of this general system would be species-specific. The expression of *Distal-less* displays the dynamic patterning process well. *Distal-less* is initially expressed in the midline areas of each wing cell, and then expands at the proximal ends of these midline rays to form small circles. *Distal-less* forms the eyespot pre-pattern in those wing cells that will later form eyespots, and its expression is lost from those wing cells that do not form eyespots. I'll return to a broader consideration of the issue of gene co-option in summing up.

THE ECHINODERM RADIAL BODY PLAN

The genetic and developmental systems that underlie evolutionary changes within body plans are now becoming accessible. They are not known, however, for transformations between phylum-level body plans, and even descriptive histories of these events remain obscure. Although they are among the most peculiar of animals, echinoderms offer the prospect of tracing such a history. The story presented below is a work in progress, but I have included it to illustrate an approach that should be feasible for investigations of other body plan transformations.

The living echinoderms share a body plan that includes an obviously pentamerally symmetrical adult. Echinoderms possess a unique ring-shaped coelomic water vascular system that generally has extensions into each of the five ambulacra (the echinoderm arm or radius and its associated podia). Hydraulic pressure generated in the water vascular system powers the many podia, or tube feet, with which the animal walks and manipulates food. Echinoderms possess a mesodermal endoskeleton composed of individual plates, each a single crystal of calcite. Each plate is built as an open meshwork, or stereom, that is a characteristic of echinoderm skeletal elements. These animals possess a circulatory, or hemal, system. A gut is usually present, but excretory organs are absent. The central nervous system (with some complexities that I'll describe below) consists of a central nerve ring from which radial nerves project into the ambulacra. (The details of the complex anatomy of echinoderms can be found in Brusca and Brusca.) These features define the body plan of the phylum as it exists today. In contrast to that of whales, the fossil record gives few clues about the appearance of this body plan, because unambiguously recognizable echinoderms first appear in the Cambrian radiation. As with other phyla, we have to infer the events that produced the echinoderm body plan by other methods. The expression of conserved pattern-regulating genes in development offers an approach for tracing the major body plan transformations.

When one looks at the most ancient echinoderms of the Cambrian, it's evident that few of the familiar pentamerally symmetrical classes were present. The predominant Cambrian echinoderms fell into nonpentameral, asymmetrical, and nearly bilaterally symmetrical classes (see Sprinkle and Sprinkle and Kier for general reviews of ancient echinoderms). Seven classes arose during the early and middle Cambrian. Of these, the Eocrinoidea, Crinoidea, and Edrioasteroidea appeared in the record with pentameral symmetry and rapidly improved the regulation of their pentamery. The organizations of the other classes, the Stylophora, Homoiostelea, Homostelea, Ctenocystoidea, and Helicoplacoidea, ranged from asymmetrical to nearly bilaterally symmetrical. More pentameral classes, including some now-extinct Paleozoic ones as well as the familiar living classes—the sea urchins, brittle stars, and holothurians—appeared some 30–50 million years later during Ordovician times. Campbell and Marshall have pointed out that there is little observable convergence between classes as one goes back in time. Features do not merge, and the classes are well defined at the time of their first appearance in the record. Despite phylogenetic analyses such as that of Paul and Smith, the relationship between the ancient nonpentameral echinoderms and the pentameral groups remains obscure. Paul and Smith place the nonpentameral groups as sister groups to the pentameral classes. That inference is consistent with the other elements of the echinoderm body plan having originated before pentameral symmetry, as cartooned in figure 12.7. This hypothesis of a stepwise addition of body plan elements is analogous to the evolution of the vertebrate body plan by addition of elements to an earlier evolved chordate body plan. It is important to keep in mind that the body plans of the ancient

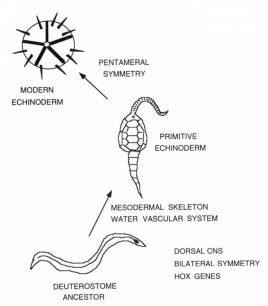

MODERN
ECHINODERM

PENTAMERAL
SYMMETRY

PRIMITIVE
ECHINODERM

MESODERMAL SKELETON
WATER VASCULAR SYSTEM

DORSAL CNS
BILATERAL SYMMETRY
HOX GENES

DEUTEROSTOME
ANCESTOR

Figure 12.7. A schematic history of the origins of features of the modern echinoderm body plan. A primitive deuterostome ancestor gave rise to chordates, hemichordates, and echinoderms. It was probably bilateral in symmetry and had a dorsal nerve cord. A number of early Paleozoic echinoderms had bilateral or nearly bilateral symmetry; thus, features characteristic of the phylum arose in nonpentameral early forms. Pentameral symmetry was added to that suite of features. Note that the arrows do not indicate direct ancestry, only the direction of transformation of features.

members of a phylum are not necessarily identical to the body plans of the living members. Older taxa have died out and ancestral features have been lost. In echinoderms, only pentameral groups remain. In characterizing the Cambrian echinoderm body plan, we would not list pentameral symmetry, but would list the water vascular system and calcitic stereom endoskeleton.

Because the nonpentameral echinoderm classes are so long extinct, DNA sequence phylogeny offers little hope for placing them with respect to the pentameral groups. However, as I'll outline below, other molecular tools involving the deployment of patterning genes offer a powerful means for discovering how symmetry was modified during the evolution of the echinoderm body plan.

Symmetry and design among the Cambrian echinoderms was downright odd (fig. 12.8). For example, the helicoplacoids were covered with calcareous plates in a spiral arrangement. A forked ambulacral tract wound around the body. The mouth was apparently located on the side, at the junction of the ambulacral tracts. Other nonpentameral echinoderms included the nearly

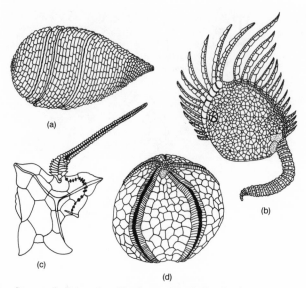

Figure 12.8. Some early Paleozoic echinoderms: *(A)* a helicoplacoid; *(B)* a paracrinoid; *(C)* a mitrate stylophoran; *(D)* a primitive echinoid. (From Raff and Kaufman 1983; drawn by E. C. Raff. Reprinted with permission of Indiana University Press.)

bilaterally symmetrical stylophorans discussed by Parsely. These old-timers, particularly the stylophorans, look very strange, and there has been a long debate over their structure and relationships. The most divergent view is that of Jefferies, who has interpreted the long arm as a tail containing structures homologous to the notochord and dorsal nerve cord of vertebrates. He has vigorously argued that these animals are the sister group of chordates. Most students of echinoderms disagree, and instead interpret these animals as primitive nonpentameral echinoderms. The stylophorans have a calcite skeleton, echinoderm-like respiratory pores, an anal pyramid, and a tail that bears the structures of an ambulacral tract with a food groove and cover plates.

Pentameral symmetry is certainly not primitive in echinoderms. As I discussed in chapter 4, echinoderms are the sister group of the hemichordates and the chordates. The ancestor of these clades was most likely bilaterally symmetrical and cephalized, and in possession of a differentiated anterior end and a dorsal nerve cord. How was this linear body plan transformed in a radially symmetrical one?

A key to the transformation of the linear body plan shared by the ancestors of all deuterostome phyla lies in the topology of the central nervous system. The central nervous system is deeply integrated into the body plan of animal groups and develops as part of the phylotypic stage. Gene expression patterns in the nervous system are conserved, as is development of its morphological features. The connections made by the neurons of arthropods, for example, are still retained among modern clades that separated during the Paleozoic.

Despite the elaborate behavior of which echinoderms are capable, their nervous systems can hardly be described as highly centralized or cephalized. Living echinoderms have three nervous systems (summarized by Hyman). The first is the oral system, arranged as a pentagonal ring surrounding the esophagus. Five radial nerves radiate from the nerve ring, each down the center of an ambulacral tract. The radial nerves are apparently primarily sensory in function. The second system is the hyponeural system. It is usually present as five radial tracts running along one side of the radial nerves of the oral system, and is primarily locomotory in function. The third system is the aboral system. It resembles the oral system in its organization, but its ring is located at the aboral pole of the animal. In most living echinoderm classes, this system is very poorly developed, and the oral and hyponeural systems provide the innervation of the animal. Crinoids are the exception. They possess a circumoral nerve ring, but the aboral system is their primary nervous system. The role of that system was illustrated by a draconian experiment conducted in the 1880s by Marshall (cited by Hyman). He eviscerated comatulid crinoids (stalkless free-swimming crinoids descended from stalked ancestors), thus removing the oral and hyponeural nerves. The eviscerated animals swam and behaved normally.

In vertebrate development, the Hox genes exhibit a pattern of expression in the central nervous system that is collinear with their organization in the Hox cluster. Since both the collinearity and the nervous system expression of these genes are conserved, the expression pattern of the Hox genes in the developing sea urchin nervous system should reveal the topological transformation. Three hypothetical transformations are presented in figure 12.9 along with their predicted consequences for Hox gene expression patterns. In the first, the radial nerves represent the ancestral dorsal nerve cord. The five radial nerves of the echinoderm would have arisen by replication of the original single nerve cord, followed by the evolution of a ring commissure to link the nerve cords together. This hypothesis predicts that the anterior-to-posterior pattern of Hox gene expression would be replicated in each radial nerve.

The other two hypotheses propose that the ancestral nerve cord was itself transformed into a ring. In the second hypothesis, the ring is suggested to have arisen by means of a head-to-tail fusion. In that case, the Hox genes would be expected to be expressed around the ring, but in one direction, like the mythical hoop snake holding its tail in its mouth. The third hypothesis suggests that the ancestral nerve cord consisted of bilaterally paired strands. A split of the center portion would have produced a ring, but one with a different Hox expression polarity than the head-to-tail ring. In both of these

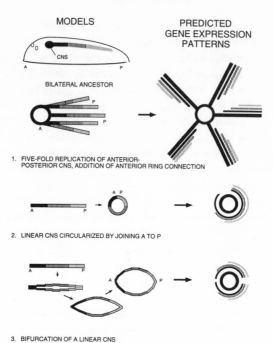

Figure 12.9.   Three models for the origin of the echinoderm radial nervous system and its predicted consequences for Hox gene expression. Models of topological transformation are shown to the left; predicted patterns of Hox gene expression are shown to the right. Hox genes are expected to be offset in their anterior boundaries, as in vertebrates (e.g., fig. 10.2). (After Raff and Popodi 1995.)

scenarios, the radial nerves would have arisen subsequent to the origin of the ring, as they do in development. With the cloning of most of the sea urchin Hox genes in our laboratory and others, an examination of expression patterns by means of in situ hybridization of Hox gene probes to developing sea urchin nervous systems is now feasible. These experiments are in progress. The few available gene expression data are consistent with the first model of figure 12.9; that is, the five axes of living echinoderms represent the multiplication of the original single body axis. The anterior ends of these axes surround the mouth. The stylophoran of figure 12.8C has a single body axis. The helicoplacoid (figure 12.8A) has three axes that converge on a mouth, and may thus represent an early experiment in multiplication of body axes. Living clades settled on five.

It's only fair to note that our diagrammed hypotheses might be incorrect. As there is more than one ring nervous system in echinoderms, we have to account for the origin of similar but differently located radial nervous systems. Do they represent homeotic duplications of the ring nervous system once it was established, or did they arise independently? Independent origins of ring nerves from a linear nervous system are suggested by the existence of ring nerves in a related phylum, the hemichordates (see Hyman). There is a major dorsal nerve cord running the length of the body in the hemichordate *Saccoglossus*. A major ventral nerve is present in the trunk, but not in the anterior part of the animal, the large proboscis. A mid-body nerve ring connects the dorsal and ventral nerve cords. A second nerve ring lies at the anterior end of the animal. That ring derives from the dorsal nerve cord and serves as a base for the many nerves that run anteriorly to innervate the proboscis. Alternatively, the echinoderm nervous system might not have arisen from an ancestral linear nerve cord at all, but from a preexisting diffuse nerve net. That appears improbable, but if correct, it would mean that the echinoderm central nervous system is not homologous to that of related deuterostomes. If this is the case, the genetic determination of its patterning should be unrelated to that of other nervous systems.

RATES AND FOSSILS

Evolution has not been consistent in rate, and unfortunately for us as observers of nature and reconstructors of evolutionary history, most of the exciting and large-scale evolutionary events have occurred rapidly relative to the resolution of the geologic record. Frustratingly, the majority of geologic time seems to be taken up by stasis, the long season of a species' life between episodes of change. Most of the fossil record consists of well-established species. Darwin discussed the problem of the rarity of transitions in the fossil record in *The Origin of Species*. He noted that transitions between species might be short in duration by comparison with the life of an established species. Simpson also suggested that periodic bouts of rapid evolution were responsible for major evolutionary changes. Eldredge and Gould (see also Gould and Eldredge) gave a more explicit formulation of the idea in their concept of punctuated equilibrium. Most of a species' duration sees little change. Change, when it occurs, is at the genetic periphery, and is rapid, giving rise to new species. The Cambrian radiation of animal body plans fits this pattern, and so do many other evolutionary episodes. The example of early whale evolution discussed above shows how rapidly body plan elements can be reorganized. Our considerations of the interaction between development and evolution are affected by the antithetical phenomena of long-term stability and rapid change.

Notwithstanding the nature and elusiveness of rapid innovation, some major body plan transformations can be observed in the fossil record. These allow us some compelling looks at patterns of change. Fossils illustrate the origin of the tetrapod limb, the transformation of the reptilian skull to its mammalian condition, the origin of the whale body plan, the evolution of the modern human skull from that of primitive hominids, and the evolution of the appendages of insects. The fossil record lets us see primitive states that can be inferred very imperfectly from living forms. The multiramous nature of the primitive insect appendage is a case in point. Without the fossils described by Kukalová-Peck, the inconspicuous exites still retained on the legs and palps of a few living insects would most likely be ignored, and we would consider insect legs as primitively uniramous. The most convincing evidence for the origin of insect wings from leg exites also derives from the fossil record. Our understanding of the origins and evolution of insect appendages has been enormously enhanced by the use of fossil evidence in conjunction with developmental genetics. The fossil record gives us crucial information on primitive states and the direction and rates of evolution. When it is combined with evidence from anatomy, molecular biology, and developmental genetics, we obtain our clearest glimpses of the origins of novel features in evolution.

EXPLORING BODY PLAN EVOLUTION

This book is my attempt to construct a triton, a merman composed of evolutionary and developmental parts. Like the triton swimming forever in the wine-dark sea that circles an ancient Greek vase, my construct must be a mosaic of harmoniously fused parts of very different intellectual origins.

Although this may be a bit startling to readers who have grown up in a tradition of genetics or developmental biology, the history of life matters, and that is why I have devoted the first part of this book to an exploration of phylogeny and to the current approaches to tracing the histories of animal body plans. The importance of phylogeny is often unappreciated by biologists whose research traditions have ignored evolutionary perspectives. Without knowledge of relationships, we cannot know directions of change in developmental processes. Without knowledge of vanished states in extinct organisms, we cannot trace the origins of the features that define living body plans. Without a knowledge of phylogeny, we cannot map the histories of master regulatory genes. At present, the histories of some of the most important transitions remain shrouded. Biologists trained in reductionist disciplines may be surprised to find that many of the relationships among phyla and other major groups remain unresolved.

Seeking the basis of the establishment and evolution of body plans and understanding the evolution of animal form requires us to integrate disciplines that were until recently largely distinct in their intellectual foundations, problems, and approaches. The task requires that we integrate three kinds of data: phylogenetic, functional, and comparative.

Through most of the twentieth century, evolution has been assumed to be irrelevant to developmental biology. However, with the discovery of the widespread role of Hox genes in animal development, evolutionary relationships have regained their century-old fascination. The reengagement of the two disciplines will be more difficult than has been commonly assumed because of a basic difference in their concepts of causality and because evolutionary and developmental biologists have focused on different problems for so long. To what degree the two fields will be altered by their reengagement is unclear. A substantial change is taking place among developmental biologists, but a full appreciation of phylogenetic information and the importance of homology to understanding developmental problems has still to be achieved. Developmental biologists, until recently, have not been concerned with issues of homology because they sought an understanding of process over any evolutionary connections. The wide phylogenetic expression of Hox and other master regulatory genes has forced a renewed focus on homology because it has become necessary to determine just what the pattern of expression of these genes among dissimilar organisms means.

Without phylogenetic information, we can compare organisms, but we cannot map the direction of change, nor estimate its rate. Of course, we are limited by the quality of our phylogenetic information. If the phylogenetic tree we use is in error, so may be our inferences. Even if a correct phylogeny is available, we must be cautious in the inferences we draw about the primitive states of developmental, morphological, or genetic features. The ideal use of phylogenetic trees or cladograms is to map changes in features and to determine ancestral traits. Unfortunately, life is never so convenient as we would like, and information about primitive states cannot be simply read off a cladogram prepared from living taxa, even if (as pointed out by Frumhoff and Reeve and by Marshall and Schultze) the tree itself is correct.

Phylogenetic information may be degraded in three important ways. First, information is lost as a consequence of the extinction of basal clades. For example, among echinoderms, only derived pentameral forms remain, but several basal lineages were different in symmetry and other features. Second, information is lost as a consequence of the subsequent evolutionary remodeling of a feature to the point at which the primitive state is no longer recognizable. The number of digits of the tetrapod hand and foot is an example. All cladograms derived from living tetrapods will agree that five digits was the primitive state because no living tetrapod has more than five digits, either as an adult or during development. Until the discovery of wellpreserved *Acanthostega* skeletons, the dogma was that the ancestral tetrapod hand or foot had five digits. In other examples, one entire order of sea urchins, the echinothurioids, has only direct development; the pluteus has been lost by the entire clade. If these were the only living sea urchins, we would have an incorrect notion of their larval evolution. The loss of old gene sequences through multiple base substitutions shows the same effect. The gene is still homologous to that of other clades, but may be low in base sequence similarity. The third cause of information loss is the reversal of states of features. The earlier discussion of Dollo's law illustrates the point. Certain relatively labile features such as life history traits will be readily reversible, as will other kinds of traits in development. Reversals can scramble what would have been a consistent suite of features, and the resulting homoplasies may confound a correct reading of phylogenetic information. Clearly the broadest range of information, from fossils to genes, needs to be used whenever possible in order to get as robust a phylogeny as possible.

Functional biology includes molecular biology, genetics, and developmental biology, as well as physiology and functional morphology. These approaches need little justification from me. They after all constitute much of the heart of modern biology. These disciplines provide our knowledge of the mechanisms of development and of the genes that underlie developmental phenomena. Without a solid grasp of the pertinent functional biology, evolutionary inference becomes essentially rampant speculation. Discussions, for example, of the evolution of sex determination would be pretty thin without the kinds of detailed mechanistic studies discussed in chapter 11, and would be positively misleading in the absence of the molecular details.

Most studies of functional developmental biology use intensively studied model systems. Model systems are vital to establishing a detailed mechanistic understanding, but their selection and use generally lack any evolutionary perspective. Gaining an evolutionary perspective requires that related non-model organisms be examined. Similarities and differences in mechanism can then be placed in a phylogenetic context, and evolutionary questions posed. An experimental evolutionary developmental biology was not possible until recently. Now organisms and phenomena can be chosen from nature and explicitly used in comparisons with model systems and with one another.

It is important to note that the testing of hypotheses is not limited to experiments. Hypotheses can be effectively tested by comparative studies as well. This can be done in two ways. In the first, regularities are sought among organisms to ask whether suspected functionally important elements are conserved. This is often done by molecular biologists seeking cis-acting control regions in the promoters of genes. Their underlying hypothesis is that functionally important elements will be more conserved in evolution than surrounding DNA. Second, predictions can be explicitly tested. For example, in 1978 Lewis proposed that homeotic genes had evolved to regulate the segmental specialization patterns seen in insects. This idea has been explicitly tested by the experiments of Warren and co-workers discussed earlier in this chapter. The Lewis hypothesis is not correct. Genes such as *Ubx* are older than the arthropods, and Carroll and his co-workers have observed that *Ubx* and *Abd-A* are expressed in the same segments of the primitive wingless insect *Thermobia* as in winged insects. The regulation of segmental identities and whether wings are produced on them is dependent not only on which Hox gene is expressed, but also on how its activity is read by downstream genes.

By themselves, functional studies carried out in model systems do not lead inevitably to good evolutionary interpretations. The extension of the findings of functional studies to nonmodel organisms has been made feasible by powerful techniques in molecular biology and by the existence of deep genetic homologies among phyla. Such extensions are needed to furnish comparative studies and ultimately produce evolutionary conclusions. The comparative method in biology is an old and productive one. Although it was comparative anatomy that made the great nineteenth-century evolutionary generalizations possible, that approach to biology lost favor through much of the twentieth century as experimental biology gained in prominence. Ironically, the advent of gene cloning and sequencing, the ease with which gene sequences can be compared, and the wide distribution of major regulatory genes have reawakened an appreciation of homology and of the importance of comparative data. Comparative data in themselves do not grant evolutionary insights any more than do functional data. In many cases, comparative studies (for example, in comparative physiology) have been used to give insight into functional aspects of biology. These studies don't in themselves require an evolutionary or phylogenetic framework. However, comparisons are required to gauge the range of mechanisms and the kinds of evolutionary alterations that have occurred. When combined with robust phylogenies, they provide the evolutionary perspective to functional data, and they supply the information needed for inferring homologies. Comparative studies lie at the heart of evolutionary biology.

Some long-standing assumptions and hypotheses have dominated our thinking about the interaction between evolution and development, and these need reexamination. Perhaps the most conspicuous such idea is that most evolutionary changes are the outcomes of heterochronic mechanisms. This hypothesis supplied the earliest coherent mechanistic explanation for a large range of phenomena, and it has been forcefully propounded in influential books, first by de Beer and more recently by Gould. I have attempted in chapter 8 to put heterochrony in a more balanced perspective. Heterochrony is clearly an element of evolutionary dissociation of developmental events, but not the only one. Because development flows along a time axis, almost any alteration will manifest itself as a change in timing. However, observed changes in timing do not necessarily mean that timing mechanisms per se

are involved. Many kinds of mutations in regulatory genes affect development, but few of these genes primarily control timing. Thus, heterochrony
is more likely to be a result than a mechanism at the lower levels in the
hierarchy of developmental processes. However, it may still serve as an
evolutionary mechanism at other hierarchical levels, such as life history strategy, and certainly is useful in the interpretation of trends in the fossil record.
Still, we should be cautious about accepting heterochronic "explanations"
for evolutionary changes because this facile and popular assignment of cause
can inhibit a more penetrating analysis.

A second long-standing and important theoretical conception of the relationship between development and evolution is that of developmental constraints. The idea that developmental rules can direct or constrain the course
of evolution has two origins. A number of evolutionists, particularly in the
generation following Darwin, took antiselectionist positions, and posited that
internal forces direct evolution and produce long-term trends independent of
the external environment. That is not a tenable position, but neither is extreme
selectionism. Internal genetic and developmental constraints of various kinds
must exist, but as I've documented in chapter 9, they are diverse and poorly
understood. Yet if internal factors constrain evolution, they are hardly a
minor issue. The acceptance of internal constraints does not mean that Darwinian selection is unimportant, but it does mean that the variation presented
to selection is not random. Rather, developmental constraints should be envisaged as resulting from both the modular structure of ontogeny and the internal
evolutionary processes that change developmental modules and the relationships among them.

Modularity may be one of the few general rules of development, and it
provides a mechanistic basis for evolutionary modifications. Modularity can
be recognized in many guises in development. Morphogenetic fields were
stressed by early-twentieth-century embryologists and, as argued by Gilbert
and co-workers, may be reappearing in new genetic formulations. Modularity
extends to most developing features, whether maternally localized regions of
eggs, axes, individual embryonic cells, fields, limb buds, organ rudiments,
or serially repeated units such as vertebrae, somites, or segments. The ever
more obvious conservation of "standard parts," both genetic and morphological, is a weighty clue pointing resolutely to modularity as a fundamental
aspect of the organization of ontogeny. The internal evolutionary processes
of dissociation, duplication and divergence, and co-option all operate within
developmental systems that exhibit modular organization. The disparate examples of the evolution of novel features presented in chapter 10 all exhibit
modular organizations, and they illustrate the operation of the internal evolutionary processes on that modularity.

GENES, HOMOLOGY, AND EVOLUTION

There are some significant generalities that allow us to envision both the nature of developmental constraints and the apparently paradoxical mixing of conservation and change observed in the evolution of ontogenies. Consideration of both body plans and regulatory genes reveals an exquisite tension between conservation and change. The patterns of the phylotypic stages that underlie body plans are conserved, yet dramatic evolution has occurred within body plans and, as we saw in chapters 6 and 7, even within the developmental trajectories leading to conserved phylotypic stages. An exploration of the evolution of early development reveals that it is enormously labile, and that changes in modes of embryonic and larval development have occurred frequently and rapidly in a large number of lineages. The persistence of phylotypic patterns in the face of changed developmental trajectories calls for an explanation. I have suggested a metaphor in the form of the developmental hourglass, and have offered a very preliminary mechanistic hypothesis of the varying roles of developmental modules and global inductive interactions over the course of ontogeny.

Conservation is not limited to phylotypic stages. It also lies in the evolutionary history of many regulatory genes. Figure 12.10 presents a diagram of the possible outcomes of regulatory gene evolution. The prediction that new master regulators would accompany morphological evolution would have been favored not so long ago. Yet it does not appear that new regulators are

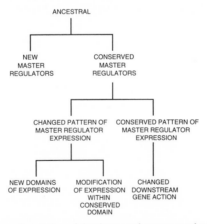

Figure 12.10.   Pathways of evolution of developmental master regulatory genes. Macroevolutionary changes result from changes in gene action. Major changes have been hypothesized to result from the evolution of new master regulators, but research on developmental regulators indicates that master regulatory genes are conserved over great phylogenetic distances. Co-option of these genes allows a great range of modifications in their actions.

the rule in evolution. Master regulators are conserved even between phyla. The amazing conclusion of the past decade of research is that conserved master regulatory genes have produced divergent ontogenies. The examples of regulatory gene and morphological evolution discussed in this chapter exemplify the kinds of changes diagrammed in figure 12.10. Conserved patterns of master regulator expression require changes in response by downstream genes. This is the situation observed for *Ubx* expression in two-winged and four-winged insects. Cases in which patterns of master regulator expression are modified are also becoming apparent, as in the fields destined to give rise to the prolegs of caterpillars and in the specification of vertebral identity.

The central problem is finding the mechanisms that connect genes and developmental processes to morphological evolution. Master regulatory genes, such as Hox genes and others involved in patterning, are apparently universally distributed among metazoans, including those at the base of the phylogenetic tree. A large number of homeobox-containing and other regulatory genes have now been listed by Hans Bode as present in *Hydra,* a cnidarian. It is a very curious thing that the broad sites of expression of these genes are frequently conserved as well, even though the structures in which they are expressed have no apparent homology between phyla. Thus, some Hox genes are expressed in the anterior ends of animals, but the resulting heads are very different. *Pax-6* regulates morphologically distinct eyes in various phyla. Just as remarkably, it has been shown by Grindley and coworkers to regulate development of the vertebrate nasal placode, admittedly a head structure but hardly homologous to anything in any other phylum. Somehow deep underlying genetic systems have remained in place in some developmental programs even though the structures constructed by genes acting downstream have undergone extensive modification. It's a bit like the archaeological discovery that the great late medieval cathedrals were successively built on older church foundations and evolved in place, with Gothic arches rising over buried Norman columns and Roman pavements. Again there is a contrast to another pattern of evolutionary behavior: conserved regulatory genes sometimes show up in new functions, having been co-opted, although their molecular structures and mechanisms are conserved.

What then do these phenomena tell us about homology? Shared underlying deep molecular processes seem to undermine our confidence in the classic structural definition of homology. Conserved underlying genetic patterns allow us to examine changes in body plans at the level of gene regulation. The conservation of Hox clusters or *wnt* genes is not in itself remarkable: these and other genes encode proteins that play very basic roles in cellular function. For example, the Hox genes fall into a large class of transcription factors. The remarkable property of these conserved genes is that they function in similar regulatory suites in development over great phylogenetic spans. Co-option of regulatory genes is a remarkably parsimonious process. Regulatory components, the pathways in which they participate, and their sites of operation are all often conserved.

Increasingly, we envision morphological structures as expressions of underlying regulatory genes coupled to downstream structural genes and epigenetic events. Some have suggested that development serve as the keystone of homology. But that idea is impaired by the divergent developmental paths that can produce admittedly homologous anatomical structures. The broad conservation of regulatory genes and regulatory pathways provides another kind of homology, one of genetic process. One might propose that we substitute this genetic homology for structural homology. This would seem to have the advantage of a direct continuity of information. In some instances, this may be a good idea; in others, it is not. There are apparently four kinds of genetic homology. The first is a direct correspondence between genetic process and morphological homology. The second is one in which genetic homology truly indicates a deep evolutionary homology, even when this is difficult to trace morphologically. The eye is a case in point. Vertebrate and insect eyes are morphologically so different that they must have had long separate histories, and must have acquired much of their structure by convergence. Yet their sharing of *Pax-6* as a master regulator suggests a deep initial homology of photoreceptors in a common ancestor. The third kind of genetic homology is one in which regional conservation is apparent. The shared role of Hox genes in body axis determination from hydra to human is the most obvious case. The "heads" formed in these two organisms can hardly be considered homologous. They are independently evolved manifestations of polarity. Similarly, the gene *Distal-less* appears in the limb fields of insects and of vertebrates. The appendages produced are not homologous, but there is a geographic commonality. Perhaps *Distal-less* evolved as part of a very ancient genetic system controlling any projection from the body axis. Any claim of homology for the products of geographically conserved gene regulation has to be made cautiously. Finally, some common regulatory gene systems function in completely disparate developmental processes, for example, *Sonic hedgehog* in the vertebrate limb bud and in the neural tube. In these instances, regulatory genes have been co-opted to serve very different developmental ends in unrelated morphogenetic pathways. Evidently, conserved gene programs can be tied to quite distinct downstream processes. The genes are indisputably homologous, but to claim any homology above the gene level for the regions and structures in which they function would be fruitless.

The examples paraded in chapter 11 show that the metazoan genome is notably fluid, and that co-option of genes can occur by two general mechanisms. The first is through changes in cis-regulatory domains of gene promoters or enhancers. Such modifications allow expression in different cell types or at different times in development. Such control shifts may be relatively crude, as in the recruitment of enzymes as lens proteins, or very intricate, as in the patterned repression of Hox gene action in presumptive caterpillar leg rudiments. These shifts involve changes in one gene at a time. A second, potentially more pervasive mechanism involves change in the transcriptional machinery of particular cell types. The result is co-option of a program consisting of many genes, which can confer a coordinated change in modular identity—that is, a homeotic transformation can occur. A simple gene change can lie at the root of such transformations if the difference in expression involves a high-level regulatory gene. Examples such as the hypothesized origin of serial homologues (e.g., the limbs of tetrapods, the lobe fins of coelacanths, or the multiple wings of Paleozoic insects) indicate that homeotic transformations of various kinds may have played significant roles in the origins of some novelties. Some classic heterochronies, such as the neoteny of axolotls or the heterochronies induced in nematode cell lineages by the *lin-14* system, fall into a similar category of shifts through a sort of overall timing change in an entire genetic module caused by a change in an upstream regulator.

Conserved genes and regulatory systems offer us important tools for tracing body plan transformations. These applications have been discussed in various places in this book, and include deducing the course of the remodeling of insect appendages, tracing the transformations that occurred during the evolution of chordate axial features, and pursuing the origin of the echinoderm radial body plan. As the evolution of conserved regulatory genes, such as the Hox genes, is defined and their expression in metazoan phyla more broadly documented, the evolution of very basic features of metazoan architecture will be better understood. The most obvious use of conserved genes is to learn directly about the evolution of the expression patterns of the genes themselves. The more interesting use is to trace the history of regulatory systems and developmental patterns. Perhaps the most desired such history would be that of regulatory genes in the metazoan radiation. It is tempting to assume that the Cambrian radiation and divergence of regulatory genes go hand in hand. It seems more likely that most of the evolution of regulatory genes and their patterns of expression predate the Cambrian radiation and even most of the radiation of body plans. The Hox gene-body axis theme certainly predates the Cambrian, and most body plans. Body plan origins involved regulatory evolution, but superimposed on basic metazoan regulatory patterns. Although some of the regulatory systems that arose in the metazoan radiation may be deeply entrenched, it is likely that regulatory evolution in the metazoan radiation was similar in kind to what we observe in major evolutionary steps within phylum-level body plans.

The study of the evolution of insect appendages is particularly important because it has encompassed such a wide spectrum of approaches, and because it has revealed with particular clarity the roles of modularity, duplication and divergence, dissociation, and co-option in the evolution of novel body plan features. Arthropods, with their complete body segmentation, are the paradigmatic modular organisms. These modules are patterned by the developmental cascade of gene action that culminates in the expression of the Hox complex genes and the downstream genes they control. Duplication of segmental elements occurred very early in the history of the arthropods. The most primitive known arthropods (such as early Cambrian trilobites or possibly the Precambrian fossil *Spriggina*) are built of a large number of similar segments that bear similar appendages. Divergence of individual segments and appendages has been going on in arthropods ever since. Selection has expanded and refined the functional roles of appendages that differ along the body axis (legs surrounding the mouth have acquired forms and functions different from those of legs farther down the body). Developmental dissociation has been a consequence of functional dissociation of serially homologous structures.

The specialization of body segments (tagmosis) has been a central theme in arthropod evolution, and it has been carried to a spectacular level of refinement by modern insects. The fossil record shows that dissociation of appendage function has been frequent. It was probably favored because there was an excess of appendages in primitive insects, a much more leggy group of creatures than their modern descendants. Dissociation was accompanied by co-option of leg elements for other functions. The consequence has been reduction and specialization of legs and exites. Legs or leg elements have been recruited to serve as antennae, jaws, palps, larval gills, wings, genitalia, and cerci. The broad pattern of gene expression has been conserved, with modifications in details of Hox gene expression, alterations in regulation of downstream genes, and co-option of genes for new functions. Novel developmental features in butterflies and other insects thus arose from minimal genetic changes that had profound morphological effects. Iridescent butterflies come from commonplace genes.

Because insect developmental genetics has been studied so intensively, genetic events in evolution seem particularly striking in that group. However, dissociation, duplication and divergence, and co-option all occur widely, although the details of their involvement may differ among taxa. Carroll and co-workers have shown that in insects, co-option of conserved genes has predominated. However, in vertebrates, gene duplication and divergence have assumed major roles. Thus, the Hox genes are conserved in insects, and changes in which they play a role have involved intricate regulation of their expression rather than their duplication. In contrast, vertebrates have duplicated members of the Hox family with abandon and have reassigned diverged members to new functions. There also has been a great deal of duplication and divergence of other genes within vertebrates. The evolution of vertebrate body plans differs from that in insects in a second important way. The vertebrae vary in shape (cervical, thoracic, and so on), and the numbers of vertebrae of particular types vary among vertebrates. The Hox code of the somites might correlate with identity of vertebrae or with position along the axis, that is, with number. Both Gaunt and Burke and co-workers, however, have found that Hox gene expression correlates with identity, but not with number. This regulatory relationship differs from that found in insects. There, evolution has changed the downstream interpretation of an invariant Hox pattern. Body plan differences between phyla may have been accompanied by distinct styles of evolution of shared regulatory gene families.

This book has encountered two aspects of animal body plans: their immense stability through time and the dramatic evolution of animal form that has occurred during the past half billion years. The stability of body plans may result from the external ecological partitioning of the world, or, perhaps more interestingly, from the operation of internal constraints. Body plans as they now exist are not readily transformable one to another. The developmental hourglass suggests a kind of internal mechanistic constraint. But why should metazoan body plans beyond the most simple ever have arisen at all, let alone with the rapidity of the events leading to the Cambrian radiation?

The answer comes from looking up into the branching pattern of the phylogenetic tree, rather than across the branch tips. Interconversion really isn't the issue for body plan origins. The initial decision as to whether the central nervous system was to lie dorsal or ventral to the gut was probably not of large consequence in the primitive flatworm-grade bilaterian ancestor. However, once the choice was made, it entrained other choices leading to complex architectures that cannot be transformed, and indeed, may be too tightly integrated in the phylotypic stages to be disintegrated to produce a new body plan.

The tension between change and constancy remains an intriguing challenge. Significant evolution has been accommodated within existing phylum-level body plans, despite the enduring stability of body plans themselves. Understanding the striking and profound changes that have produced whales, insect wings, and human brains should keep us occupied for some time to come.

References

Abbott, E. A. 1983. *Flatland: A Romance of Many Dimensions.* Harper Collins, New York.

Abele, L. G., W. Kim, and B. E. Felgenhauer. 1989. Molecular evidence for inclusion of the phylum Pentastomida in the Crustacea. *Mol. Biol. Evol.* 6:685–91.

Adams, D. 1982. *Life, the Universe and Everything.* Harmony Books, New York.

Adoutte, A., and H. Philippe. 1993. The major lines of metazoan evolution: Summary of traditional evidence and lessons from ribosomal RNA sequences analysis. In *Comparative Molecular Neurobiology.* Y. Pichon (ed.). Birkhäuser Verlag, Basel. Pp. 1–30.

Agassiz, L. 1857. Essay on classification. In *Contributions to the Natural History of the United States.* E. Lurie (ed.). Harvard University Press, Cambridge, Mass.

Ahlberg, P. E. 1991. Tetrapod or near-tetrapod fossils from the Upper Devonian of Scotland. *Nature* 354:298–301.

Ahlberg, P. E. 1992. Coelacanth fins and evolution. *Nature* 358:459.

Alberch, P. 1982. Developmental constraints in evolutionary processes. In *Evolution and Development.* J. T. Bonner (ed.). Springer-Verlag, Berlin. Pp. 313–32.

Alberch, P. 1985. Problems with the interpretation of developmental sequences. *Syst. Zool.* 34:46–58.

Alberch, P. 1987. Evolution of a developmental process: Irreversibility and redundancy in amphibian metamorphosis. In *Development as an Evolutionary Process.* R. A. Raff and E. C. Raff (eds.). Alan R. Liss, New York. Pp. 23–46.

Alberch, P., and E. Gale. 1983. Size dependency during the development of the amphibian foot: Colchicine induced digital loss and reduction. *J. Embryol. Exp. Morphol.* 76:177–97.

Alberch, P., and E. Gale. 1985. A developmental analysis of an evolutionary trend: Digit reduction in amphibians. *Evolution* 39:8–23.

Alberch, P., S. J. Gould, G. F. Oster, and D. B. Wake. 1979. Size and shape in ontogeny and phylogeny. *Paleobiology* 5:296–317.

Alexander, R. McN. 1989. *Dynamics of Dinosaurs and Other Extinct Giants.* Columbia University Press, New York.

Allard, M. A., and R. L. Honeycutt. 1992. Nucleotide sequence variation in the mitochondrial 12S rRNA gene and the phylogeny of African mole-rats (Rodentia: Bathyergidae). *Mol. Biol. Evol.* 9:27–40.

Allen, G. E. 1978. *Thomas Hunt Morgan: The Man and His Science.* Princeton University Press, Princeton.

Allison, P. A., and D. E. G. Briggs. 1991. The taphonomy of soft-bodied animals. In *The Process of Fossilization.* S. K. Donovan (ed.). Columbia University Press, New York. Pp. 120–40.

Altangerel, P., M. A. Norell, L. M. Chiappe, and J. M. Clark. 1993. Flightless bird from the Cretaceous of Mongolia. *Nature* 362:623–26.

Ambros, V. 1989. A hierarchy of regulatory genes controls a larva-to-adult developmental switch in *C. elegans. Cell* 57:49–57.

Ambros, V., and W. Fixsen. 1987. Cell lineage variation among nematodes. In *Development as an Evolutionary Process.* R. A. Raff and E. C. Raff (eds.). Alan R. Liss, New York. Pp. 139–59.

Ambros, V., and H. R. Horvitz. 1984. Heterochronic mutants of the nematode *Caenorhabditis elegans. Science* 226:409–16.

Ambros, V., and H. R. Horvitz. 1987. The *lin-14* locus of *Caenorhabditis elegans* controls the time of expression of specific developmental events. *Genes Dev.* 1: 398–414.

Ambros, V., and E. G. Moss. 1994. Heterochronic genes and the temporal control of *C. elegans* development. *Trends Genet.* 10:123–27.

Amemiya, C. T., and G. W. Litman. 1991. Early evolution of immunoglobin genes. *Am. Zool.* 31:558–69.

Amemiya, C. T., Y. Ohta, R. T. Litman, J. P. Rast, R. N. Haire, and G. W. Litman. 1993. VH gene organization in a relict species, the coelacanth *Latimeria chalumnae:* Evolutionary implications. *Proc. Natl. Acad. Sci. USA* 90:6661–65.

Amemiya, S., and R. B. Emlet. 1992. The development and larval form of an echinothurioid echinoid, *Asthenosoma ijimai,* revisited. *Biol. Bull.* 182:15–30.

Amero, S. A., R. H. Kretsinger, N. D. Moncrief, K. R. Yamamoto, and W. R. Pearson. 1992. The origin of nuclear receptor proteins: A single precursor distinct from other transcription factors. *Mol. Endocrinol.* 6:3–7.

Anderson, D. T. 1963. The embryology of *Dacus tyroni.* 2. Development of imaginal discs in the embryo. *J. Embryol. Exp. Morphol.* 11:339–51.

Anderson, D. T. 1969. On the embryology of the cirripede crustaceans *Tetraclita rosea* (Krauss), *T. pururascenes* (Wood), *Chthamalus antennatus* (Darwin) and *Chamaesipho columna* (Spengler) and some considerations of crustacean phylogenetic relationships. *Phil. Trans. R. Soc. Lond.* B 256:183–235.

Anderson, D. T. 1973. *Embryology and Phylogeny in Annelids and Arthropods.* Pergamon Press, Oxford.

Anderson, D. T. 1979. Embryos, fate maps, and the phylogeny of arthropods. In *Arthropod Phylogeny.* A. P. Gupta (ed.). Van Nostrand Reinhold, New York. Pp. 59–105.

Anderson, M. M., and S. Conway Morris. 1982. A review, with descriptions of four unusual forms, of the soft-bodied fauna of the Conception and St. John's Groups (late Precambrian), Avalon Peninsula, Newfoundland. In *Proceedings, Third North American Paleontological Convention.* B. Mamet and M. J. Copeland (eds.). Pp. 1–8.

Appel, T. A. 1987. *The Cuvier-Geoffroy Debate: French Biology in the Decades Before Darwin*. Oxford University Press, New York.

Applebury, M. L., and P. A. Hargrave. 1986. Molecular biology of the visual pigments. *Vision Res.* 26:1881–95.

Arasu, P., B. Wightman, and G. Ruvkun. 1991. Temporal regulation of *lin-14* by the antagonistic action of two other heterochronic genes, *lin-4* and *lin-28*. *Genes Dev.* 5:1825–33.

Archie, J. W. 1989. A randomization test for phylogenetic information in systematic data. *Syst. Zool.* 38:239–54.

Armstrong, E. 1982. Mosaic evolution in the primate brain: Differences and similarities in the hominid thalamus. In *Primate Brain Evolution*. E. Armstrong and D. Falk (eds.). Plenum Press, New York. Pp. 131–61.

Arn, P., H. Chen, C. M. Tuck-Muller, C. Mankinen, G. Wachtel, S. Li, C.-C. Shen, and S. S. Wachtel. 1994. SRVX, a sex reversing locus in Xp21.2-p22.11. *Hum. Genet.* 93:389–93.

Arnold, A. P. 1980. Sexual differences in the brain. *Am. Sci.* 68:165–73.

Arthur, W. 1988. *A Theory of the Evolution of Development*. John Wiley & Sons, Chichester.

Avivi, A., A. Yayon, and D. Givol. 1993. A novel form of FGF receptor-3 using an alternative exon in the immunoglobulin domain III. *FEBS Lett.* 330:249–52.

Ax, P. 1984. *Das phylogenetische System*. G. Fischer Verlag, Stuttgart.

Ax, P. 1985. The position of the Gnathostomulida and Platyhelminthes in the phylogenetic system of the Bilateria. In *The Origins and Relationships of Lower Invertebrates*. S. Conway Morris, J. D. George, R. Gibson, and H. M. Platt (eds.). Clarendon Press, Oxford. Pp. 168–80.

Ax, P. 1989. Basic phylogenetic systematization of the Metazoa. In *The Hierarchy of Life: Molecules and Morphology in Phylogenetic Analysis*. B. Fernholm, K. Bremer, and H. Jörnvall (eds.). Elsevier Science Publishers, Amsterdam. Pp. 229–45.

Babcock, R. C., C. N. Mundy, and D. Whitehead. 1994. Sperm diffusion models and in situ confirmation of long-distance fertilization in the free-spawning asteroid *Acanthaster planci. Biol. Bull.* 186:17–28.

Baier, H., and F. Bonhoffer. 1992. Axon guidance by gradients of a target-derived component. *Science* 255:472–75.

Baker, B. S. 1989. Sex in flies: The splice of life. *Nature* 340:521–24.

Baldauf, S. L., and J. D. Palmer. 1993. Animals and fungi are each other's closest relatives: Congruent evidence from multiple proteins. *Proc. Natl. Acad. Sci. USA* 90:11558–62.

Ballard, J. W. O., G. J. Olsen, D. P. Faith, W. A. Odgers, D. M. Rowell, and P. W. Atkinson. 1992. Evidence from 12S ribosomal RNA sequences that onychophorans are modified arthropods. *Science* 258:1345–48.

Baltzer, F. 1967. *Theodor Boveri: Life and Work of a Great Biologist*. University of California Press, Berkeley.

Bambach, R. K. 1983. Ecospace utilization and guilds in marine communities through the Phanerozoic. In *Biotic Interactions in Recent and Fossil Benthic Communities.*

M. J. S. Tevesz and P. L. McColl (eds.). Plenum Press, New York. Pp. 719–46.

Bambach, R. K. 1985. Classes and adaptive variety: The ecology of diversification in marine faunas through the Phanerozoic. In *Phanerozoic Diversity Patterns*. J. W. Valentine (ed.). Princeton University Press, Princeton. Pp. 191–253.

Bambach, R. K. 1986. Phanerozoic marine communities. In *Patterns and Processes in the History of Life*. D. M. Raup and D. Jablonski (eds.). Springer-Verlag, Berlin. Pp. 407–28.

Bardoni, B., E. Zanaria, S. Guioli, G. Floridia, K. C. Worley, G. Tonini, E. Ferrante, G. Chiumello, E. R. B. McCabe, M. Fraccaro, O. Zuffardi, and G. Camerino. 1994. A dosage sensitive locus at chromosome Xp21 is involved in male to female sex reversal. *Nature Genet.* 7:497–501.

Barnes, J. 1989. *A History of the World in 10½ Chapters*. Jonathan Cape, London.

Bartels, J., M. T. Murtha, and F. H. Ruddle. 1993. Multiple Hox/HOM-class homeoboxes in Platyhelminthes. *Mol. Phyl. Evol.* 2:143–51.

Barthel, K. W., N. H. M. Swinburne, and S. Conway Morris. 1990. *Solnhofen: A Study in Mesozoic Palaeontology*. Cambridge University Press, Cambridge.

Bartnik, E., and K. Weber. 1989. Widespread occurrence of intermediate filaments in invertebrates; common principles and aspects of diversion. *Eur. J. Cell Biol.* 50:17–33.

Barton, M. K., and J. Kimble. 1990. *fog-1*, a regulatory gene required for specification of spermatogenesis in the germ line of *Caenorhabditis elegans. Genetics* 125:29–39.

Basler, K., and G. Struhl. 1994. Compartment boundaries and the control of *Drosophila* limb pattern by hedgehog protein. *Nature* 368:208–14.

Bateson, W. 1894. *Materials for the Study of Variation*. Macmillan and Co., London.

Bautzmann, H., J. Holtfreter, H. Spemann, and O. Mangold. 1932. Versuche zur Analyse der Induktionsmittel in der Embryonalentwicklung. *Naturwissenschaften* 20:971–74.

Baverstock, P. R., M. Adams, R. W. Polkinghorne, and M. Gelder. 1982. A sex-linked enzyme in birds: Z-chromosome conservation but no dosage compensation. *Nature* 296:763–66.

Baverstock, P. R., R. Fielke, A. M. Johnson, R. A. Bray, and I. Beveridge. 1991. Conflicting phylogenetic hypotheses for the parasitic platyhelminths tested by partial sequencing of 18S ribosomal RNA. *Int. J. Parisitol.* 21:329–39.

Beachy, P. A. 1990. A molecular view of the Ultrabithorax homoeotic gene of *Drosophila. Trends Genet.* 6:46–51.

Beeman, R. W. 1987. A homoeotic gene cluster in the red flour beetle. *Nature* 327:247–49.

Behringer, R. R., T. M. Lewin, C. J. Quaife, R. D. Palmiter, R. L. Brinster, and A. J. D'Ercole. 1990. Expression of insulin-like growth factor I stimulates normal somatic growth in growth hormone-deficient transgenic mice. *Endocrinology* 127:1033–40.

Bell, L. R., J. I. Horabin, P. Schedl, and T. W. Cline. 1991. Positive autoregulation of *Sex-lethal* by alternative splicing maintains the female determined state in *Drosophila. Cell* 65:229–39.

Bender, W. 1983. Molecular genetics of the bithorax complex in *Drosophila melanogaster*. *Science* 221:23–29.

Benezra, R., R. L. Davis, D. Lockshon, D. L. Turner, and H. Weintraub. 1990. The protein Id: A negative regulator of helix-loop-helix DNA binding proteins. *Cell* 61:49–59.

Bengtson, S. 1990. The solution to a jigsaw puzzle. *Nature* 345:765–66.

Bengtson, S. 1991. Oddballs from the Cambrian start to get even. *Nature* 351:184–85.

Bengtson, S. 1994. The advent of animal skeletons. In *Early Life on Earth*. S. Bengston (ed.). Columbia University Press, New York. Pp. 412–25.

Bengtson, S., and S. Conway Morris. 1992. Early radiation of biomineralizing phyla. In *Origin and Early Evolution of the Metazoa*. J. H. Lipps and P. W. Signor (eds.). Plenum Press, New York. Pp. 447–81.

Bennett, A. F., and J. A. Rubin. 1986. The metabolic and thermoregulatory status of therapsids. In *The Ecology and Biology of Mammal-like Reptiles*. N. Hotton III, P. D. MacLean, J. J. Roth, and E. C. Roth (eds.). Smithsonian Institution Press, Washington, D.C. Pp. 207–18.

Benoit, R., D. Sassoon, B. Jacq, W. Gehring, and M. Buckingham. 1989. *Hox-7*, a mouse homeobox gene with a novel pattern of expression during embryogenesis. *EMBO J.* 8:91–100.

Benton, M. J. 1995. Diversification and extinction in the history of life. *Science* 268:52–58.

Benus, A. P. 1988. Sedimentological context of a deep water Ediacaran fauna (Mistaken Point Formation, Avalon Zone, Eastern Newfoundland). In *Trace Fossils, Small Shelly Fossils and the Precambrian-Cambrian Boundary: Proceedings*. E. Landing, G. M. Narbonne, and P. Myron (eds.). *Bull. New York State Museum*, 463:8–9.

Bergquist, P. R. 1985. Poriferan relationships. In *The Origins and Relationships of Lower Invertebrates*. S. Conway Morris, J. D. George, R. Gibson, and H. M. Platt (eds.). Clarendon Press, Oxford. Pp. 14–27.

Berkner, L. V., and L. C. Marshall. 1964. The history of oxygen concentration in the earth's atmosphere. *Disc. Faraday Soc.* 37:122–41.

Berrill, N. J. 1935. Studies on tunicate development. Part III. Differential retardation and acceleration. *Phil. Trans. R. Soc. Lond.* B 225:255–326.

Berta, A. 1994. What is a whale? *Science* 263:180–81.

Beverly, S. M., and A. C. Wilson. 1984. Molecular evolution in *Drosophila* and the higher Diptera. II. A time scale for fly evolution. *J. Mol. Evol.* 21:1–13.

Boag, P. T., and P. R. Grant. 1981. Intense natural selection in a population of Darwin's finches (Geospizinae) in the Galápagos. *Science* 214:82–85.

Bocharova-Messner, O. M. 1959. Development of the wing in the early post-embryonic stage in the ontogeny of dragonflies (order Odonata). *Trudy Inst. Morf. Zhiv. Im. Severtsova* 27:187–200.

Bolker, J. A. 1995. The choice and consequences of model systems in developmental biology. *BioEssays* 17:451–55.

Bonnichsen, R., M. T. Beatty, M. D. Turner, and M. Stoneking. 1994. What can be learned from hair? A hair record from the Mammoth Meadow Locus, Southwestern

Montana. In *Prehistoric Mongoloid Dispersals*. T. Akazawa and E. Szathmary (eds.). Oxford University Press, Oxford.

Bonnichsen, R., C. W. Bolen, M. Turner, J. C. Turner, and M. T. Beatty. 1992. Hair from Mammoth Meadow II, Southwestern Montana. *Curr. Res. Pleistocene* 9:75–78.

Bordereau, C., and S. H. Han. 1986. Stimulatory influence of the queen and king on soldier differentiation in the higher termites *Nasutitermes lujae* and *Cubitermes fungifaber*. *Insectes Sociaux* (Paris) 33:296–305.

Bottjer, D. J., J. K. Schubert, and M. L. Droser. 1995. Comparative evolutionary paleoecology: Assessing the changing ecology of the past. In *Biotic Recoveries from Mass Extinctions*. M. Hart (ed.). Geological Society of London Special Publication, London.

Boveri, T. 1907. Zellenstudien VI: Die Entwicklung dispermer Seeigelier. Ein Beitrag zur Befruchtungslehre und zur Theorie des Kernes. *Jena. Zeitschr. Naturw.* 43: 1–292.

Bowerman, B., B. W. Draper, C. C. Mello, and J. R. Priess. 1993. The maternal gene *skn-1* encodes a protein that is distributed unequally in early *C. elegans* embryos. *Cell* 74:443–52.

Bowring, S. A., J. P. Grotzinger, C. E. Isachsen, A. H. Knoll, S. M. Pelechaty, and P. Kolosov. 1993. Calibrating rates of early Cambrian evolution. *Science* 261:1293–98.

Brenner, S., G. Elgar, R. Sandford, A. Macrae, B. Venkatesh, and S. Aparicio. 1993. Characterization of the pufferfish *(Fugu)* genome as a compact model verte-brate genome. *Nature* 366:265–68.

Brew, K., and R. L. Hill. 1975. Lactose biosynthesis. *Rev. Physiol. Biochem. Phar-macol.* 72:105–58.

Bridges, T. S. 1993. Reproductive investment in four developmental morphs of *Streblospio* (Polychaeta: Spionidae). *Biol. Bull.* 184:144–52.

Briggs, D. E. G., D. H. Erwin, and F. J. Collier. 1994. *The Fossils of the Burgess Shale*. Smithsonian Institution Press, Washington, D.C.

Briggs, D. E. G., R. A. Fortey, and M. A. Wills. 1992. Morphological disparity in the Cambrian. *Science* 256:1670–73.

Bromage, T. G. 1985. Taung facial remodeling: A growth and development study. In *Hominid Evolution: Past, Present and Future*. P. V. Tobias (ed.). Alan R. Liss, New York. Pp. 239–45.

Bronner-Fraser, M. 1995. Patterning of the vertebrate neural crest. *Perspect. Dev. Neurobiol.* 3 (1): 53–62.

Brown, S. D. M., and A. J. Greenfield. 1985. Possible functions of the homeobox. *Nature* 313:185–86.

Brusca, R. C., and G. J. Brusca. 1990. *Invertebrates*. Sinauer Associates, Sunder-land, Mass.

Bruton, D. L. 1991. Beach and laboratory experiments with the jellyfish *Aurelia* and remarks on some fossil "medusoid" traces. In *The Early Evolution of Metazoa and the Significance of Problematic Taxa*. A. M. Simonetta and S. Conway Morris (eds.). Cambridge University Press, Cambridge. Pp. 125–29.

Bryant, P. J., and P. Simpson. 1984. Intrinsic and extrinsic control of growth in developing organs. *Q. Rev. Biol.* 59:387–415.

Buckland, W. 1824. *Reliquiae Diluvianae; or Observations on the Organic Remains Contained in Caves, Fissures, and Diluvial Gravel, and on Other Geological Phenomena, Attesting the Action of an Universal Deluge.* John Murray, London.

Budd, G. 1993. A Cambrian gilled lobopod from Greenland. *Nature* 364:709–11.

Bulfinch, T. 1994. *The Golden Age of Myth and Legend.* Studio Editions, London.

Burke, A. C. 1989. Development of the turtle carapace: Implications for the evolution of a novel bauplan. *J. Morphol.* 199:363–78.

Burke, A. C., C. E. Nelson, B. A. Morgan, and C. Tabin. 1995. Hox genes and the evolution of vertebrate axial morphology. *Development* 121:333–46.

Burke, R. D., R. L. Myers, T. L. Sexton, and C. Jackson. 1991. Cell movements during the initial phase of gastrulation in the sea urchin embryo. *Dev. Biol.* 146: 542–57.

Burnet, T. 1690/91. *Sacred Theory of the Earth.* Reprinted by Southern Illinois University Press (1965), Carbondale.

Buss, L. W., and A. Seilacher. 1994. The phylum Vendobionta: A sister group of the Eumetazoa? *Paleobiology* 20:1–4.

Butterfield, N. J. 1994. Burgess Shale-type fossils from a lower Cambrian shallow-shelf sequence in northwestern Canada. *Nature* 369:477–79.

Byrne, M. 1991. Developmental diversity in the starfish genus *Patiriella* (Asteriodea: Asterinidae). In *Biology of Echinodermata.* T. Yanagisawa, I. Yasumasu, C. Oguro, N. Suzuki, and T. Motokawa (eds.). A. A. Balkema, Rotterdam. Pp. 499–508.

Byrne, M. 1992. Reproduction of sympatric populations of *Patiriella calcar, P. exigua,* and *P. gunnii,* asterinid sea urchins with abbreviated development. *Mar. Biol.* 114:297–316.

Byrne, M., and M. J. Anderson. 1994. Hybridization of sympatric *Patiriella* species (Echinodermata: Asteroidea) in New South Wales. *Evolution* 48:564–76.

Calvino, I. 1968. *Cosmicomics.* Harcourt Brace Jovanovich, San Diego.

Calvino, I. 1969. *t zero.* Harcourt Brace Jovanovich, San Diego.

Cameron, J. L., F. S. McEuen, and C. M. Young. 1988. Floating lecithotrophic eggs from the bathyal echinothuriid sea urchin *Araeosoma fenestratum.* In *Echinoderm Biology.* R. D. Burke, P. V. Mladenov, P. Lambert, and R. L. Parsley (eds.). A. A. Balkema, Rotterdam. Pp. 177–80.

Cameron, R. A., R. J. Britten, and E. H. Davidson. 1993. The embryonic ciliated band of the sea urchin, *Strongylocentrotus purpuratus* derives from both oral and aboral ectoderm. *Dev. Biol.* 160:369–76.

Cameron, R. A., S. E. Fraser, R. J. Britten, and E. H. Davidson. 1989. The oral-aboral axis of a sea urchin is specified by first cleavage. *Development* 106:641–47.

Cameron, R. A., S. E. Fraser, R. J. Britten, and E. H. Davidson. 1991. Macromere cell fates during sea urchin development. *Development* 113:1085–91.

Cameron, R. A., B. R. Hough-Evans, R. J. Britten, and E. H. Davidson. 1987. Lineage and fate of each blastomere of the eight-cell sea urchin embryo. *Genes Dev.* 1:75–85.

Campbell, K. S. W., and C. R. Marshall. 1987. Rates of evolution among Paleozoic echinoderms. In *Rates of Evolution*. K. S. W. Campbell and M. F. Day (eds.). Allen and Unwin, London. Pp. 61–100.

Cann, R. L., M. Stoneking, and A. C. Wilson. 1987. Mitochondrial DNA and human evolution. *Nature* 325:31–36.

Cano, R. J., H. N. Poinar, N. J. Pieniazek, A. Acra, and G. O. Poinar. 1993. Amplification and sequencing of DNA from a 120–135-million-year-old weevil. *Nature* 363:536–38.

Carpenter, K., K. F. Hirsch, and J. R. Horner. 1994. *Dinosaur Eggs and Babies*. Cambridge University Press, Cambridge.

Carroll, R. L. 1988. *Vertebrate Paleontology and Evolution*. W. H. Freeman, New York.

Carroll, S. B. 1995. Homeotic genes and the evolution of arthropods and chordates. *Nature* 376:479–85.

Carroll, S. B., J. Gates, D. Keys, S. W. Paddock, G. F. Panganiban, J. Selegue, and J. A. Williams. 1994. Pattern formation and eyespot determination in butterfly wings. *Science* 265:109–14.

Carroll, S. B., S. D. Weatherbee, and J. A. Langeland. 1995. Homeotic genes and the regulation and evolution of insect wing number. *Nature* 375:58–61.

Cavalier-Smith, T. 1985a. Cell volume and the evolution of eukaryotic genome size. In *The Evolution of Genome Size*. T. Cavalier-Smith (ed.). John Wiley & Sons, Chichester. Pp. 105–84.

Cavalier-Smith, T. 1985b. Eukaryotic gene numbers, non-coding DNA and genome size. In *The Evolution of Genome Size*. T. Cavalier-Smith (ed.). John Wiley & Sons, Chichester. Pp. 69–103.

Cavalli-Sforza, L. L. 1991. Genes, peoples and languages. *Sci. Am.* 265 (5): 104–10.

Cavener, D. R. 1992. Transgenic animal studies on the evolution of genetic regulatory circuitries. *BioEssays* 14:237–44.

Chambers, R. 1844. *Vestiges of the Natural History of Creation*. John Churchill, London.

Charig, A. J. 1982. Systematics in biology: A fundamental comparison of some major schools of thought. In *Problems of Phylogenetic Reconstruction*. K. A. Joysey and A. E. Friday (eds.). Academic Press, London. Pp. 363–440.

Chatwin, B. 1987. *The Songlines*. Viking, New York.

Chellaiah, A. T., D. G. McEwen, S. Werner, J. Xu, and D. M. Ornitz. 1994. Fibroblast growth factor receptor (FGFR)3. *J. Biol. Chem.* 269:11620–27.

Chen, J., and D. D. Erdtmann. 1991. Lower Cambrian lagerstatte from Chengjiang, Yunnan, China: Insights for reconstructing early metazoan life. In *The Early Evolution of Metazoa and the Significance of Problematic Taxa*. A. M. Simonetta and S. Conway Morris (eds.). Cambridge University Press, Cambridge. Pp. 57–76.

Cherry-Garrard, A. 1922. *The Worst Journey in the World*. Chatto and Windus, London.

Chevallier, A., M. Kieny, A. Mauger, and P. Sengel. 1977. Developmental fate of the somitic mesoderm in the chick embryo. In *Vertebrate Limb and Somite Morphogenesis*. D. A. Ede, J. R. Hinchliffe, and M. Balls (eds.). Cambridge University Press, Cambridge. Pp. 421–32.

Chiou, S. H. 1988. A novel crystallin from octopus lens. *FEBS Lett.* 241:261–64.

Chisaka, O., and M. Capecchi. 1991. Regionally restricted developmental defects resulting from targeted disruption of the mouse homoeobox gene *Hox-1.5. Nature* 350:473–79.

Cho, K. W. Y., B. Blumberg, H. Steinbeisser, and E. M. De Robertis. 1991. Molecular nature of Spemann's organizer: The role of the *Xenopus* homeobox gene *goosecoid. Cell* 67:1111–20.

Chouinard, S., and T. C. Kaufman. 1991. Control of the expression of the homoeotic *labial (lab)* locus of *Drosophila melanogaster:* Evidence for both positive and negative autogenous regulation. *Development* 113:1267–80.

Chow, K. L., and S. W. Emmons. 1994. HOM-C/Hox genes and four interacting loci determine the morphogenetic properties of single cells in the nematode male tail. *Development* 120:2579–93.

Christen, R., A. Ratto, A. Baroin, R. Perasso, K. G. Grell, and A. Adoutte. 1991. An analysis of the origin of metazoans, using comparisons of partial sequences of the 28S RNA, reveals an early emergence of triploblasts. *Eur. Mol. Biol. Org. J.* 10:499–503.

Clark, R. B. 1964. *Dynamics in Metazoan Evolution: The Origin of the Coelom and Segments.* Clarendon Press, Oxford.

Clark, R. B. 1979. Radiation of the metazoa. In *The Origin of Major Invertebrate Groups.* M. R. House (ed.). Academic Press, London. Pp. 55–102.

Clarkson, E. N. K. 1975. The evolution of the eye in trilobites. *Fossils and Strata* 4:7–31.

Cleveland, L. R. 1926. Symbiosis among animals with special reference to termites and their intestinal flagellates. *Q. Rev. Biol.* 1:51–60.

Cleveland, L. R., S. R. Hall, E. P. Sanders, and J. Collier. 1934. The wood-feeding roach *Cryptocercus,* its Protozoa, and the symbiosis between Protozoa and the roach. *Mem. Am. Acad. Arts Sci.* 17:185–342.

Cline, T. W. 1984. Autoregulatory functioning of a *Drosophila* gene product that establishes and maintains the sexually determined state. *Genetics* 107:231–77.

Cline, T. W. 1988. Evidence that *sisterless-a* and *sisterless-b* are two of several discrete "numerator elements" of the X/A sex determination signal in *Drosophila* that switch *Sxl* between two alternative stable expression states. *Genetics* 119: 829–62.

Cloud, P. E. 1949. Some problems and patterns of evolution exemplified by fossil invertebrates. *Evolution* 2:322–50.

Coates, M. 1991. New palaeontological contributions to limb ontogeny and phylogeny. In *Developmental Patterning of the Vertebrate Limb.* J. R. Hinchliffe, J. M. Huerle, and D. Summerbell (eds.). Plenum Press, New York. Pp. 328–38.

Coates, M. 1994. The origin of vertebrate limbs. *Development* (Suppl.): 169–80.

Coates, M. I., and J. A. Clack. 1990. Polydactyly in the earliest tetrapod limbs. *Nature* 347:66–69.

Coates, M. I., and J. A. Clack. 1991. Fish-like gills and breathing in the earliest known tetrapod. *Nature* 352:234–36.

Coelho, C. N. D., K. M. Krabbenhoft, W. B. Upholt, J. F. Fallon, and R. A. Kosher. 1991. Altered expression of the chicken homeobox-containing genes *GHox-7* and

GHox-8 in the limb buds of limbless mutant chick embryos. *Development* 113: 1487–93.

Coelho, C. N. D., W. B. Upholt, and R. A. Kosher. 1993. The expression pattern of the chicken homeobox-containing gene *GHox-7* in developing polydactylous limb buds suggests its involvement in apical ectodermal ridge-directed outgrowth of limb mesoderm and in programmed cell death. *Differentiation* 52:129–37.

Cohen, B., A. A. Simcox, and S. M. Cohen. 1993. Allocation of the thoracic imaginal primordia in the *Drosophila* embryo. *Development* 117:597–608.

Cohn, M. J., J. C. Izpisúa-Belmonte, H. Abud, J. K. Heath, and C. Tickle. 1995. Fibroblast growth factors induce additional limb development from the flank of chick embryos. *Cell* 80:739–46.

Colbert, E. H. 1961. *Dinosaurs: Their Discovery and Their World.* E. P. Dutton, New York.

Colbert, E. H. 1968. *Men and Dinosaurs: The Search in Field and Laboratory.* E. P. Dutton, New York.

Colbert, E. H. 1995. *The Little Dinosaurs of Ghost Ranch.* Columbia University Press, New York.

Comins, N. F. 1993. *What if the Moon Didn't Exist.* Harper Collins, New York.

Compston, W., I. S. Williams, J. L. Kirschvink, Z. Zichao, and M. Guogan. 1992. Zircon U-Pb ages for the early Cambrian time-scale. *J. Geol. Soc. Lond.* 149: 171–84.

Compton, A. W., and F. A. Jenkins Jr. 1979. Origin of mammals. In *Mesozoic Mammals: The First Two-Thirds of Mammalian History.* J. A. Lillegraven, Z. Kielan-Jaworowska, and W. A. Clemens (eds.). University of California, Berkeley. Pp. 59–73.

Conklin, E. G. 1905. Organization and cell lineage of the ascidian egg. *J. Natl. Acad. Sci.* (Phila.) 13:1–119.

Conway Morris, S. 1986. The community structure of the middle Cambrian Phyllopod Bed (Burgess Shale). *Palaeontology* 29:423–67.

Conway Morris, S. 1989. Burgess Shale faunas and the Cambrian explosion. *Science* 246:339–46.

Conway Morris, S. 1993a. Ediacaran-like fossils in Cambrian Burgess Shale-type faunas of North America. *Palaeontology* 36:593–635.

Conway Morris, S. 1993b. The fossil record and the early evolution of the Metazoa. *Nature* 361:219–25.

Conway Morris, S. 1994. Why molecular biology needs palaeontology. *Development* (Suppl.): 1–13.

Conway Morris, S., and J. S. Peel. 1990. Articulated halkieriids from the Lower Cambrian of north Greenland. *Nature* 345:802–5.

Conway Morris, S., and J. S. Peel. 1995. Articulated halkieriids from the Lower Cambrian of North Greenland and their role in early protostome evolution. *Phil. Trans. R. Soc. Lond.* B 347:305–58.

Coomans, A. 1981. Phylogenetic implications of the photoreceptor structure. In *Origine dei Grandi Phyla dei Metazoi.* Accademia Nationale Dei Lincei, Rome. Pp. 23–68.

Cooper, A., C. Mourer-Chauviré, G. K. Chambers, A. von Haeseler, A. C. Wilson, and S. Pääbo. 1992. Independent origins of New Zealand moas and kiwis. *Proc. Natl. Acad. Sci. USA* 89:8741–44.

Count, E. W. 1947. Brain and body weight in man: Their antecedents in growth and evolution. *Ann. N.Y. Acad. Sci.* 46:993–1122.

Cracraft, J. 1990. The origin of evolutionary novelties: Pattern and process at different hierarchical levels. In *Evolutionary Innovations.* M. H. Nitecki (ed.). University of Chicago Press, Chicago. Pp. 21–44.

Crews, D. 1994. Temperature, steroids and sex determination. *J. Endocrinol.* 142:1–8.

Crews, D., J. M. Bergeron, J. J. Bull, D. Flores, A. Tousignant, J. K. Skipper, and T. Wibbels. 1994. Temperature-dependent sex determination in reptiles: Proximate mechanisms, ultimate outcomes, and practical applications. *Dev. Genet.* 15: 297–312.

Crichton, M. 1990. *Jurassic Park.* Alfred A. Knopf, New York.

Crimes, T. P. 1992a. Changes in the trace fossil biota across the Proterozoic-Phanerozoic boundary. *J. Geol. Soc. Lond.* 149:637–46.

Crimes, T. P. 1992b. The record of trace fossils across the Proterozoic-Cambrian boundary. In *Origin and Early Evolution of the Metazoa.* J. H. Lipps and P. W. Signor (eds.). Plenum Press, New York. Pp. 177–202.

Crompton, A. W., and F. A. J. Jenkins. 1979. Origin of mammals. In *Mesozoic Mammals: The First Two-Thirds of Mammalian History.* J. A. Lillegraven, Z. Kielan-Jaworwska, and W. A. Clemens (eds.). University of California, Berkeley. Pp. 59–73.

Crowther, R. J., and J. R. Whittaker. 1986. Developmental autonomy of presumptive notochord cells in partial embryos of an ascidian. *Int. J. Invert. Reprod. Dev.* 9:253–61.

Cummings, E. G., P. Bringas Jr., M. S. Grodin, and H. S. Slavkin. 1981. Epithelial-directed mesenchyme differentiation in vitro. *Differentiation* 20:1–9.

Cvekl, A., C. M. Sax, E. H. Bresnick, and J. Piatigorsky. 1994. A complex array of positive and negative elements regulates the chicken αA-crystallin gene: Involvement of Pax-6, USF, CREB and/or CREM, and AP_1 proteins. *Mol. Cell. Biol.* 14:7363–76.

Darwin, C. 1860. *A Naturalist's Voyage Round the World in H.M.S. "Beagle."* Anchor Books, New York.

Darwin, C. 1872. *The Origin of Species by Means of Natural Selection.* John Wanamaker, Philadelphia.

Darwin, C.. 1874. *The Descent of Man, and Selection in Relation to Sex.* A. L. Burt, New York.

Davidson, E. H. 1986. *Gene Activity in Early Development.* Academic Press, Orlando.

Davidson, E. H. 1989. Lineage-specific gene expression and the regulative capacities of the sea urchin embryo. *Development* 105:421–45.

Davis, R. L., H. Weintraub, and A. B. Lassar. 1987. Expression of a single transfected cDNA converts fibroblasts to myoblasts. *Cell* 51:987–1000.

Dean, M. C., and B. A. Wood. 1984. Phylogeny, neoteny and growth of the cranial base in hominoids. *Folia Primatol.* 43:157–80.

de Beer, G. 1930. *Embryology and Evolution.* Clarendon Press, Oxford.

de Beer, G. 1958. *Embryos and Ancestors.* Oxford University Press, Oxford.

de Beer, G. 1971. *Homology, an Unsolved Problem.* Oxford Biology Readers. J. J. Head and O. E. Lowenstein (eds.). Oxford University Press, London.

Deflande-Rigaud, M. 1946. Vestiges microscopiques des larves d'Echinoderms de l'Oxfordien de Villers-sur-Mer. *C. R. Acad. Sci.* 222:908–10.

de Jong, W. W., W. Hendriks, J. W. Mulders, and H. Bloemendal. 1989. Evolution of eye lens crystallins: The stress connection. *Trends Biochem. Sci.* 14:365–68.

Delaye, M., and A. Tardieu. 1983. Short-range order of crystallin proteins accounts for eye lens transparency. *Nature* 302:415–17.

del Pino, E. M. 1983. Progesterone induces incubatory changes in the brooding pouch of the frog *Gastrotheca riobambae* (Fowler). *J. Exp. Zool.* 227:159–63.

del Pino, E. M. 1989a. Marsupial frogs. *Sci. Am.* 260 (5): 110–18.

del Pino, E. M. 1989b. Modifications of oogenesis and development in marsupial frogs. *Development* 107:169–87.

del Pino, E. M., and R. P. Elinson. 1983. A novel developmental pattern for frogs: Gastrulation produces an embryonic disk. *Nature* 306:589–91.

del Pino, E. M., and B. Escobar. 1981. Embryonic stages of *Gastrotheca riobambae* (Fowler) during maternal incubation and comparison with that of other egg-brooding hylid frogs. *J. Morphol.* 167:277–95.

del Pino, E. M., and A. A. Humphries Jr. 1978. Multiple nuclei during early oogenesis in *Flectonotus pygmaeus* and other marsupial frogs. *Biol. Bull.* 154:198–212.

del Pino, E. M., and G. Sanchez. 1977. Ovarian structure of the marsupial frog *Gastrotheca riobambae* (Fowler). *J. Morphol.* 153:153–62.

Deno, T., and Satoh, N. 1984. Studies on the cytoplasmic determinant for muscle cell differentiation in ascidian embryos: An attempt at transplantation of the myoplasm. *Dev. Growth Differ.* 26:43–48.

Dent, J. N. 1968. Survey of amphibian metamorphosis. In *Metamorphosis.* W. Etkin and L. I. Gilbert (eds.). Appleton-Century-Crofts, New York. Pp. 271–311.

De Robertis, E. M., E. A. Morita, and K. W. Y. Cho. 1991. Gradient fields and homeobox genes. *Development* 112:669–78.

Derry, L. A., A. J. Kaufman, and S. B. Jacobsen. 1992. Sedimentary cycling and environmental change in the late Proterozoic: Evidence from stable and radiogenic isotopes. *Geochim. Cosmochim. Acta.* 56:1317–29.

DeSalle, R., J. Gatesy, W. Wheeler, and D. Grimaldi. 1992. DNA sequences from a fossil termite in Oligo-Miocene amber and their phylogenetic implications. *Science* 257:1933–36.

Des Marais, D. J., H. Strauss, R. E. Summons, and J. M. Hayes. 1992. Carbon isotope evidence for the stepwise oxidation of the Proterozoic environment. *Nature* 359:605–9.

Desmond, A. 1982. *Archetypes and Ancestors: Paleontology in Victorian London 1850–1875.* University of Chicago Press, Chicago.

Dhouailly, D. 1975. Formation of cutaneous appendages in dermo-epidermal recombinations between reptiles, birds and mammals. *Roux's Arch. Dev. Biol.* 177:323–40.

Dickinson, W. J. 1988. On the architecture of regulatory systems: Evolutionary insights and implications. *BioEssays* 8:204–8.

Dodd, J., and T. M. Jessell. 1988. Axon guidance and the patterning of neural projections in vertebrates. *Science* 242:692–99.

Dodd, M. H. I., and J. M. Dodd. 1976. The biology of metamorphosis. In *Physiology of the Amphibia*. B. Lofts (ed.). Academic Press, New York. Pp. 467–597.

Dodson, P. 1990. Counting dinosaurs: How many kinds were there? *Proc. Natl. Acad. Sci. USA* 87:7608–12.

Dodson, S. 1989. Predator-induced reaction norms. *BioScience* 39:447–52.

Doe, C. Q., and C. S. Goodman. 1993. Early events in insect neurogenesis. *Dev. Biol.* 111:193–205.

Dollé, P., J. C. Izpisua-Belmonte, E. Boncinelli, and D. Duboule. 1991. The *Hox-4.8* gene is localized at the 5' extremity of the Hox-4 complex and is expressed in the most posterior parts of the body during development. *Mech. Dev.* 36:3–13.

Dollé, P., J. C. Izpisua-Belmonte, H. Falkenstein, A. Renucci, and D. Duboule. 1989. Coordinate expression of the murine Hox-5 complex-containing genes during limb pattern formation. *Nature* 342:767–72.

Donoghue, M. J., J. A. Doyle, J. Gauthier, A. G. Kluge, and T. Rowe. 1989. The importance of fossils in phylogeny reconstruction. *Annu. Rev. Ecol. Syst.* 20:431–60.

Doolittle, R. F. 1983. Probability and the origin of life. In *Scientists Confront Creationism*. L. R. Godfrey (ed.). W. W. Norton, New York. Pp. 85–97.

Doolittle, W. F., and C. Sapienza. 1980. Selfish genes, the phenotype paradigm and genome evolution. *Nature* 285:618–20.

Douglas, S. E., C. A. Murphy, D. F. Spencer, and M. W. Gray. 1991. Cryptomonad algae are evolutionary chimaeras of two phylogenetically distinct unicellular eukaryotes. *Nature* 350:148–51.

Dover, G. A. 1986. Molecular drive in multigene families: How biological novelties arise, spread, and are assimilated. *Trends Genet.* 2:159–65.

Drager, B. J., M. A. Harkey, M. Iwata, and A. H. Whiteley. 1989. The expression of embryonic primary mesenchyme genes of the sea urchin, *Strongylocentrotus purpuratus,* in the adult skeletogenic tissues of this and other species of echinoderms. *Dev. Biol.* 133:14–23.

Duboule, D. 1994. Temporal colinearity and the phylotypic progression: A basis for the stability of a vertebrate Bauplan and the evolution of morphologies through heterochrony. *Development* (Suppl.): 135–42.

Ducibella, T. 1974. The occurrence of biochemical metamorphic events without anatomical metamorphosis in the axolotl. *Dev. Biol.* 38:175–86.

Duellman, W. E., and L. Trueb. 1986. *Biology of the Amphibia*. McGraw-Hill, New York.

Duncan, T., and T. F. Stuessy. 1984. *Cladistics: Perspectives on the Reconstruction of Evolutionary History*. Columbia University Press, New York.

Dung, V. V., P. M. Giao, N. N. Chinh, D. Tuoc, P. Arctander, and J. MacKinnon. 1993. A new species of living bovid from Vietnam. *Nature* 363:443–45.

Eakin, R. M. 1970. A third eye. *Am. Sci.* 58:73–79.

Eakin, R. M. 1979. Evolutionary significance of photoreceptors: In retrospect. *Am. Zool.* 19:647–53.

Echelard, Y., D. J. Epstein, B. St-Jacques, L. Shen, J. Mohler, J. A. McMahon, and A. P. McMahon. 1993. Sonic hedgehog, a member of a family of putative signaling molecules, is implicated in the regulation of CNS polarity. *Cell* 75: 1417–30.

Edgecombe, G. D., and B. D. E. Chatterton. 1987. Heterochrony and the Silurian radiation of encrinurine trilobites. *Lethaia* 20:337–51.

Eernisse, D. J., J. S. Albert, and F. E. Anderson. 1992. Annelida and Arthropoda are not sister taxa: A phylogenetic analysis of spiralian metazoan morphology. *Syst. Biol.* 41:305–30.

Ehlers, U. 1985. Phylogenetic relationships within the platyhelminthes. In *The Origins and Relationships of Lower Invertebrates*. S. Conway Morris, J. D. George, R. Gibson, and H. M. Platt (eds.). Clarendon Press, Oxford. Pp. 143–58.

Eigenmann, J. E., A. Amador, and D. F. Patterson. 1988. Insulin-like growth factor I levels in proportionate dogs, chondrodystrophic dogs and in giant dogs. *Acta Endocrinol.* 118:105–8.

Eigenmann, J. E., D. F. Patterson, and E. R. Froesch. 1984. Body size parallels insulin-like growth factor I levels but not growth hormone secretory capacity. *Acta Endocrinol.* 106:448–53.

Eisele, J. 1994. *Survival and detection of blood residues on stone tools*. Technical Report 94-1. Department of Anthropology, University of Nevada, Reno. 46 pp.

Eldredge, N., and S. J. Gould. 1972. Punctuated equilibria: An alternative to phyletic gradualism. In *Models in Paleobiology*. T. J. M. Schopf (ed.). Freeman, Cooper and Company, San Francisco. Pp. 82–115.

Elinson, R. P. 1987. Changes in developmental patterns: Embryos of amphibians with large eggs. In *Development as an Evolutionary Process*. R. A. Raff and E. C. Raff (eds.). Alan R. Liss, New York. Pp. 1–21.

Elinson, R. P. 1989. Egg evolution. In *Complex Organismal Functions: Integration and Evolution in Vertebrates*. D. B. Wake and G. Roth (eds.). John Wiley & Sons, New York. Pp. 251–62.

Elinson, R. P., and E. M. del Pino. 1985. Cleavage and gastrulation in the egg-brooding marsupial frog, *Gastrotheca riobambae*. *J. Embryol. Exp. Morphol.* 90: 223–32.

Emlet, R. B. 1995. Larval spicules, cilia, and symmetry as remnants of indirect development in the direct developing sea urchin *Heliocidaris erythrogramma*. *Dev. Biol.* 167:405–15.

Emlet, R. B., L. R. McEdward, and R. R. Strathmann. 1987. Echinoderm larval ecology viewed from the egg. In *Echinoderm Studies 2*. A. A. M. Jangoux and J. M. Lawrence (eds.). Balkema, Rotterdam. Pp. 55–136.

Endler, J. A. 1995. Multiple-trait coevolution and environmental gradients in guppies. *Trends Ecol. Evol.* 10:22–29.

Enlow, D. H. 1966. A comparative study of facial growth in *Homo* and *Macaca*. *Am. J. Phys. Anthropol.* 24:293–307.

Ephrussi, A., and R. Lehmann. 1992. Induction of germ cell formation by *Oscar*. *Nature* 358:387–92.

Erickson, J. W., and T. W. Cline. 1991. Molecular nature of the *Drosophila* sex determination signal and its link to neurogenesis. *Science* 251:1071–74.

Erwin, D. H. 1993a. *The Great Paleozoic Crisis: Life and Death in the Permian.* Columbia University Press, New York.

Erwin, D. H. 1993b. The origin of metazoan development: A paleontological perspective. *Biol. J. Linn. Soc.* 50:255–74.

Erwin, D. H., J. W. Valentine, and J. J. Sepkoski Jr. 1987. A comparative study of diversification events: The early Paleozoic versus the Mesozoic. *Evolution* 41: 1177–86.

Estes, R., Z. V. Spinar, and E. Nevo. 1978. Early Cretaceous pipid tadpoles from Israel (Amphibia: Anura). *Herpetologica* 34:374–93.

Etkin, W. 1970. The endocrine mechanism of amphibian metamorphosis: An evolutionary achievement. *Mem. Soc. Endocrinol.* 18:137–55.

Ettensohn, C. A. 1984. Primary invagination of the vegetal plate during sea urchin gastrulation. *Am. Zool.* 24:571–88.

Evans, R. M. 1988. The steroid and thyroid hormone receptor superfamily. *Science* 240:889–95.

Ewart, C. 1921. The nestling feathers of the mallard, with observations on the composition, origin, and history of feathers. *Proc. Zool. Soc. Lond.* 1921:609–42.

Falk, D. 1987. Brain lateralization in primates and its evolution in hominids. *Yrbk. Phys. Anthropol.* 30:107–25.

Fallon, J. F., A. Lopez, M. A. Ros, M. P. Savage, B. B. Olwin, and B. K. Simandl. 1994. FGF-2: Apical ectodermal ridge growth signal for chick limb development. *Science* 264:104–7.

Farris, J. S. 1983. The logical basis of phylogenetic analysis. In *Advances in Cladistics.* Vol. 2. N. I. Platnick and V. A. Funk (eds.). Columbia University Press, New York. Pp. 1–36.

Fedonkin, M. A. 1992. Vendian faunas and the early evolution of metazoa. In *Origin and Early Evolution of the Metazoa.* J. H. Lipps and P. W. Signor (eds.). Plenum Press, New York. Pp. 87–129.

Fedonkin, M. A. 1994. Vendian body fossils and trace fossils. In *Early Life on Earth.* S. Bengston (ed.). Columbia University Press, New York. Pp. 370–88.

Fedonkin, M. A., and B. N. Runnegar. 1992. Proterozoic metazoan trace fossils. In *The Proterozoic Biosphere.* J. W. Schopf and C. Klein (eds.). Cambridge University Press, Cambridge. Pp. 389–95.

Feduccia, A. 1980. *The Age of Birds.* Harvard University Press, Cambridge, Mass.

Felsenstein, J. 1978. Cases in which parsimony and compatibility methods will be positively misleading. *Syst. Zool.* 27:401–10.

Felsenstein, J. 1985. Confidence limits on phylogenies: An approach using the bootstrap. *Evolution* 39:783–91.

Felsenstein, J., and H. Kishino. 1993. Is there something wrong with the bootstrap on phylogenies? A reply to Hillis and Bull. *Syst. Biol.* 42:193–200.

Felts, W. J. L. 1959. Transplantation studies of factors in skeletal organogenesis: The subcutaneously implanted immature long-bone of the rat and mouse. *Am. J. Phys. Anthropol.* 17:201–13.

Feng, H., and B. P. Brandhorst. 1994. Evolution of actin gene families of sea urchins. *J. Mol. Evol.* 39:347–56.

Fernholm, B., K. Bremer, and H. Jörnvall. 1989. *The Hierarchy of Life: Molecules and Morphology in Phylogenetic Analysis*. Excerpta Medica, Amsterdam.

Field, K. G., G. J. Olsen, D. J. Lane, S. J. Giovannoni, M. T. Ghiselin, E. C. Raff, N. R. Pace, and R. A. Raff. 1988. Molecular phylogeny of the animal kingdom. *Science* 239:748–53.

Field, K. G., J. M. Turbeville, R. A. Raff, and B. A. Best. 1990. Evolutionary relationships of phylum Cnidaria inferred from 18S rRNA sequence data. *Fourth International Congress of Systematic and Evolutionary Biology*, College Park, Maryland. Abstract.

Fink, W. L. 1988. Phylogenetic analysis and the detection of ontogenetic patterns. In *Heterochrony in Evolution: A Multidisciplinary Approach*. M. L. McKinney (ed.). Plenum Press, New York. Pp. 71–91.

Fischer, A. L. 1965. Fossils, early life, and atmospheric history. *Proc. Natl. Acad. Sci. USA* 53:1205–13.

Flynn, T. T., and J. P. Hill. 1939. The development of the Monotremata. IV. Growth of the ovarian ovum, maturation, fertilization and early cleavage. *Trans. Zool. Soc. Lond.* 24:445–662.

Flynn, T. T., and J. P. Hill. 1942. The later stages of cleavage and the formation of the primary germ layers in the Monotremata (preliminary communication). *Proc. Zool. Soc. Lond.* A 111:233–53.

Foote, M. 1992. Paleozoic record of morphological diversity in blastozoan echinoderms. *Proc. Natl. Acad. Sci. USA* 89:7325–29.

Foote, M., and S. J. Gould. 1992. Cambrian and recent morphological disparity. *Science* 258:1816.

Ford, E. B., and J. S. Huxley. 1921. The nestling feathers of the mallard, with observations on the composition, origin, and history of feathers. *Roux's Arch. Dev. Biol.* 117:67–79.

Forey, P. L. 1988. Golden jubilee for the coelacanth *Latimeria chalumnae*. *Nature* 336:727–32.

Forey, P. L., and P. Janvier. 1993. Agnathans and the origin of jawed vertebrates. *Nature* 361:129–34.

Fowles, J. 1969. *The French Lieutenant's Woman*. Little Brown & Co., Boston.

Franco, B., S. Guioli, A. Pragliola, B. Incerti, B. Bardoni, R. Tonlorenzi, R. Carrozzo, E. Maestrini, M. Pieretti, P. Taillon-Miller, C. J. Brown, H. F. Willard, C. Lawrence, M. G. Persico, G. Camerino, and A. Ballabio. 1991. A gene deleted in Kallmann's syndrome shares homology with neural cell adhesion and axonal path-finding molecules. *Nature* 353:529–36.

Frazzetta, T. H. 1975. *Complex Adaptations in Evolving Populations*. Sinauer Associates, Sunderland, Mass.

Freeman, G., and J. W. Lundelius. 1992. Evolutionary implications of the mode of D quadrant specification in coelomates with spiral cleavage. *J. Evol. Biol.* 5: 205–47.

French, V. 1993. The long and the short of it. *Nature* 361:400–401.

Frumhoff, P. C., and H. K. Reeve. 1994. Using phylogenies to test hypotheses of adaptations: A critique of some current proposals. *Evolution* 48:172–80.

Frumkin, A., Z. Rangini, A. Ben-Yehuda, Y. Gruenbaum, and A. Fainsod. 1991.

A chicken caudal homolog, *CHoc-cad,* is expressed in the epiblast with posterior localization and in the early endodermal lineage. *Development* 112:207–19.

Fuenzalida, M., R. Lemus, S. Romero, R. Fernandez-Valencia, and D. Lemus. 1990. Behavior of rabbit dental tissues in heterospecific association with embryonic quail ectoderm. *J. Exp. Zool.* 256:264–72.

Futuyma, D. J. 1986. *Evolutionary Biology.* Sinauer Associates, Sunderland, Mass.

Galaburda, A. M., and D. N. Pandya. 1982. Role of architectonics and connections in the study of primate brain evolution. In *Primate Brain Evolution.* E. Armstrong and D. Falk (eds.). Plenum Press, New York. Pp. 203–16.

Gans, C., and R. G. Northcutt. 1983. Neural crest and the origin of vertebrates: A new head. *Science* 220:268–74.

Garcia-Fernàndez, J., and P. W. H. Holland. 1994. Archtypal organization of the amphioxus Hox gene cluster. *Nature* 370:563–66.

Gardner, B. G. 1982. Tetrapod classification. *Zool. J. Linn. Soc.* 74:207–32.

Garstang, W. 1922. The theory of recapitulation: A critical restatement of the biogenetic law. *Zool. J. Linn. Soc. Lond., Zool.* 35:81–101.

Garstang, W. 1928. The morphology of the Tunicata, and its bearing on the phylogeny of the Chordata. *Q. J. Microsc. Soc.* 75:51–187.

Gaunt, S. J. 1994. Conservation in the Hox code during morphological evolution. *Int. J. Dev. Biol.* 38:549–52.

Gaunt, S. J., and P. B. Singh. 1990. Homeogene expression patterns and chromosomal imprinting. *Trends Genet.* 6:208–12.

Gauthier, J., D. Cannatella, K. De Queiroz, A. G. Kluge, and T. Rowe. 1989. Tetrapod phylogeny. In *The Hierarchy of Life: Molecules and Morphology in Phylogenetic Analysis.* B. Fernholm, K. Bremer, and H. Jörnvall (eds.). Excerpta Medica, Amsterdam. Pp. 337–53.

Gauthier, J., A. G. Kluge, and T. Rowe. 1988. Amniote phylogeny and the importance of fossils. *Cladistics* 4:105–209.

Gavis, E. R., and R. Lehmann. 1992. Localization of *nanos* RNA controls embryonic polarity. *Cell* 71:301–13.

Geetha-Habib, M., R. Noiva, H. A. Kaplan, and W. J. Lennarz. 1988. Glycosylation site binding protein, a component of oligosaccharide transferase, is highly similar to three other 57 kd luminal proteins of the ER. *Cell* 54:1053–60.

Gehling, J. G. 1986. Algal binding of siliciclastic sediments: A mechanism in the preservation of Ediacaran fossils. *12th International Sedimentological Congress.* Abstr. 117.

Gehling, J. G. 1987. Earliest known echinoderm: A new Ediacaran fossil from the Pound Subgroup of South Australia. *Alcheringa* 11:337–45.

Gehling, J. G. 1991. The case for Ediacaran fossil roots to the metazoan tree. *Geological Society of India Memoir No. 20.* Pp. 181–224.

Geist, V. 1986. The paradox of the great Irish stags. *Nat. Hist.* 3:54–65.

Gendron-Maguire, M., M. Mallo, M. Zhang, and T. Gridley. 1993. *Hoxa-2* mutant mice exhibit homoeotic transformation of skeletal elements derived from cranial neural crest. *Cell* 75:1317–31.

Ghiselin, M. T. 1974. A radical solution to the species problem. *Syst. Zool.* 23: 536–44.

Gibbs, H. L., and P. H. Grant. 1987. Oscillating selection on Darwin's finches. *Nature* 327:511–13.

Gilbert, S. F. 1994. *Developmental Biology*. Sinauer Associates, Sunderland, Mass.

Gilbert, S. F., J. M. Opitz, and R. A. Raff. 1995. Resynthesizing evolutionary and developmental biology. *Dev. Biol*. In press.

Gill, P., P. L. Ivanov, C. Kimpton, P. Romelle, N. Benson, G. Tully, I. Evett, E. Hagelberg, and K. Sullivan. 1994. Identification of the remains of the Romanov family by DNA analysis. *Nature Genet.* 6:130–35.

Gingerich, P. D. 1983. Rates of evolution: Effects of time and temporal scaling. *Science* 222:159–61.

Gingerich, P. D., S. M. Raza, M. Arif, M. Anwar, and X. Zhou. 1994. New whale from the Eocene of Pakistan and the origin of cetacean swimming. *Nature* 368:844–47.

Gingerich, P. D., B. H. Smith, and E. L. Simons. 1990. Hind limbs of Eocene *Basilosaurus:* Evidence of feet in whales. *Science* 249:154–57.

Gingerich, P. D., N. A. Wells, D. E. Russell, and S. M. I. Shah. 1983. Origin of whales in epicontinental remnant seas: New evidence from the early Eocene of Pakistan. *Science* 220:403–6.

Glaessner, M. F. 1961. Pre-Cambrian animals. *Sci. Am.* 204 (3): 72–78.

Glaessner, M. F. 1984. *The Dawn of Animal Life: A Biohistorical Study*. Cambridge University Press, Cambridge.

Glaessner, M. F., and M. Wade. 1966. The Late Precambrian fossils from Ediacara, South Australia. *Paleontology* 9:599–628.

Globus, M., and S. Vethamany-Globus. 1976. An in vitro analogue of early chick limb bud outgrowth. *Differentiation* 6:91–96.

Goldschmidt, R. 1938. *Physiological Genetics*. McGraw-Hill, New York.

Goldschmidt, R. 1940. *The Material Basis of Evolution*. Yale University Press, New Haven.

Golenberg, E. M., D. E. Giannis, M. T. Clegg, C. J. Smiley, M. Durbin, D. Henderson, and G. Zurawski. 1990. Chloroplast DNA sequence from a Miocene *Magnolia* species. *Nature* 344:656–58.

Goodman, C. S., and C. J. Shatz. 1993. Developmental mechanisms that generate precise patterns of neuronal connectivity. *Cell 72/Neuron 10* (Suppl.): 77–98.

Goodwin, B. C. 1982. Development and evolution. *J. Theor. Biol.* 97:43–55.

Goodwin, B. C. 1985. What are the causes of morphogenesis? *BioEssays* 3:32–36.

Goodwin, B. C. 1990. The evolution of generic forms. In *Organizational Constraints on the Dynamics of Evolution*. J. Maynard Smith and G. Vida (eds.). Manchester University Press, Manchester. Pp. 107–17.

Gorr, T., T. Kleinschmidt, and H. Fricke. 1991. Close tetrapod relationships of the coelacanth *Latimeria* indicated by hemoglobin sequences. *Nature* 351:394–97.

Gosse, P. H. 1857. *Omphalos: An Attempt to Untie the Geological Knot*. John Van Voorst, London.

Gould, S. J. 1970. Dollo on Dollo's Law: Irreversibility and the status of evolutionary laws. *J. Hist. Biol.* 3:189–212.

Gould, S. J. 1974. The origin and function of "bizarre" structures: Antler size and skull size in the "Irish Elk," *Megaloceros giganteus. Evolution* 28:191–220.

Gould, S. J. 1977. *Ontogeny and Phylogeny.* The Belknap Press of Harvard University Press, Cambridge, Mass.

Gould, S. J. 1981. *The Mismeasure of Man.* W. W. Norton, New York.

Gould, S. J. 1988. The uses of heterochrony. In *Heterochrony in Evolution: A Multidisciplinary Approach.* M. L. McKinney (ed.). Plenum Press, New York. Pp. 1–13.

Gould, S. J. 1989. *Wonderful Life: The Burgess Shale and the Nature of History.* W. W. Norton, New York.

Gould, S. J. 1991. The disparity of the Burgess Shale arthropod fauna and the limits of cladistic analysis: Why we must strive to quantify morphospace. *Paleobiology* 17:411–23.

Gould, S. J., and N. Eldredge. 1993. Punctuated equilibrium comes of age. *Nature* 366:223–27.

Gould, S. J., and R. C. Lewontin. 1979. The spandrels of San Marcos and the Panglossian paradigm: A critique of the adaptationist program. *Proc. R. Soc. Lond.* B 205:581–98.

Gould, S. J., and E. S. Vrba. 1982. Exaptation: A missing term in the science of form. *Paleobiology* 8:4–15.

Gouy, M., and W.-H. Li. 1989. Molecular phylogeny of the kingdoms Animalia, Plantae, and Fungi. *Mol. Biol. Evol.* 6:109–22.

Grant, M. 1962. *Myths of the Greeks and Romans.* World Publishing Co., Cleveland.

Grant, P. 1986. *Ecology and Evolution of Darwin's Finches.* Princeton University Press, Princeton.

Grant, S. G., and V. M. Chapman. 1988. Mechanisms of X-chromosome regulation. *Annu. Rev. Genet.* 22:199–233.

Grant, S. W. F. 1990. Shell structure and distribution of *Cloudina,* a potential index fossil for the terminal Proterozoic. *Am. J. Sci.* 290-A:261–94.

Gray, J., and W. Shear. 1992. Early life on land. *Am. Sci.* 80:444–56.

Greene, E. 1989. A diet-induced developmental polymorphism in a caterpillar. *Science* 243:643–46.

Gregory, W. K. 1951. *Evolution Emerging.* MacMillan, New York.

Grell, K. 1982. Placozoa. In *Synopsis and Classification of Living Organisms.* Vol. 1. S. Parker (ed.). McGraw-Hill, New York. P. 639.

Grindley, J. C., D. R. Davidson, and R. E. Hill. 1995. The role of *Pax-6* in eye and nasal development. *Development* 121:1433–42.

Grobben, K. 1908. Die systematische Einteilung des Tierreisches. *Verh. Zool. Bot. Ges. Wien.* 58:491–511.

Hadfield, K. A., B. J. Swalla, and W. R. Jeffery. 1995. Multiple origins of anural development in ascidians inferred from rDNA sequences. *J. Mol. Evol.* 40:413–27.

Hadzi, J. 1953. An attempt to reconstruct the system of animal classification. *Syst. Zool.* 2:145–54.

Hadzi, J. 1963. *The Evolution of the Metazoa.* Macmillan, New York.

Hafner, J. C., and M. S. Hafner. 1988. Heterochrony in rodents. In *Heterochrony in Evolution: A Multidisciplinary Approach.* M. L. McKinney (ed.). Plenum Press, New York. Pp. 217–35.

Hagelberg, E., M. Thomas, C. Cook Jr., A. Sher, G. Baryshnikov, and A. Lister. 1994. DNA from ancient mammoth bones. *Nature* 370:333–34.

Hahn, J.-H., J. C. Kissinger, and R. A. Raff. 1995. Structure and evolution of CyI cytoplasmic actin-encoding genes in the indirect- and direct-developing sea urchins *Heliocidaris tuberculata* and *Heliocidaris erythrogramma*. *Gene* 153:219–24.

Halanych, K. M. 1993. The phylogenetic position of the lophophorates based on 18S ribosomal gene sequence data. *Am. Zool.* 33:288. Abstract.

Halanych, K. M., J. D. Bacheller, A. M. Aguinaldo, S. M. Liva, D. M. Hillis, and J. A. Lake. 1995. 18S rDNA evidence that the lophophorates are protostome animals. *Science* 267:1641–43.

Haldane, J. B. S. 1932. The time of action of genes and its bearing on some evolutionary problems. *Am. Nat.* 66:5–24.

Halder, G., P. Callaerts, and W. J. Gehring. 1995. Induction of ectopic eyes by targeted expression of the *eyeless* gene in *Drosophila*. *Science* 267:1788–92.

Hall, B. K. 1984. Developmental mechanisms underlying the formation of atavisms. *Biol. Rev.* 59:89–124.

Hall, B. K. 1992. *Evolutionary Developmental Biology*. Chapman and Hall, London.

Halliday, T. R., and P. Verrell. 1986. Salamanders and newts. In *The Encyclopedia of Reptiles and Amphibians*. T. R. Halliday and K. Adler (eds.). Facts on File, New York. Pp. 18–29.

Hambrey, M. B., and W. B. Harland. 1981. *Earth's Pre-Pleistocene Glacial Record*. Cambridge University Press, Cambridge.

Hamburger, V. 1988. *The Heritage of Experimental Embryology: Hans Spemann and the Organizer*. Oxford University Press, Oxford.

Hampé, A. 1960. Le compétition entre les éléments osseux du zeugopode de poulet. *J. Embryol. Exp. Morphol.* 8:241–45.

Handt, O., M. Richards, M. Trommsdorff, C. Kilger, J. Simanainen, O. Grergiev, K. Bauer, A. Stone, R. Hedges, W. Schaffner, G. Utermann, B. Sykes, and S. Pääbo. 1994. Molecular genetic analysis of the Tyrolian ice man. *Science* 264: 1775–78.

Hanken, J. 1985. Morphological novelty in the limb skeleton accompanies miniaturization in salamanders. *Science* 229:871–74.

Hanken, J. 1986. Developmental evidence for amphibian origins. *Evol. Biol.* 20: 389–417.

Hanken, J. 1993. Model systems versus outgroups: Alternative approaches to the study of head development and evolution. *Am. Zool.* 33:448–56.

Hanken, J., W. Klymkowsky, C. H. Summers, D. W. Seufert, and N. Ingebrigtsen. 1992. Cranial ontogeny in the direct-developing frog, *Eleutherodactylus coqui* (Anura: Leptodactylidae), analyzed using whole-mount immunohistochemistry. *J. Morphol.* 211:95–118.

Hanken, J., and D. B. Wake. 1993. Miniaturization of body size: Organismal consequences and evolutionary significance. *Annu. Rev. Ecol. Syst.* 24:501–19.

Hänni, C., L. Vincent, and P. Taberlet. 1994. Tracking the origins of the cave bear *(Ursus spelaeus)* by mitochondrial DNA sequencing. *Proc. Nat. Acad. Sci. USA* 91:12336–40.

Haqq, C. M., C.-Y. King, P. K. Donahoe, and M. A. Weiss. 1993. SRY recognizes conserved DNA sites in sex-specific promoters. *Proc. Natl. Acad. Sci. USA* 90: 1097–1101.

Hardin, J. 1989. Local shifts in position and polarized motility drive cell rearrangements during sea urchin gastrulation. *Dev. Biol.* 136:430–45.

Hardin, J., and L. Y. Cheng. 1986. The mechanisms and mechanics of archenteron elongation during sea urchin gastrulation. *Dev. Biol.* 115:490–501.

Hardin, J., and D. R. McClay. 1991. Target recognition by the archenteron during sea urchin gastrulation. *Dev. Biol.* 142:86–102.

Harkey, M. A., K. Klueg, P. Sheppard, and R. A. Raff. 1995. Structure, expression and extracellular targeting of PM27, a skeletal protein associated specifically with growth of the sea urchin larval spicule. *Dev. Biol.* 168:549–66.

Harrelson, A. L., and C. S. Goodman. 1988. Growth cone guidance in insects: Fasciculin II is a member of the immunoglobulin superfamily. *Science* 242: 700–708.

Harris, C. W., and L. Glover. 1988. The regional extent of the ca. 600 Ma Virgilina deformation: Implications for stratigraphic correlation in the Carolina terrane. *Geol. Soc. Am. Bull.* 100:200–217.

Hasegawa, M., T. Hashimoto, J. Adachi, N. Iwabe, and T. Miyata. 1993. Early branchings in the evolution of eukaryotes: Ancient divergence of entamoeba that lacks mitochondria revealed by protein sequence data. *J. Mol. Evol.* 36:270–81.

Hasegawa, M., Y. Iida, T. Yano, F. Takaiwa, and M. Iwabuchi. 1985. Phylogenetic relationships among eukaryotic kingdoms inferred from ribosomal RNA sequences. *J. Mol. Evol.* 22:32–38.

Heitzler, P., and P. Simpson. 1991. The choice of cell fate in the epidermis of *Drosophila. Cell* 64:1083–92.

Hendriks, L., R. DeBaere, Y. Van de Peer, J. Neefs, and A. Goris. 1991. The evolutionary position of rhodophyte *Pophyra umbilicalis* and the basidiomycete *Leucosporidium scottii* among other eukaryotes as deduced from complete sequences of small ribosomal subunit RNA. *J. Mol. Evol.* 32:167–77.

Hendy, M. D., and D. Penny. 1989. A framework for the quantitative study of evolutionary trees. *Syst. Zool.* 38:297–309.

Henry, J. J., and R. M. Grainger. 1987. Inductive interactions in the spatial and temporal restriction of lens-forming potential in embryonic ectoderm of *Xenopus laevis. Dev. Biol.* 124:200.

Henry, J. J., K. M. Klueg, and R. A. Raff. 1992. Evolutionary dissociation between cleavage, cell lineage and embryonic axes in sea urchin embryos. *Development* 114:931–38.

Henry, J. J., and R. A. Raff. 1990. Evolutionary change in the process of dorsoventral axis determination in the direct developing sea urchin, *Heliocidaris erythrogramma. Dev. Biol.* 141:55–69.

Henry, J. J., and R. A. Raff. 1992. Development and evolution of embryonic axial systems and cell determination in sea urchins. *Semin. Dev. Biol.* 3 (1): 35–42.

Henry, J. J., and R. A. Raff. 1994. Progressive determination of cell fates along the dorsoventral axis in the sea urchin *Heliocidaris erythrogramma. Roux's Arch. Dev. Biol.* 204:62–69.

Henry, J. J., G. A. Wray, and R. A. Raff. 1990. The dorsoventral axis is specified prior to first cleavage in the direct developing sea urchin *Heliocidaris erythrogramma*. *Development* 110:875–84.

Herr, W. 1988. The POU domain: A large conserved region in the mammalian pit-1, oct-1, oct-2, and *Caenorhabditis elegans* unc-86 gene products. *Genes Dev.* 2: 1513–16.

Hertweck, A. 1972. Georgia coastal region, Sapelo Island, U.S.A.: Sedimentology and biology. V. Distribution and environmental significance of Lebensspuren and in situ skeletal remains. *Senckenbergiana Maritima* 4:125–67.

Herzog, M., and L. Maroteaux. 1986. Dinoflagellate 17S rRNA sequence inferred from the gene sequence: Evolutionary implications. *Proc. Natl. Acad. Sci. USA* 83:8644–48.

Higuchi, R. G., B. Bowman, M. Freiberger, O. A. Ryder, and A. C. Wilson. 1984. DNA sequences from the quagga, an extinct member of the horse family. *Nature* 312:282–84.

Hill, R. E., P. F. Jones, A. R. Rees, C. M. Sime, M. J. Justice, N. R. Copeland, N. A. Jenkins, E. Graham, and D. R. Davidson. 1989. A new family of mouse homeobox-containing gene: Molecular structure, chromosomal location, and developmental expression of *Hox-7.1*. *Genes Dev.* 3:26–37.

Hillis, D. M., J. J. Bull, M. E. White, M. Badgett, and I. J. Molineux. 1992. Experimental phylogenetics: Generation of a known phylogeny. *Science* 255:589–92.

Hillis, D. M., J. J. Bull, M. E. White, M. Badgett, and I. J. Molineux. 1993. Experimental approaches to phylogenetic analysis. *Syst. Biol.* 42:90–92.

Hillis, D. M., and C. W. Huelsenbeck. 1994. Application and accuracy of molecular phylogenies. *Science* 264:671–77.

Hillis, D. M., and J. P. Huelsenbeck. 1992. Signal, noise, and reliability in molecular phylogenetic analyses. *J. Hered.* 83:189–95.

Hinchliffe, J. R., and D. R. Johnson. 1980. *The Development of the Vertebrate Limb.* Clarendon Press, Oxford.

Hitchcock, E. 1858. *Ichnology of New England: A Report on the Sandstone of the Connecticut Valley, Especially Its Fossil Footmarks.* Commonwealth of Massachusetts, Boston.

Hodgkin, J. 1987. Sex determination and dosage compensation in *Caenorhabditis elegans*. *Annu. Rev. Genet.* 21:133–54.

Hodgkin, J. 1990. Sex determination compared in *Drosophila* and *Caenorhabditis*. *Nature* 344:721–28.

Hodgkin, J. 1992. Genetic sex determination mechanisms and evolution. *BioEssays* 14:253–61.

Hodin, J., and G. Wistow. 1993. 5'-RACE PCR of mRNA for three taxon-specific crystallins: For each gene one promoter controls both lens and non-lens expression. *Biochim. Biophys. Res. Commun.* 190:391–96.

Hofmann, H. J. 1992. Proterozoic carbonaceous films. In *The Proterozoic Biosphere*. J. W. Schopf and C. Klein (eds.). Cambridge University Press, Cambridge. Pp. 349–57.

Hofmann, H. J. 1994. Proterozoic carbonaceous compressions ("metaphytes" and "worms"). In *Early Life on Earth*. S. Bengston (ed.). Columbia University Press, New York. Pp. 342–57.

Holder, N. 1983. Developmental constraints and the evolution of vertebrate limb patterns. *J. Theor. Biol.* 104:451–71.

Holland, P. W. H. 1990. Homeobox genes and segmentation: Co-option, co-evolution, and convergence. *Semin. Dev. Biol.* 1:135–45.

Holland, P. W. H. 1991. Cloning and evolutionary analysis of *msh*-like homeobox genes from mouse, zebrafish and ascidian. *Gene* 98:253–57.

Holland, P. W. H., J. Garcia-Fernandez, N. A. Williams, and A. Sidow. 1994. Gene duplications and the origins of vertebrate development. *Development* (Suppl.): 125–33.

Holland, P. W. H., A. M. Hacker, and N. A. Williams. 1991. A molecular analysis of the phylogenetic affinities of *Saccoglossus cambrensis* Brambell & Cole (Hemichordata). *Phil. Trans. R. Soc. Lond.* B 332:185–89.

Holland, P. W. H., and B. L. M. Hogan. 1986. Phylogenetic distribution of *Antennapedia*-like homoeoboxes. *Nature* 321:251–53.

Holland, P. W. H., L. Z. Holland, N. A. Williams, and N. D. Holland. 1992. An amphioxus homeobox gene: Sequence conservation, spatial expression during development and insight into vertebrate evolution. *Development* 116:653–61.

Holland, P. W. H., P. Ingham, and S. Krauss. 1992. Mice and flies head to head. *Nature* 358:627–28.

Holt, A. B., D. B. Cheek, E. D. Mellits, and D. E. Hill. 1975. Brain size and the relation of the primate to the nonprimate. In *Fetal and Postnatal Cellular Growth: Hormones and Nutrition.* D. B. Cheek (ed.). John Wiley, New York. Pp. 23–44.

Holtfreter, J. 1934. Ueber die Verbreitung induzierender Substanzen und ihre Leistungen im Triton-Keim. *Arch. Entw. Mech. Org.* 132:307–83.

Horridge, G. A. 1978. The separation of visual axes in apposition compound eyes. *Phil. Trans. R. Soc. Lond.* B 285:1–59.

Hörstadius, S. 1973. *Experimental Embryology of Echinoderms.* Clarendon Press, Oxford.

Horwitz, J. 1992. α-Crystallin can function as a molecular chaperone. *Proc. Natl. Acad. Sci. USA* 89:10449–53.

Höss, M., and S. Pääbo. 1993. DNA extraction from Pleistocene bones by a silica-based purification method. *Nucleic Acids Res.* 21:3913–14.

Höss, M., S. Pääbo, and N. K. Vereshchagin. 1994. Mammoth DNA sequences. *Nature* 370:333.

Hou, X., and J. Bergström. 1994. Palaeoscolecid worms may be nematomorphs rather than annelids. *Lethaia* 27:11–17.

Hoyle, H. E., and E. C. Raff. 1990. Two *Drosophila* beta tubulin isoforms are not functionally equivalent. *J. Cell Biol.* 111:1009–26.

Hulbert, R. C. J. 1993. Taxonomic evolution in North American Neogene horses (subfamily Equinea): The rise and fall of an adaptive radiation. *Paleobiology* 19: 216–34.

Hülskamp, M., and D. Tautz. 1991. Gap genes and gradients: The logic behind the gaps. *BioEssays* 13:261–68.

Huneycutt, R. L. 1992. Naked mole-rats. *Am. Sci.* 80:43–53.

Hunt, P., and R. Krumlauf. 1991. Deciphering the Hox code: Clues to patterning branchial regions of the head. *Cell* 66:1075–78.

Hunt, P., J. Whiting, I. Muchamore, H. Marshall, and R. Krumlauf. 1991. Homeobox genes and models for patterning the hindbrain and branchial arches. *Development* (Suppl.) 1:187–96.

Huxley, A. 1939. *After Many a Summer Dies the Swan*. Harper and Brothers, New York.

Huxley, J. S. 1932. *Problems of Relative Growth*. Methuen and Company, London.

Hyman, L. H. 1940. *The Invertebrates. Protozoa through Ctenophora*. Vol. I. McGraw-Hill, New York.

Hyman, L. H. 1951. *The Invertebrates: Platyhelminthes and Rhynchocoela. The Acoelomate Bilateria*. McGraw-Hill, New York.

Hyman, L. H. 1955. *The Invertebrates: Echinodermata. The Coelomate Bilateria*. McGraw-Hill, New York.

Hyman, L. H. 1959. *The Invertebrates: Smaller Coelomate Groups*. McGraw-Hill, New York.

Ingham, P. W., and A. Martinez-Arias. 1992. Boundaries and fields in early embryos. *Cell* 68:221–35.

Insel, T. R., Z.-X. Wang, and C. F. Ferris. 1994. Patterns of brain vasopressin receptor distribution associated with social organization in microtine rodents. *J. Neurosci.* 14:5381–92.

Izpisua-Belmonte, J.-C., C. Tickle, P. Dollé, L. Wolpert, and D. Duboule. 1991. Expression of the homeobox *Hox-4* genes and the specification of position in chick wing development. *Nature* 350:585–89.

Jaanusson, V. 1981. Functional thresholds in evolutionary progress. *Lethaia* 14:251–60.

Jablonski, D. 1986a. Background and mass extinctions: The alteration of macrevolutionary regimes. *Science* 231:129–33.

Jablonski, D. 1986b. Larval ecology and macroevolution in marine invertebrates. *Bull. Mar. Sci.* 39:565–87.

Jablonski, D., and D. J. Bottjer. 1990. The ecology of evolutionary innovation: The fossil record. In *Evolutionary Innovations*. M. H. Nitecki (ed.). University of Chicago Press, Chicago. Pp. 253–88.

Jablonski, D., and R. A. Lutz. 1983. Larval ecology of marine invertebrates: Paleobiological implications. *Biol. Rev.* 58:21–89.

Jacob, F. 1977. Evolution and tinkering. *Science* 196:1161–66.

Jacob, F., and J. Monod. 1961. On the regulation of gene activity. *Cold Spring Harb. Symp. Quant. Biol.* 26:193–211.

Jacobs, D. K. 1990. Selector genes and the Cambrian radiation of Bilateria. *Proc. Natl. Acad. Sci. USA* 87:4406–10.

Jacobs, W. P. 1994. *Caulerpa. Sci. Am.* 271 (6): 100–105.

Jacobson, A. G. 1966. Inductive processes in embryonic development. *Science* 152:25–34.

Jan, Y. N., and L. Y. Jan. 1993. HLH proteins, fly neurogenesis, and vertebrate myogenesis. *Cell* 75:827–30.

Janczewski, D. N., N. Yuhki, D. A. Gilbert, G. T. Jefferson, and S. J. O'Brien. 1992. Molecular phylogenetic inference from saber-toothed cat fossils of Rancho La Brea. *Proc. Natl. Acad. Sci. USA* 89:9769–73.

Janzen, F., and G. Paukstis. 1991. Environmental sex determination in reptiles: Ecology, evolution, and experimental design. *Q. Rev. Biol.* 66:149–79.

Jarvis, J. U. M. 1981. Eusociality in a mammal: Cooperative breeding in naked mole-rat colonies. *Science* 212:571–73.

Jarzembowski, E. A. 1981. An early Cretaceous termite from southern England (Isoptera: Hodotermitidae). *Syst. Entomol.* 6:91–96.

Jeannotte, L., M. Lemieux, J. Cherron, F. Poirer, and E. J. Robertson. 1993. Specification of axial identity in the mouse: Role of the *Hoxa-5 (Hox 1.3)* gene. *Genes Dev.* 7:2085–96.

Jefferies, R. P. S. 1986. *The Ancestry of the Vertebrates.* British Museum of Natural History, London.

Jefferson, T. 1780–81. *Notes on the State of Virginia.* University of North Carolina Press, Chapel Hill.

Jeffery, W. R., and S. Meier. 1984. Ooplasmic segregation of the myoplasmic actin network in stratified ascidian eggs. *Roux's Arch. Dev. Biol.* 193:257–62.

Jeffery, W. R., and B. J. Swalla. 1992. Evolution of alternate modes of development in ascidians. *BioEssays* 14:219–26.

Jenkins, R. J. F. 1985. The enigmatic Ediacaran (late Precambrian) genus *Rangea* and related forms. *Paleobiology* 11:336–55.

Jenkins, R. J. F. 1992. Functional and ecological aspects of Ediacaran assemblages. In *Origin and Early Evolution of the Metazoa.* J. H. Lipps and P. W. Signor (eds.). Plenum Press, New York. Pp. 131–76.

Jeram, A. J., P. A. Selden, and D. Edwards. 1990. Land animals in the Silurian: Arachnids and myriapods from Shropshire, England. *Science* 250:658–61.

Jerison, H. J. 1982. Allometry, brain size, cortical surface, and convolutedness. In *Primate Brain Evolution.* E. Armstrong and D. Falk (eds.). Plenum Press, New York. Pp. 77–84.

Jiang, Z.-W. 1992. The lower Cambrian fossil record of China. In *Origin and Early Evolution of the Metazoa.* J. H. Lipps and P. W. Signor (eds.). Plenum Press, New York. Pp. 311–33.

Johnson, D. E., and L. T. Williams. 1993. Structural and functional diversity in the FGF receptor multigene family. *Adv. Cancer Res.* 60:1–41.

Johnson, M. S., and J. R. G. Turner. 1979. Absence of dosage compensation for a sex-linked gene in butterflies *(Heliconius). Heredity* 43:71–77.

Jones, D. S. 1988. Sclerochronology and the size versus age problem. In *Heterochrony in Evolution: A Multidisciplinary Approach.* M. L. McKinney (ed.). Plenum Press, New York. Pp. 93–108.

Jongens, T. A., B. Hay, L. Y. Jan, and Y. N. Jan. 1992. The germ cell-less gene product: A posteriorly localized component necessary for germ cell development in *Drosophila. Cell* 70:569–84.

Kafka, F. 1915. *The Metamorphosis.* Translated by S. Corngold, 1972. Bantam Books, Toronto.

Kammerer, P. 1924. *The Inheritance of Acquired Characteristics.* Boni and Liverlight, New York.

Kauffman, S. A. 1989. Cambrian explosion and Permian quiescence: Implications of rugged fitness landscapes. *Evol. Ecol.* 3:274–81.

Kauffman, S. A. 1993. *The Origins of Order: Self-Organization and Selection in Evolution.* Oxford University Press, Oxford.

Kaufman, T. C., M. A. Seeger, and G. Olsen. 1990. Molecular and genetic organization of the *Antennapedia* gene complex of *Drosophila melanogaster. Adv. Genet.* 27:309–62.

Keller, R. E. 1986. The cellular basis of amphibian gastrulation. In *Developmental Biology: A Comprehensive Synthesis.* Vol. 2. *The Cellular Basis of Morphogenesis.* L. Browder (ed.). Plenum Press, New York. Pp. 241–327.

Kelly, P. A., S. Ali, M. Rozakis, L. Goujon, M. Nagano, I. Pellagrini, D. Gould, J. Djiane, M. Edery, and J. Finidori. 1993. The growth hormone/prolactin receptor family. *Recent Prog. Horm. Res.* 48:123–64.

Kemphues, K. J., J. R. Priess, D. G. Morton, and N. Cheng. 1988. Identification of genes required for cytoplasmic localization in early *C. elegans* embryos. *Cell* 52:311–20.

Kenyon, C., and B. Wang. 1991. A cluster of Antennapedia-class homeobox genes in a nonsegmented animal. *Science* 253:516–17.

Kessel, M., and P. Gruss. 1990. Murine developmental control genes. *Science* 249: 374–79.

Kettlewell, H. B. D. 1973. *The Evolution of Melanism.* Clarendon Press, Oxford.

Keyes, L. N., T. W. Cline, and P. Schedl. 1992. The primary sex determination signal of *Drosophila* acts at the level of transcription. *Cell* 68:933–43.

Kidwell, M. G. 1990. Evolutionary aspects of hybrid dysgenesis in *Drosophila. Can. J. Zool.* 68:1716–26.

Kidwell, M. G. 1992. Horizontal transfer of P-elements and other short inverted repeat transposons. *Genetica* 86:275–86.

Kieny, M., and A. Chevallier. 1979. Autonomy of tendon development in the embryonic chick wing. *J. Embryol. Exp. Morphol.* 49:153–65.

Kimble, J. 1994. An ancient molecular mechanism for establishing embryonic polarity? *Science* 266:577–78.

Kimble, J., and J. White. 1981. On the control of germ-line development in *Caenorhabditis elegans. Dev. Biol.* 81:208–19.

Kimura, M. 1983. *The Neutral Theory of Molecular Evolution.* Cambridge University Press, New York.

King, M.-C., and A. C. Wilson. 1975. Evolution at two levels in humans and chimpanzees. *Science* 188:107–16.

Kingsley, D. M. 1994. What do BMPs do in mammals? Clues from the mouse *short-ear* mutation. *Trends Genet.* 10:16–21.

Kingsley, D. M., A. E. Bland, J. M. Grubber, P. C. Marker, L. B. Russell, N. G. Copeland, and N. A. Jenkins. 1992. The mouse *short ear* skeletal morphogenesis locus is associated with defects in a bone morphogenetic member of the TGF-B superfamily. *Cell* 71:399–410.

Kissinger, C. R., B. Liu, E. Martin-Bianco, T. B. Kornberg, and C. O. Pabo. 1990. Crystal structure of an *engrailed* homeodomain-DNA complex at 2.8 A resolution: A framework for understanding homeodomain-DNA interactions. *Cell* 63:579–90.

Klingenberg, C. P., and J. R. Spence. 1993. Heterochrony and allometry: Lessons from the water strider genus *Limnoporus. Evolution* 47:1834–53.

Kluge, A. G., and R. E. Strauss. 1985. Ontogeny and systematics. *Annu. Rev. Ecol. Syst.* 16:247–68.

Knoll, A. H. 1992a. Biological and biogeochemical preludes to the Ediacaran radiation. In *Origin and Early Evolution of the Metazoa.* J. H. Lipps and P. W. Signor (eds.). Plenum Press, New York. Pp. 53–84.

Knoll, A. H. 1992b. The early evolution of eukaryotes: A geological perspective. *Science* 256:622–27.

Knoll, A. H. 1994. Neoproterozoic evolution and environmental change. In *Early Life on Earth.* S. Bengston (ed.). Columbia University Press, New York. Pp. 439–49.

Knoll, A. H., and M. R. Walter. 1992. Latest Proterozoic stratigraphy and earth history. *Nature* 356:673–78.

Koelle, M., W. S. Talbot, W. A. Segraves, M. T. Bender, P. Cherbas, and D. S. Hogness. 1991. The *Drosophila EcR* gene encodes an ecdysone receptor, a new member of the steroid receptor superfamily. *Cell* 67:59–77.

Koestler, A. 1971. *The Case of the Midwife Toad.* Random House, New York.

Kojima, S., T. Hashimoto, M. Hasegawa, S. Murata, S. Ohta, H. Seki, and N. Okada. 1993. Close phylogenetic relationship between Vestimentifera (tube worms) and Annelida revealed by the amino acid sequence of elongation factor-1α. *J. Mol. Evol.* 37:66–70.

Kollar, E. J., and C. Fisher. 1980. Tooth induction in chick epithelium: Expression of quiescent genes for enamel synthesis. *Science* 207:993–95.

Koopman, P., J. Gubbay, N. Vivian, P. Goodfellow, and R. Lovell-Badge. 1991. Male development of chromosomally female mice transgeneic for *Sry. Nature* 351:117–21.

Korschelt, E., and K. Heider. 1900. *Textbook of the Embryology of the Invertebrates.* Macmillan, New York.

Krauss, S., J.-P. Concordet, and P. W. Ingham. 1993. A functionally conserved homolog of the *Drosophila* segment polarity gene *hh* is expressed in tissues with polarizing activity in zebrafish embryos. *Cell* 75:1431–44.

Kristensen, N. P. 1989. Insect phylogeny based on morphological evidence. In *The Hierarchy of Life: Molecules and Morphology in Phylogenetic Analysis.* B. Fernholm, K. Bremer, and H. Jörnvall (eds.). Excerpta Medica, Amsterdam. Pp. 295–306.

Kristensen, R. M. 1983. Loricifera, a new phylum with Aschelminthes characters from the meiobenthos. *Z. Zool. Syst. Evolutionsforsch.* 21:163–80.

Krumlauf, R. 1992. Evolution of the vertebrate Hox homeobox genes. *BioEssays* 14:245–52.

Kukalová, J. 1968. Permian mayfly nymphs. *Psyche* 75:310–27.

Kukalová-Peck, J. 1978. Origin and evolution of insect wings and their relation to metamorphosis, as documented by the fossil record. *J. Morphol.* 156:53–126.

Kukalová-Peck, J. 1983. Origin of the insect wing and wing articulation from the arthropodan leg. *Can. J. Zool.* 61:1618–69.

Kukalová-Peck, J. 1987. New Carboniferous Diplura, Monura, and Thusanura, the hexapod ground plan, and the role of thoracic side lobes in the origin of wings (Insecta). *Can. J. Zool.* 65:2327–45.

Kukalová-Peck, J. 1991. Fossil history and the evolution of hexapod structures. In *Insects of Australia*. Melbourne University Press, Melbourne. Pp. 141–79.

Kumé, M., and K. Dan. 1988. *Invertebrate Embryology*. Garland, New York.

Kurtén, B. 1963. Return of a lost structure in the evolution of the felid dentition. *Finska Vetenshaps-Societeten, Helsingfors. Commentationes Biol.* 26:1–12.

Kurtén, B. 1986. *How to Deep-Freeze a Mammoth*. Columbia University Press, New York.

Labandeira, C. C., B. S. Beall, and F. M. Hueber. 1989. Early insect diversification: Evidence from Lower Devonian bristletail from Quebec. *Science* 242:913–16.

LaBarbera, M. 1981. Water flow patterns in and around three species of articulate brachiopods. *J. Exp. Mar. Biol. Ecol.* 55:185–206.

Lacalli, T. C., N. D. Holland, and J. E. West. 1994. Landmarks in the anterior central nervous system of amphioxus larvae. *Phil. Trans. R. Soc.* B 344:165–85.

Laitman, J. T. 1985. Evolution of the hominid upper respiratory tract: The fossil evidence. In *Hominid Evolution: Past, Present and Future*. P. V. Tobias (ed.). Alan R. Liss, New York. Pp. 281–86.

Laitman, J. T., and R. C. Heimbuch. 1984. A measure of basicranial flexion in *Pan paniscus*, the pygmy chimpanzee. In *The Pygmy Chimpanzee: Evolutionary Biology and Behaviour*. R. L. Susman (ed.). Plenum, New York. Pp. 49–64.

Lake, J. A. 1987. Rate-independent technique for analysis of nucleic acid sequences: Evolutionary parsimony. *Mol. Biol. Evol.* 4:167–91.

Lake, J. A. 1990. Origin of the Metazoa. *Proc. Natl. Acad. Sci. USA* 87:763–66.

Land, M. F. 1984. Molluscs (Eyes). In *Photoreception and Vision in Invertebrates*. M. A. Alii (ed.). Plenum Press, New York. Pp. 699–725.

Lande, R. 1987. Quantitative genetics and evolutionary theory. In *Proceedings of the Second International Conference on Quantitative Genetics*. B. S. Wier, E. J. Eisen, M. M. Goodman, and G. Namkoong (eds.). Sinauer Associates, Sunderland, Mass. Pp. 71–84.

Landing, E. 1994. Precambrian-Cambrian boundary global stratotype ratified and a new perspective of Cambrian time. *Geology* 22:179–82.

Langeland, J. A., and S. B. Carroll. 1993. Conservation of regulatory elements controlling *hairy* pair-rule stripe formation. *Development* 117:585–96.

Lanyon, S. 1985. Detecting internal inconsistencies in distance data. *Syst. Zool.* 34:397–403.

Larsen, W. J. 1992. *Human Embryology*. Churchill Livingstone, New York.

Larson, A., and P. Chippindale. 1993. Molecular approaches to the evolutionary biology of plethodontid salamanders. *Herpetologica* 49:204–15.

Lauder, G. V. 1994. Homology, form, and function. In *Homology: The Hierarchical Basis of Comparative Biology*. B. K. Hall (ed.). Academic Press, San Diego. Pp. 151–96.

Lauder, G. V., and K. F. Liem. 1989. The role of historical factors in the evolution of complex organismal functions. In *Complex Organismal Functions: Integration and Evolution in Vertebrates*. D. B. Wake and G. Roth (eds.). John Wiley & Sons, Chichester. Pp. 63–78.

Lawrence, P. A. 1992. *The Making of a Fly: The Genetics of Animal Design*. Blackwell Scientific Publications, Oxford.

LeDouarin, N. M., P. Cochard, M. Vincent, J. L. Duband, G. C. Tucker, M.-A. Teillet, and J.-P. Thiery. 1984. Nuclear, cytoplasic, and membrane markers to follow neural crest cell migration: A comparative study. In *The Role of the Extracellular Matrix in Development.* R. L. Trelstad (ed.). Alan R. Liss, New York. Pp. 373–98.

Lee, D. C., P. Gonzalez, and G. Wistow. 1994. z-Crystallin: A lens-specific promoter and the gene recruitment of an enzyme as a crystallin. *J. Mol. Biol.* 236:669–78.

Lee, J. J., R. J. Shott, S. J. Rose III, T. L. Thomas, R. J. Britten, and E. H. Davidson. 1984. Sea urchin actin gene subtypes: Gene number, linkage and evolution. *J. Mol. Biol.* 172:149–76.

Lee, R. C., R. Feinbaum, and V. Ambros. 1993. The *C. elegans* heterochronic gene *lin-14* encodes small RNAs with antisense complementarity to *lin-14. Cell* 75:843–54.

Lefeuve, P., and C. Bordereau. 1984. Soldier formation regulated by a primer pheromone from the soldier frontal gland in a higher termite, *Nasutitermes lujae. Proc. Natl. Acad. Sci. USA* 81:7665–68.

Lehmann, R., and C. Rongo. 1993. Germ plasm formation and germ cell determination. *Semin. Dev. Biol.* 4:149–59.

Leid, M., P. Kastner, R. Lyons, H. Nakshatri, M. Saunders, T. Zacharewski, J.-Y. Chen, A. Staub, J.-M. Garnier, S. Mader, and P. Chambon. 1992. Purification, cloning, and RXR identity of the HeLa cell factor with which RAR or TR heterodimerizes to bind target sequences efficiently. *Cell* 68:377–95.

Lem, S. 1983. *His Master's Voice.* Harcourt Brace Jovanovich, New York.

Le Mouellic, H., Y. Lallemand, and P. Brulet. 1992. Homoeosis in the mouse induced by a null mutation in the *Hox-3.1* gene. *Cell* 69:251–64.

Lemus, D., L. Coloma, M. Fuenzalida, J. Illanes, Y. Paz De La Vega, A. Ondarza, and M. J. Blanquez. 1986. Odontogenesis and amelogenesis in interacting lizard-quail tissue combinations. *J. Morphol.* 189:121–29.

Levin, L. A., and T. S. Bridges. 1994. Control and consequences of alternative developmental modes in a poecilogonous polychaete. *Am. Zool.* 34:323–32.

Levinton, J. 1988. *Genetics, Palentology, and Macroevolution.* Cambridge University Press, Cambridge.

Levi-Setti, R. 1993. *Trilobites.* 2d ed. University of Chicago Press, Chicago.

Levitan, D. R., M. A. Sewall, and F.-S. Chia. 1992. How distribution and abundance influence fertilization success in the sea urchin *Strongylocentrotus purpuratus. Ecology* 73:248–54.

Lewin, R. 1993. *The Origin of Modern Humans.* Scientific American Library, New York.

Lewis, E. B. 1963. Genes and developmental pathways. *Am. Zool.* 3:33–56.

Lewis, E. B. 1978. A gene complex controlling segmentation in *Drosophila. Nature* 276:565–70.

Lewis-Williams, J. D. 1986. Cognitive and optical illusions in San rock art research. *Curr. Anthropol.* 27:171–77.

Lewis-Williams, J. D., and T. A. Dowson. 1988. The signs of all times: Entoptic phenomena in Upper Paleolithic art. *Curr. Anthropol.* 29:201–45.

Li, X., and M. Noll. 1994. Evolution of distinct developmental functions of three

Drosophila genes by acquisition of different cis-regulatory regions. *Nature* 367: 83–87.

Liem, K. F. 1990. Key evolutionary innovations, differential diversity, and symecomorphosis. In *Evolutionary Innovations*. M. H. Nitecki (ed.). University of Chicago Press, Chicago. Pp. 147–70.

Light, S. F. 1944. Experimental studies on ectohormonal control of the development of supplementary reproductives in the termite genus *Zootermopsis* (formerly *Termopsis*). *Univ. Calif. Publ. Zool.* 43:413–54.

Light, S. F., and F. M. Weesner. 1951. Further studies on the production of supplementary reproductives in *Zootermopsis* (Isoptera). *J. Exp. Zool.* 117:397–414.

Lillie, F. R. 1895. The embryology of the Unionidae. *J. Morphol.* 10:1–100.

Lillie, F. R. 1898. Adaptation in cleavage. In *Biological Lectures of the Marine Biological Laboratory of Woods Hole, Mass.* Ginn and Company, Boston. Pp. 43–67.

Lillie, F. R. 1927. The gene and the ontogenetic process. *Science* 66:361–68.

Lindahl, T. 1993. Instability and decay of the primary structure of DNA. *Nature* 362:709–15.

Lindberg, D. R. 1988. Heterochrony in gastropods: A neontological view. In *Heterochrony in Evolution: A Multidisciplinary Approach*. M. L. McKinney (ed.). Plenum Press, New York. Pp. 197–216.

Little, C. 1990. *The Terrestrial Invasion: An Ecophysiological Approach to the Origins of Land Animals*. Cambridge University Press, Cambridge.

Lively, C. M. 1986. Predator-induced shell dimorphism in the acorn barnacle *Chthamalus anisopoma*. *Evolution* 40:232–42.

Logan, G. A., J. M. Hayes, G. B. Hieshima, and R. E. Summons. 1995. Terminal Proterozoic reorganization of biogeochemical cycles. *Nature* 376:53–56.

Lombard, R. E., and D. B. Wake. 1976. Tongue evolution in the lungless salamanders, family Plethodontidae. I. Introduction, theory and a general model of dynamics. *J. Morphol.* 148:265–86.

Lombard, R. E., and D. B. Wake. 1977. Tongue evolution in the lungless salamanders, family Plethodontidae. II. Function and evolutionary diversity. *J. Morphol.* 153:39–80.

Lombard, R. E., and D. B. Wake. 1986. Tongue evolution in the lungless salamanders, family Plethodontidae. IV. Phylogeny of plethodontid salamanders and the evolution of feeding dynamics. *Syst. Zool.* 35:532–51.

Lord, E. M., and J. P. Hill. 1987. Evidence for heterochrony in the evolution of plant form. In *Development as an Evolutionary Process*. R. A. Raff and E. C. Raff (eds.). Alan R. Liss, New York. Pp. 47–70.

Lorenzen, S. 1985. Phylogenetic aspects of pseudocoelomate evolution. In *The Origins and Relationships of Lower Invertebrates*. S. Conway Morris, J. D. George, R. Gibson, and H. M. Platt (eds.). Clarendon Press, Oxford. Pp. 210–23.

Lowenstein, J. M. 1980. Species-specific proteins in fossils. *Naturwissenschaften* 67:343–46.

Lowenstein, J. M. 1981. Immunological reactions from fossil material. *Phil. Trans. R. Soc. Lond.* B 292:143–49.

Lowenstein, J. M. 1988. Immunological methods for determining phylogenetic relationships. In *Molecular Evolution and the Fossil Record: Short Courses in Paleontology #1*. B. Runnegar and J. W. Schopf (eds.). The Paleontological Society, Knoxville. Pp. 12–19.

Lowenstein, J. M., V. M. Sarich, and B. J. Richardson. 1981. Albumin systematics of the extinct mammoth and Tasmanian wolf. *Nature* 291:409–11.

Loy, T. H. 1983. Prehistoric blood residues: Detection on tool surfaces and identification of species of origin. *Science* 220:1269–71.

Loy, T. H. 1993. The artifact as site: An example of the biomolecular analysis of organic residues on prehistoric tools. *World Archeol.* 25:44–63.

Lucchesi, J. C., and J. E. Manning. 1987. Gene dosage and compensation in *Drosophila melanogaster*. *Adv. Genet.* 24:371–429.

Lufkin, T., A. Dierich, M. LeMeur, M. Mark, and P. Chambon. 1991. Disruption of the *Hox-1.6* homeobox gene results in defects in a region corresponding to its rostral domain of expression. *Cell* 66:1105–19.

Lüscher, M. 1955. Zur Frage der Übertragung sozialer Wirkstoff bei termiten. *Naturwissenschaften* 42:186.

Luykx, P. 1985. Genetic relations among castes in lower termites. In *Caste Differentiation in Social Insects*. J. A. L. Watson, B. M. Okot-Kotber, and C. Noirot (ed.). Pergamon Press, Oxford. Pp. 17–25.

Lyell, C. 1835. *Principles of Geology: Being an Inquiry How Far the Former Changes of the Earth's Surface Are Referable to Causes Now in Operation*. John Murray, London.

Lyngstadaas, S. P., S. Risnes, H. Noedbo, and A. G. Floves. 1990. Amelogenin gene similarity in vertebrates: DNA sequences encoding amelogenin seem to be conserved during evolution. *J. Comp. Physiol.* B 160:469–72.

Lynn, W. G. 1942. The embryology of *Eleutherodactylus nubicola*, an anuran which has no tadpole stage. *Carnegie Inst. (Wash.) Contrib. Embryol.* 30:27–62.

Lyons, K. M., R. W. Pelton, and B. L. M. Hogan. 1990. Organogenesis and pattern formation in the mouse: RNA distribution patterns suggest a role for bone morphogenetic protein-2A (BMP-2A). *Development* 109:833–44.

Mabee, P. M. 1989. An empirical rejection of the ontogenetic polarity criterion. *Cladistics* 5:409–16.

Macgregor, H. C., and E. M. del Pino. 1982. Ribosomal gene amplification in multinucleate oocytes of the egg brooding hylid frog *Flectonotus pygmaeus*. *Chromosoma* (Berl.) 85:475–88.

Maddison, D. R., M. Ruvolo, and D. L. Swofford. 1992. Geographic origins of human mitochondrial DNA: Phylogenetic evidence from control region sequences. *Syst. Biol.* 41:111–24.

Maden, M. 1993. The homeotic transformation of tails into limbs in *Rana temporaria* by retinoids. *Dev. Biol.* 159:379–91.

Malacinski, G. M. 1989. Amphibian somite development: Contrasts of morphogenetic and molecular differentiation patterns between the laboratory archetype species *Xenopus* (anuran) and axolotl (urodele). *Zool. Sci.* 6:1–14.

Malakhov, V. V. 1994. *Nematodes: Structure, Development, Classification, and Phylogeny*. Smithsonian Institution Press, Washington, D.C.

Manak, J. R., and M. P. Scott. 1994. A class act: Conservation of homeodomain protein functions. *Development* (Suppl.): 61–71.

Mangold, O. 1961. Grundzüge der Entwicklungsphysiologie der Wireltiere mit besonder Berücksichtigung der Missbildungen auf grund experimenteller Arbeiten an Urodelen. *Acta Genet. Med. Gmellol.* 10:1–49.

Manton, S. M. 1977. *The Arthropods: Habits, Functional Morphology and Evolution.* Oxford University Press, Oxford.

Manton, S. M., and D. T. Anderson. 1979. Polyphyly and the evolution of arthropods. In *The Origin of Major Invertebrate Groups.* M. R. House (ed.). Academic Press, London. Pp. 269–321.

Marden, J. H., and M. G. Kramer. 1994. Surface-skimming stoneflies: A possible intermediate in insect flight evolution. *Science* 266:427–30.

Marshall, C. R. 1990. Confidence intervals on stratigraphic ranges. *Paleobiology* 16:1–10.

Marshall, C. R., E. C. Raff, and R. A. Raff. 1994. Dollo's law and the death and resurrection of genes. *Proc. Natl. Acad. Sci. USA* 91:12283–87.

Marshall, C., and H.-P. Schultze. 1992. Relative importance of molecular, neontological and paleontological data in understanding the biology of the vertebrate invasion of land. *J. Mol. Evol.* 35:93–101.

Martin, R. L., C. Wood, W. Baehr, and M. L. Applebury. 1986. Visual pigment homologies revealed by DNA hybridization. *Science* 232:1266–69.

Maynard Smith, J., R. Burian, S. Kauffman, P. Alberch, J. Campbell, B. Goodwin, R. Lande, D. Raup, and L. Wolpert. 1985. Developmental constraints and evolution. *Q. Rev. Biol.* 60:265–87.

Maynard Smith, J., and K. C. Sondhi. 1960. The genetics of a pattern. *Genetics* 45:1039–50.

Mayr, E. 1960. The emergence of evolutionary novelties. In *Evolution after Darwin.* Vol. I. *The Evolution of Life.* S. Tax (ed.). University of Chicago Press, Chicago. Pp. 349–80.

Mayr, E. 1982. *The Growth of Biological Thought: Diversity, Evolution, and Inheritance.* Belknap Press of Harvard University Press, Cambridge, Mass.

McEdward, L. R., and D. A. Janies. 1993. Life cycle evolution in asteroids: What is a larva? *Biol. Bull.* 184:255–68.

McElreavey, K., E. Vilain, N. Abbas, I. Herskowitz, and M. Fellows. 1993. A regulatory cascade hypothesis for mammalian sex determination: SRY represses a negative regulator of male development. *Proc. Natl. Acad. Sci. USA* 90:3368–72.

McFall-Ngai, M. J., and M. K. Montgomery. 1990. The anatomy and morphology of the adult bacterial light organ of *Euprymna scolopes* Berry (Cephalopoda: Sepionidae). *Biol. Bull.* 179:332–39.

McGinnis, W., R. L. Garber, J. Wirz, A. Kuroiwa, and W. J. Gehring. 1984. A homologous protein-coding sequence in *Drosophila* homoeotic genes and its conservation in other metazoans. *Cell* 37:403–8.

McGinnis, W., and R. Krumlauf. 1992. Homeobox genes and axial patterning. *Cell* 68:283–302.

McGinnis, W., M. S. Levine, E. Hafen, A. Kuroiwa, and W. Gehring. 1984. A conserved DNA sequence in homoeotic genes of the *Drosophila* Antennapedia and Bithorax complexes. *Nature* 308:428–33.

McKinney, M. L. 1986. Ecological causation of heterochrony: A test and implications for evolutionary theory. *Paleobiology* 12:282–89.

McKinney, M. L. 1988. Classifying heterochrony: Allometry, size, and time. In *Heterochrony in Evolution: A Multidisciplinary Approach*. M. L. McKinney (ed.). Plenum Press, New York. Pp. 17–34.

McKinney, M. L., and K. J. McNamara. 1991. *Heterochrony: The Evolution of Ontogeny*. Plenum Press, New York.

McLaren, A. 1993. Germ cell sex determination. *Semin. Dev. Biol.* 4:171–77.

McMahon, A. P. 1991. Pattern regulation in the vertebrate embryo: The role of the wnt-family of putative signalling molecules. *Semin. Dev. Biol.* 2:425–33.

McMenamin, M. A. S., and D. L. S. McMenamin. 1990. *The Emergence of Animals: The Cambrian Breakthrough*. Columbia University Press, New York.

McMillan, W. O., R. A. Raff, and S. R. Palumbi. 1992. Population genetic consequences of developmental evolution in sea urchins (Genus *Heliocidaris*). *Evolution* 46:1299–1312.

McNamara, K. J. 1986. A guide to the nomenclature of heterochrony. *J. Paleontol.* 60:4–13.

McNamara, K. J. 1988a. The abundance of heterochrony in the fossil record. In *Heterochrony in Evolution: A Multidisciplinary Approach*. M. L. McKinney (ed.). Plenum Press, New York. Pp. 287–325.

McNamara, K. J. 1988b. Heterochrony and the evolution of echinoids. In *Echinoderm Phylogeny and Evolutionary Biology*. C. R. C. Paul and A. B. Smith (eds.). Clarendon Press, Oxford. Pp. 149–63.

McShea, D. W. 1993. Arguments, tests, and the Burgess Shale: A commentary on the debate. *Paleobiology* 19:399–402.

Meedle, T. H. 1992. Development of the ascidian embryo: Cell fate specification by autonomous and inductive processes. In *Morphogenesis: An Analysis of the Development of Biological Form*. E. F. Rossomand and S. Alexander (eds.). Marcel Dekker, New York. Pp. 263–317.

Mello, C. C., B. W. Draper, M. Krause, H. Weintraub, and J. R. Priess. 1992. The *pie-1* and *mex-1* genes and maternal control of blastomere identity in early *C. elegans* embryos. *Cell* 70:163–76.

Melville, H. 1851. *Moby Dick*. Bantam Books (1981), New York.

Mercola, M., and C. D. Stiles. 1988. Growth factor superfamilies and mammalian embryogenesis. *Development* 102:451–60.

Merimee, T. J., J. Zapf, B. Hewlett, and L. L. Cavalli-Sforza. 1987. Insulin-like growth factors in pygmies: The role of puberty in determining final stature. *N. Engl. J. Med.* 316:906–11.

Meyer, A., and S. I. Dolvin. 1992. Molecules, fossils, and the origin of tetrapods. *J. Mol. Evol.* 35:102–13.

Meyer, A., and A. C. Wilson. 1990. Origin of tetrapods inferred from their mitochondrial DNA affiliation to lungfish. *J. Mol. Evol.* 31:359–64.

Meyer, A., and A. C. Wilson. 1991. Coelacanth's relationships. *Nature* 353:219.

Miklos, G. L. G., and K. S. W. Campbell. 1994. From protein domains to extinct phyla: Reverse engineering approaches to the evolution of biological complexities. In *Early Life on Earth*. S. Bengston (ed.). Columbia University Press, New York. Pp. 501–16.

Miklos, G. L. G., K. S. W. Campbell, and D. R. Kankel. 1994. The rapid emergence of bio-electronic novelty, neural architecture and behavioral performance. In *Flexibility and Constraint in Behavioral Systems*. Wiley, Chichester. In press.

Mikulic, D. G., D. E. G. Briggs, and J. Kluessendorf. 1985a. A new exceptionally preserved biota from the Lower Silurian of Wisconsin, U.S.A. *Phil. Trans. R. Soc. Lond.* B 311:75–85.

Mikulic, D. G., D. E. G. Briggs, and J. Kluessendorf. 1985b. A Silurian soft-bodied biota. *Science* 228:715–17.

Mileikovsky, S. A. 1971. Types of larval development in marine bottom invertebrates, their disribution and ecological significance: A re-evaluation. *Mar. Biol.* 10:193–213.

Miller, L. M., J. D. Plenefisch, L. P. Casson, and B. J. Meyer. 1988. *xol-1:* A gene that controls the male modes of both sex determination and dosage compensation in *C. elegans. Cell* 55:167–83.

Minot, S. 1981. *Surviving the Flood*. Atheneum, New York.

Mittenthal, J. E. 1989. Physical aspects of the organization of development. In *Complex Systems: SFI Studies in the Sciences of Complexity*. D. Stein (ed.). Addison-Wesley Longman, Reading, Mass. Pp. 225–74.

Mittenthal, J. E., and A. G. Jacobson. 1990. The mechanics of morphogenesis in multicellular embryos. In *Biomechanics of Active Movement and Deformation*. N. Akkas (ed.). Springer-Verlag, Heidelberg. Pp. 295–401.

Mivart, S. G. 1871. *Genesis of Species*. Appleton and Co., New York.

Mohanty-Hejmadi, P., S. K. Dutta, and P. Mahapatra. 1992. Limbs generated at site of tail amputation in marbled balloon frog after vitamin A treatment. *Nature* 355:352–53.

Montagu, M. F. A. 1981. *Growing Young*. McGraw-Hill, New York.

Montgomery, M. K., and M. J. McFall-Ngai. 1992. The muscle-derived lens of a squid bioluminescent organ is biochemically convergent with the ocular lens: Evidence for recruitment of aldehyde dehydrogenase as a predominant structural protein. *J. Biol. Chem.* 267:20999–21003.

Moore, W. J., and C. L. B. Lavelle. 1974. *Growth of the Facial Skeleton in the Hominoidea*. Academic Press, New York.

Morgan, B. A., J.-C. Izpisúa-Belmonte, D. Duboule, and C. J. Tabin. 1992. Targeted misexpression of *Hox-4.6* in the avian limb bud causes apparent homeotic transformations. *Nature* 358:236–39.

Morgan, T. H. 1927. *Experimental Embryology*. Columbia University Press, New York.

Morgan, T. H. 1932. *The Scientific Basis of Evolution*. W. W. Norton, New York.

Morgan, T. H. 1934. *Embryology and Genetics*. Columbia University Press, New York.

Morris, J. 1973. *Heaven's Command: An Imperial Progress*. Harcourt Brace Jovanovich, New York.

Morris, P. J. 1993. The developmental role of the extracellular matrix suggests a monophyletic origin of the kingdom Animalia. *Evolution* 47:152–65.

Morris, V. B. 1995. A pluteal development of the sea urchin *Holopneustes purpurescens* Agassiz (Echinodermata: Echinoidea: Euechinoidea). *Zool. J. Linn. Soc.* In press.

Moses, K. 1991. The role of transcription factors in the developing *Drosophila* eye. *Trends Genet.* 7:250–55.

Mousseau, T. A., and H. Dingle. 1991. Maternal effects in insect life histories. *Annu. Rev. Entomol.* 36:511–34.

Müller, F. 1864. *Für Darwin.* Translated into English by W. S. Dallas as *Facts and Arguments for Darwin* (1869). J. Murray, London.

Müller, G. B. 1989. Ancestral patterns in bird limb development: A new look at Hampé's experiment. *J. Evol. Biol.* 2:31–47.

Müller, G. B. 1990. Developmental mechanisms at the origin of morphological novelty: A side-effect hypothesis. In *Evolutionary Innovations.* M. H. Nitecki (ed.). University of Chicago Press, Chicago. Pp. 99–130.

Muller, H. J. 1939. Reversibility in evolution considered from the standpoint of genetics. *Biol. Rev.* 14:261–80.

Müller, K. J., and D. Walossek. 1986. Arthropod larvae from the Upper Cambrian of Sweden. *Trans. R. Soc. Edinburgh: Earth Sci.* 77:157–79.

Mundy, G. R. 1989. Local factors in bone remodeling. *Recent Prog. Horm. Res.* 45:507–31.

Murtha, M. T., J. F. Leckman, and F. H. Ruddle. 1991. Detection of homeobox genes in development and evolution. *Proc. Natl. Acad. Sci. USA* 88:10711–15.

Nagy, L. M., and S. B. Carroll. 1994. Conservation of *wingless* patterning functions in the short-germ embryos of *Tribolium castaneum. Nature* 367:460–63.

Nathans, J., D. Thomas, and D. S. Hogness. 1986. Molecular genetics of human color vision: Genes encoding blue, green, and red pigments. *Science* 232:193–202.

Needham, J. 1933. On the dissociability of the fundamental process in ontogenesis. *Biol. Rev.* 8:180–223.

Needham, J., C. H. Waddington, and D. Needham. 1934. Physico-chemical experiments on the amphibian organizer. *Proc. R. Soc. Lond.* B 114:393–423.

Nelsen, O. E. 1953. *Comparative Embryology of the Vertebrates.* Blakiston, New York.

Nelson, G. J. 1978. Ontogeny, phylogeny, paleontology and the biogenetic law. *Syst. Zool.* 27:324–45.

Newman, S. A. 1993. Is segmentation generic? *BioEssays* 15:277–83.

Newman, S. A., and W. D. Comper. 1990. "Generic" physical mechanisms of morphogenesis and pattern formation. *Development* 110:1–18.

Nijhout, H. F. 1991. *The Development and Evolution of Butterfly Wing Patterns.* Smithsonian Institution Press, Washington, D.C.

Nijhout, H. F., and D. E. Wheeler. 1982. Juvenile hormone and the physiological basis of insect polymorphisms. *Q. Rev. Biol.* 57:109–33.

Nilsson, A., J. Isgaard, A. Lindahl, A. Dahlström, A. Skottner, and O. G. Isaksson. 1986. Regulation by growth hormone of number of chondrocytes containing IGF-I in rat growth plate. *Science* 233:571–74.

Nilsson, D.-E., and S. Pelger. 1994. A pessimistic estimate of the time required for an eye to evolve. *Proc. R. Soc. Lond.* B 256:53–58.

Nishida, H. 1994. Localization of determinants for formation of the anterior-posterior axis in eggs of the ascidian *Halocynthia roretzi*. *Development* 120:3093–3104.

Nishikata, T., I. Mita-Miyazawa, T. Deno, K. Takamura, and N. Satoh. 1987. Expression of epidermis-specific antigens during embryogenesis of the ascidian, *Halocynthia roretzi*. *Dev. Biol.* 121:408–16.

Niswander L., and G. M. Martin. 1993. FGF-4 and BMP-2 have opposite effects on limb growth. *Nature* 361:68–71.

Niswander, L., C. Tickle, A. Vogel, I. Booth, and G. R. Martin. 1993. FGF-4 replaces the apical ectodermal ridge and directs outgrowth and patterning of the limb. *Cell* 75:579–87.

Nohno, T., S. Noji, E. Koyama, K. Ohyama, F. Myokai, A. Kuroiwa, T. Saito, and S. Taniguchi. 1991. Involvement of the *Chox-4* chicken homeobox genes in determining of anteroposterior axial polarity during limb development. *Cell* 64: 1197–1205.

Noirot, C. 1985. Pathways of caste development in the lower termites. In *Caste Differentiation in Social Insects*. J. A. L. Watson, B. M. Okot-Kotber, and C. Noirot (eds.). Pergamon Press, Oxford. Pp. 41–57.

Noji, S., T. Nohno, E. Koyama, K. Muto, K. Ohyama, Y. Aoki, K. Tamura, K. Ohsugi, H. Ide, S. Taniguchi, and T. Saito. 1991. Retinoic acid induces polarizing activity but is unlikely to be a morphogen in the chick limb bud. *Nature* 350:83–86.

Nonet, M. L., and B. J. Meyer. 1991. Early aspects of *Caenorhabditis elegans* sex determination and dosage compensation are regulated by a zinc-finger protein. *Nature* 351:65–68.

Normark, B. B., A. R. McCune, and R. G. Harrison. 1991. Phylogenetic relationships of neopterygian fishes, inferred from mitochondrial DNA sequences. *Mol. Biol. Evol.* 8:819–34.

Norris, D. O., and W. A. Gern. 1976. Thyroxine-induced activation of hypothalmo-hypophysial axis in neotenic salamander larvae. *Science* 194:525–27.

Nottebohm, F. 1991. Reassessing the mechanisms and origins of vocal learning in birds. *Trends Neural Sci.* 14:206–11.

Nottebohm, F., and A. P. Arnold. 1976. Sexual dimorphism in vocal control areas of the songbird brain. *Science* 194:211–13.

Novacek, M. J. 1994. Whales leave the beach. *Nature* 368:807.

Nüsslein-Volhard, C. 1991. Determination of the embryonic axes of *Drosophila*. *Development* (Suppl.): 1–10.

Odin, G. S., N. H. Gale, B. Auvray, M. Bielski, F. Doré, J.-R. Lancelot, and P. Pasteels. 1983. Numerical dating of Precambrian-Cambrian boundary. *Nature* 301:21–23.

Ohno, S. 1970. *Evolution by Gene Duplication*. Springer-Verlag, New York.

Olsen, R. R., J. L. Cameron, and C. M. Young. 1993. Larval development (with observations on spawning) of the pencil urchin *Phyllacanthus imperialis:* A new intermediate larval form? *Biol. Bull.* 185:77–85.

Orgel, L. H., and F. H. C. Crick. 1980. Selfish DNA: The ultimate parasite. *Nature* 284:604–7.

Osorio, D. 1991. Patterns of function and evolution in the arthropod optic lobe. In *Evolution of the Eye and Visual System*. J. R. Cronly-Dillon and R. L. Gregory (eds.). Boca-Raton: CRC Press. Pp. 203–28.

Ossowski, I. V., G. Hausner, and P. C. Loewen. 1993. Molecular evolutionary analysis based on the amino acid sequence of catalase. *J. Mol. Evol.* 37:71–76.

Oster, G., J. Murray, and M. Miani. 1985. A model for chondrogenic condensations in the developing limb: The role of extracellular matrix and cell tractions. *J. Embryol. Exp. Morphol.* 89:93–112.

Otting, G., Y. Q. Qian, M. Billeter, M. Muller, M. Affolter, W. J. Gehring, and K. Wuthrich. 1990. Protein-DNA contacts in the structure of a homeodomain-DNA complex determined by nuclear magnetic resonance spectroscopy in solution. *EMBO J.* 9:3085–92.

Owen, R. 1846. Report on the archetype and homologies of the vertebrate skeleton. *Report of the British Association for the Advancement of Science* (Southampton Meeting). Pp. 169–340.

Pääbo, S. 1989. Ancient DNA: Extraction, characterization, molecular cloning, and enzymatic amplification. *Proc. Natl. Acad. Sci. USA* 86:1939–43.

Pääbo, S. 1990. Amplifying ancient DNA. In *PCR Protocols: A Guide to Methods and Applications*. M. A. Innis, D. H. Gelfand, J. J. Sninsky, and T. J. White (eds.). Academic Press, San Diego. Pp. 159–66.

Pääbo, S., R. G. Higuchi, and A. C. Wilson. 1989. Ancient DNA and the polymerase chain reaction: The emerging field of molecular archaeology. *J. Biol. Chem.* 264: 9709–12.

Pääbo, S., D. M. Irwin, and A. C. Wilson. 1990. DNA damage promotes jumping between templates during enzymatic amplification. *J. Biol. Chem.* 265:4718–21.

Pabst, D. A. 1993. Intramuscular morphology and tendon geometry of the epaxial swimming muscles of dolphins. *J. Zool.* (Lond.) 230:159–76.

Pace, N. R., D. A. Stahl, D. J. Lane, and G. J. Olsen. 1985. Analyzing natural microbial populations by rRNA sequences. *Am. Soc. Microbiol. News* 51:4–12.

Packard, A. 1972. Cephalopods and fish: The limits of convergence. *Biol. Rev.* 47:241–307.

Page, D. C., R. Mosher, E. M. Simpson, E. M. C. Fisher, G. Mardon, J. Pollack, B. McGillivray, A. de la Chapelle, and L. G. Brown. 1987. The sex-determining region of the human Y chromosome encodes a finger protein. *Cell* 51:1091–1104.

Pagel, M., and R. A. Johnstone. 1992. Variation across species in the size of the nuclear genome supports the junk-DNA explanation for the C-value paradox. *Proc. R. Soc. Lond.* B 249:119–24.

Paley, W. 1837. *Paley's Natural Theology*. Charles Knight, London.

Palumbi, S. R., and B. D. Kessing. 1991. Population biology of the trans-Arctic exchange: MtDNA sequence similarity between Pacific and Atlantic sea urchins. *Evolution* 45:1790–1805.

Palumbi, S. R., and E. C. Metz. 1991. Strong reproductive isolation between closely related tropical sea urchins (Genus *Echinometra*). *Mol. Biol. Evol.* 8:227–39.

Palumbi, S. R., and A. C. Wilson. 1990. Mitochondrial DNA similarity in the sea urchins *Strongylocentrotus purpuratus* and *S. droebachiensis*. *Evolution* 44: 403–15.

Panganiban, G., L. Nagy, and S. Carroll. 1994. The evolution and patterning of insect limbs. *Curr. Biol.* 4:671–75.

Parks, A. L., B. W. Bisgrove, G. A. Wray, and R. A. Raff. 1989. Direct development in the sea urchin *Phyllacanthus parvispinus* (Cidaroidea): Phylogenetic history and functional modification. *Biol. Bull.* 177:96–109.

Parks, A. L., B. A. Parr, J.-E. Chin, D. S. Leaf, and R. A. Raff. 1988. Molecular analysis of heterochronic changes in the evolution of direct developing sea urchins. *J. Evol. Biol.* 1:27–44.

Parr, B. A., and A. P. McMahon. 1995 Dorsalizing signal *Wnt-7a* required for normal polarity of D-V and A-P axes of mouse limbs. *Nature* 374:350–53.

Parsley, R. L. 1988. Feeding and respiratory strategies in Stylophora. In *Echinoderm Phylogeny and Evolutionary Biology*. C. R. C. Paul and A. B. Smith (eds.). Clarendon Press, Oxford. Pp. 347–61.

Partanen, J., T. P. Makela, E. Eerola, J. Korhonen, H. Hirvonen, L. Claesson-Welch, and K. Alitalo. 1991. FGFR-4, a novel acidic fibroblast growth factor receptor with a distinct expression pattern. *EMBO J.* 10:1347–54.

Passingham, R. E. 1981. Broca's area and the origins of human vocal skill. *Phil. Trans. R. Soc. Lond.* B 292:167–75.

Passingham, R. E. 1985. Rates of brain development in mammals including man. *Brain Behav. Evol.* 26:167–75.

Passingham, R. E., C. A. Heywood, and P. D. Nixon. 1986. Reorganization in the human brain as illustrated by the thalamus. *Brain Behav. Evol.* 29:68–76.

Patel, N. H. 1994. Developmental evolution: Insights from studies of insect segmentation. *Science* 266:581–90.

Patel, N. H., E. E. Ball, and C. S. Goodman. 1992. Changing role of *even-skipped* during the evolution of insect pattern formation. *Nature* 357:339–42.

Patel, N. H., E. Martin-Blanco, K. Coleman, S. J. Poole, M. C. Ellis, T. B. Kornberg, and C. S. Goodman. 1989. Expression of engrailed proteins in arthropods, annelids, and chordates. *Cell* 58:955–68.

Patstone, G., E. B. Pasquale, and P. A. Maher. 1993. Different members of the fibroblast growth factor receptor family are specific to distinct cell types in the developing chick embryo. *Dev. Biol.* 155:107–23.

Pattatucci, A. M., and T. C. Kaufman. 1991. The homoeotic gene *Sex combs reduced* of *Drosophila melanogaster* is differentially regulated in the embryonic and imaginal states of development. *Genetics* 129:443–61.

Patterson, C. 1981. Significance of fossils in determining evolutionary relationships. *Annu. Rev. Ecol. Syst.* 12:195–223.

Patterson, C. 1982. Morphological characters and homology. In *Problems of Phylogenetic Reconstruction*. K. A. Joysey and A. E. Friday (eds.). Academic Press, New York. Pp. 21–74.

Patterson, C. 1983. How does phylogeny differ from ontogeny? In *Development and Evolution*. B. C. Goodwin, N. Holder, and C. Wylie (eds.). Cambridge University Press, Cambridge. Pp. 1–31.

Patterson, C. 1989. Phylogenetic relations of major groups: Conclusions and prospects. In *The Hierarchy of Life: Molecules and Morphology in Phylogenetic Analy-*

sis. B. Fernholm, K. Bremer, and H. Jörnvall (eds.). Elsevier Science Publishers, Amsterdam. Pp. 471–88.

Patterson, P. H. 1992. Neuron-target interactions. In *An Introduction to Molecular Neurobiology*. Z. Hall (ed.). Sinauer Associates, Sunderland, Mass. Pp. 428–59.

Paul, C. R. C., and A. B. Smith. 1984. The early radiation and phylogeny of echinoderms. *Biol. Rev.* 59:443–81.

Paulus, H. F. 1979. Eye structure and the monophyly of the Arthropoda. In *Arthropod Phylogeny*. A. P. Gupta (ed.). Van Nostrand Reinhold, New York. Pp. 299–383.

Peifer, M., F. Karch, and W. Bender. 1987. The bithorax complex: Control of segmental identity. *Genes Dev.* 1:891–98.

Pendleton, J. W., B. K. Nagai, M. T. Murtha, and F. H. Ruddle. 1993. Expansion of the Hox gene family and the evolution of chordates. *Proc. Natl. Acad. Sci. USA* 90:6300–6304.

Pennington, J. T. 1985. The ecology of fertilization of echinoid eggs: The consequences of sperm dilution, adult aggregation and synchronous spawning. *Biol. Bull.* 169:417–30.

Penny, D., M. D. Hendy, and M. A. Steele. 1992. Progress with methods for constructing evolutionary trees. Pp. 73–79.

Perrett, D. I., K. A. May, and S. Yoshikawa. 1994. Facial shape and judgements of female attractiveness. *Nature* 368:239–42.

Peters, K., D. Ornitz, S. Werner, and L. Williams. 1993. Unique expression pattern of the FGF receptor 3 gene during mouse organogenesis. *Dev. Biol.* 155:423–30.

Pettigrew, J. B. 1908. *Design in Nature*. Longmans, Green, and Co., London.

Pfennig, D. W. 1992. Polymorphism in sparefoot toad tadpoles as a locally adjusted evolutionarily stable strategy. *Evolution* 46:1408–20.

Pfizenmayer, E. W. 1939. *Siberian Man and Mammoth*. Blackie & Son, London.

Pflug, H. D. 1970. Zur Fauna der Nama-Schichten in Sudwest-Afrika. 1. Pteridinia, Bau und systematische Zugerhorigkeit. *Palaeontogr. Abt. A* 134:226–62.

Piatigorsky, J., and G. Wistow. 1991. The recruitment of crystallins: New functions precede gene duplication. *Science* 252:1078–79.

Plenefisch, J. D., L. DeLong, and B. Meyer. 1989. Genes that implement the hermaphrodite mode of dosage compensation in *Caenorhabditis elegans*. *Genetics* 121:57–76.

Poiner, G. O. J. 1993. Still life in amber. *The Sciences,* March/April: 34–39.

Poiner, G. O. J., and R. Hess. 1982. Ultrastructure of 40-million-year-old insect tissue. *Science* 215:1241–42.

Priess, J. R., H. Schnabel, and R. Schnabel. 1987. The *glp-1* locus and cellular interactions in early *C. elegans* embryos. *Cell* 51:601–11.

Qian, Y. Q., M. Billeter, G. Otting, M. Muller, W. J. Gehring, and K. Wuthrich. 1989. The structure of the Antennapedia homeodomain determined by NMR spectroscopy in solution: Comparison with prokaryotic repressors. *Cell* 59:573–80.

Quiring, R., U. Walldorf, U. Kloter, and W. J. Gehring. 1994. Homology of the *eyeless* gene of *Drosophila* to the *Small eye* gene in mice and *Aniridia* in humans. *Science* 265:785–89.

Rabinow, L., and W. J. Dickinson. 1986. Complex cis-acting regulators and locus structure of *Drosophila* tissue-specific ADH variants. *Genetics* 112:523–37.

Radice, G. P., A. W. Neff, Y. H. Shim, J.-J. Brustis, and G. M. Malacinski. 1989. Developmental histories in amphibian myogenesis. *Int. J. Dev. Biol.* 33:325–43.

Raff, E. C., and R. A. Raff. 1985. Possible functions of the homeobox. *Nature* 313:185.

Raff, R. A. 1987. Constraint, flexibility, and phylogenetic change in the evolution of direct development in sea urchins. *Dev. Biol.* 119:6–19.

Raff, R. A. 1988. Direct developing sea urchins: A system for the study of developmental processes in evolution. In *Echinoderm Biology*. R. D. Burke, P. V. Mladenov, P. Lambert, and R. L. Parsley (eds.). A. A. Balkema, Rotterdam. Pp. 63–69.

Raff, R. A. 1992. Direct-developing sea urchins and the evolutionary reorganization of early development. *BioEssays* 14:211–18.

Raff, R. A. 1994. Developmental mechanisms in the evolution of animal form: Origins and evolvability of body plans. In *Early Life on Earth*. S. Bengston (ed.). Columbia University Press, New York. Pp. 489–500.

Raff, R. A., K. G. Field, M. T. Ghiselin, D. J. Lane, G. J. Olsen, N. R. Pace, A. L. Parks, B. A. Parr, and E. C. Raff. 1988. Molecular analysis of distant phylogenetic relationships in echinoderms. In *Echinoderm Phylogeny and Evolutionary Biology*. C. R. C. Paul and A. B. Smith (eds.). Clarendon Press, Oxford. Pp. 29–41.

Raff, R. A., L. Herlands, V. B. Morris, and J. Healy. 1990. Evolutionary modification of echinoid sperm correlates with developmental mode. *Dev. Growth Differ.* 32:283–91.

Raff, R. A., and T. C. Kaufman. 1983. *Embryos, Genes, and Evolution*. Macmillan, New York. Reprinted 1991 by Indiana University Press, Bloomington.

Raff, R. A., C. R. Marshall, and J. M. Turbeville. 1994. Using DNA sequences to unravel the Cambrian radiation of the animal phyla. *Annu. Rev. Ecol. Syst.* 25:351–75.

Raff, R. A., B. A. Parr, A. L. Parks, and G. A. Wray. 1990. Heterochrony and other mechanisms of radical evolutionary change in early development. In *Evolutionary Innovations*. M. H. Nitecki (ed.). University of Chicago Press, Chicago. Pp. 71–98.

Raff, R. A., and E. M. Popodi. 1995. Evolutionary approaches to analyzing development. In *Molecular Zoology: Advances, Stategies and Protocols*. J. D. Ferraris and S. R Palumbi (eds.). John Wiley and Sons, New York. In press.

Raff, R. A., and E. C. Raff. 1970. Respiratory mechanisms and the metazoan fossil record. *Nature* 228:1003–5.

Raff, R. A., and G. A. Wray. 1989. Heterochrony: Developmental mechanisms and evolutionary results. *J. Evol. Biol.* 2:409–34.

Raff, R. A., G. A. Wray, and J. J. Henry. 1991. Implications of radical evolutionary changes in early development for concepts of developmental constraint. In *New Perspectives on Evolution*. L. Warren and H. Koprowski (eds.). Wiley-Liss, New York. Pp. 189–207.

Raikow, R. J., S. R. Borecky, and S. L. Berman. 1979. The evolutionary reestablishment of a lost ancestral muscle in the bower bird assemblage. *Condor* 81:203–6.

Rainey, W. E., J. M. Lowenstein, V. M. Sarich, and D. M. Magor. 1984. Sirenian molecular systematics: The extinct Steller's sea cow *(Hydrodamalis gigas)*. *Naturwissenschaften* 67:343–46.

Ramirez-Solis, R., H. Zheng, J. Whiting, R. Krumlauf, and A. Bradley. 1993. *Hoxb-4 (Hox 2.6)* mutant mice show homoeotic transformations of a cervical vertebra and defects in the closure of the sternal rudiments. *Cell* 73:279–94.

Ramsköld, L., and X. Hou. 1991. New early Cambrian animal and onychophoran affinities of enigmatic metazoans. *Nature* 351:225–28.

Randazzo, F. M., D. L. Cribbs, and T. C. Kaufman. 1991. Rescue and regulation of *proboscipedia:* A homoeotic gene of the Antennapedia Complex. *Development* 113:257–71.

Randazzo, F. M., M. A. Seeger, C. A. Huss, M. A. Sweeney, J. K. Cecil, and T. C. Kaufman. 1993. Structural changes in the Antennapedia Complex of *Drosophila pseudoobscura*. *Genetics* 133:319–30.

Rasmussen, N. 1987. A new model of developmental constraints as applied to the *Drosophila* system. *J. Theor. Biol.* 127:271–99.

Raup, D. M. 1968. Theoretical morphology of echinoid growth. In *Paleobiological Aspects of Growth and Development: A Symposium.* D. B. Macurda Jr. (ed.). Paleontological Society Memoirs. Society of Economic Paleontologists and Mineralogists, Tulsa. Pp. 50–63.

Raup, D. M. 1979. Size of the Permo-Triassic bottleneck and its evolutionary implications. *Science* 206:217–18.

Raup, D. M. 1991. *Extinction: Bad Genes or Bad Luck?* W. W. Norton, New York.

Raup, D. M., and A. Michelson. 1965. Theoretical morphology of the coiled shell. *Science* 147:1294–95.

Raup, D. M., and J. J. Sepkoski Jr. 1984. Periodicity of extinctions in the geologic past. *Proc. Natl. Acad. Sci. USA* 81:801–5.

Ready, D. F. 1989. A multifaceted approach to neural development. *Trends Neural Sci.* 12:102–10.

Regier, J. C., and N. S. Vlahos. 1988. Heterochrony and the introduction of novel modes of morphogenesis during the evolution of moth choriogenesis. *J. Mol. Evol.* 28:19–31.

Reid, D. G. 1989. The comparative morphology, phylogeny and evolution of the gastropod family Littorinidae. *Phil. Trans. R. Soc. Lond.* B 342:1–110.

Reiter, R. S., and M. Solursh. 1982. Mitogenic property of the apical ectodermal ridge. *Dev. Biol.* 93:28–35.

Renucci, A., V. Zappavigna, J. Zakany, J.-C. Izpisua-Belmonte, K. Bürkl, and D. Duboule. 1992. Comparison of mouse and human HOX-4 complexes defines conserved sequences involved in the regulation of *Hox-4.4. EMBO J.* 11:1459–68.

Retallack, G. J. 1993. Were the Ediacaran fossils lichens? *Paleobiology* 20:523–44.

Rhoads, D. C., and J. W. Morse. 1971. Evolutionary and ecologic significance of oxygen deficient marine basins. *Lethaia* 4:413–28.

Richards, R. J. 1992. *The Meaning of Evolution.* University of Chicago Press, Chicago.

Richardson, J., A. Cvekl, and G. Wistow. 1995. *Pax-6* is essential for lens-specific expression of z-crystallin. *Proc. Natl. Acad. Sci. USA* 92:4676–80.

Richardson, M. K. 1996. Heterchrony and the phylotypic stage. *Dev. Biol.* In press.

Riddiford, L. M. 1994. Cellular and molecular actions of juvenile hormone. I. General considerations and premetamorphic actions. *Adv. Insect Physiol.* 24:213–74.

Riddle, R. D., R. L. Johnson, E. Laufer, and C. Tabin. 1993. *Sonic hedgehog* mediates the polarizing activity of the ZPA. *Cell* 75:1401–16.

Riedl, R. 1978. *Order in Living Organisms: A Systems Analysis of Evolution.* John Wiley & Sons, Chichester.

Rieger, R. M. 1986. Über den Ursprung der Bilateria: Die Bedeutung der Ultrastructurforschung für ein neues Verstehen der Metazoanevolution. *Verh. Dtsch. Zool. Ges.* 79:31–50.

Rijli, F. M., M. Mark, S. Lakkaraju, A. Dierich, P. Dollé, and P. Chambon. 1993. A homoeotic transformation is generated in the rostral branchial region of the head by disruption of *Hoxa-2*, which acts as a selector gene. *Cell* 75:1333–49.

Rimoin, D. L., T. J. Merimee, and V. A. McKusick. 1966. Growth-hormone deficiency in man: An isolated, recessive inherited defect. *Science* 152:1635–37.

Riutort, M., K. G. Field, J. M. Turbeville, R. A. Raff, and J. Bguna. 1992. Enzyme electrophoresis, 18S rRNA sequences, and levels of phylogenetic resolution among several species of freshwater planarians (Platyhelminthes, Tricladida, Paludicola). *Can. J. Zool.* 70:1425–39.

Robert, B., G. Lyons, B. K. Simandi, A. Kuroiwa, and M. Buckingham. 1991. The apical ectodermal ridge regulates *Hox-7* and *Hox-8* gene expression in developing chick limb buds. *Genes Dev.* 5:2363–74.

Roberts, C. W. M., J. R. Shutter, and S. J. Korsmeyer. 1994. *Hox11* controls the genesis of the spleen. *Nature* 368:747–49.

Robertson, R. M., K. G. Pearson, and H. Reichert. 1982. Flight interneurons in the locust and the origin of insect wings. *Science* 217:177–79.

Robinson, G. E. 1992. Regulation of division of labor in insect societies. *Annu. Rev. Entomol.* 37:637–65.

Rosen, D. E. 1982. Do current theories of evolution satisfy the basic requirements of explanation? *Syst. Zool.* 31:76–85.

Rosen, V., and S. Thies. 1992. The BMP proteins in bone formation and repair. *Trends Genet.* 8:97–102.

Roth, G., J. Blanke, and D. B. Wake. 1994. Cell size predicts morphological complexity in the brains of frogs and salamanders. *Proc. Natl. Acad. Sci. USA* 91: 4796–4800.

Roth, G., and A. Schmidt. 1993. The nervous system of plethodontid salamanders: Insight into the interplay between genome, organism, behavior, and ecology. *Herpetologica* 49:185–94.

Roth, V. L. 1988. The biological basis of homology. In *Ontogeny and Systematics.* C. J. Humphries (ed.). Columbia University Press, New York. Pp. 1–26.

Roth, V. L. 1994. Within and between organisms: Replicators, lineages, and homologies. In *Homology: The Hierarchical Basis of Comparative Biology.* B. K. Hall (ed.). Academic Press, San Diego. Pp. 301–37.

Roughgarden, J. 1989. The evolution of marine life cycles. In *Mathematical Evolutionary Biology.* M. Feldman (ed.). Princeton University Press, Princeton. Pp. 270–300.

Rouse, G. A. F., and K. Fitzhugh. 1994. Broadcasting fables: Is external fertilization really primitive? Sex, size, and larvae in sabellid polychaetes. *Zool. Scripta* 23: 271–312.

Roux, W. 1894. The problems, methods and scope of developmental mechanics. An introduction to the "Archiv für Entwicklungsmechanik der Organismen," translated by W. M. Wheeler. In *Biological Lectures of the Marine Biological Laboratory of Woods Hole, Mass.* (1895). Ginn and Company, Boston. Pp. 149–90.

Rozanov, A. Y., and A. Y. Zhuravlev. 1992. The lower Cambrian fossil record of the Soviet Union. In *Origin and Early Evolution of the Metazoa*. J. H. Lipps and P. W. Signor (eds.). Plenum Press, New York. Pp. 205–82.

Rubock, M. J., Z. Larin, M. Cook, N. Papalopulu, R. Krumlauf, and H. Lehrach. 1990. A yeast artificial chromosome containing a mouse homoeobox cluster *Hox-2*. *Proc. Natl. Acad. Sci. USA* 87:4751–55.

Ruddle, F. H., K. L. Bentley, M. T. Murtha, and N. Risch. 1994. Gene loss and gain in the origin of the vertebrates. *Development* (Suppl.): 155–61.

Rudwick, M. J. S. 1976. *The Meaning of Fossils: Episodes in the History of Paleontology*. Neale Watson, New York.

Rudwick, M. J. S. 1992. *Scenes from Deep Time: Early Pictorial Representations of the Prehistoric World*. University of Chicago Press, Chicago.

Ruiz i Altaba, A., and D. A. Melton. 1990. Axial patterning and the establishment of polarity in the frog embryo. *Trends Genet.* 6:57–64.

Runnegar, B. N. 1982a. The Cambrian explosion: Animals or fossils? *J. Geol. Soc. Aust.* 29:395–411.

Runnegar, B. N. 1982b. Oxygen requirements, biology and phylogenetic significance of the late Precambrian worm *Dickinsonia*, and the evolution of the burrowing habit. *Alcheringa* 6:223–39.

Runnegar, B. N. 1986. Molecular paleontology. *Palaeontology* 29:1–24.

Runnegar, B. N. 1987. Rates and modes of evolution in the Mollusca. In *Rates of Evolution*. K. S. W. Campbell and M. F. Day (eds.). Allen and Unwin, London. Pp. 39–60.

Runnegar, B. N. 1992. Proterozoic fossils of soft-bodied metazoans (Ediacara faunas). In *The Proterozoic Biosphere*. J. W. Schopf and C. Klein (eds.). Cambridge University Press, Cambridge. Pp. 999–1007.

Runnegar, B. N. 1994. Proterozoic eukaryotes: Evidence from biology and geology. In *Early Life on Earth*. S. Bengston (ed.). Columbia University Press, New York. Pp. 287–97.

Runnegar, B. N., and G. B. Curry. 1992. Amino acid sequences of hemerythrins from *Lingula* and a priapulid worm and the evolution of oxygen transport in the metazoa. *Int. Geol. Congr. Kyoto* 2:346.

Runnegar, B. N., and M. A. Fedonkin. 1992. Proterozoic metazoan body fossils. In *The Proterozoic Biosphere*. J. W. Schopf and C. Klein (eds.). Cambridge University Press, Cambridge. Pp. 369–88.

Ruppert, E. E. 1991. Introduction to aschelminth phyla: A consideration of mesoderm, body cavities, and cuticle. In *Microscopic Anatomy of Invertebrates*. Vol. 4. *Aschelminthes*. F. W. Harrison and E. E. Ruppert (eds.). Wiley-Liss, New York. Pp. 1–17.

Russell, E. S. 1916. *Form and Function: A Contribution to the History of Animal Morphology.* John Murray, London.

Ruvkun, G., and J. Giusto. 1989. The *Caenorhabditis elegans* heterochronic gene *lin-14* encodes a nuclear protein that forms a temporal developmental switch. *Nature* 338:313–19.

Sagan, C. 1983. *Cosmos.* Random House, New York.

Salser, S. J., and C. Kenyon. 1994. Patterning, *C. elegans* homeotic cluster genes, cell fates and cell migrations. *Trends Genet.* 10:159–64.

Salvini-Plawen, L. V., and E. Mayr. 1977. On the evolution of photoreceptors and eyes. In *Evolutionary Biology.* Vol. 10. M. K. Hecht, W. C. Steere, and B. Wallace (eds.). Plenum, New York. Pp. 207–63.

Sánchez-Herrero, E., I. Vernos, R. Marco, and G. Morata. 1985. Genetic organization of *Drosophila* bithorax complex. *Nature* 313:108–13.

Sarich, V. M., and A. C. Wilson. 1967. Immunological time scale for hominid evolution. *Science* 158:1200–1203.

Satoh, N., and W. R. Jeffery. 1995. Chasing tails in ascidians: Developmental insights into the origin and evolution of chordates. *Trends Genet.* In press.

Saunders, J. W. Jr., and M. T. Gasseling. 1968. Ectodermal-mesodermal interactions in the origin of limb symmetry. In *Epithelial-Mesenchymal Interactions.* R. Fleischmajer and R. E. Billingham (eds.). Williams and Wilkins, Baltimore. Pp. 78–97.

Scalan, B. E., L. R. Maxon, and W. E. Duellman. 1980. Albumin evolution in marsupial frogs (Hylidae: Gastrotheca). *Evolution* 34:222–29.

Schaller, D., C. Wittmann, A. Spisher, F. Müller, and H. Tobler. 1990. Cloning and analysis of three new homeobox genes from the nematode *Caenorhabditis elegans. Nucleic Acids Res.* 18:2033–36.

Schank, J. C., and W. C. Wimsatt. 1986. Generative entrenchment and evolution. *Phil. Sci. Assoc.* 2:33–60.

Schartl, M., I. Schlupp, A. Schartl, M. K. Meyer, I. Nanda, M. Schmid, J. T. Epplen, and J. Parzefall. 1991. On the stability of dispensible constituents of the eukaryotic genome: Stability of coding sequences versus truly hypervariable sequences in a cloned vertebrate, the Amazon molly *Poecilia formosa. Proc. Natl. Acad. Sci. USA* 88:8759–63.

Schatt, P. 1985. *Development et Croissance embryonaire de L'oursin incubant Abatus cordatus (Echinoidea: Spatangoida).* Ph.D. dissertation, Pierre and Marie Curie University, National Museum of Natural History, Paris.

Schedl, P., and J. Kimble. 1988. *fog-2,* a germ-line-specific sex determination gene required for hermaphrodite spermatogenesis in *Caenorhabditis elegans. Genetics* 119:43–61.

Scheltema, A. H. 1993. Aplacophora as progenetic aculiferans and the coelomic origin of mollusks as the sister taxon of Sipuncula. *Biol. Bull.* 184:57–78.

Schierwater, B., M. Murtha, M. Dick, F. H. Ruddle, and L. W. Buss. 1991. Homeoboxes in cnidarians. *J. Exp. Zool.* 260:413–16.

Schlinger, B. A., and A. P. Arnold. 1992. Plasma sex steroids and tissue aromatization in hatching zebra finches: Implications for the sexual differentiation of singing behaviour. *Endocrinology* 130:289–99.

Schnabel, R., and H. Schnabel. 1990. Early determination in the *C. elegans* embryo:

A gene, *cib-1*, required to specify a set of stem-cell-like blastomeres. *Development* 108:107–19.

Schopf, J. W., and C. Klein. 1992. *The Proterozoic Biosphere: A Multidisciplinary Study.* Cambridge University Press, Cambridge.

Schopf, T. J. M. 1978. Fossilization potential of an intertidal fauna: Friday Harbor, Washington. *Paleobiology* 4:261–70.

Schram, F. R. 1991. Cladistic analysis of metazoan phyla and the placement of fossil problematica. In *The Early Evolution of Metazoa and the Significance of Problematic Taxa.* A. M. Simonetta and S. Conway Morris (eds.). Cambridge University Press, Cambridge. Pp. 35–46.

Scott, M. P. 1992. Vertebrate homeobox gene nomenclature. *Cell* 71:551–53.

Scott, M. P., and A. J. Weiner. 1984. Structural relationships among genes that control development: Sequence homology between the Antennapedia, Ultrabithorax, and fushi tarazu loci of *Drosophila. Proc. Natl. Acad. Sci. USA* 81: 4115–19.

Sechrist, J., G. N. Serbedzija, T. Scherson, S. E. Fraser, and M. Bronner-Fraser. 1993. Segmental migration of the hindbrain neural crest does not arise from its segmental generation. *Development* 118:691–703.

Segraves, K. L. 1975. *The Great Dinosaur Mistake.* Beta Books, San Diego.

Seilacher, A. 1974. Fabricational noise in adaptive morphology. *Syst. Zool.* 22: 451–65.

Seilacher, A. 1979. Constructional morphology of sand dollars. *Paleobiology* 5: 191–221.

Seilacher, A. 1984a. Constructional morphology of bivalves: Evolutionary pathways in primary vs. secondary soft-bottom dwellers. *Paleontology* 27:207–37.

Seilacher, A. 1984b. Late Precambrian and early Cambrian metazoa: Preservational or real extinctions. In *Patterns of Change in Earth Evolution.* H. D. Holland and A. F. Trendall (eds.). Springer-Verlag, Berlin. Pp. 159–68.

Seilacher, A. 1989. Vendozoa: Organismic construction in the Proterozoic biosphere. *Lethaia* 22:229–39.

Seilacher, A. 1994. Early multicellular life: Late Proterozoic fossils and the Cambrian explosion. In *Early Life on Earth.* S. Bengston (ed.). Columbia University Press, New York. Pp. 389–400.

Sepkoski, J. J. Jr. 1984. A kinetic model of Phanerozoic taxonomic diversity. III. Post-Paleozoic families and mass extinction. *Paleobiology* 10:246–67.

Sessions, S. K., and A. Larson. 1987. Developmental correlations of genome size in plethodontid salamanders and their implications for genome evolution. *Evolution* 41:1239–51.

Shaffer, H. B. 1984. Evolution in a paedomorphic lineage. I. An electrophoretic analysis of the Mexican ambystomatid salamanders. *Evolution* 38:1194–1206.

Sharp, P. M., A. T. Lloyd, and D. G. Higgins. 1991. Coelacanth's relationships. *Nature* 353:218–19.

Sharpton, V. L., K. Burke, A. Camargo-Zanoguera, S. A. Hall, D. S. Lee, L. E. Marín, G. Suárez-Reynoso, J. M. Quezada-Muxeton, P. D. Spudis, and J. Urrutia-Fucugauchi. 1993. Chicxulub multiring impact basin: Size and other characteristics derived from gravity analysis. *Science* 261:1564–67.

Shawlot, W., and R. R. Behringer. 1995. Requirement for *Lim1* in head-organizer function. *Nature* 374:425–30.

Shea, B. T. 1988. Heterochrony in primates. In *Heterochrony in Evolution: A Multi-disciplinary Approach.* M. L. McKinney (ed.). Plenum Press, New York. Pp. 237–66.

Shea, B. T. 1989. Heterochrony in human evolution: The case for neoteny reconsidered. *Yrbk. Phys. Anthropol.* 32:69–101.

Shea, B. T. 1992. Developmental perspective on size change and allometry in evolution. *Evol. Anthropol.* 1:125–34.

Shea, B. T., R. E. Hammer, and R. L. Brinster. 1987. Growth allometry of the organs in giant transgenic mice. *Endocrinology* 121:1924–30.

Shea, B. T., R. E. Hammer, R. L. Brinster, and M. R. Ravosa. 1990. Relative growth of the skull and postcranium in giant transgenic mice. *Genet. Res. Camb.* 56:21–34.

Shear, W. A. 1991. The early development of terrestrial ecosystems. *Nature* 351: 283–89.

Shear, W. A. 1993. One small step for an arthropod. *Nat. Hist.* 102 (3):47–51.

Shear, W. A., and J. Kukalová-Peck. 1990. The ecology of Paleozoic terrestrial arthropods: The fossil evidence. *Can. J. Zool.* 68:1807–34.

Shenk, M. A., and R. E. Steele. 1993. A molecular snapshot of the metazoan "Eve." *Trends Biochem. Sci.* 18:459–63.

Sherman, P. W., J. U. M. Jarvis, and R. D. Alexander (eds.). 1991. *The Biology of the Naked Mole-Rat.* Princeton University Press, Princeton.

Sherman, P. W., J. U. M. Jarvis, and S. H. Braude. 1992. Naked mole rats. *Sci. Am.* 267:72–78.

Shiang, R., L. M. Thompson, Y. Z. Zhu, D. M. Church, T. J. Fielder, M. Bocian, S. T. Winokur, and J. J. Wasmuth. 1994. Mutations in the transmembrane domain of FGFR3 cause the most common genetic form of dwarfism, achondroplasia. *Cell* 78:335–42.

Shields, G. F., and A. C. Wilson. 1987. Calibration of mitochondrial DNA evolution in geese. *J. Mol. Evol.* 24:212–17.

Shoshani, J., J. M. Lowenstein, D. A. Walz, and M. Goodman. 1985. Proboscidean origins of mastodon and woolly mammoth demonstrated immunologically. *Paleobiology* 11:429–37.

Shubin, N. H. 1994. History, ontogeny, and the evolution of the archetype. In *Homology: The Hierarchical Basis of Comparative Biology.* B. K. Hall (ed.). Academic Press, San Diego. Pp. 249–71.

Shubin, N. H., and P. Alberch. 1986. A morphogenetic approach to the origin and basic organization of the tetrapod limb. *Evol. Biol.* 20:319–87.

Sidow, A., A. C. Wilson, and S. Pääbo. 1991. Bacterial DNA in Clarkia fossils. *Phil. Trans. R. Soc. Lond.* B 333:429–33.

Signor, P. W., and J. H. Lipps. 1992. Origin and early radiation of the Metazoa. In *Origin and Early Evolution of the Metazoa.* J. H. Lipps and P. W. Signor (eds.). Plenum Press, New York. Pp. 3–23.

Simcox, A. A., E. Hersperger, A. Shearn, J. R. S. Whittle, and S. M. Cohen. 1991.

Establishment of imaginal discs and histoblast nests in *Drosophila. Mech. Dev.* 34:11–20.

Simcox, A. A., I. J. H. Roberts, E. Hersberger, M. C. Gribbin, A. Shearn, and J. R. S. Whittle. 1989. Imaginal discs can be recovered from cultured embryos mutant for the segment polarity genes *engrailed, naked* and *patched* but not from *wingless. Development* 107:715–22.

Simeone, A., D. Acampora, M. Gulisano, A. Stornaiuolo, and E. Boncinelli. 1992. Nested expression domains of four homeobox genes in developing rostral brain. *Nature* 358:687–90.

Simpson, G. G. 1934. *Attending Marvels: A Patagonian Journal.* Macmillan, New York.

Simpson, G. G. 1953. *The Major Features of Evolution.* Columbia Univerity Press, New York.

Simpson, G. G. 1959. The history of life. In *Evolution after Darwin.* Vol. I. *The Evolution of Life.* S. Tax (ed.). University of Chicago Press, Chicago. Pp. 117–96.

Simpson, G. G. 1978. *Concession to the Improbable.* Yale University Press, New Haven.

Simpson, G. G. 1984. *Discoverers of the Lost World.* Yale University Press, New Haven.

Sinclair, A. H., P. Berta, M. Palmer, J. R. Hawkins, B. L. Griffiths, M. J. Smith, J. W. Foster, A.-M. Frischauf, R. Lovell-Badge, and P. N. Goodfellow. 1990. A gene from the human sex-determining region encodes a protein with homology to a conserved DNA-binding motif. *Nature* 346:240–44.

Sinervo, B. 1993. The effects of offspring size on physiology and life history. *BioScience* 43:210–18.

Sinervo, B., and R. B. Huey. 1990. Allometric engineering: An experimental test of the causes of interpopulational differences in performance. *Science* 248:1106–9.

Sinervo, B., and L. R. McEdward. 1988. Developmental consequences of an evolutionary change in egg size: An experimental test. *Evolution* 42:885–99.

Skiba, F., and E. Schierenberg. 1992. Cell lineages, developmental timing, and spatial pattern formation in embryos of free-living soil nematodes. *Dev. Biol.* 151:597–610.

Slack, J. M. W., P. W. H. Holland, and C. F. Graham. 1993. The zootype and the phylotypic stage. *Nature* 361:490–92.

Smith, A. B. 1984. *Echinoid Paleobiology.* Allen and Unwin, London.

Smith, A. B. 1988. Phylogenetic relationship, divergence times, and rates of molecular evolution for camarodont sea urchins. *Mol. Biol. Evol.* 5:345–65.

Smith, A. B. 1990. Evolutionary diversification of echinoderms during the early Paleozoic. In *Major Evolutionary Radiations.* P. D. Taylor and G. P. Larwood (eds.). Clarendon Press, Oxford. Pp. 265–86.

Smith, E. P., J. Boyd, G. R. Frank, H. Takahashi, R. M. Cohen, B. Speckler, T. C. Williams, D. B. Lubahn, and K. S. Korach. 1994. Estrogen resistance caused by a mutation in the estrogen-receptor gene in a man. *N. Engl. J. Med.* 331:1056–61.

Smith, H. W. 1932. *Kamongo.* Viking Press, New York.

Smith, J. P. S. I., S. Teyler, and R. M. Rieger. 1986. Is the Turbellaria polyphyletic? *Hydrobiologia* 132:13–21.

Smith, M. J., J. D. G. Boom, and R. A. Raff. 1990. Single copy DNA distances between two congeneric sea urchin species exhibiting radically different modes of development. *Mol. Biol. Evol.* 7:315–26.

Smothers, J. F., C. D. von Dohlen, L. H. Smith, and R. D. Spall. 1994. Molecular evidence that the myxozoan protists are metazoans. *Science* 265:1719–21.

Sober, E. 1993. Experimental tests of phylogenetic inference methods. *Syst. Biol.* 42:85–89.

Sogin, M. L. 1994. The origin of eukaryotes and evolution into major kingdoms. In *Early Life on Earth.* S. Bengston (ed.). Columbia University Press, New York. Pp. 181–92.

Sogin, M. L., U. Edman, and H. Elwood. 1989a. A single kingdom of eukaryotes. In *The Hierarchy of Life: Molecules and Morphology in Phylogenetic Analysis.* B. Fernholm, K. Bremer, and H. Jörnvall (eds.). Excerpta Medica, Amsterdam. Pp. 133–43.

Sogin, M. L., J. H. Gunderson, H. J. Elwood, R. A. Alonso, and D. A. Peattie. 1989b. Phylogenetic meaning of the kingdom concept: An unusual ribosomal RNA from *Giardia lamblia. Science* 243:75–77.

Soltis, P. S., and D. E. Soltis. 1993. Ancient DNA: Prospects and limitations. *New Zealand J. Bot.* 31:203–9.

Soltis, P. S., D. E. Soltis, and C. J. Smiley. 1992. An rbcL sequence from a Miocene *Taxodium* (bald cypress). *Proc. Natl. Acad. Sci. USA* 89:449–51.

Sommer, R. J., L. K. Carta, and P. W. Sternberg. 1994. The evolution of cell lineage in nematodes. *Development* (Suppl.): 85–95.

Sommer, R. J., and P. W. Sternberg. 1994. Changes of induction and competence during the evolution of vulva development in nematodes. *Science* 265:114–18.

Sommer, R. J., and P. W. Sternberg. 1995. Evolution of cell lineage and pattern formation in the vulval equivalence group of rhabditid nematodes. *Dev. Biol.* 167: 61–74.

Sommer, R. J., and D. Tautz. 1991. Segmentation gene expression in the housefly *Musca domestica. Development* 113: 419–30.

Sommer, R. J., and D. Tautz. 1993. Involvement of an ortholog of the *Drosophila* pair-rule gene *hairy* in segment formation of the short germ band embryo of *Tribolium* (Coleoptera). *Nature* 361:448–50.

Sordino, P., F. van der Hoeven, and D. Duboule. 1995. Hox gene expression in teleost fins and the origin of vertebrate digits. *Nature* 375:687–91.

Sprinkle, J. 1992. Radiation of Echinodermata. In *Origin and Early Evolution of the Metazoa.* J. H. Lipps and P. W. Signor (eds.). Plenum Press, New York. Pp. 375–98.

Sprinkle, J., and P. M. Kier. 1987. Phylum Echinodermata. In *Fossil Invertebrates.* R. S. Boardman, A. H. Cheetham, and A. J. Rowell (eds.). Blackwell Scientific Publications, Palo Alto. Pp. 550–611.

Stanley, S. M. 1976. Ideas on the timing of metazoan diversification. *Paleobiology* 2:209–19.

Stasek, C. R. 1972. The molluscan framework. *Chem. Zool.* 12:1–44.

Stearns, S. C. 1989. The evolutionary significance of phenotypic plasticity. *BioScience* 39:436–45.

Stearns, S. C. 1992. *The Evolution of Life Histories*. Oxford University Press, Oxford.

Steinmann-Zwicky, M. 1992a. How do germ cells choose their sex? *Drosophila* as a paradigm. *BioEssays* 14:513–18.

Steinmann-Zwicky, M. 1992b. Sex determination of *Drosophila* germ cells. *Semin. Dev. Biol.* 3:341–47.

Sternberg, P. W., and H. R. Horvitz. 1981. Gonadal cell lineages of the nematode *Panagrellus redivivus* and implications for evolution by modification of cell lineages. *Dev. Biol.* 88:147–66.

Sterrer, W., M. Mainitz, and R. M. Rieger. 1985. Gnathostomulida: Enigmatic as ever. In *The Origins and Relationships of Lower Invertebrates*. S. Conway Morris, J. D. George, R. Gibson, and H. M. Platt (eds.). Clarendon Press, Oxford. Pp. 181–99.

Stock, D. W., and D. L. Swofford. 1991. Coelacanth's relationships. *Nature* 353: 217–18.

Storm, E. E., T. V. Huynh, N. G. Copeland, N. A. Jenkins, D. M. Kingsley, and S.-J. Lee. 1994. Limb alterations in *brachypodism* mice due to mutations in a new member of the TGFβ-superfamily. *Nature* 368:639–43.

Strathmann, R. R. 1978. The evolution and loss of feeding larval stages of marine invertebrates. *Evolution* 32:894–906.

Strathmann, R. R. 1988. Larvae, phylogeny, and von Baer's law. In *Echinoderm Phylogeny and Evolutionary Biology*. C. R. C. Paul and A. B. Smith (eds.). Clarendon Press, Oxford. Pp. 51–68.

Strathmann, R. R., and M. F. Strathmann. 1982. The relationship between adult size and brooding in marine invertebrates. *Am. Nat.* 119:91–101.

Strickberger, M. W. 1976. *Genetics*. Macmillan, New York.

Stringer, C., and C. Gamble. 1993. *In Search of the Neanderthals: Solving the Puzzle of Human Origins*. Thames and Hudson, New York.

Struhl, G., and K. Basler. 1993. Organizing activity of wingless protein in *Drosophila*. 72:527–40.

Stuart, J. J., S. J. Brown, R. W. Beeman, and R. E. Denell. 1991. A deficiency of the homoeotic complex of the beetle *Tribolium*. *Nature* 350:72–74.

Stuart, J. J., S. J. Brown, R. W. Beeman, and R. E. Denell. 1992. The *Tribolium* homoeotic gene *Abdominal* is homologous to *abdominal-A* of the *Drosophila* Bithorax complex. *Development* 117:233–43.

Sturtevant, M. H. 1923. Inheritance of direction of coiling in *Limnaea*. *Science* 58:269–70.

Summerbell, D. 1979. The zone of polarizing activity: Evidence for a role in abnormal chick limb morphogenesis. *J. Embryol. Exp. Morphol.* 33:621–43.

Summerbell, D., J. H. Lewis, and L. Wolpert. 1973. Positional information in chick limb morphogenesis. *Nature* 224:492–96.

Suttie, J. M., P. F. Fennessy, I. D. Corson, F. J. Laas, S. F. Crosbie, J. H. Butles, and P. D. Gluckman. 1989. Pulsatile growth hormone, insulin-like growth factors and antler development in red deer *(Cervas elaphus scoticus)* stags. *J. Endocrinol.* 121:351–60.

Swalla, B. J., M. A. Badgett, and W. R. Jeffery. 1991. Identification of a cytoskeletal protein localized in the myoplasm of ascidian eggs: Localization is modified during anural development. *Development* 111:425–36.

Swalla, B. J., and W. R. Jeffery. 1990. Interspecific hybridization between an anural and urodele ascidian: Differential expression of urodele features suggests multiple mechanisms control anural development. *Dev. Biol.* 142:319–34.

Swalla, B. J., K. W. Makabe, N. Satoh, and W. R. Jeffery. 1993. Novel genes expressed differentially in ascidians with alternate modes of development. *Development* 119:307–18.

Swofford, D. L., and G. J. Olsen. 1990. Phylogeny reconstruction. In *Molecular Systematics*. D. M. Hillis and C. Moritz (eds.). Sinauer Associates, Sunderland. Pp. 411–501.

Syren, R. M., and P. Luykx. 1981. Geographic variation of sex-linked translocation heterozygosity in the termite *Kalotermes approximatus* Snyder (Insecta: Isoptera). *Chromosoma* (Berl.). 82:65–88.

Tabin, C. J. 1991. Retinoids, homeoboxes, and growth factors: Towards molecular models for limb development. *Cell* 66:199–217.

Tabin, C. J. 1992. Why we have (only) five fingers per hand: Hox genes and the evolution of paired limbs. *Development* 116:289–96.

Tabin, C. J. 1995. The initiation of the limb bud: Growth factors, Hox genes, and retinoids. *Cell* 80:671–74.

Tabin, C., and E. Laufer. 1993. Hox genes and serial homology. *Nature* 361:692–93.

Takada, S., K. L. Stark, M. J. Shea, G. Vassileva, J. A. McMahon, and A. P. McMahon. 1994. *Wnt-3a* regulates somite and tailbud formation in the mouse embryo. *Genes Dev.* 8:174–89.

Takashima, T., T. Hamamoto, K. Otazai, W. D. Grant, and K. Horikoshi. 1990. *Haloarcula japonica* sp. nov., a new triangular halophilic archaebacterium. *Syst. Appl. Microbiol.* 13:177–81.

Telford, M. J., and P. W. H. Holland. 1993. The phylogenetic affinities of the chaetognaths: A molecular analysis. *Mol. Biol. Evol.* 10:660–76.

Templeton, A. R. 1991. Human origins and analysis of mitochondrial DNA sequences. *Science* 255:737.

Templeton, A. R. 1993. The "Eve" hypothesis: A genetic critique and reanalysis. *Am. Anthropol.* 95:51–72.

Thaller, C., and G. Eichele. 1987. Identification and spatial distribution of retinoids in the developing chick limb bud. *Nature* 327:625–28.

Themerson, S. 1975. *Professor Mmaa's Lecture*. The Overlook Press, Woodstock.

Thewissen, J. G. M., and S. T. Hussain. 1993. Origin of underwater hearing in whales. *Nature* 361:444–45.

Thewissen, J. G. M., S. T. Hussain, and M. Arif. 1994. Fossil evidence for the origin of aquatic locomotion in Archaeocete whales. *Science* 263:210–12.

Thiery, J. P., J. L. Duband, and A. Delouvée. 1982. Pathways and mechanisms of avian trunk neural crest cell migration and localization. *Dev. Biol.* 93:324–43.

Thomas, J. B., M. J. Bastiani, M. Bate, and C. S. Goodman. 1984. From grasshopper to *Drosophila:* A common plan for neuronal development. *Nature* 310:203–7.

Thomas, R. D. K., and E. C. Olson. 1980. *A Cold Look at the Warm-Blooded Dinosaurs*. Westview Press, Boulder, Colo.

Thomas, R. H., W. Schaffner, A. C. Wilson, and S. Pääbo. 1989. DNA phylogeny of the extinct marsupial wolf. *Nature* 340:465–67.

Thomson, K. S. 1991. Parallelism and convergence in the horse limb: The internal-external dichotomy. In *New Perspectives on Evolution*. L. Warren and H. Koprowski (eds.). Wiley-Liss, New York. Pp. 101–22.

Thorne, B. L., and J. M. Carpenter. 1992. Phylogeny of the Dictyoptera. *Syst. Entomol.* 17:253–68.

Thornhill, R. 1991. Female preference for the pheromone of males with low fluctuating asymmetry in the Japanese scorpionfly (*Panorpa japonica:* Mecoptera). *Behav. Ecol.* 3:277–83.

Thornhill, R. 1992. Fluctuating asymmetry, interspecific aggression and male mating tactics in two species of Japanese scorpionflies. *Behav. Ecol. Sociobiol.* 30:357–63.

Thorson, G. 1950. Reproductive and larval ecology of marine bottom invertebrates. *Biol. Rev.* 25:1–45.

Tickle, C. 1994. On making a skeleton. *Nature* 368:587–88.

Tickle, C., B. Alberts, L. Wolpert, and J. Lee. 1982. Local application of retinoic acid to the limb bud mimics the action of the polarizing region. *Nature* 296:564–66.

Tickle, C., J. Lee, and G. Eichele. 1985. A quantitative analysis of the effects of all-*trans*-retinoic acid on the pattern of chick wing development. *Dev. Biol.* 109:82–95.

Tickle, C., D. Summerbell, and L. Wolpert. 1975. Positional signaling and specification of digits in chick limb morphogenesis. *Nature* 296:564–66.

Tiegs, O. W., and S. M. Manton. 1958. The evolution of the Arthropoda. *Biol. Rev. Camb. Phil. Soc.* 33:255–337.

Tiersch, T. R., M. J. Mitchell, and S. S. Wachtel. 1991. Studies on the phylogenetic conservation of the *SRY* gene. *Hum. Genet.* 87:571–73.

Tiong, S. Y. K., J. R. S. Whittle, and M. C. Gribbin. 1987. Chromosomal continuity in the abdominal region of the bithorax complex of *Drosophila* is not essential for its contribution to metameric identity. *Development* 101:135–42.

Tolkien, J. R. R. 1965. *The Lord of the Rings*. Houghton Mifflin, Boston.

Tomarev, S. I., R. D. Zinovieva, and J. Piatigorsky. 1991. *J. Biol. Chem.* 266: 24226–31.

Tompkins, R. 1978. Genic control of axolotl metamorphosis. *Am. Zool.* 18:313–19.

Tong, P. Y., S. E. Tollefsen, and S. Kornfeld. 1988. The cation-independent mannose 6-phosphatase receptor binds insulin-like growth factor II. *J. Biol. Chem.* 263:2585–88.

Towe, K. M. 1970. Oxygen-collagen priority and the early metazoan fossil record. *Proc. Natl. Acad. Sci. USA* 65:781–88.

Treisman, J., P. Gönczy, M. Vashishta, E. Harris, and C. Desplan. 1989. A single amino acid can determine the DNA binding specificity of homeodomain proteins. *Cell* 59:553–62.

Treisman, J., E. Harris, D. Wilson, and C. Desplan. 1992. The homeodomain: A new face for the helix-turn-helix. *BioEssays* 14:145–50.

Trinkaus, E., and P. Shipman. 1993. *The Neanderthals: Changing the Image of Mankind.* Alfred A. Knopf, New York.

Tschanz, K. 1988. Allometry and heterochrony in the growth of the neck of the Triassic prolacertiform reptiles. *Palaeontology* 31:997–1011.

Turbeville, J. M. 1986. An ultrastructural analysis of coelomogenesis in the hoplonemertine *Prosorhochmus americanus* and the polychaete *Magelona* sp. *J. Morphol.* 187:51–60.

Turbeville, J. M. 1991. Nemertina. In *Microscopic Anatomy of Invertebrates.* Vol. 3. *Platyhelminthes and Nemertinea.* F. W. Harrison and B. J. Bogitsh (eds.). Wiley-Liss, New York. Pp. 285–328.

Turbeville, J. M., K. G. Field, and R. A. Raff. 1992. Phylogenetic position of Phylum Nemertini, inferred from 18S rRNA sequences: Molecular data as a test of morphological character homology. *Mol. Biol. Evol.* 9:235–49.

Turbeville, J. M., D. M. Pfeifer, K. G. Field, and R. A. Raff. 1991. The phylogenetic status of arthropods, as inferred from 18S rRNA sequences. *Mol. Biol. Evol.* 8:669–86.

Turbeville, J. M., J. R. Schulz, and R. A. Raff. 1993. Deuterostome phylogeny and the sister group of the chordates: Evidence from molecules and morphology. *Mol. Biol. Evol.* 11:648–55.

Turing, A. M. 1952. The chemical basis of morphogenesis. *Phil. Trans. R. Soc.* B 237:37–72.

Turner, B., and H. R. Steeves. 1989. Induction of spermatogenesis in an all-female fish species by treatment with an exogenous androgen. In *Evolution and Ecology of Unisexual Vertebrates.* R. M. Dawley and J. P. Bogart (eds.). The New York State Museum, Albany. Pp. 113–22.

Twitty, V. C. 1966. *Of Scientists and Salamanders.* W. H. Freeman, San Francisco.

Twitty, V. C., and J. L. Schwind. 1931. The growth of eyes and limbs transplanted heteroplastically between two species of *Ambystoma. J. Exp. Zool.* 59:61–86.

Vachon, G., B. Cohen, C. Pfeifle, M. E. McGuffin, J. Botas, and S. M. Cohen. 1992. Homeotic genes of the Bithorax complex repress limb development in the abdomen of the *Drosophila* embryo through the target gene *Distal-less. Cell* 71: 437–50.

Valentine, J. W., A. G. Collins, and C. P. Meyer. 1994. Morphological complexity increase in metazoans. *Paleobiology* 20:131–42.

Valentine, J. W., and D. H. Erwin. 1987. Interpreting great developmental experiments: The fossil record. In *Development as an Evolutionary Process.* R. A. Raff and E. C. Raff (eds.). Alan R. Liss, New York. Pp. 71–107.

Vandekerckhove, J., and K. Weber. 1984. Chordate muscle actins differ distinctly from invertebrate muscle actins: The evolution of the different vertebrate muscle actins. *J. Mol. Biol.* 179:391–413.

Van de Peer, Y., J.-M. Neefs, P. De Rijk, and R. De Wachter. 1993. Reconstructing evolution from eukaryotic small-ribosomal-subunit RNA sequences: Calibration of the molecular clock. *J. Mol. Evol.* 37:221–32.

Van Valen, L. 1982. Homology and causes. *J. Morphol.* 173:305–12.

Veblen, T. 1899. *The Theory of the Leisure Class.* Macmillan, New York.

Venuti, J. M., and W. R. Jeffery. 1989. Cell lineage and determination of cell fate in ascidian embryos. *Int. J. Dev. Biol.* 33:197–212.

Vergnaud, G., D. C. Page, M.-C. Simmler, L. Brown, F. Rouyer, B. Noel, D. Botstein, A. de la Chapelle, and J. Weissenbach. 1986. A deletion map of the human Y chromosome based on DNA hybridization. *Am. J. Hum. Genet.* 38: 109–24.

Vermeij, G. J. 1987. *Evolution and Escalation: An Ecological History of Life.* Princeton University Press, Princeton.

Vigilant, L., M. Stoneking, H. Harpending, K. Hawkes, and A. C. Wilson. 1991. African populations and the evolution of human mitochondrial DNA. *Science* 253: 1503–7.

Vogel, A., and C. Tickle. 1993. FGF-4 maintains polarizing activity of posterior limb bud cells in vivo and in vitro. *Development* 119:199–206.

Vossbrinck, C. R., J. V. Maddox, S. Friedman, B. A. Debrunner-Vossbrinck, and C. R. Woese. 1987. Ribosomal RNA sequence suggests microsporidia are extremely ancient eukaryotes. *Nature* 326:411–14.

Wada, H., and N. Satoh. 1994. Details of the evolutionary history from invertebrates to vertebrates, as deduced from the sequences of 18S rDNA. *Proc. Natl. Acad. Sci. USA* 91:1801–4.

Waddington, C. H. 1942. Canalization of development and the inheritance of acquired characters. *Nature* 150:563–65.

Waddington, C. H., J. Needham, and J. Brachet. 1936a. Studies on the nature of the amphibian organization center. III. The activation of the evocator. *Proc. R. Soc. Lond.* B 120:172–98.

Waddington, C. H., J. Needham, W. W. Nowinski, R. Lemberg, and A. Cohen. 1936b. Studies on the nature of the amphibian organization center. IV. Further experiments on the chemistry of the evocator. *Proc. R. Soc. Lond.* B 120:198–208.

Wagner, G. P. 1994. Homology and the mechanisms of development. In *Homology: The Hierarchical Basis of Comparative Biology.* B. K. Hall (ed.). Academic Press, San Diego. Pp. 274–99.

Wainright, P. O., G. Hinkle, M. L. Sogin, and S. K. Stickel. 1993. Monophyletic origins of the metazoa: An evolutionary link with fungi. *Science* 260:340–42.

Wake, D. B., and J. Hanken. 1995. Evolution of direct development in the lungless salamanders (family Plethodontidae), the most common reproductive mode in urodeles. *Int. J. Dev. Biol.* In press.

Wakimoto, B. T., F. R. Turner, and T. C. Kaufman. 1984. Defects in embryogenesis in mutants associated with the Antennapedia gene complex of *Drosophila melanogaster. Dev. Biol.* 102:147–72.

Walcott, C. D. 1910. Abrupt appearance of the Cambrian Fauna on the North American continent. Cambrian Geology and Paleontology, II. *Smithsonian Miscellaneous Collections* 57:1–16.

Walossek, D. 1993. The Upper Cambrian *Rehbachiella* and the phylogeny of Branchiopoda and Crustacea. *Fossils and Strata* 32:1–202.

Walossek, D., and K. J. Müller. 1994. Pentastomid parasites from the Lower Paleozoic of Sweden. *Trans. R. Soc. Edinburgh: Earth Sci.* 85:1–37.

Walossek, D., K. J. Müller, and R. M. Kristensen. 1994. A more than half a billion years old stem-group tardigrade from Siberia. *Sixth International Symposium on Tardigrada, Selwyn College, Cambridge.* Abstract.

Walsby, A. E. 1980. A square bacterium. *Nature* 283:69–71.

Walter, M. R. 1994. Stromatolites: The main source of information on the evolution of the early benthos. In *Early Life on Earth.* S. Bengston (ed.). Columbia University Press, New York. Pp. 270–86.

Wanek, N., D. M. Gardiner, K. Mueoka, and S. V. Bryant. 1991. Conversion by retinoic acid of anterior cells into ZPA cells in the chick wing bud. *Nature* 350:8–83.

Wang, C., and R. Lehmann. 1991. *Nanos* is the localized posterior determinant in *Drosophila. Cell* 66:637–47.

Wanyonyi, K. 1974. The influence of the juvenile hormone analogue ZR 512 (Zoecon) on caste development in *Zootermopsis nevadensis* (Hagen) (Isoptera). *Insectes Sociaux* (Paris) 21:35–44.

Warren, R., L. Nagy, J. Selegue, J. Gates, and S. Carroll. Evolution of homeotic gene regulation and function in flies and butterflies: Testing the Lewis hypothesis. *Nature* 372:458–61.

Wassersug, R. J., and W. E. Duellman. 1984. Oral structures and their development in egg-brooding hylid frog embryos and larvae: Evolutionary and ecological implications. *J. Morphol.* 182:1–37.

Wayne, R. K. 1986a. Cranial morphology of domestic and wild canids: The influence of development on morphological change. *Evolution* 40:243–61.

Wayne, R. K. 1986b. Developmental constraints on limb growth in domestic and some wild canids. *J. Zool.* (Lond.) A 210:381–99.

Webb, J. F., and D. M. Noden. 1993. Ectodermal placodes: Contributions to the development of the vertebrate head. *Am. Zool.* 33:434–47.

Wedeen, C. J., and D. A. Weisblat. 1991. Segmental expression of an *engrailed*-class gene during early development and neurogenesis in an annelid. *Development* 113: 805–14.

Weiner, J. 1994. *The Beak of the Finch.* Alfred A. Knopf, New York.

Weintraub, H. 1993. The MyoD family and myogenesis: Redundancy, networks, and thresholds. *Cell* 75:1241–44.

Weisblat, D. A., C. J. Wedeen, and R. G. Kostrikan. 1994. Evolution of developmental mechanisms: Spatial and temporal modes of rostrocaudal patterning. *Curr. Top. Dev. Biol.* 29:101–34.

West-Eberhard, M. J. 1989. Phenotypic plasticity and the origins of diversity. *Annu. Rev. Ecol. Syst.* 20:249–78.

Westerfield, M., M.-A. Akimenko, M. Ekker, and A. Püschel. 1993. Eyes, ears, and homeobox genes in zebrafish embryos. In *Molecular Basis of Morphogenesis.* M. Bernfield (ed.). Wiley-Liss, New York. Pp. 69–77.

Weston, J. A. 1963. A radioautographic analysis of the migration and localization of trunk neural crest cells in the chick. *Dev. Biol.* 6:274–310.

Weygoldt, P. 1986. Arthropod interrelationships: the phylogenetic-systematic approach. *Z. Zool. Syst. Evol.* 24:19–35.

Wheeler, W. C., P. Cartwright, and C. Y. Hayashi. 1993. Arthropod phylogeny: A combined approach. *Cladistics* 9:1–39.

Whitcomb, J. C., and H. M. Morris. 1961. *The Genesis Flood*. Baker, Grand Rapids.

White, B. A., and C. S. Nicoll. 1981. Hormonal control of amphibian metamorphosis. In *Metamorphosis: A Problem in Developmental Biology*. L. I. Gilbert and E. Frieden (eds.). Plenum Press, New York. Pp. 363–96.

Whiting, M. F., and W. C. Wheeler. 1994. Insect homeotic transformation. *Nature* 368:696–97.

Whitington, P. M., D. Leach, and R. Sandeman. 1993. Evolutionary change in neural development within the arthropods: Axonogenesis in the embryos of two crustaceans. *Development* 118:449–61.

Whitington, P. M., T. Meier, and P. King. 1991. Segmentation, neurogenesis and formation of early axonal pathways in the centipede, *Ethmostigmus rubripes* (Brandt). *Roux's Arch. Dev. Biol.* 199:349–63.

Whittaker, J. R. 1980. Acetylcholinesterase development in extra cells caused by changing the distribution of myoplasm in ascidian embryos. *J. Embryol. Exp. Morphol.* 55:343–54.

Whittaker, J. R. 1982. Muscle lineage cytoplasm can change the developmental expression in epidermal lineage cells of ascidian embryos. *Dev. Biol.* 93:463–70.

Whittaker, J. R., and T. H. Meedle. 1989. Two histospecific enzyme expressions in the same cleavage-arrested one-celled ascidian embryos. *J. Exp. Zool.* 250:168–75.

Whittington, H. B. 1979. Early arthropods, their appendages and relationships. In *The Origin of Major Invertebrate Groups*. M. R. House (ed.). Academic Press, London. Pp. 253–68.

Whittington, H. B. 1985. *The Burgess Shale*. Yale University Press, New Haven.

Whorf, B. L. 1940. Science and linguistics. *Technol. Rev.* 42:229–31, 247–48.

Wibbels, T., and D. Crews. 1994. Putative aromatase inhibitor induces male sex determination in a female unisexual lizard and in a turtle with temperature-dependent sex determination. *J. Endocrinol.* 141:295–99.

Wigglesworth, V. B. 1973. Evolution of insect wings and flight. *Nature* 246:127–29.

Wightman, B., T. R. Buerglin, J. Gatto, P. Arasu, and G. Ruvkun. 1991. Negative regulatory sequences in the *lin-14* 3'-untranslated region are necessary to generate a temporal switch during *Caenorhabditis elegans* development. *Genes Dev.* 5:1813–24.

Williams, D. H. C., and D. T. Anderson. 1975. The reproductive system, embryonic development, larval development, and metamorphosis of the sea urchin *Heliocidaris erythrogramma* (Val.) (Echinoidea: Echinometridae). *Aust. J. Zool.* 23:371–403.

Williams, J. A., and S. B. Carroll. 1993. The origin, patterning and evolution of insect appendages. *BioEssays* 15:567–77.

Williams, J. A., S. W. Paddock, and S. B. Carroll. 1993. Pattern formation in a secondary field: A hierarchy of regulatory genes subdivides the developing *Drosophila* wing into discrete subregions. *Development* 117:571–84.

Williams, J. A., S. W. Paddock, K. Vorwerk, and S. B. Carroll. 1994. Organization

of wing formation and induction of a wing-patterning gene at the dorsal/ventral compartment boundary. *Nature* 368:299–305.

Willmer, P. 1990. *Invertebrate Relationships: Patterns in Animal Evolution.* Cambridge University Press, Cambridge.

Wills, M. A., D. E. G. Briggs, and R. A. Fortey. 1994. Disparity as an evolutionary index: A comparison of Cambrian and Recent arthropods. *Paleobiology* 20:93–130.

Wilson, A. C., L. R. Maxson, and V. M. Sarich. 1974. Two types of molecular evolution: Evidence from studies of intraspecific hybridization. *Proc. Natl. Acad. Sci. USA* 71:2843–47.

Wilson, E. B. 1904. Experimental studies in germinal localization. II. Experiments on the cleavage-mosaic in *Patella* and *Dentalium. J. Exp. Zool.* 1:197–268.

Wilson, E. O. 1971. *The Insect Societies.* Harvard University Press, Cambridge, Mass.

Wilson, E. O. 1985. The sociogenesis of insect colonies. *Science* 228:1489–95.

Wilson, E. O. 1988. The current state of biological diversity. In *Biodiversity.* E. O. Wilson and F. M. Peter (eds.). National Academy Press, Washington, D.C. Pp. 3–18.

Wimsatt, W. C., and J. C. Schank. 1988. Two constraints on the evolution of complex adaptations and the means for their avoidance. In *Evolutionary Progress.* M. H. Nitecki (ed.). University of Chicago Press, Chicago. Pp. 231–73.

Winslow, J., N. Hastings, C. Carter, and T. Insel. 1993. A role for central vasopressin in pair bonding in monogamous prairie voles. *Nature* 365:545–48.

Winsor, M. P. 1991. *Reading the Shape of Nature: Comparative Zoology at the Agassiz Museum.* University of Chicago Press, Chicago.

Wistow, G. 1993. Lens crystallins: Gene recruitment and evolutionary dynamism. *Trends Biochem. Sci.* 18:301–6.

Wistow, G., A. Anderson, and J. Piatigorsky. 1990. Evidence for neutral and selective processes in the recruitment of enzyme-crystallins in avian lenses. *Proc. Natl. Acad. Sci. USA* 87:6277–80.

Woese, C. R., O. Kandler, and M. L. Wheelis. 1990. Towards a natural system of organisms: Proposal for the domains Archaea, Bacteria, and Eucaria. *Proc. Natl. Acad. Sci. USA* 87:4576–79.

Wolpert, L. 1969. Positional information and the spatial pattern of cellular differentiation. *J. Theor. Biol.* 25:1–47.

Wolpert, L. 1971. Positional information and pattern formation. *Curr. Top. Dev. Biol.* 6:183–224.

Wootton, R. J., and C. P. Ellington. 1991. Biomechanics and the origin of insect flight. In *Biomechanics in Evolution.* J. M. V. Rayner and R. J. Wootton (eds.). Cambridge University Press, Cambridge. Pp. 99–112.

Wozney, J. M., V. Rosen, A. J. Celeste, L. M. Mitsock, M. J. Whitters, R. W. Kriz, R. M. Hewick, and E. A. Wang. 1988. Novel regulators of bone formation: Molecular clones and activities. *Science* 242:1528–34.

Wray, G. A. 1992. The evolution of larval morphology during the post-Paleozoic radiation of echinoids. *Paleobiology* 18:258–87.

Wray, G. A. 1994. The evolution of cell lineage in echinoderms. *Am. Zool.* 34:353–63.

Wray, G. A., and A. E. Bely. 1994. The evolution of echinoderm development is driven by several distinct factors. *Development* (Suppl.): 97–106.

Wray, G. A., and D. R. McClay. 1989. Molecular heterochronies and heterotopies in early echinoid development. *Evolution* 43:803–13.

Wray, G. A., and R. A. Raff. 1989. Evolutionary modification of cell lineage in the direct-developing sea urchin *Heliocidaris erythrogramma*. *Dev. Biol.* 132:458–70.

Wray, G. A., and R. A. Raff. 1990. Novel origins of lineage founder cells in the direct-devloping sea urchin *Heliocidaris erythrogramma*. *Dev. Biol.* 141:41–54.

Wray, G. A., and R. A. Raff. 1991. The evolution of developmental strategy in marine invertebrates. *Trends Ecol. Evol.* 6:45–50.

Wright, S. 1934a. An analysis of variability of digits in an inbred strain of guinea pigs. *Genetics* 19:506–36.

Wright, S. 1934b. The results of crosses between inbred strains of guinea pigs, differing in number of digits. *Genetics* 19:537–51.

Yokobori, S., M. Hasegawa, T. Ueda, N. Okada, K. Nishikawa, and K. Watanabe. 1994. Relationships among coelacanths, lungfishes, and tetrapods: A phylogenetic analysis based on mitochondrial cytochrome oxidase I gene sequences. *J. Mol. Evol.* 38:602–9.

Yokouchi, Y., H. Sasaki, and A. Kuroiwa. 1991. Homoeobox gene expression correlated with the bifuracation process of limb cartilage development. *Nature* 353:443–45.

Yokoyama, T., N. G. Copeland, N. A. Jenkins, C. A. Montgomery, F. F. B. Elder, and P. A. Overbeek. 1993. Reversal of left-right asymmetry: A *situs inversus* mutation. *Science* 260:679–82.

Younger-Shepherd, S., H. Vaessin, E. Bier, L. Y. Jan, and Y. N. Jan. 1992. *deadpan*, an essential pan-neural gene encoding an HLH protein, acts as a denominator in *Drosophila* sex determination. *Cell* 70:911–22.

Zhang, X.-G., and B. R. Pratt. 1994. Middle Cambrian arthropod embryos with blastomeres. *Science* 266:637–39.

Index

Abatus cordatus, 227

Abdominal A gene, 414, 416

Abdominal B gene, 413, 414

absolute dating, 66

acanthocephalans: lacking fossil record as far back as the Cambrian, 175; as pseudocoelomates, 39, 118, 119

Acanthodes, 342

Acanthostega, 162, 355–56, 425

acceleration (peramorphosis), 262, 265, 290

Acetabularia, 85

acetyl cholinesterase, 245

achaete scute complex, 339, 371

Acheta domestica, 408

acoelomate animals: examples of diversity of body plans, 40; flatworms as last remnant of, 118; in Hyman's metazoan phylogeny, 39; intermediate filaments in, 135; principal elements of body plan organization, 41. *See also* platyhelminth flatworms

acorn worm, 40

acron, 126

actin gene family, 234–35

adaptation, 8; exaptation, 340; preadaptation, 340

adaptationism, 294

additive distance methods, 107

adult body plan, 195–96

Aepyornis, 250

AER (apical ectodermal ridge), 346–47, 349

aesthetes, 377

African rock art, 321

Agassiz, Louis, 5

agmatans, 89

agnathans, 342, 354; lampreys, 379

Albumares, 79

albumin, 165, 248

algae, 73–74; *Acetabularia*, 85; stromatolites, 74

allometric engineering, 252

allometric growth, 264, 297, 319

allometry: defined, 16, 252; as a developmental constraint, 319–20; heterochrony and, 263–65, 273; mechanisms underlying, 278–82; as result of local controls, 280

ambers, ancient DNA from, 169–70

Ambulocetus natans, 401, 402, 403

Ambystoma, 267

Ambystoma punctatum, 281

ambystomatid salamanders, 391

Ambystoma tigrinum, 281

Ameghino, Florentino, 398

Ameria, 45–46

Amiskwia, 96

ammonites, 176, 251

amniotes, 49, 51, 52–53, 198, 317

amphibians: genomic constraints in, 305; neoteny in, 272; notochord, 200; number of cell types in, 307; as primitive pattern of tetrapods, 214; spadefoot toads, 219–20. *See also* frogs; salamanders

Amphioxus (*Branchiostoma*), 185, 197, 198–99, 379

anagenesis, 58–59

analogy, 37–38

ancestral metazoans hypothesis, 71, 72

Anfesta, 79

animals: Cambrian fauna, 98–99; Chengjiang fauna, 91, 380; colonial, 369–70; contingency in history of, 67; as eukaryotes, 61; the first animals, 68–73; fungi and, 113; Modern fauna, 99–100; number of species of, 60; Paleozoic fauna, 99, 405–13, 420; South American, 397–99; three great evolutionary faunas, 98–99. *See also* Burgess Shale fauna; Ediacaran fauna; metazoans; terrestrial animals

annelids (Annelida): in alternative metazoan phylogenies, 46; arthropods and, 44, 122, 123, 124, 125, 130, 132, 133, 183, 190; burrowing, 123; cleavage in, 43; common ancestry with mollusks and brachiopods, 93, 117–18, 121–22; Ediacaran fauna as, 82, 84; *engrailed* gene in, 183; as eucoelomates, 39; eyes, 377, 381; halkieriids as link between brachiopods and, 102; miniaturized, 270–71; in molecular phylogenetic tree of metazoans, 112; motility of, 123; in protostome coelomate clade, 113; as protostomes, 40; segmentation in, 42; skeletonized debut of, 89; as terrestrial, 178. *See also* polychaetes

Anomalocaris, 95, 96, 99, 135, 380

Anser, 158

antelopes, 59

Antennapedia complex, 183, 186–87, 190, 307–9, 339, 413

Antennapedia gene, 186, 307, 413, 416

anterior-posterior axis, 184, 186, 189, 209, 414

ants, 60

apes: chimpanzees, 286, 288, 331; divergence from humans, 157; foramen magnum, 288; gorillas, 286; growth rates compared with humans, 286; humans as fetalized descendants of, 256; humans resembling young apes more than adults, 24; skull, 287; upper respiratory tract and associated features, 287

apical ectodermal ridge (AER), 346–47, 349

appendages: dissociation of function in arthropods, 433; in *Drosophila,* 352–53; fins, 354–55, 388; head appendages in living and Cambrian arthropods, 128, 136; insects, 409, 433; segments and appendages of arthropods, 127. *See also* limbs; wings

appositional compound eye, 378

apterous gene, 414, 415, 417

Arabidopsis thaliana, 213

archaebacteria, 61, 62, 73

archaeocetes, 401–4

Archaeopteryx, 270, 323, 392

archenteron elongation, 223–24, 233

archeocyathids, 89

archetypes, 33, 34

archosaurs, 52, 53

Arkarua, 79

arrow worms. *See* chaetognaths

arthropleurids, 177

arthropodization, 123, 126, 129, 135

arthropods (Arthropda): in alternative metazoan phylogenies, 46; and annelids, 44, 122, 123, 124, 125, 130, 132, 133, 183, 190; appearing in fossil record with jointed legs, 352; arthropodization, 123, 126, 129, 135; brain evolution, 88; in Burgess Shale, 126, 136, 162; *Canadaspis,* 95; centipedes, 134, 177, 178, 296; compound eye, 129, 378, 381, 382; cuticle, 123; disparity of Cambrian, 135–37; early divergence from coelomate protostomes, 121; *Emeraldella,* 128; as eucoelomates, 39; eyes, 129, 376, 378, 381, 382; first fossils of, 87, 101; five molecular phylogenies for, 133; gnathobase, 126, 127–28; head appendages in living and Cambrian, 128, 136; hemocoel, 124; homologous homeotic genes in mammals and, 182; hypotheses for origins of, 124; image formation, 376; intermediate filaments absent in, 135; *Leanchoilia,* 128; legs, 352; mandibles, 125; millipedes, 178; as modular, 433; in molecular phylogenetic tree of metazoans, 112; molecular phylogenies of, 132–35; monophyly versus polyphyly in, 123–32; as most diverse animal phylum, 123; myriapods, 130, 131, 133–34; nervous system, 124; as paraphyletic, 112; pentastomids and, 134; as plankton feeders, 99; post-Cambrian arthropods lacking developmental flexibility, 203; principal elements of body plan organization, 41; as protostomes, 44; pycnogonids, 10; scorpions, 178; segmentation in, 42, 123–24; segments and appendages of, 127; serial homology in, 337–38; shared anatomical features of, 124; *Sidneyia,* 128; spiralian ancestry of, 243; tagmosis, 123, 126–27, 203, 337–38, 415, 433; teloblast, 124; terrestrial, 177, 178; trigonotarbids, 177; *Waptia,* 95. *See also* chelicerates; crustaceans; onychophorans; trilobites; unirames

articulates (Articulata), 6, 42, 122, 132

artiodactyls: antelopes, 59; herbivory in, 97; Irish elk, 319–20; moose, 320; whales related to, 400

aschelminths. *See* pseudocoelomates

ascidians, 244–47; as colonial animals, 369; cytoplasm in embryos of, 14; as deuterostomes, 40; developmental modules in, 327;

direct developing, 192–93; evolutionary relationship to chordates, xvi, 8–9, 139–40, 197, 218; fertilization, 246; larvae, xvi, 9, 244–45; *Molgula occulta* and *Molgula oculata*, 245–47; mosaic development in, 334; muscle actin genes, 198; pharyngula stage in, 196; pre-displacement in, 275
Asthenosoma ijimai, 226
atavisms, 386–87
Atdabanian stage, 87
attractors, 328–29
Australopithecus, 59
autapomorphies, 52, 108
Avalon Peninsula (Newfoundland), 69
axolotl, 263, 272, 337, 391
axoneme, 330, 336
axon guidance, 289
Aysheaia, 178

bacteria, 314–15
bacteriophage T7, 154–55
barnacles: cleavage in, 130; larvae revealing as crustaceans, xv, 42–43; orientation depending on presence of predators, 220
bases, alignment of, 105
base substitution, 105, 152
Basilosaurus, 401
basin of attraction, 328
Bateson, William, 16, 186
Bathyergidae, 369
bats: birds compared with, 51; wings, 207–8
Bauplan. See body plan
beetles: divergence from flies, 412; *Tribolium,* 202, 412–13; weevils, 61, 170
behavior, 87, 100, 150
Beresovka mammoth, 168
biblical record, 63
bicoid gene, 188, 357
bicoid proteins, 310–11
bilateral symmetry, 38, 41
Bilateria: base of radiation of, 115–18; common body plan of, 285–86; examples of diversity of body plans, 40; eyes, 381; in molecular phylogenetic tree of metazoans, 112; as primitive metazoans, 38–39; principal elements of body plan organization, 41
biochemistry, 12–13
biogenetic law, 2
biology: comparative anatomy, 293–94, 427; ecology, 2; functional biology, 426; paleontology, 63, 65, 102, 140; philosophical differences within, 212. *See also* developmental biology; embryology; evolutionary biology; genetics; molecular biology; morphology; systematics
birds: *Aepyornis,* 250; *Anser,* 158; *Archaeopteryx,* 270, 323, 392; as archosaurs, 52; bats compared with, 51; *Branta,* 158; chicks, 194, 195, 345, 347, 392; crystallins in, 383; and dinosaurs, 146; emperor penguin, 2–4; Galápagos Island finches, 323; kiwis, 168; limbs, 392–93; mitochondrial DNA evolution, 158; moas, 168; neural revolution in, 100; pseudodontorn, 388; sex determination in, 374; songbirds, 279–80; teeth in, 388, 393; wing evolution as constraint on further development, 301
Bithorax complex, 16–17, 183, 186–87, 190, 307–9, 339, 413–14, 415
black-mound termites, 365
blastophore, 44
blastozoans, 97
BMPs (bone morphogenic protein families), 281, 338–39, 349, 351–53
body plan (*Bauplan*): adult body plan, 195; age of basic plans, xiii; arising from transformation of single ancestral plan, 185, 200; building similar animals in different ways, 211–54; in Cuvier's four embranchements, 6; as deeply immanent in ontogeny, 33; defined, xiv, 30–31, 400; developmental basis of, 173–210; and developmental biology, 173–74; developmental body plan, 196; developmental constraints on during Cambrian radiation, 301; Ediacaran, 77, 81; evolution within, 180; evolving new plans, 397–434; examples of diversity of, 40; extinct Cambrian plans, 96; and function for Cuvier and Geoffroy, 7; homology and, 33–38; Hox genes and, 185–90; hypotheses on stability of, 179–81; interpreting lost plans, 148–49; of land plants, 178; of larval forms, 196; macroevolution and, 30–33; metazoan, 30–62; missing links in, 67; number of basic plans, xiii, 32; number of and rate of evolution, 94–98; origin of concept of, 33–34; origins of, 63; Owen's archetypes compared with, 33, 34; paradox about, xiv, 197; of phyla, 31, 38; phylogeny necessary for understanding, 47–50, 148; principal elements of, 41; stasis in, 32, 179–81; tension between conservation and change in, 429;

body plan (*continued*)
  three ideas entangled in concept of, 195–96;
  three kinds of data for understanding, 424–
  28; transformability, 180–81, 182; why
  plans beyond the simplest should have
  arisen, 434; Woodger's *Bauplan* concept,
  196; zootype, 186
bone, fossil, 167–68
bone morphogenic protein families (BMPs),
  281, 338–39, 349, 351–53
bone remodeling, 282, 287
Bonner, John Tyler, 18
bony fishes: coelacanths and, 161; Haeckel's
  diagram of development in, 194; muscle actin genes, 198; notochord development, 200;
  tail development, 5; teleosts, 200, 213–14
bootstrapping, 109–110
Boveri, Theodor, 14
Bowers, H. R., 1
box jellies (Cubomedusae), 376–77
brachiopods (Brachiopoda): in alternative
  metazoan phylogenies, 46; ancestors of, 94;
  in the Cambrian fauna, 98; common ancestors with mollusks and annelids, 93, 121–
  22; as dominant filter feeders before Permian extinction, 98; as eucoelomates, 39;
  halkieriids as link between annelids and,
  102; as lophophorates, 122; lophophore, 42;
  in molecular phylogenetic tree, 112; in Paleozoic fauna, 99; in Permian mass extinction, 176; in protostome coelomate clade,
  113; as protostomes, 40, 122–23; shells as
  basic functional component of, 90; skeletonized debut of, 89
Brachydanio rerio. See zebrafish
brachypodism mutation, 351–52
brain: arthropod, 88; chimp, 288; giant frog,
  279; human, 286, 287, 288–89; mammal,
  100; Neanderthal, 150, 287; vertebrate, 88
branch-and-bound method, 109
branchial skeleton, 206
branching events: cladistics on, 51, 52; cladogram representation of, 45, 54, 56; in early
  metazoans, 39; genetic distance and
  branching patterns, 106; mapping evolution
  of developmental features onto branching
  patterns, 171; model systems ignoring
  branching order, 213; in phylogenetic trees,
  48, 50, 58, 59; the phylogeny of animals as
  branching, 173; time between, 95

Branchiostoma (*Amphioxus*), 185, 197,
  198–99, 379
Branta, 158
Brisaster latifrons, 226
brooding, 221
brooding frogs, 247–49
Brooks, William Keith, 10
bryophytes, 178
bryozoans: as colonial animals, 369; as eucoelomates, 39; as lophophorates, 123. *See also*
  entoprocts
Buckland, William, 64–65
burden, 316, 317
Burgess Shale fauna, 94–96; *Anomalocaris*,
  95, 96, 99, 135, 380; arthropods, 126,
  136, 162; *Canadaspis*, 95; *Choia*, 95; eyes
  in, 379–80; fossils of soft-bodied animals
  in, 87, 91; *Hallucigenia*, 96; *Opabina*, 40,
  95, 96, 380; *Ottoia*, 95; possible fossil
  ctenophore in, 175; priapulids in, 119; Vendian descendants in, 78; Walcott's discovery of, 69, 95; *Waptia*, 95; *Wiwaxia*, 92, 122
Burnet, Thomas, 64
burrowing, 77, 87–88, 90, 101, 120, 123
butterflies, 415–17

Caenorhabditis elegans: cell fates in, 238–40;
  developmental modules in, 332–33; heterochronic mutation, in 275, 276–77, 282–83;
  Hox gene cluster in, 309; intense study of,
  26; maternal gene action in, 243; as model
  system, 119, 213; ray development in, 359;
  sex determination in, 371, 372, 374–75;
  vulva development in, 205
calcareous skeletons, 75
Cambrian period: *Amiskwia*, 96; *Anomalocaris*, 95, 96, 99, 135, 380; arthropods of,
  135–37; Atdabanian stage, 87; the Cambrian fauna, 98–99; Chengjiang fauna, 91,
  380; *Canadaspis*, 95; echinoderms of, 418;
  Emeraldella, 128; fauna as soft-bodied, 87;
  Halkieria, 91–93; *Hallucigenia*, 96; *Leanchoilia*, 128; length of, 68, 93; Manykaian
  stage, 87; *Microdictyon*, 91, 96; modern animal phyla appearing in, 68, 95; no new
  phyla since, xiv, 32, 173, 174–75; *Odontogriphus*, 96; *Opabinia*, 40, 95, 96, 380; *Pikaia*, 88; planktotrophic larvae appearing
  in, 222; the Precambrian-Cambrian boundary, 68, 86, 87, 89, 101; *Sidneyia*, 128;

skeletonized animals appearing in, 89;
Thaumaptilon, 78, 86; Tommotian stage,
87, 89, 90. *See also* Burgess Shale fauna
Cambrian radiation of metazoans: apparent
suddenness of, 86; appearance of phyla in,
31–32, 95–97, 102, 173; base of, 113–15;
dating of, 74, 93; developmental constraints
during, 301; dissecting by molecular phylogeny, 103–41; extinct body plans produced in, 96; factors in, 73; loss of data
making understanding difficult, 149; molecular biology and, 140–41; molecular phylogenetic tree revealing base of, 114; oxygen
levels and, 75–76; patterns of development
in, 181–85; quickening evolution just before, 101–2; regulatory genes and, 432;
reliable phylogenies required to infer structure of, 47–48, 95; structuralism and,
313; as under way by 530 million years
ago, 68
Canadaspis, 95
canalization, 253
canids, 265; dogs, 264–65, 272, 278
cappuccino gene, 357
carbon, 76
carnivores, 100, 220
carpoids, 139
cartilage, 343, 345
cartilage concentration, zones of, 350
cascades of genes. *See* gene cascades
Cassiopeia, 79
caste differentiation in termites, 364–68
catalase, 113
catastrophes, 21, 65, 66, 67, 99
catastrophism, 65, 66
cats: *Lynx*, 393–94; saber-toothed cats, 167
Caulerpa, 85
causality, 20, 21
cave painting, 321
ceh13 gene, 309
cell fate determination, 239
cell lineages: as approximations to phylogeny,
22; autonomy of, 268; in *Caenorhabditis
elegans*, 238–40, 275, 276–77, 283, 432;
changes in early development, 231; developmental modules including, 327; in direct-
developing ascidians, 193, 245, 247, 254;
early studies of, 14, 19; in *Heliocidaris
erythrogramma*, 237; heterochronies, 290;
mapping of, 218–19; in mollusks, 240,

242, 243; in soil nematode phylotypic stage,
209; as temporally connected series of modules, 332
cell size, 304
cell types, 254, 306–7, 328–29, 330
centaurs, 295, 322
centipedes, 134, 177, 178, 296
Cephalobus, 277
cephalochordates: *Amphioxus*, 185, 197,
198–99, 379; in deuterostome phylogeny,
137; eyes, 379; Hox clusters, 354; muscle
actin genes, 198
cephalopods: ammonites, 176, 251; eyes, 376,
377, 381, 384; nautiloids, 176; *Nautilus*,
40, 377; neural revolution in, 100; octopuses, 377, 385; in Permian extinction, 176;
spiral cleavage absent in, 243; squids, 377,
385
cetaceans. *See* whales
chaetognaths (arrow worms): as deuterostomes, 44, 137, 138; as eucoelomates, 39;
lacking fossil record as far back as the Cambrian, 175; as protostomes, 138
Chambers, Robert, 146
character conflicts, 103–4, 108
Charnia, 77
Charniodiscus, 70, 78, 79, 86
chelicerates (Chelicerata): in arthropod phylogeny, 124, 133; head appendages, 128,
136; limbs of, 125, 126; pycnogonids and,
10; segments and appendages of, 127
Chengjiang fauna, 91, 380
Cherry-Garrard, Apsley, 1–4
chicks, 195, 196, 345, 347, 392
chimpanzees, 286, 288, 331
chitin, 89, 138, 378
choanocyte, 114
choanoflagellates, 113, 114
Choia, 95
chondrichthyans (Chondrichthyes), 160, 161;
sharks, 160, 161, 200
chordates (Chordata): basic body organization
of, 31, 197; as deuterostomes, 44, 138–40;
echinoderms and, 420; as eucoelomates, 39;
evolutionary relationship to ascidians, xvi,
8–9, 139–40, 197, 218; first appearance of,
88; in molecular phylogenetic tree of metazoans, 112; as monophyletic, 140; neural
crest cells, 198, 207, 328, 342–44; notochord, 197, 200, 328; pharyngula stage in,

chordates (Chordata) (*continued*)
196; *Pikaia,* 88; somites, 200; traditional
hypothesis for origins of, 217–18. *See also*
cephalochordates; urochordates; vertebrates
chromosomal sex determination, 370, 371, 374
chromosomes, 14
cilia, 330
ciliary photoreceptors, 380–81
circulating hormones, 278, 279
cis-acting control regions of promoters, 426,
431–32
clades, 31, 50–51, 400
cladistic methods, 108–111; distance-based
methods compared to, 105, 106, 111; maximum likelihood methods, 110, 153; parsimony methods, 108–10, 151, 152, 153,
155; transversion parsimony methods, 109
cladistics, xvi, 50–55; autopomorphies in, 52,
108; clades, 31, 50–51, 400; in determining
evolutionary relationships between phyla,
45; exposing problems but not providing unequivocal solutions, 103; fossil record dismissed by some cladists, 404; nested pattern
of synapomorphies assumed by, 395; not all
similarities used in, 51–52; not all traits
conforming to favored tree, 53; outgroups
used in, 55–56; pattern cladists, 37; shared
primitive features as unhelpful in, 47; symplesiomorphies in, 52; synapomorphies in,
52, 54–55, 108, 395. *See also* cladistic
methods; cladograms
cladogenesis, 58–59
cladograms, 54–55; defined, 45; four major
characteristics of, 56; ideal use of, 425; and
phylogenetic trees, 56–60
clams, 98, 176; *Unio,* 43, 241–43
Clarkia beds (Idaho), 169
classes (taxonomic), 159
cleavage, 43–44, 130, 251, 312–13; spiral
cleavage, 43–44, 240, 243
cloning, 26, 165
Cloudina, 76, 90
cluster analysis, 106–7, 155
Clypeaster rosaceus, 226
cnidarians (Cnidaria): as colonial animals,
369; depth in molecular phylogenetic tree,
111–13; Ediacaran fauna as, 45, 77, 79, 85;
eyes, 376–77; fungi and, 112; *Hydra,* 430;
intermediate filaments in, 135; lacking complex nervous system, 39; medusa, 40, 78,
376; in metazoan tree, 115; as most primi-

tive metazoans, 45; photosynthetic endo-
symbiodonts, 78; possessing only tissue
level of organization, 38; sea pens, 79–80;
skeletonized debut of, 89; *Thaumaptilon,*
78, 86; Trilobozoa, 79, 89. *See also* jellyfishes
coelacanths, 159, 160–62, 163, 354
coelom, 39, 77, 118, 120
coelomate animals: appearance at beginning of
Cambrian, 102; disparity of, 120; Ediacaran
fauna as, 77; in emergence of metazoans,
73; as first competent burrowers, 101, 120;
in Hyman's metazoan phylogeny, 39; in
molecular phylogenetic tree, 112, 113; morphogenetic regulatory genes widely shared
in, 184; as not having long hidden history,
93. *See also* deuterostomes; protostomes
coelomate protostomes, 120–22; disparity of,
121; early divergence from arthropods, 121;
segmentation in, 183. *See also* annelids;
brachiopods; mollusks; nemerteans; pogonophorans; sipunculans
Coelophysis, 264
collagen, 75
collinearity hypothesis, 308
colonial animals, 369–70
color vision, 379
comatulids (feather stars), 180
comparative anatomy, 293–94, 427
complexity, constraint by, 315–19
compound eye, 129, 378, 381, 382, 384
concentricycloids, 62
concerted evolution, 297
Conklin, Edwin, 14
Conomedusites, 79
convergent evolution, 49, 53
co-option, 340–41; defined, 326; of eye structures and genes, 382–86; of a growth factor
system, 282; incipient structures explained
by, 340; of leg elements in insects, 433; as
making evolution of ontogeny possible,
325; of regulatory genes, 341, 431; in vertebrate jaw development, 341–44
copulation, 119
cornea, 379
corpora allata, 367
correlated progression model, 115
creationism, modern, 66, 315
Creation of Genesis, 63–65
Crepidula, 43
Cretaceous mass extinction, 175, 176

criminal man, 387

crinoids (Crinoidea), 176, 180, 418, 421

crocodiles, 383

Cro-Magnon painters, 321

crustaceans (Crustacea): in arthropod phylogeny, 124, 133; barnacles as, xv, 42–43; *Eurytemora*, 191; *Gammarus chevreuxi*, 15–16; head appendages, 128, 136; limbs, 125–26; lobsters, 264; mandibles, 126, 127–28; naupilius larvae, 131; nervous system development, 130, 134; pentastomids and, 134; *Rehbachiella*, 191; segments and appendages of, 127; unirames and, 130–31, 133. *See also* barnacles

Cryptobatrachus, 248

crystallins, eye lens, 383, 385, 386

Crystal Palace dinosaurs, 142–47

Ctenocystoidea, 418

Ctenophalides felis, 408

ctenophores (Ctenophora): as diploblastic, 77; in metazoan tree, 115; in molecular phylogenetic tree, 112; possessing only tissue level of organization, 38; possible Cambrian fossil, 175

Cubomedusae (box jellies), 376–77

cuticle, 123

Cuvier, Georges, 6–7, 65, 293–94

C-value paradox, 304, 305

cytoplasm, 14

Dacus tyroni, 411–12

Daphnia, 220

Darwin, Charles: on embryos for tracing evolutionary relationships, xv; on eyes, 380; as first to construct phylogenetic tree, 59; Haeckel and, 2; on homology, 34; on incompleteness of fossil record, 158; and Lyell, 25, 145; Mivart's attack on, 340; on Precambrian life, 69; on recapitulation, 8; synthesis of, 28–29; variation recognized as critical by, 213, 293

daughterless gene, 339

deadpan gene, 371

de Beer, Gavin, xiii, 23–24, 36, 129, 217, 255, 258, 259

de Beerian heterochrony, 261, 266

decapentaplegic gene, 353, 411, 412, 414

deep time, 63–68; the Flood as connecting to human time scale, 64; metazoan origins and, 63–102; ordered and subdivided, 66.

See also Cambrian period; Precambrian period

Deformed gene, 309

Delta gene, 332

deuterostomes (Deuterostomia), 137–40; the ancestral deuterostome, 140; characteristics of, 44–45; examples of diversity in body plans, 40; intermediate filaments in, 135; lophophorates as, 122; in molecular phylogenetic tree of metazoans, 112, 113; as monophyletic, 138; photoreceptors, 380–81; principal elements of body plan organization, 41; three phylogenies for, 139. *See also* chordates; echinoderms; hemichordates; urochordates

development: canalization, 253; in colonial animals, 369–70; conservation and change in early, 222–23; as consisting of independent chains of causally linked processes, 260; developmental basis of body plans, 173–210; developmental hourglass, 208–10, 318; developmental regulation, 253–54; developmental sequence, 255; dissociability of elements in, 23, 203, 282, 290; early development as evolutionarily flexible, 211–12; egg size and, 251–52; in eusocial species, 370; evolution of after phylotypic stage, 205–8; evolution's relationship to, xiii, xiv, xv–xvi, xviii, 9, 17–18, 425, 427–28; experimental approaches to evolution of early, 238–50; generality of evolutionary changes in early, 250–51; generative entrenchment, 192; genes underlying processes of, xvi–xvii, 17; heredity and, 13–15; heterochronies in early, 267–68; homeobox-containing genes in, 26; life history and, 219–22; as modular, 203–4, 428; mosaic development, 253, 334–35; nymphs, 364, 366; ontogeny recapitulating phylogeny, 1–29; organogenesis, 204, 268; patterns of in metazoan radiation, 181–85; phylogeny and the evolution of, 171–72; polarity of change in, 171, 181–82; quantitative changes in modularity in during, 204; radical evolutionary changes in early, 192, 222–23, 317; regulative development, 253–54, 334–35; stability of larval forms, 190–92; underlying flexibility of, 252–54; using related organisms to study, 217. *See also* developmental biology; developmental constraints; developmental reca-

development (*continued*)
pitulation; direct development; eggs; embryos; growth; indirect development; larvae;
morphogenesis; ontogeny; phylotypic stage
developmental biology: body plans and, 173–
74; causality in, 20; diversity avoided in,
22; and evolutionary biology, xv, 4, 18, 19,
20–23, 216, 425; examples in, 344–45; favored organisms in, 213–14; in functional
biology, 426; genes assuming primacy in,
22, 310; genetics and, 13–15, 17–18;
hypotheses in, 20; implicit and real phylogenies in model systems in, 214; life history
not included in, 215, 317; model systems
as norm in, 213; operating demands of the
discipline, 215; reductionism in, 212; transmission genetics and, 14. *See also* development; evolutionary developmental biology
developmental body plan, 196
developmental constraints, 292–320; allometry as, 319–20; biomechanical constraints,
299–300; complexity as, 315–19; constructional or architectural constraints, 297; genetic constraints, 300, 307–10; genomic
constraints, 304–6; internal organizational
constraints, 300–301; limits of structure im-
posing, 310–15; natural selection and,
297–99, 428; nature of existing developmental systems as, 294–96; number of cell
types as, 306–7; ontogenetic rules as, 301;
phyletic constraints, 301; sources of, 299–
302; vertical and horizontal, 317–18
developmental genetics, xvi, 28, 215, 217, 315
developmental hourglass, 208–10, 318
developmental modules. *See* modularity
developmental program, 331
developmental recapitulation, 5–8; Darwin
on, 8; embryologists abandon, 20;
Haeckel's embryo diagrams as evidence for,
194; in Haeckel's mechanism of evolution,
2, 9–10; heterochrony and, 259; in popular
culture, 255; von Baer's refutation of, 6
developmental regulation, 253–54
developmental sequence, 255
dichotomies and model systems, 212–14
Dickinsonia, 70, 75, 77, 81, 82, 83, 86
diffusion mode of gas exchange, 75
digits, 355–56, 425
dinosaurs: as archosaurs, 52, 53; birds and,
146; *Coelophysis*, 264; the Crystal Palace
dinosaurs, 142–47; Dinosauria group first

conceived by Owen, 143; estimated number
of genera, 147; extinction in late Cretaceous, 175; *Hylaeosaurus*, 144; *Iguanodon*,
143, 144–45, 146; *Jurassic Park*, 163–64;
Megalosaurus, 143, 144, 146; *Mononychus*, 147; on Noah's ark, 66; skin of, 145;
warm-bloodedness of, 147
diploblastic animals, 38, 40, 41, 77, 100
dipterans (Diptera): conservation of pattering
in, 202; *Musca domestica*, 202; Strepsiptera
and, 189. *See also* fruit flies
direct development: in ascidians, 192–93; as
degenerative evolution, 229; ecological
context of, 220–22; egg size and, 251–52;
in frogs, 192–93, 221–22, 249–50, 275;
heterochrony and, 274–75; as radical evolutionary change in early development, 192–
93; in salamanders, 250; in sea urchins,
193, 221, 224–37, 251–52, 275, 425–26;
structures lost in, 391
disparity, 60–62, 97, 159
dissociated heterochronies, 267
dissociation, 334–37; of appendage function
in insects, 433; defined, 326; dissociability
of developmental processes required for
evolution, 27; dissociability of elements in
development, 23, 203, 282, 290; in evolution, 282; heterochrony as, 282, 290, 326;
as making evolution of ontogeny possible,
325
Distal-less gene, 185, 333, 353, 411, 413,
414, 416, 417, 431
distal tip cell, 239, 333
distance-based methods, 106–8; additive distance methods, 107; cladistic methods compared with, 105, 106, 111; cluster analysis,
106–7, 155; data loss in, 153
diversity: developmental biology avoiding, 22;
and disparity, 60–62, 97; idealistic conceptions of, 293; as inconvenience to nonevolutionary approach, 213; as lower at higher
systematic levels, 97; molecular biology for
studying, 25; as original issue of evolutionary biology, 21
divine plenitude, principle of, 293
divine watchmaker, the, 292
divisions (taxonomic), 178
Dlx-1 gene, 184
DNA: alignment of bases, 105; base substitution, 105, 152; degradation over time, 164,
165–66; in fossils, 163–71; four bases in,

50; junk DNA, 304; mitochondrial DNA, 151, 156, 158, 165, 166; phylogenetic information in, 104; really old sequences of, 168–69; recovering by polymerase chain reaction technique, 166–67; selfish DNA, 304; structure of, 17. *See also* genes

dogs, 264–65, 272, 278

Dollo, Louis, 143, 387

Dollo's law, 386–89, 426

dosage compensation, 370–71

doublesex gene, 371

D quadrant determination, 240

dragonflies, 405, 412

Driesch, Hans, 14

Drosophila: Antennapedia complex, 183, 186–87, 190, 307–9, 339, 413; appendage development, 352–53; Bithorax complex, 16–17, 183, 186–87, 190, 307–9, 339, 413–14, 415; cascade of genes determining Hox gene expression, 188; cells in nervous system development, 332; developmental modules in, 327; diversity of species in, 61; duplication and redundancy in, 344; eye development, 129, 184–85, 358; gap genes in, 188, 201, 202, 310–11, 319; growth control in, 280; head segmentation, 199; *hedgehog* gene, 216; helix-loop-helix transcription factors in, 339, 371; as highly modified and streamlined, 406–7; homeotic mutation in, 186–89; Hox clusters in, 185; larvae, 407; left/right symmetry in, 80–81, 302–3; limb and wing development, 411, 412, 413–14; limb fields in, 333; as long germ band insect, 201, 407; *msh* gene, 198; pair-rule genes in, 188, 201, 202, 319; parallel processing in, 335–36; segmentation in, 16, 26, 132, 183–84, 186–87, 327, 357–58; segment polarity genes in, 183, 188, 201, 202, 319; sex determination in, 370–71, 372, 375; spermatogenesis, 336; tagmosis studied in, 123; transposable elements affecting mutation rates in, 300

Drosophila melanogaster, 16, 26, 213, 308–9, 408

Drosophila pseudoobscura, 308–9

duplication and divergence, 337–40; defined, 326; evolutionary changes facilitated by, 203; genetic complexity arising from, 315; as making evolution of ontogeny possible, 325; novel features created by, 339–40; producing families of related genes, 152, 156,

337; redundancy and, 344; in vertebrates, 433–34

dwarfism, 278

early diploblasts hypothesis, 72, 78

ecdysone, 368

echinoderms (Echinodermata): blastozoans, 97; body plans, 97–98, 417–23; Cambrian, 418; carpoids, 139; chordates and, 420; concentricycloids, 62; crinoids, 176, 180, 418, 421; Ctenocystoidea, 418; as deuterostomes, 42, 44, 137, 138–40; developmental hourglass as diagram for, 209; Eocrinoidea, 418; as eucoelomates, 39, 42; extinctions making molecular phylogeny difficult, 162; eyespots, 377; helicoplacoids, 418, 419, 420; hemichordates and, 420, 423; homeobox-containing genes in, 187; Homoiostelea, 418; Homostelea, 418; Hox gene expression in, 421–22; in molecular phylogenetic tree of metazoans, 112; nervous system, 418, 420–23; in Paleozoic fauna, 99, 420; paracrinoids, 420; skeleton, 89, 418; Stylophora, 418, 420; *Tribrachidium* as ancestral, 79. *See also* edrioasteroids; sea urchins

echiurians, 39, 43, 175

ecology, 2

ecospace utilization, 179

Ediacara, 77, 79

Ediacaran fauna, 68–71; *Albumares,* 79; ancestral metazoans hypothesis, 71, 72; *Anfesta,* 79; as animals, 77–86; as annelids, 82, 84; body forms, 77, 81; *Charnia,* 77; *Charniodiscus,* 70, 78, 79, 86; as cnidarians, 45, 77, 79, 85; *Conomedusites,* 79; *Dickinsonia,* 70, 75, 77, 81, 82, 83, 86; difficulties in interpreting, 148–49; discovery of, 69; diversity of, 69; early diploblasts hypothesis, 72, 78; *Ediacara,* 77, 79; Ediacaran medusoids, 78–79; environmental changes favoring, 74–75; *Ernietta,* 80; fossils of, 70, 76, 77, 81, 83, 85; Garden of Ediacara hypothesis, 72, 78, 85; as harboring photosynthetic endosymbionts, 78; hypotheses regarding, 71–72; integumental compartmentalization in, 83; as lichens, 85–86, 100; *Parvancorina,* 70; photoreception, 380–81; *Pterdinium,* 80; quilted construction of, 83, 85; *Rangea,* 80; relationship to metazoans, 71, 72, 81, 100, 102; as

Ediacaran fauna (*continued*)
representing the radiation of diploblastic animals, 100; segmentation in, 80, 83;
Spriggina, 70, 77, 80, 84–85, 433; *Tribrachidium,* 70, 79; *Vendia,* 80; as Vendobionta, 85, 100; vendozoan hypothesis, 72,
86
Ediacaran medusoids, 78–79
edrioasteroids (Edrioasteroidea): as deuterostomes, 40; symmetry, 418; *Tribrachidium*
as resembling, 79
eggs: *Aepyornis,* 250; amniotic, 317; ascidian,
246; frogs, 249–50; oogenesis, 372; sea urchin, 225–26, 231–32; size of and development, 225–26, 251–52; terrestrial vertebrates, 251
18S ribosomal RNA, 111–13, 140
elephants, 165, 168
Eleutherodactylus, 249–50
Eleutherodactylus nubicola, 275
elongation factor-1α, 113, 118
embryology: applied to arthropod phylogeny,
131; as experimental science, 11, 19, 212,
213, 334; recapitulation abandoned by, 20
embryonic ectoderm, 43
embryos: as blocked out in early development,
211–12; ciliary bands in, 330; cleavage,
43–44, 130, 251, 312–13; in the developmental hourglass, 208; developmental recapitulation in, 5–6, 8; as fields in which domains are generated, 312; induction, 11,
12–13, 205; maternally encoded information in, 231; as modular, 204; mosaic and
regulative, 253–54, 334–35; physical location of modules in, 331; regulation in, 253;
size of and development, 252; spiral cleavage, 43–44, 240, 243; in tracing evolutionary relationships, xv; vertebrate embryos
passing through typical phylotypic stage,
193–94. *See also* gastrulation
Emeraldella, 128
emperor penguin, 2–4
empty spiracle gene, 199
Emx gene, 199
engrailed gene, 183, 188, 201, 202, 352
enterocoely, 44
entoprocts: as pseudocoelomates, 39, 118;
lacking fossil record as far back as the Cambrian, 175; lophophore, 42
environmental sex determination, 370, 374
Eocene radiation of aquatic mammals, 403

Eocrinoidea, 418
Eosaccharomyces, 73
epidermal growth factor superfamily, 281
Equus hemionus, 168
Ernietta, 80
essentialism, 212, 213
estrogen, 279, 280, 373
eubacteria, 61, 73
eucoelom, 39
eucoelomates: examples of body plans, 40;
major bilateral phyla as, 39; principal elements of body plan organization, 41
eukaryotes, 61, 73
Eurytemora, 191
eusocial species, 370
eutely, 119
even-skipped gene, 202
evolution: adaptation and homology unified in,
8; Agassiz's opposition to, 5; within body
plans, 180; anagenesis, 58–59; change in
number of units as common in, 207; changes
facilitated by duplication of genes, 203;
cladogenesis, 58–59; concerted evolution,
297; convergent evolution, 49, 53; as descent
with modification, 53–54; development's relationship to, xiii, xiv, xv–xvi, xviii, 9,
17–18, 425, 426–27; directionality in,
24–25, 48, 98, 183; dissociability of developmental processes required in, 27; dissociations in, 282; Dollo's law, 386–89, 426;
early development as evolutionarily flexible,
211–12; embryos in tracing evolutionary relationships, xv; evolutionary reversals, 386–
95; evolving new body plans, 397–434; experimental approaches to evolution of early
development, 238–50; four elements making
evolution of ontogeny possible, 325; generality of changes in early development, 250–51;
Haeckel's mechanism of, 2; heterochrony as
mechanism of, xviii, 24; homology as basis
for all evolutionary comparisons, 37; industrial melanism, 54; internal organization as
factor in, xviii, 32, 324; larval forms conserved in, 190–92; mechanisms of, 323–25;
models of evolutionary change, 152; morphology transformed in, 23; mosaic evolution, 323, 403; neutral evolution, 297; number of body plans and rate of, 94–98;
orthogenesis, 25; parsimony in, 53; phylogeny required for understanding, 18, 32; preadaptation, 340; principles of evolvability,

325–26; and progress, 98–100; punctuated equilibrium, 21, 271, 423; rate of, 94–98, 323, 423–24; selfish DNA evolution, 297; as tinkering with existing structures, 324; transformation as early concept of, 5, 33. *See also* evolutionary biology; macroevolution; microevolution; natural selection; phylogeny; variation

evolutionary biology: causality in, 20; comparison as fundamental for, 18; and developmental biology, xv, 4, 18, 19, 20–23, 216, 425; differences between developmental biology and, 21; diversity as driving issue in, 21; historical focus of, 21–22; hypotheses in, 20; microevolution as concern of, 30; population genetics as tool for, 14, 21; time frame in, 21. *See also* evolution; evolutionary developmental biology

evolutionary developmental biology: conceptual framework for, 216–19; context of, 4–5; developmental genetics compared with, 28; dichotomies of perspective retarding, 213; experimental discipline for, 19; heterochrony as vital concept for, 24; issues for, 23–29; phylogenetics as vital to, 47; three sources of data for, 216

evolutionary parsimony, 109

evolutionary systematics, 50, 103

exaptation, 340

experimentalist tradition, 213

extinction: contingency in surviving, 67, 182; extinct lineages affecting molecular phylogenies, 159–63; heterochronies of extinct animals, 268–70; large-scale extinctions, 68; more than a single catastrophic, 65; principle of plenitude and, 293; pseudoextinction, 58. *See also* mass extinctions

extramachrochaetae gene, 339

eyeless gene, 184, 359, 384

eyes, 375–82; aesthetes, 377; in annelids, 377, 381; in arthropods, 129, 376, 378, 381, 382; in bilaterians, 381; in Burgess Shale fauna, 379–80; in cephalochordates, 379; in cephalopods, 376, 377, 381; in cnidarians, 376–77; color vision, 379; compound eye, 129, 378, 381, 382, 384; co-option of eye structures and genes, 382–86; cornea, 379; in *Drosophila*, 129, 184–85, 358; in flatworms, 377; in honeybees, 375; image formation, 376; infrared eyes, 376; lens, 381, 382–83; in mollusks, 377–78,

381; multiple origins for, 382; in onychophorans, 378; parietal eyes, 379; photoreception, 129, 376, 378, 380–81; pineal eyes, 379; in pseudocoelomates, 377; radio wave eyes, 376; retina, 379; retina as inducer of lens in, 336–37; schizochroal eye, 378, 382; in snakes, 207, 376; in trilobites, 129, 378, 382; in vertebrates, 184–85, 376, 378–79, 381; vision, 375

eyespots, 377

fair game model, 176

falsifiability, 153–54

families (taxonomic), 159

fate maps, 130, 131, 218, 229, 230

fauna. *See* animals

feathers, 4, 340

feather stars (comatulids), 180

fecal pellets, 94, 369

feeding modes, 99

FGF (fibroblast growth factors), 281, 345, 349

fibroblast growth factor receptors (FGFR), 281–82

fibroblast growth factors (FGF), 281, 345, 349

field of bullets hypothesis, 176

finches, Galápagos Island, 323

fins, 354–55, 388

fishes: *Acanthodes,* 342; agnathans, 342, 354, 379; experimental reversals in, 390; fins, 354–55, 388; guppies, 323; lungfishes and, 159–60; neural revolution in, 100; parthenogenesis in, 390; ray-finned fishes, 161, 342; whales compared with, 49, 388. *See also* bony fishes; chondrichthyans; lobe-finned fishes; zebrafish

Fitch-Margoliash method, 107

FitzRoy, Captain Robert, 397

flatworms. *See* platyhelminth flatworms

Flectonotus, 248, 249

flies: butterflies and, 415–17; dragonflies, 405, 412; emergence of, 412; mayflies, 190–91, 405, 407, 409; scorpion flies, 303; stoneflies, 411. *See also* dipterans

Flood, the, 63–67, 397

fog gene, 372

forelimbs: in the earliest tetrapods, 354; fin folds as origin of, 49, 354; limb bud development, 345; as serially homologous to hindlimbs, 36–37; in snakes, 207; in whales, 401; wings evolving from, 324

fossilization, 86–87

fossil record: accuracy of molecular clock resting on, 157–59; of Cambrian period, 68,
94; cladists dismissing as unimportant, 404;
new groups appearing in, 271; numerous
extinctions indicated by, 65; phyla with no
record as far back as the Cambrian, 175;
rate of evolution in, 323, 423–24
fossils: arthropods, 87, 101; ascribed to the
Flood, 64; of bone, 167–68; DNA in, 163–
71; Ediacaran, 70, 76, 77, 81, 83, 85; the
first metazoan fossils, 68; fossil heterochronies, 268–70; fossilization, 86–87; importance in understanding metazoan history,
100–102; index fossils, 65; from La Brea
tar pits, 167; larvae, 190–91; Neanderthal,
149; small shelly fossils of the Phanerozoic
eon, 89–90; of soft-bodied animals, 83,
86–87, 91; trace fossils, 76–77, 87–88,
100, 101
free oxygen levels, 75–76
Fritziana, 248
frogs, 247–50; brooding frogs, 247–49;
Cryptobatrachus, 248; direct-developing,
192–93, 221–22, 249–50, 275; *Eleutherodactylus*, 249–50; *Eleutherodactylus nubicola*, 275; *Flectonotus*, 248, 249; fossil
pipid frog tadpoles, 191; *Fritziana*, 248;
Gastrotheca riobambae, 247–49; genomic
constraints in, 305; *Hemiphractus*, 248;
indirect-developing, 275; induction in, 11;
marsupial frogs, 247–49; oldest fossils from
lower Jurassic, 247; pipid frogs, 191; pre-
displacement in, 275; *Rana*, 275; regulative
development in, 334; salamander development compared to, 200; *Stefania*, 248; thyroxin in metamorphosis of, 279; tropical
frogs, 215, 247; twenty-nine reproductive
modes for, 247; *Xenopus*, 249; *Xenopus
laevis*, 213, 275
fruit flies: *Dacus tyroni*, 411–12. *See also*
Drosophila
function: and body plan for Cuvier and Geoffroy, 7; in Cuvier's four embranchements,
6; duplication and, 344; genetic co-option
and, xviii; selection molding homologies to
meet demands of, 8
functional biology, 426
fungi, 61, 112, 113

Galápagos Island finches, 323
Gammarus chevreuxi, 15–16

gap genes: bicoid protein establishing pattern
of expression of, 310–11; conservation of,
202; encoding transcription factors, 188,
201; *hunchback* gene, 188, 310–11; *Krüppel* gene, 202; mutations in, 319
Garden of Ediacara hypothesis, 72, 78, 85
Garstang, Walter, xvi, 23, 255, 259
gas exchange, diffusion mode of, 75
Gastrotheca riobambae, 247–49
gastrotichs: eyespots, 377; lacking fossil record as far back as the Cambrian, 175; as
pseudocoelomates, 39, 118
gastrulation: cleavage and, 251; in frogs and
salamanders, 193, 200; in *Heliocidaris
erythrogramma*, 233; organogenesis beginning in, 268; in sea urchins, 209, 224
GDF-5 gene, 351–52
Gdf proteins, 339
gene cascades: in *Drosophila*, 188, 201, 316;
in early development, 192, 193; as highly
conserved, 318; regulatory genes as, 331–
32; in sexual differentiation, 368; in short
germ band insects, 202; in somatic sex determination, 372
gene duplication. *See* duplication and divergence
gene knockout techniques, 189, 315
generative entrenchment, 192
genes: in arthropod eye development, 129; assuming primacy in developmental biology,
22, 310; chromosomes, 14; co-option for
functions, xviii, 341; co-option of eye structures and, 382–86; in developmental and
evolutionary biology compared, 21; differential gene expression as underlying morphological differences, 234; in evolutionary
reversals, 389–95; evolving at different
rates, 157; and the execution of pattern,
350–52; fossil genes, 163–71; gene silencing, 389–90; genetic constraints on development, 300; genetic homology, 431; genetic organization of modules, 327–29;
genetics of heterochrony, 276–77; and induction, 12–13; in insect leg and wing development, 414–15; as molecular clocks,
107; and morphology, 310, 312; operon theory, 17; organization as developmental constraint, 307–10; paired-box genes, 344,
384; pathways in development, 329; rate
genes, 15–16; structural genes, 234, 331,
431; switch genes, 190, 333, 356–59; as

underlying developmental processes, xvi–xvii, 17. *See also* duplication and divergence; gene cascades; gene sequences; genomes; mutations; regulatory genes; segmentation genes; *and particular genes*

gene sequences: and cleavage, 312–13; comparison of, 17–18; extinct lineages affecting phylogenetic conclusions based on, 159–63; gaps in, 106; homology of, xviii, 18, 34, 105; in inferring molecular phylogenies, xvi, 25, 104–11, 152–55; in phylogeny of humans, 151; for phylum-level phylogenies, 140; in producing reliable phylogenies, 47; rate of evolution, 155–56

Genesis, Creation of, 63–65

genetics: and development, 13–15, 17–18; developmental genetics, xvi, 28, 215, 217, 315; in functional biology, 426; Mendelian genetics, 9, 13, 20, 23, 25; molecular genetics, 215–16; physiological genetics, 16; population genetics, 13–14, 21, 296; quantitative genetics, 296; transmission genetics, 13, 14

gene trees, 155–56

genomes: consequences of fluid, 395–96; genomic constraints on evolution, 304–6; opportunism of, 362–96; size in animals, 304

Geoffroy Saint-Hilaire, Etienne, 5, 34

Geospiza fortis, 323

germ cell-less gene, 357

giant mammals, Pleistocene, 397–99

gill arches, 342, 354

global heterochronies, 261–66; in human evolution, 285, 289; versus local heterochronies, 266–67. *See also* paedomorphosis; peramorphosis

glochidium, 241–42

Glyptodon, 397

gnathobase, 126, 127–28

gnathostomulids: flatworms and, 116; lacking fossil record as far back as the Cambrian, 175; as pseudocoelomates, 39, 118

Goldschmidt, Richard, 12, 15, 16

Gollum, 386

gorillas, 286

grades, 52

granulosa cells, 373

grasshoppers, 304; *Schistocerca*, 202

Great Chain of Being, 6, 293

Great Flood, 63–67, 397

Grobben, K., 44

growth: allometric, 264, 297, 319; differential, 252, 272, 280; dwarfism, 278; growth hormones, 278–82; isometric, 264; time and growth control, 278–82. *See also* allometry

growth factor superfamilies, 281

growth hormone gene, 278

growth hormones, 278–82

Grypania, 73

guinea pigs, 207, 394

guppies, 323

Hadorn, Ernst, 16

Haeckel, Ernst: on arthropod phylogeny, 125; continued relevance of, 18; diagrams of vertebrate embryos, 193–94; and heterochrony, 259; on interaction of phylogeny and ontogeny, xv, 2, 3, 9–10, 42; program of rejected, 19–20

hair, 168

hairy gene, 202

Haldane, J. B. S., 16

Halkiera, 91–93

halkieriids, 89, 102, 121–22

Hallucigenia, 96

Haloarcula, 314

halteres, 186, 189, 412, 413, 415, 416

Hawkins, Waterhouse, 143–44

head, vertebrate, 198–99

heart urchins, 180, 227, 283

hedgehog gene, 216, 348, 353

helicoplacoids (Helicoplacoidea), 418, 419, 420

Heliocidaris erythrogramma: burden and, 317; cell types of, 254, 329; direct development in, 227–37; egg size in, 226, 252; heterochrony in, 275; time to shift to direct development, 391

Heliocidaris tuberculata, 227–28, 234, 235, 236, 329

Heliocidaris wnt-5A gene, 236

helix-loop-helix transcription factors, 339, 371, 395

hemerythrin, 121

hemichordates (Hemichordata): acorn worm, 40; as deuterostomes, 44, 137, 138–40; echinoderms and, 420, 423; in molecular phylogenetic tree of metazoans, 112; *Saccoglossus*, 423

Hemiphractus, 248

hemipterans, 407, 408, 409

hemocoel, 124
hemoglobin, 160
hemoglobin genes, 156
Hennig, Willi, 51
Henson's node, 348, 359
heparin-binding growth factor superfamily, 281
herbivory, 97
heredity: development and, 13–15; Lamarckian, 9, 20, 294, 324. *See also* genes
her-1 gene, 372
heterochrony, 255–91; in *Caenorhabditis elegans*, 275, 276–77, 282–83; classes of, 258–66; classic manifestations of, 261–66; de Beerian heterochrony, 261, 266; defined, xiv, 255; dissociated heterochronies, 267; as dissociations in processes in time, 282, 290, 326, 335; in early development, 267–68; as evolutionary process that affects development, 276, 282; in evolution of direct development in sea urchins, 229, 267, 283; in extinct species, 268–70; genetics of, 276–77; Haeckel's coining of term, 2; in humans, 285–90; local heterochronies, 266–67, 289; regulatory genes and, 432; as a result not a process, 282–84, 290, 427–28; as unifying ontogeny and phylogeny, 24; as universal mechanism of developmental evolution, xviii, 24, 256, 258. *See also* global heterochronies
heterotopy, 260, 335
higher termites, 364
hindlimbs: fin folds as origin of, 49, 354; as serially homologous to forelimbs, 36–37, 353–55; in whales, 401, 403, 404
history, 21, 27
Hitchcock, Edward, 146–47
Holopneustes purpurescens, 237
Holtfreter, Johannes, 11
homeobox-containing genes: in arthropods and vertebrates, 182; duplication and, 344; in eye development in *Drosophila* and vertebrates, 184–85; in flatworms, 88; in forebrain development in mice, 199; in *Hydra,* 430; in insects and vertebrates, xvii, 27, 187; in nonsegmental phyla, 187; orphans, 188; phylogenetic distribution of, 27; regulating head organization in vertebrates, 198; as regulatory genes, 26–27; and segmentation, 26, 187
homeobox-containing proteins, 187

homeodomain, 187–88, 384
homeodomain protein superfamily, 339
homeosis (homoeosis): discovery by Bateson, 16, 186; in evolution of insect body plans, 189; homeotic mutation for generating hopeful monsters, 189
homeotic genes: Antennapedia complex, 183, 186–87, 190, 307–9, 339, 413; Bithorax complex, 16–17, 183, 186–87, 190, 307–9, 339, 413–14, 415; as homologous in arthropods and mammals, 182; mutations in, 319; regulating segmentation in *Drosophila,* 16, 183–84, 186–87; as selector switches, 16, 186; transcription factors encoded by, 186
homeotic selector genes, 312, 359
hominids, 59
Homo, 59
homoeosis. *See* homeosis
Homoiostelea, 418
homology: as basis for all evolutionary comparisons, 37; biological homology concept, 35; and body plan, 33–38; classic concept of questioned, 35; criteria for establishing, 35–36; Darwin on, 34; different kinds of homologous similarities, 51; for establishing correspondence of structures, 35; evolutionary continuity of information in, 34; existing at any level of organization, 8; of gene sequences, xviii, 18, 34, 105; genetic homology, 431; in Geoffroy's approach to anatomy, 7, 34; of gill arches and jaws, 342; high-linkage phylogeny and, 48; historical homology concept, 35; of homeotic genes in arthropods and mammals, 182; and homoplasy, 49–50; Hox genes renewing focus on, 425; identifying homologous features required for phylogenetic trees, 49; molded by selection to meet functional demands, 8; molecular processes and, 430–31; morphological homology concept, 35; Owen on, 34, 35; as recording inherited contingencies of evolutionary history, 8; of scales and feathers, 4; structuralist definition of, 313; in study of development, 218; three concepts of, 35; underlying evolutionary transformations within taxonomic group, 196. *See also* serial homology
homoplasy, 49–50, 54, 108–9, 155, 395
Homostelea, 418
honeybees, 375

hormones: circulating hormones, 278, 279; ecdysone, 368; growth hormones, 278–82; juvenile hormones in termites, 367–68; thyroxin, 279, 337, 391. *See also* sex
hormones
horses: digits, 355; *Equus hemionus,* 168; litopterns, 398; pattern of evolution in, 25;
phylogeny of, 58–59; quaggas, 165
Hox A cluster, 350
HoxA genes, 189
HoxB genes, 189, 343, 345
Hox C-8 gene, 189
Hox clusters: in *Caenorhabditis elegans,* 309;
cascade of genes determining expression of,
188; conservation of, 430; constraint in,
308; deletion in *Tribolium,* 413; in determining anterior-posterior body axes, 186,
189; in *Drosophila,* 309; evolutionary relationships among, 184, 185; in limb development, 350–52, 354; original function of,
190; single cluster as primitive state, 187
HoxD cluster, 351, 354–55
HoxD genes, 346, 351
Hox-11 gene, 358
Hox genes: in annelids and arthropods, 183;
in arthropod segmentation, 433; body plans
and, 185–90; conservation of, 216, 307,
431; constraints on phylotypic stage of vertebrates due to, 209; as distributed among
metazoans, 430; in *Drosophila* segmentation, 327; duplication of in vertebrates,
433–34; in echinoderms, 421–22; homeotic
mutations of, 186; in insect wing and leg
expression, 414, 415–17; in mammalian
jaw development, 342–44; in nonsegmented taxa, 310; as predating the Cambrian, 432; renewing focus on homology,
425; as switch genes, 190; as transcription
factors, 430. *See also* Hox clusters; *and particular genes*
Hox-7 gene, 185
Hox 7.1 gene, 198
humans: animal-human hybrids conceived by,
322; art appearing in, 321–22; brain, 286,
287, 288–89; chimp genetics and morphology compared with, 331; digits, 355; divergence from apes, 157; DNA from mummies
of, 165; foramen magnum, 288; fossil species required for phylogenetic inferences,
56; growth rates compared with apes', 286;
Haeckel's diagram of embryo of, 194; het-

erochrony, 285–90; hominids, 59; *Homo,*
59; mitochondrial Eve, 151; Neanderthals
and modern, 149–51, 287; neoteny in, 256,
285–89; number of cell types in, 307; "out
of Africa" scenario, 150–51; phylogenetic
tree for, 150–51; pygmies, 278; resembling
young apes more than adults, 24; skull
shape, 286–88; stages in development of,
196; Taung skull, 287; Tyrolean ice man,
167; upper respiratory tract and associated
features, 287. *See also* Neanderthals
hunchback gene, 188, 310–11
Hutton, James, 65
Huxley, Aldous, 256–58
Huxley, J. S., 15, 16
Hydra, 430
Hylaeosaurus, 144
Hyman, Libbie, 38, 39, 44, 45, 113, 122
hypermorphosis, 262, 263–64, 267, 273, 285,
286
hypotheses, 20, 426–27

ichthyosaurs, 388
Ichthyostega, 356
IGF-I (insulin-like growth factor-I), 278–79
Iguanodon, 143, 144–45, 146
image formation, 376
imaginal discs, 303, 328, 352, 407, 411, 417
immunoglobulin genes, 161
incipient structures, 340
Incisitermes schwarzi, 364–65
index fossils, 65
indirect development: direct development versus a larval feeding stage, 220–22; in frogs,
275; in sea urchins, 233, 224–28, 237,
253–54. *See also* larvae
inducers, 11–12
induction: in frogs, 11; genes and, 12–13; inducers, 11–12; neural induction, 11–12, 13;
in vulva development, 205
industrial melanism, 54
infrared radiation, 376
innovation. *See* novelty
insects: alternative pathways to phylotypic
stage in, 201–3; ants, 60; appendages, 409,
433; in arthropod phylogeny, 133; butterflies, 415–17; classic hypothesis for evolution of body plan, 415; developmental modules in, 328; dissociation of appendage
function in, 433; grasshoppers, 202, 304;
halteres, 186, 189, 412, 413, 415, 416;

insects (*continued*)
heads of newly hatched, 408; hemipterans, 407, 408, 409; homeobox-containing genes in development of, xvii; homeotic transformation in evolution of body plans of, 189; honeybees, 375; imaginal discs, 303, 328, 352, 407, 411, 417; legs, 407, 414–15, 433; limb and wing primordia, 411–12; long germ band insects, 201, 202; *Lymantria dispar*, 15; multiramous limbs of primitive, 129; myriapods and, 131, 133; *Nemora arizonaria*, 220; Neoptera, 405; nervous system development, 130, 134; neural revolution in, 100; orthopterans, 408; Palaeodictyoptera, 405–6; Paleoptera, 405; Paleozoic, 405–13; phylotypic stage of, 201; *Precis coenia*, 416–17; Protocoleoptera, 412; protowings, 410–11; pterygotes, 411; segmentation in, 183; siphonapterans, 408; social insects, 100, 368; Strepsitera, 189; tagmosis in, 433; as terrestrial, 177, 178; thysanurans, 408; two developmental classes of, 201; undiscovered species of, 60; wing development, 352, 404–15. *See also* beetles; dipterans; flies; termites
insulin-like growth factor-I (IGF-I), 278–79
insulin-like growth factor superfamily, 281
integumental compartmentalization, 83, 85
intermediate filaments, 135
internal organization: as constraint on development, 300–301; in evolution, xviii, 32, 324
invertebrates: brooding in, 221; direct development versus larval stage in, 221; evolutionary connection to vertebrates, 8; miniaturization in marine, 270. *See also invertebrate phyla by name*
Irish elk, 319–20

jackknifing, 110, 154
Jacob, F., 17
jaws, 294, 341–44, 369
Jefferson, Thomas, 293
jellyfishes; box jellies, 376–77; *Cassiopeia*, 79; Ediacaran medusoids as, 78–79; fossils, 83
junk DNA, 304
juvenile hormone system in termites, 367–68

Kammerer, Paul, 20
Kaufman, Thomas, xiv, xviii
Kerygmachela kierkegaardi, 135
kinorhynchs: eyespots, 377; lacking fossil record as far back as the Cambrian, 175; as pseudocoelomates, 39, 118
kiwis, 168
Koestler, Arthur, 20
Krox20 gene, 358
Krüppel gene, 202

labial gene, 309
La Brea tar pits, 167
α-lactalbumin, 339–40
lactose, 339–40
Lamarckian heredity, 9, 20, 294, 324
lampreys, 379
land animals. *See* terrestrial animals
language, 362
larvae: ascidians, xvi, 9, 244–45; barnacles, xv, 42–43; body plan of larval forms, 196; bypassing in favor of direct development, 192–93; direct development versus a larval stage, 220–22; evolutionary conservation of larval forms, 190–92; fossils, 191–92; glochidium, 241–42; heterochronies in, 268; mayflies, 405, 407, 409; natural selection operating on, 23; nauplii, 42–43, 131, 134, 190, 191; Paleozoic insects, 407, 409, 410; photoreception, 376; polychaetes, 225, 392; salamanders, 206; sea urchin plutei, 191, 221, 224–28; tadpoles, 191, 219–20, 247–50; termites, 364; trochophore, 44, 121, 130, 225, 242
Leanchoilia, 128
left/right symmetry, 80–81, 302–4
legs: arthropods, 352; insects, 407, 414–15, 433; trilobites, 125; unirames, 407; whales, 323, 401. *See also* limbs
lens, 381, 382–83
Lewis, E. B., 16
Leydig cells, 373
lichens, 85–86, 100
life history, 215, 219–22, 317, 424
Lillie, Frank, 15, 241–43
limb buds: apical ectodermal ridge, 346–47, 349; genes and evolution of pattern in, 350–52; modular evolution of, 344–47; molecules signaling between modules of, 347–49; progress zone, 346–47; two cell sheets forming, 331; zone of polarizing activity, 346–48, 356
limbless mutant, 349
Lim1 gene, 198, 358
limbs: in birds, 392–93; commonalities of de-

velopment in insects and vertebrates, 352–
53; constraint on number in tetrapods,
295–96, 301; digits, 355–56, 425; limb
fields, 333; limb loss, 393; as serial homologues, 353–55. *See also* forelimbs; hindlimbs; legs; limb buds
linear transcriptional hypothesis, 307–8
lin-4 gene, 276
lin-14 gene, 276, 282–83, 333, 432
lin-28 gene, 276
lin-29 gene, 276
Lipalian interval, 71
litopterns, 398
lizards, 379, 390
lobe-finned fishes (rhipidistians); coelacanths,
159, 160–62, 163, 354; extinction of most
lineages of, 160, 162; metapterygial axis,
355. *See also* lungfishes
lobopods, 91, 96, 135, 178
lobsters, 264
local heterochronies, 266–67, 289
Loch Ness monster, 144
Lombroso, Cesare, 387
long germ band insects, 201, 202
lophophorates, 122–23; brachiopod ancestors
as, 94; as deuterostomes, 137, 138; intermediate filaments in, 135; in metazoan phylogeny, 42; photoreceptors, 380–81. *See also*
brachiopods; bryozoans; phoronids
lophophore, 42, 94, 122
lophotrochozoans, 122
loriciferans: discovery of, 62; lacking fossil
record as far back as the Cambrian, 175; as
pseudocoelomates, 39, 118
lower termites, 364
lungfishes: fishes compared with, 159–60; genome size, 304; metapterygial axis, 355; in
tetrapod evolution, 159–62, 163
lunules, 273
lycopodophytes, 178
Lyell, Charles, 25, 66, 145
Lymantria dispar, 15
Lynx, 393–94
lysozyme, 340
Lytechinus pictus, 234

macroevolution: and body plans, 30–33; issues of, 21; sufficiency of microevolution
for, 324
mago nashi gene, 357
mammals: bats, 51, 207–8; brain, 100; canids,

265; cats, 167, 393–94; dogs, 264–65, 272,
278; elephants, 165, 168; Eocene radiation
of aquatic, 403; estrogen in growth regulation of, 280; growth hormone gene, 278;
herbivore orders, 97; homologous homeotic
genes in arthropods and, 182; Hox clusters
in, 185; jaws, 294, 342–44; left/right symmetry in, 303; mammary glands, 339–40;
mammoths, 165, 168, 293; marsupials, 55,
193–94, 251, 165, 398; *Megatherium*, 397;
middle ear, 36, 147, 341, 343; monotremes,
193–94, 251; myoD family in, 339; of New
Zealand, 62; neural revolution in, 100;
number of species of, 60; opossums, 55;
pandas, 341; pantodonts, 97; placental
mammals, 55, 100, 251; Pleistocene giants,
397–99; sea cows, 165; sex determination
in, 371, 372–74; synapsids, 52; therapsids,
147. *See also* apes; artiodactyls; horses; humans; rodents; whales
mammary glands, 339–40
mammoths, 165, 168, 293
Mangold, Hilde, 11
Mantell, Gideon, 143
Manx gene, 246
Manykaian stage, 87
marine plants, 178
marsupial frogs, 247–49
marsupials, 193–94, 251, 398; opossums, 55;
thylacines, 165
mass extinctions, 175–77; catastrophes in
causing, 21, 99, 175–76; Cretaceous mass
extinction, 175, 176; cyclic pattern of, 175–
76; hypothetical mechanisms of, 176; introduction of higher-level taxa more frequent
following, 179. *See also* Permian mass extinction
master regulatory genes: as conserved between
phyla, 430; for coordinating dosage compensation and sex determination, 371; *eyeless* as, 384; in mouse head development,
358; pathways of evolution of, 429–30;
Pax-6 as, 358, 384, 431; phylogeny required to map histories of, 424
Mastotermes dawiniensis, 169
Mastotermes electrodominicus, 169
maximum likelihood methods, 110, 153
maximum parsimony, 108
mayflies, 190–91, 405, 407, 409
Meckel's cartilage, 343
medusa, 40, 78, 376

medusoids, Ediacaran, 78–79
Mendelian genetics, 9, 13, 20, 23, 25
Megalosaurus, 143, 144, 146
Megatherium, 397
meristic changes, 337
mesentery, 41
mesentoblast cell, 43, 44
messenger RNA hypothesis, 17
mesoderm, 38–39, 43, 44, 345–46
mesoglea, 38, 83
mesonychids, 400, 402, 403
mesothoracic wings, 406
mesozoans, 175
Messel Shale (Germany), 83
metapterygial axis, 355
metathoracic wings, 406
metazoans (Metazoa): alternative phylogenies
    of, 46; Ameria, 45–46; common genetic
    strategy of, 331; diploblastic animals, 38,
    40, 41, 77, 100; Ediacaran fauna and, 71,
    72, 81, 100, 102; emergence and radiation
    of, 71–73; events near Precambrian-
    Cambrian boundary, 101; extracellular ma-
    trix of, 114; first fossils of, 68, 100; genome
    of, 304; Hox genes distributed among, 430;
    initial radiation in Vendian, 76–77; as
    monophyletic, 115, 185; the most primitive,
    38; number of cell types in, 306–7; Oligo-
    meria, 45–46; origins in deep time, 63–102;
    Phycodes pedum zone, 68; phyla and body
    plans, 30–62; phyla originating 800 million
    years ago, 93–94; Polymeria, 45–46; in
    Precambrian, 73–77; relationship to fungi
    and plants, 113; serial homology in, 337;
    sexual reproduction in, 370; sponges as,
    113–15; standard parts in organization of,
    330; triploblastic animals, 40, 41, 73, 76,
    93, 101. *See also* animals; Cambrian radia-
    tion of metazoans; phyla
methylene blue, 12
mice: BMP family in development of, 339;
    brachypodism mutation, 351–52; growth
    hormone experiments with, 278, 279; head
    development in, 358; homeobox-containing
    genes in forebrain development, 199;
    Hox-11 knockout in, 358; limb development
    in, 350, 351–52, 355; *Mus musculus,* 213;
    sex determination in, 373; *short-ear* muta-
    tion, 352; *wnt* genes, 236, 338, 350
Microdictyon, 91, 96
microevolution: issues of, 30; as sufficient for

macroevolutionary patterns, 324; time scale
    of, 21
micromeres, 231, 243, 254, 327
microtubules, 330, 336
middle ear, mammalian, 36, 147, 341, 343
millipedes, 178
miniaturization, 270–71
missing phenotypes, 298–99
mitochondrial DNA, 151, 156, 158, 165, 166
mitochondrial Eve, 151
Mivart, St. George, 340
moas, 168
mobergellans, 89
model systems: dichotomies and, 212–14; im-
    plicit and real phylogenies in, 214; limits
    of, 214–16; known regulatory genes from,
    236; as the norm in developmental biology,
    213
Modern fauna, 99–100
modularity, 326–34; arthropods as modular,
    433; connectivity between modules, 331–
    32; defined, 325–26; development as modu-
    lar, 203–4, 428; embryos as modular, 204;
    in evolution, 322; genetic organization of
    modules, 327–29; in hybrid monsters, 322;
    limb buds as modules, 344–52; modules as
    dynamic, 326, 332; modules set up by
    switch genes, 356–59; morphogenetic fields
    compared with modules, 333; physical loca-
    tion within developing system, 331; proper-
    ties of modules, 326–27; standard parts,
    330–31; temporal transformations of mod-
    ules, 332–34
molecular biology: in functional biology, 426;
    and the metazoan radiation, 140–41; mor-
    phology complementing, 105; as opening up
    the diversity of species, 25; paleontology as
    crucial to, 140; two themes in, 17
molecular clocks, 107, 157–59, 248, 308
molecular genetics, 215–16
molecular phylogeny, 103–41; for arthropods,
    132–35; extinct lineages affecting, 159–63;
    gene-sequence data for inferring, xvi, 25,
    104–11; methods for inferring, 103–11;
    outlining a molecular phylogeny of the liv-
    ing phyla, 111–13
molecular processes: and homology, 430–31;
    underlying development, xvi–xvii, 17
molecular systematics, 94, 139, 154
Molgula occulta, 245–47
Molgula oculata, 245–47

mollusks (Mollusca): in alternative metazoan phylogenies, 46; cleavage in, 43, 240; common ancestry with annelids and brachiopods, 93, 117–18; *Crepadula*, 43; defense against predation in shells of, 100; as embranchement of Cuvier, 6; as eucoelomates, 39; evolution of early development in, 240–44; eyes, 377–78, 381; glochidium, 241–42; *Halkieria* as early, 92–93; in molecular phylogenetic tree of metazoans, 112; monoplacophoran mollusks, 46; *Patella*, 243; phylogenetic problems of, 46–47; platyhelminth flatworms and, 116, 117–18; progenetic males, 263; in protostome coelomate clade, 113; shell morphologies not found, 298; skeletonized debut of, 89; as terrestrial, 178. *See also* cephalopods; clams

Mononychus, 147

monoplacophoran mollusks, 46

monotremes, 193–94, 251

monsters; hybrid, 322; teratological, 6

moose, 320

Morgan, Thomas Hunt, 10–11, 14–15

morphogenesis: changes in, 233; of limb buds, 345; morphogenetic fields, 333, 428; regulatory genes in, 184; structuralism on, 312; transcription factors in, 203

morphogens, 347–48, 353

morphology: as complementing molecular biology, 105; differential gene expression as underlying differences in, 234; four properties of morphological data, 103; genes and, 310, 312; as most comprehensive database for phylogenies, 154; structuralist view of, 312; transformed in evolution, 23. *See also* morphogenesis

mosaic development, 253, 334–35

mosaic evolution, 323, 403

mouths, 77–78

msh gene, 185, 198, 349

msx genes, 349

Müller, Fritz, 9

Müllerian duct, 373

multicellularity, 73, 93, 100, 114

Musca domestica, 202

muscle actin genes, 198

Mus musculus, 213

mutations: affecting developmental pathways, 272, 329; in *Caenorhabditis elegans,* 238; canalization resisting, 253; in develop-

mental genetics, 28, 217; in dissecting development processes, 22; in early development, 254, 260, 316, 319; and evolutionary reversal, 388–89; heterochronic, 275, 276–77, 282–83; homeotic, 186–89, 300; multiple, 106, 107, 109; physiological genetics and, 16; in regulatory genes, 428; single-gene, 239; in spermatogenesis, 336; transposable elements affecting rate of, 300

myf-5, 339

myoD, 332, 339

myogenin, 339

myriapods, 130, 131, 133–34

Mysticeti, 402

myxozoans, 116

naked mole rats, 368–69

nanos gene, 188, 357–58

nasal placode, 197

Nasutitermes, 367

naturalist tradition, 212

natural selection: as a cause in evolutionary biology, 20; developmental constraints and, 296–99, 428; directionality and, 24–25; in evolutionary change, xviii; the eye arising from, 380; internal versus external selection, 324; in mass extinctions, 176; molding homologies to meet functional demands, 8; ontogeny as product of, xiv; as operating on larvae, 23; operating on microevolutionary phenomena, 30, 324; pattern and process included in, 324–25; seen as providing all order in biological systems, 324; variant phenotypes sorted by, 323–25

nauplii, 42–43, 131, 134, 190, 191

nautiloids, 176

Nautilus, 40, 377

Neanderthals, 149–51; art lacking in, 321; brain, 150, 287; DNA analysis of tools of, 168; older restorations of, 149; recent restorations of, 149–50; speech, 150

Nectocaris, 380

Needham, Joseph, 12, 13, 23, 260

neighbor-joining method, 107

nematodes: cell lineage autonomy in, 268; *Cephalobus,* 277; diversity of, 119; experimental approaches to evolution of early development in, 238–40; eyespots, 377; heterochronies in, 276–77; homeobox-containing genes in, 187; lacking fossil record as far back as the Cambrian, 175;

nematodes (*continued*)
mosaic development in, 334; *Panagrellus redivivus*, 238–39, 333; parasitizing humans, 119; phylotypic stage of, 209–10; as pseudocoelomates, 39, 118; *Rhabditis terricola*, 277; as soil dwelling, 178; vulva development, 205. *See also Caenorhabditis elegans*
nematomorphs: eyespots, 377; possible Cambrian fossil, 175; as pseudocoelomates, 39, 118, 119; as soil dwelling, 178
nemerteans (Nemertea; ribbon worms): as eucoelomates, 39; in molecular phylogenetic tree of metazoans, 112; phylogenetic placement of, 116–17; in protostome coelomate clade, 113; as terrestrial, 178
Nemora arizonaria, 220
Neoptera, 405
neoteny, 263; in amphibians, 272; as category of heterochrony, 262; defined, 255–56; in humans, 256, 285–89; in salamanders, 263, 272, 337, 391; in sea urchins, 267
nervous system: appearing in platyhelminth flatworms, 39; in arthropods, 124; in the Cambrian, 88; in crustaceans, 130, 134; Cuvier on, 6; in *Drosophila*, 332; in echinoderms, 418, 420–23; in insects, 130, 134; neural crest cells, 197, 207, 328, 342–44; neural revolution, 100; neurons, 289; in plethodontid salamanders, 305–6. *See also* brain
neural crest cells, 197, 207, 328, 342–44
neural induction, 11–12, 13
neural revolution, 100
neurons, 289
neutral evolution, 297
New Zealand, mammalian fauna of, 62
Noah's ark, 63–67
Notch gene, 332
notochord, 197, 200, 328
novelty: defined, 399–400; duplication and divergence creating, 339–40; evolving new body plans, 397–434; novel patterns occurring in early development of sea urchins, 209; paedomorphosis associated with, 270, 271; regulatory genes conserved and recombined in novel pathways, 203, 360; serial homology introducing, 337–39
nymphs, 364, 366

Occam's razor, 53
octopuses, 377, 385

Odontoceti, 402
Odontogriphus, 96
Oligomeria, 45–46
Oncopeltus fasciatus, 408
ontogenetic criterion, 56
ontogenetic trajectories, 265–66
ontogeny: body plan as deeply immanent in, 33; complexity of making evolution seem impossible, 325; environmental features changing, 220; evolutionary data revealing perspectives on, 28; four elements making evolution possible in, 325; heterochrony and, 259–60; interface with phylogeny, 23–24; as internal criterion of polarization, 55–56; as not a single process, 23; ontogenetic rules as developmental constraints, 301; phylogeny interacting with, xv, 1–29; parallel processing in, 334; as product of selection and evolutionary history, xiv; as source of transformational information, 36. *See also* development; developmental recapitulation
onychophorans (Onychophora): in arthropod phylogeny, 124, 130, 134; eyes, 378; as link between annelids and arthropods, 60; orders as not morphologically disparate, 159; peripatids, 60; segments and appendages of, 127; tagmosis patterns of, 126; as terrestrial, 178; unirames and, 125, 130
oogenesis, 372
Opabina, 40, 95, 96, 380
operon theory, 17
opossums, 55
opsins, 381
optic placode, 197
optimal tree, 109, 110
orders (taxonomic), 159, 310
organogenesis, 204, 268
organs, 38, 328
orthodenticle gene, 199
orthogenesis, 25
orthologous gene sequences, 156
orthopterans, 408
oskar gene, 357
otic placode, 197
Ottoia, 95
Otx gene, 199
outgroups: in inferring temporal order, 270; and primitive characters, 55–56
"out of Africa" scenario, 150–51
Owen, Richard, 33–34, 35, 36, 37, 143–47, 195

oxygen levels, 75–76
oysters, 271

paedomorphosis, 262–63, 270–72; defined, 261; frequency of, 261–62; in human evolution, 285–89; novel forms associated with, 270, 271; post-displacement, 262, 263, 274–75; progenesis, 262–63; as selecting between developmental modes in ancestral repertoire, 271; in trilobites, 267. *See also* neoteny
paired-box genes, 344, 384
pair-rule genes: controlling expression of segment polarity genes, 188; *even-skipped* gene, 202; gap genes controlling expression of, 201, 202; *hairy* gene, 202; mutations in, 319
Pakicetus, 401–2, 403
Palaeodictyoptera, 405–6
paleontology, 63, 65, 102, 140
Paleoptera, 405
Paleozoic fauna, 99; echinoderms, 99, 420; insects, 405–13
Paley, William, 292
Panagrellus redivivus, 238–39, 333
pandas, 341
pantodonts, 97
paracrinoids, 420
parallel processing, 334, 335–36
paralogous gene sequences, 156, 161
paranota, 404, 407
paraphyletic groups, 52
parietal eyes, 379
parsimony, 53–54, 104, 109
parsimony methods, 108–10, 151–53, 155
parthenogenesis, 390
Parvancorina, 70
Patagonia, 397
Patella, 243
pattern, genes and the execution of, 350–52
pattern cladists, 37
pattern formation, 312, 344, 345, 417
Pax-6 gene, 184, 358, 382, 384–85, 386, 430, 431
PCR (polymerase chain reaction) techniques, 111, 164, 166–67, 236
pentastomids (Pentastomida): and arthropods, 134; eyespots, 377; fossil larvae, 190; intermediate filaments lacking in, 135; in molecular phylogenetic tree of metazoans, 112;

more adult development produced by, 272; possible Cambrian fossils, 175
peramorphosis, 263–65, 272–73; acceleration, 262, 265, 290; defined, 261; frequency of, 261–62; hypermorphosis, 262, 263–64, 267, 273, 285, 286; pre-displacement, 262, 265, 274–75; in trilobites, 267
peripatids, 60
Permian mass extinction: blastozoans dying out in, 97; clams replacing brachiopods after, 98, 176; Paleozoic fauna decimated by, 99; possible causes of, 99, 176; as setting the stage for a reradiation, 176–77
Peronella japonica, 226
Petalonomae, 80
Phanerozoic eon: marine animal diversity, 98; small shelly fossils of, 89–90; Triassic period, 177, 179. *See also* Paleozoic fauna
pharyngeal arches, 342–44
pharyngula stage, 196, 197
phenotypes, missing, 298–99
phenotypic plasticity, 220
pheromones, 367
phoronids: as eucoelomates, 39; lacking fossil record as far back as the Cambrian, 175; as lophophorates, 122; lophophore, 42
photoreception, 129, 376, 378, 380–81
photosynthesis, 75–76
Phycodes pedum zone, 68
phyla: arising millions of years before first remains, 93; body plan at level of, 31, 38; classes, orders, and families in, 159; date of appearance of modern, 68; determining evolutionary relationships of, 45; as disjunct in their features, 103; and the embranchements of Cuvier, 6; hypotheses on stability of, 179–81; the living phyla, 38–47; living phyla appearing in the Cambrian, 68, 95; marine versus terrestrial, 100, 177–78; master regulators conserved between, 430; metazoan phyla and body plans, 30–62; a molecular phylogeny of the living phyla, 111–13; no new phyla since the Cambrian, xiv, 32, 173, 174–75; phylogenetically informative morphological features few at level of, 103; splitting of, 94, 95; sudden appearance in Cambrian radiation, 31–32, 95–97, 102, 173; suites of anatomical features of, xiv, 31. *See also* phylogeny; *and metazoan phyla by name*
phyletic constraints, 301

Phyllacanthus imperialis, 227, 228
Phyllacanthus parvispinis, 227
phylogenetic trees: for bacteriophage T7,
154–55; basic aim of, 58; cladograms and,
56–60; computer tools for constructing, 22,
104–5; consistency in method of constructing, 153; efficiency in constructing,
152–53; Haeckel on, 2, 9; of humans, 150–
51; ideal use of, 425; identifying homologous features required for, 49; inferential
nature of, 59; inferred in a variety of ways,
50; inferring from molecular phylogeny,
104–11; mapping innovations onto, 399;
number of possible rooted trees, 48, 109;
optimal tree, 109, 110; power of method of
constructing, 153; robustness of, 109–10,
153; rooted trees, 48, 106, 109; of sea urchins, 225, 226; testability of, 153–54; two
aspects of evolution appearing in, 50. *See
also* cladistic methods; distance-based
methods
phylogeny: alternative metazoan phylogenies,
46; and the Cambrian radiation, 47–48, 95;
degradation of phylogenetic information,
425–26; developmental regulatory genes
and, 183–85; and the evolution of development, 171–72; gene sequences for inferring,
xvi, 25, 152–55; Haeckel's coining of term,
2; information in DNA, 104; interface with
ontogeny, 23–24; lack of means for testing
phylogenies, 45; as necessary for understanding evolution, 18, 32; need for reliable
phylogenies, 47–50; ontogeny interacting
with, xv, 1–29; phylogenetics as vital to
evolutionary developmental biology, 47,
217–18; problem in existing approaches to,
49; as resting on comparative embryology,
xvi; tools of, 50–55; two aspects of, 59; for
understanding body plan evolution, 148,
424, 425. *See also* developmental recapitulation; molecular phylogeny; phylogenetic
trees
phylotypic stage: conservation in, 429; defined, 33; and the developmental hourglass,
208–10, 211, 318; evolution after, 205–8;
heterochronies in, 268; in insects, 201; low
probability of evolutionary change in, 211;
as midpoint in development, 204–5; pharyngula stage, 196, 197; stability of, 203–5;
in nematodes, 209–10; in vertebrates,
193–94, 196–97

Physarum, 85
physiological genetics, 16
Pikaia, 88
pineal eyes, 379
pineal gland, 379
pipid frogs, 191
pit vipers, 376
placental mammals, 55, 100, 251
placozoans (Placozoa), 38, 113, 115
planktotrophic larvae, 221–22
plants, 61, 113, 178
platelet-derived growth factor superfamily, 281
platyhelminth flatworms (Platyhelminthes): as
acoelomates, 39, 40, 118; in alternative
metazoan phylogenies, 46; chaetognaths
and, 138; cleavage in, 43; complex nervous
system appearing in, 39; eyespots, 377; homeobox genes in, 88; lacking fossil record
as far back as the Cambrian, 175; in molecular phylogenetic tree of metazoans, 112,
113; as monophyletic, 116; as most primitive metazoans, 45; neural evolution in, 88;
as terrestrial, 178
Pleistocene giant mammals, 397–99
plenitude, principle of divine, 293
plethodontid salamanders, 206, 250, 304–6
pogonophorans (Pogonophora): cleavage in,
43; as eucoelomates, 39; lacking fossil record as far back as the Cambrian, 175;
lenses, 381; in molecular phylogenetic tree
of metazoans, 112; in protostome coelomate
clade, 113; segmentation in, 42
polarity genes, segment. *See* segment polarity
genes
polarity of change in development, 171,
181–82
polychaetes: eyes, 377; feeding larvae, 225,
392; and *hedgehog* expression, 353; trochophore, 130, 225
polymerase chain reaction (PCR) techniques,
111, 164, 166–67, 236
Polymeria, 45–46
polymorphic technology, 363, 364–70
populational thought, 212
population genetics, 13-14, 21, 296
Porifera. *See* sponges
post-displacement, 262, 263, 274–75
POU domain, 188
preadaptation, 340
Precambrian-Cambrian boundary, 68, 86, 87,
89, 101

Precambrian period: Darwin on life in, 69; environmental changes in late, 74–75; eukaryotic evolution in late, 73; evolution during, 95; life and environments, 73–77; problematica, 76; stromatolites, 74; tectonic events in late, 76; Vendian age, 69, 74, 76–77, 87, 89, 93
Precis coenia, 416–17
predation, 78, 90, 100, 323
pre-displacement, 262, 265, 274–75
priapulans: in Burgess Shale, 119; *Ottoia*, 95; as pseudocoelomates, 39, 118; size of, 119
primary reproductives (termites), 364, 365, 366
primitive characters: outgroups and, 55–56; reappearing in more advanced living animals, 386–87
problematica, 76
proboscipedia gene, 327
progenesis, 262–63
progress, 98–100
progress zone, 346–47
promoters, 318, 383–84, 426, 431–32
proteins: bicoid proteins, 310–11; bone morphogenic protein families, 281, 338–39, 349, 351–52; crystallins, 383–86; as encoded by genes, 310; in fossils, 165; Gdf proteins, 339; homeobox-containing proteins, 187; homeodomain protein superfamily, 339; influencing growth, 278; with multiple functions, 394–95
prothoracic wings, 406
protists: algae, 73–74, 85; as base of metazoan radiation, 113; disparate groups of, 61, 73; multinucleate protists, 85; in termite digestion, 366
Protocoleoptera, 412
protostomes (Protostomia): cleavage in, 43; examples of body plans of, 40; larval mouth construction in, 44; in molecular phylogenetic tree, 112, 113; principal elements of body plan organization in, 41. *See also* arthropods; coelomate protostomes
protowings, 410–11
pseudergates, 364, 366
pseudocoelom, 39, 118, 120
pseudocoelomates (aschelminths), 118–20; anatomy of, 119; ancestors of, 119–20; examples of diversity of body plans, 40; eyespots, 377; as polyphyletic, 119; principal elements of body plan organization, 41; uncertain phylogenetic relationships of, 39

pseudodontorn, 388
pseudoextinction, 58
Pterdinium, 80
pteridophytes, 178
pterygotes, 411
punctuated equilibrium, 21, 271, 423
pycnogonids (sea spiders), 10
pygmies, 278

quaggas, 165
quantitative genetics, 296
queens (termites), 365

radial symmetry, 38
radiates, 6, 38
radio waves, 376
Rana, 275
Rangea, 80
rate genes, 15–16
rattlesnakes, 376
ray-finned fishes, 161, 342
recapitulation. *See* developmental recapitulation
recovering data from the past, 142–72; all data from the past as degraded, 148; genes from extinct lineages, 159–63; fossil genes, 163–71; interpreting lost body plans, 148–49; loss and recovery of data, 148; lost data on Neanderthals, 149–51; the molecular clock for timing molecular events, 157–59; recovering phylogeny from genes, 152–56
Redkinia, 76
reductionism, 212
redundancy, 344
regulative development, 253–54, 334–35
regulatory genes: Cambrian radiation and, 432; as cascades of genes, 331–32; cell identities depending on, 332–33; cloning of, 26; conservation of, 216; as conserved and recombined in novel pathways, 203, 360; as constraining development, 300; co-option in, 341, 431; duplication in evolution of vertebrate body plan, 198; heterochrony and, 432; in *Hydra*, 430; important events in regulatory evolution, 184; in inferring phylogenetic relationships, 47, 183–85; in insect body plan evolution, 415–17; from known models, 236; morphogenetic genes widely shared among coelomates, 184; mutations in, 428; pathways of expression of, 329; in patterning insect legs and wings,

regulatory genes (*continued*)
414; promoters, 318, 383–84, 426, 431–
32; redundancy in, 344; serial homology in,
338–39; steroid receptor family, 27, 203;
and structural genes, 234, 331, 431; tension
between conservation and change in, 429;
wide-ranging conservation and co-option
of, 27. *See also* homeobox-containing
genes; Hox genes; master regulatory genes;
transcription factors
Rehbachiella, 191
Reichert, Karl, 8
reproductives (termites), 364, 365, 366–67,
368
reptiles: crocodiles, 383; different evolutionary lineages within, 52; extinction of great
marine, 400; ichthyosaurs, 388; jaws, 294;
lizards, 379, 390; mammalian middle ear
deriving from jaw of, 36, 147, 341, 343; as
paraphyletic group, 52; sex determination
in, 374; synapsids, 52; *Tanystropheus longobardicus,* 284; therapsids, 147; turtles,
200–201. *See also* dinosaurs; snakes
restriction maps, 155
retina, 336–37, 379
retinoic acid, 13, 345, 347–48
Rhabditis terricola, 277
rhabdomeres, 129, 378, 380–81
rhiniophytes, 178
rhipidistians. *See* lobe-finned fishes
rhodopsins, 381
ribbon worms. *See* nemerteans
Riftia pachytila, 77–78
RNA sequencing, 111
rodents: guinea pigs, 207, 394; naked mole
rats, 368–69; voles, 289. *See also* mice
Rodhocetus kasrani, 402–4
rooted trees, 48, 106, 109
rotifers: eyespots, 377; lacking fossil record as
far back as the Cambrian, 175; as pseudocoelomates, 118, 119; soil dwelling, 178
rotulid sand dollars, 273, 274
Roux, Wilhelm, 10, 11, 19, 213, 334
Russian royal family, 166

Saarina, 76
saber-toothed cats, 167
Saccoglossus, 423
salamanders: *Ambystoma,* 267; *Ambystoma
punctatum* and *Ambystoma tigrinum,* 281;

ambystomatid salamanders, 391; axolotl,
263, 272, 337, 391; branchial skeleton in,
206; digits, 355; direct developing, 250;
early stages of development in, 196; gastrulation, 200; genomic constraints in, 305;
growth factors, 281; heterochrony in, 267;
loss of digits in, 302; miniaturized, 270;
neotenic, 263, 272, 391; plethodontid salamanders, 206, 250, 304–6; *Triturus,* 267
sand dollars: evolution of shape in, 283; rotulid, 273, 274
scales, 4
scalloped gene, 414, 415, 417
Schistocerca, 202
schizochroal eye, 378, 382
schizocoely, 44
Scott, Robert Falcon, 1
scorpion flies, 303
scorpions, 178
Scr gene, 416
sdc gene, 372
sea cows, 165
sea pens, 79–80; *Thaumaptilon,* 78, 86
sea spiders (pycnogonids), 10
sea urchins: *Abatus cordatus,* 227; actin gene
evolution in, 234–35; altering distribution
of nuclei in embryo of, 14; archenteron
elongation, 223–24, 233; *Asthenosoma ij-
mai,* 226; bilateral symmetry in heart urchins, 180; *Brisaster latifrons,* 226; *Clypeaster rosaceus,* 226; developmental
modes in, 223–29; developmental modules
in, 327; direct-developing, 193, 221,
224–37, 275; eggs, 225–26, 231–32; features of development of, 223; gastrulation,
224, 233; heart urchins, 180, 227, 283; *Heliocidaris tuberculata,* 227–28, 234, 235,
236, 329; heterochrony in, 229, 267, 283–
84; *Holopneustes purpurescens,* 237; Hox
gene expression in, 421–22; indirect-
developing, 233, 224–28, 237, 253–54; larvae, 191, 221, 224–28; *Lytechinus pictus,*
234; novel patterns occurring in early development of, 209; partial plutei, 391; in Permian extinction, 176–77; *Peronella japonica,* 226; *Phyllacanthus imperialis,* 227,
228; *Phyllacanthus parvispinis,* 227; phylogenetic tree for, 226, 227; pre-displacement
in, 275; regulative development in, 334;
skeleton, 224, 234; *Strongylocentrotus*

droebachiensis, 228; *Strongylocentrotus purpuratus,* 223, 234–35; studies of fertilization in, 215; vestibule, 224. *See also Heliocidaris erythrogramma;* sand dollars
secondary reproductives (termites), 364, 365, 366
segmentation: in annelids, 118; in arthropods, 123–24; in *Drosophila,* 16, 26, 132, 183–84, 186–87, 327, 344, 357–58; in Ediacarans, 80, 83; in eucoelomates, 42; homeoboxes and, 187; as inevitable given structural rules of metazoans, 182–83; in post-Cambrian orders, 310; in vertebrates, 189, 344
segmentation genes. *See* gap genes; pair-rule genes; segment polarity genes
segment polarity genes: *engrailed* gene, 183, 188, 201, 202, 352; expression of, 183, 188; mutations in, 319; *wingless* gene, 188, 344, 353, 411, 412, 413, 414, 415, 417
selection. *See* natural selection
selection limits, 296
self-assembly, 330
selfish DNA, 304
selfish DNA evolution, 297
senses, 380
serial homology, 36–37; duplication and divergence introducing, 326; of jaw and pharyngeal arches, 342–44; limbs as, 353–55; novel features and complexity introduced by, 337–39
Serres, Etienne, 6
Sertoli cells, 373
sex: appearance of, 73; copulation, 119; and polymorphic technology, 364–70; sexual differentiation in termites, 364, 366–68; spermatogenesis, 336, 372. *See also* sex determination; sex hormones
Sex combs reduced gene, 413
sex determination: in birds, 374; in *Caenorhabditis elegans,* 371, 372, 374–75; chromosomal, 370, 371, 374; dosage compensation, 370–71; in *Drosophila,* 370–71, 372, 375; environmental, 370, 374; fluidity of mechanisms of, 370–75; in mammals, 371, 372–74; in mice, 373; in naked mole rats, 368–69; in reptiles, 374; temperature-dependent, 374; in termites, 364–66; in vertebrates, 374, 375
sex hormones: developmental effects of, 279;

estrogen, 279, 280, 373; in temperature-dependent sex determination, 374; testosterone, 279; in vertebrate sex determination, 375
sex lethal gene, 371
sexual differentiation in termites, 364, 366–68
sharks, 160, 161, 200
shell, 90, 100
shelly fossils, 89–90
short-ear mutation, 352
short germ band insects, 201, 202
Sidneyia, 128
siphonapterans, 408
sipunculans (Sipunculida): cleavage in, 43; as eucoelomates, 39; lacking fossil record as far back as the Cambrian, 175; in molecular phylogenetic tree of metazoans, 112; in protostome coelomate clade, 113
sisterless gene, 371
situs inversus, 303–4
skeletons, 75, 89–90, 102, 224, 234, 418
sliding reflection, 80, 81
slime molds, 85
small shelly fossils, 89–90
Smith, William, 65
snakes: body plan of, 206–7; eyes, 207, 376; forelimbs, 207; pit vipers, 376; ribs functioning as legs in, 341
social insects, 100, 368
soldiers (termites), 364, 365, 367, 366, 368
somites, 200, 206, 207, 345, 434
songbirds, 279–80
sonic hedgehog gene, 348, 351, 431
South American fauna, 397–99
spadefoot toads, 219–20
species: defined, 61; diversity and disparity of, 60–62, 97; duration of, 423; as genetically distinct but sharing morphological features, 152; number of living, 60
species trees, 155–56
Spemann, Hans, 11
spermatogenesis, 336, 372
spermatophytes, 178
sphenophytes, 178
spherical abberation, 378
spiral cleavage, 43–44, 240, 243
spiralians, 43–44, 47, 240, 242–43, 253, 334
spire gene, 357
sponges (Porifera): as base of metazoan radiation, 113–15; *Choia,* 95; first spicules, 93;

sponges (Porifera) (*continued*)
  in molecular phylogenetic tree, 112, 113;
  primitive body organization of, 40; principal elements of body plan organization, 41;
  skeletonized debut of, 89
Spriggina, 70, 77, 80, 84–85, 433
squids, 377, 385
Sry gene, 373
SRY gene, 373–74
standard parts, 208, 330–31, 428
staufen gene, 357
Stefania, 248
Steller's sea cow, 165
Stern, Curt, 16
steroid receptor family, 27, 203
steroids, 13
stoneflies, 411
Strepsitera, 189
stromatolites, 74
Strongylocentrotus droebachiensis, 228
Strongylocentrotus purpuratus, 223, 234–35
structural genes, 234, 331, 431
structuralism, 312–15; strong version of,
  312–13; weak version of, 314
structure, constraints imposed by limits of,
  310–15
styli, thoracic, 405
Stylophora, 418, 420
superphyla, 222
superpositional compound eye, 378
switch genes: cell type and, 333; Hox genes
  as, 190; modules set up by, 356–59
Sxl gene, 371, 372
symmetry: bilateral, 38, 41; left/right, 80–81,
  302–4; radial, 38
symplesiomorphies, 52
synapomorphies, 52, 54–55, 108, 395
synapsids, 52
systematics: division of phyla, 159; evolutionary, 50, 103; molecular, 94, 139, 154;
  schools of, 50. *See also* cladistics

tadpoles, 191, 219–20, 247–50
tagmosis, 123, 126–27, 203, 337–38, 415,
  433
talpid mutant, 349, 351
Tanystropheus longobardicus, 284
tardigrades (water bears): Cambrian fossils,
  134, 175; as eucoelomates, 39; eyespots,
  377; as soil dwelling, 178
Taung skull, 287

teleosts, 200, 213–14
teloblast, 124
temperature-dependent sex determination, 374
teratological monsters, 6
terminal Cretaceous mass extinction, 175, 176
termites, 362–63; anal feeding in, 366–67;
  black-mound termites, 365; caste determination in, 364–68; corpora allata, 367; *Incisitermes schwarzi*, 364–65; juvenile hormone in, 367–68; lower and higher, 364;
  Mastotermes dawiniensis and *Mastotermes
  electrodominicus*, 169; *Nasutitermes*, 367;
  sex determination in, 364–66; sexual differentiation in, 364, 366–68
terrestrial animals (land animals), 177–79; return to aquatic habitats, 388, 400–404; tetrapod origins, 159–62, 163
tertiary reproductives (termites), 364, 365
testosterone, 279
tetrapods: *Acanthostega*, 162, 355–56, 425;
  amphibians illustrating primitive pattern of,
  214; as built from standard parts, 208; constraint on number of limbs in, 295–96, 301;
  difficult modifications in elements of, 200;
  digits, 355–56, 425; *Ichthyostega*, 356;
  limbs in the earliest, 354; origins of,
  159–62, 163; pruning of stem lineages by
  extinction, 163. *See also* amphibians; birds;
  mammals; reptiles
TGF-β superfamily (transforming growth factor-β superfamily), 281, 338, 353
Thaumaptilon, 78, 86
thecal cells, 373
therapsids, 147
Thermobia domestica, 408, 427
thermopterans, 51, 53
thoracic styli, 405
thylacine, 165
thyroxin, 279, 337, 391
thysanurans, 408
time scale, 21
tissue, 38
Tommotian stage, 87, 89, 90
tommotiids, 89
tools, ancient, 168
trace fossils, 76–77, 87–88, 100, 101
tra gene, 371
transcription factors: in cascades of genes,
  331–32; developmental control genes encoding, 184; evolutionary conservation and,
  216; families of, 339; gap genes encoding,

188, 201; gene duplication and, 203; helix-loop-helix transcription factors, 339, 371, 395; homeobox-containing proteins as, 187; homeotic genes encoding, 186; Hox genes as, 430; interaction with promoters, 318; orphan genes as, 188
transformation, 5, 33
transformer gene, 371
transforming growth factor-β superfamily (TGF-β superfamily), 281, 338, 353
transitions, 109, 110
transmission genetics, 13, 14
transversion parsimony methods, 109
transversions, 109, 110
Triassic period, 177, 179
Tribolium, 202, 412–13
Tribrachidium, 70, 79
Trichoplax adherens, 115
trigonotarbids, 177
trilobites (Trilobita): in arthropod phylogeny, 124; biramous leg of, 125; body fossils of, 87; eyes, 129, 378, 382; fossil embryos, 191; head appendages, 128, 136; heterochrony in, 267; as protostomes, 40; segments and appendages of, 127
Trilobozoa, 79, 89
trimerophytes, 178
triploblastic animals, 40, 41, 73, 76, 93, 101
tritons, 322, 424
Triturus, 267
trochophore, 44, 121, 130, 225, 242
tropical frogs, 215, 247
tudor gene, 357
Tully monster (*Tullimonstrum*), 96
turtles, 200–201
Twitty, Victor, 12
Tyrolean ice man, 167

Ubx gene, 189, 416, 427, 430
Ultrabithorax gene, 413, 414
unc-86 gene, 188
Unio, 43, 241–43
unirames (Uniramia): in arthropod phylogeny, 124; crustaceans and, 130–31, 133; developmental differences in, 131; head appendages of, 128, 136; legs of primitive, 407; limbs, 125, 126; mandibles, 127–28; onychophorans and, 125, 130; as polyphyletic, 134; segments and appendages of, 127; as terrestrial, 178. *See also* insects

urochordates: as deuterostomes, 44, 137, 139; as eucoelomates, 39; lacking fossil record as far back as the Cambrian, 175. *See also* ascidians
utahphosphids, 89

Varangian Ice Age, 74
variation: appearing nonrandom in developing systems, 324, 428; in constructing vertebrate body plan elements, 197–201; Darwin on, 213, 293; in developmental and evolutionary biology, 21, 22; natural selection sorting variant phenotypes, 323–25; sources of, 325; three processes producing nonrandom, 325
vasa gene, 357
Vendia, 80
Vendian age, 69, 74, 76–77, 87, 89, 93
Vendobiota, 85, 100
vendozoan hypothesis, 72, 86
vertebrae: Hox code and, 434; as serially homologous, 36; in snakes, 206
vertebrates: ascidians and, 139–40; basic body organization of, 31; brain evolution, 88; as deuterostomes, 137; developmental hourglass as diagram for, 209; duplication of Hox genes in, 433–34; eggs of terrestrial, 250–51; as embranchement of Cuvier, 6; evolutionary connection to invertebrates, 8; eyes, 184–85, 376, 378–79, 381; head, 197, 198–99; *hedgehog* gene, 216; homeobox-containing genes in development of, xvii, 27, 187; homeobox-containing genes regulating eye development, 184–85; homeobox genes in arthropods and, 182; Hox clusters in, 185; image formation, 376; intrinsic growth regulation in development of, 280–82; jaw development, 341–44; limb development compared with *Drosophila*, 352–53; as not inevitable, 67; Owen's archetype of, 34; pharyngula stage, 196, 197; phylotypic stage of, 193–94, 196–97; principal elements of body plan organization, 41; segmentation in, 189, 344; serial homology in, 36–37; sex determination in, 374, 375; stages in development of, 196; as terrestrial, 177, 178; variation in constructing body plan elements, 197–201; wing evolution in, 324. *See also* amphibians; birds; fishes; forelimbs; hindlimbs; mammals; reptiles; tetrapods

vestibule, 224
Vestiges of the Natural History of Creation
    (Chambers), 146
vestigial gene, 414, 415
vision, 375; color vision, 379; image forma-
    tion, 376
voles, 289
von Baer, Karl Ernst, 6, 7–8, 33, 197, 316
vulva, 205

Waddington, C. H., 12, 13, 16
Walcott, Charles, 69, 95–96
wanton extinction model, 176
Waptia, 95
watchmaker, the divine, 292
water bears. *See* tardigrades
weevils, 61, 170
Weiner, A. J., 26
whales (cetaceans), 400–404; archaeocetes,
    401–4; early whales with legs, 323, 401;
    fish compared with, 49, 388; limb loss and
    recovery in, 393; Mysticeti, 402; Odonto-
    ceti, 402; a phylogeny for, 402; specialized
    features of modern, 401
William of Occam, 53
Wilson, Edward A., 1–4
wing cells, 417
wingless gene, 188, 344, 353, 411, 412, 413,
    414, 415, 417

wings: in bats, 207–8; in birds, 301; in chicks,
    347; co-option in development of, 341;
    halteres, 186, 189, 412, 413, 415, 416;
    in insects, 352, 404–15; in vertebrates,
    324
Wiwaxia, 92, 122
wiwaxiids, 89
wnt gene family, 236, 338, 344, 353, 430
wnt-7a gene, 350
Wolffian duct, 373
Woodger, Joseph Henry, 196
Woods Hole, 10, 19
workers (termites), 364, 365, 366–67

Xenopus, 249
Xenopus laevis, 213, 275
xol-1 gene, 372

zebrafish (*Brachydanio rerio*): embryonic
    cell types, 328; as favored organism in
    developmental biology, 213–14; fins,
    354–55
zone of polarizing activity (ZPA), 346–48,
    356
zones of cartilage concentration, 350
zootype, 186
ZPA (zone of polarizing activity), 346–48,
    356

The Shape of Life

GENES, DEVELOPMENT,
AND THE EVOLUTION OF
ANIMAL FORM

Rudolf A. Raff

I n *The Shape of Life*, Raff analyzes the rise
of evolutionary developmental biology
and proposes new research questions,
hypotheses, and approaches to guide the
growth of this recently founded discipline.
Drawing on a number of key discoveries from
the past decade, Raff explains how research in
diverse disciplines has forged closer links
between developmental and evolutionary
biology. A revitalized systematics has enabled
researchers to map developmental features
onto more objectively inferred phylogenies;
paleontology has put the Cambrian radiation
of animal body plans into sharper focus; and
deeper probing of the molecular machinery
underlying development has revolutionized
understanding of the evolution of development. For instance, the discovery during the
past decade that both insects and vertebrates
use homologous homeobox-containing genes
in the development of their body plans has
revealed that fundamental genetic relationships underly the development of animals in
disparate phyla.

Raff uses the evolution of animal body
plans to exemplify the interplay between
developmental mechanisms and evolutionary